the man-made world

Engineering Concepts Curriculum Project
Polytechnic Institute of Brooklyn

McGRAW-HILL BOOK COMPANY

New York · St. Louis · San Francisco · Düsseldorf · Johannesburg
Kuala Lumpur · London · Mexico · Montreal · New Delhi · Panama
Rio de Janeiro · Singapore · Sydney · Toronto

ECCP STAFF

Co-Directors

Dr. E. E. David, Jr., Science Adviser to the President

Dr. J. G. Truxal, Polytechnic Institute of Brooklyn

Executive Director

Dr. E. J. Piel, Polytechnic Institute of Brooklyn

Staff Associate

Dr. Thomas T. Liao, Polytechnic Institute of Brooklyn

Administrative Assistant

Mrs. Jo Simonelli

Consultants

Mr. John S. Barss, Andover, Massachusetts

Mr. Edward G. Blakeway, Raleigh Schools, North Carolina

Dr. Ludwig Braun, Polytechnic Institute of Brooklyn

Mr. Henry Fraze, University of South Florida

Dr. Lois Greenfield, University of Wisconsin

Mr. Lester Hollinger, Glen Rock High School, Glen Rock, N. J.

Mr. Richard King, Staples High School, Connecticut

Mr. Dean Larsen, Jefferson County Schools, Colorado

Mr. James J. McNeary, University of Wisconsin

Dr. David Miller, National Academy of Engineering

Dr. John Nickel, Wichita State University, Kansas

This book was set in Times Roman by Typographic Sales, Inc., printed and bound by Von Hoffmann Press, Inc. The designers were Mary Gaitskill and Ted Smith; the cover design was done by Otto von Ersal with photo detail by Scala New York/Florence. The drawings were done by Typographic Sales, Inc., and RSR Associates. The editors were Marnie Hauff and Eugene Wheetley. Lee Painter supervised the production.

ECCP has been supported primarily by grants from the National Science Foundation.

ISBN 07-019502-1

PREFACE

Throughout the writing of this book, three ideas have guided the authors.

1. *Learning should be fun.* There are well over a million engineers and scientists in the United States. Most of these men and women chose their careers because they enjoy understanding and creating something worthwhile. Learning how nature and the man-made world interact is like a never-ending game: each new bit of understanding leads to a host of new questions.

2. *Subject matter should be relevant.* Naturally, there are parts of science which are concerned with fundamental questions—the structure of the atom, the origin of the solar system, and the history of life. These are really the concerns of the theoretical scientist, seeking to advance human knowledge.

Most people, however, are directly concerned with technology which surrounds them and shapes their lives. This is the man-made world: the technology which has been created to improve men's lives. This is the transportation systems, the new buildings forming the modern city, the water supply and electrical energy systems, modern communication including television and the picturephone, and the automated factory or traffic-control system. It is also the pollution, the debasing of the earth, the unwise use of natural resources. If a technical course is to be *relevant*, it should develop those concepts which lead to an understanding of such use and misuse of technology and science.

3. *Science should be easy.* In our educational system beyond junior high school, technical studies usually have the reputation of difficulty. There is no logical explanation for this. Social science is inherently a much more difficult subject than the physical or life sciences. Social systems are more complex than systems with which technology and science are concerned. The largest computer ever built is much simpler than the justice system of even a small city.

ECCP

This book is a product of the Engineering Concepts Curriculum Project (ECCP). The Project started in 1965, when it became clear that fewer and fewer students were electing the physical sciences. This situation seemed particularly alarming when we heard on the national scene that the United States was entering an "age of technology." The computer alone promised changes which would make the Industrial Revolution seem a minor development.

CONTRIBUTORS

Dr. Jack Alford
Harvey Mudd College

Professor Hugh Allen
Montclair State College

Mr. Edmund Anderson
Monroe Sr. High School, Wisconsin

Dr. E. J. Angelo
Bell Telephone Laboratories

Dr. John L. Asmuth
University of Wisconsin

Mr. N. W. Badger
Garden City High School, N. Y.

Dr. Euval S. Barrekette
IBM Research

Dr. D. L. Bitzer
University of Illinois

Professor William B. Blesser
Polytechnic Institute of Brooklyn

Dr. Joseph Bordogna
University of Pennsylvania

Dr. Manfred Brotherton
Morristown, New Jersey

Professor A. E. Bryson
Stanford University

Mr. D. R. Coffman
James Caldwell High School, N. J.

Dr. Thomas Earnshaw
Episcopal Academy
Philadelphia

Mr. Jay Emmer
North Shore High School, N. Y.

Dr. R. L. Garwin
International Business Machines Corp.

Professor J. Richard Goldgraben
Polytechnic Institute of Brooklyn

Survival of our civilization depends in part upon our ability to adapt to these technological developments and to control the changes so that they remain in the realm within which society can adjust. Technology cannot be allowed to develop in such a way that groups of people are forced to withdraw from society. In a democratic country, such control can be realized only with public understanding of the nature, the capabilities, the limitations, and the trends of technology. Furthermore, the advance of technology cannot be understood without consideration of the interplay with social, economic, political, and psychological forces.

The problems we face as a nation are monumental. We not only need to understand technology, we must also anticipate its side effects. Change today is so rapid and the visible effects are often so delayed that the wrong decisions can lead to major problems before the effects can be reversed. Unfortunately, history emphasizes how difficult the problems are.

Thus, the primary objective of the book is a first step toward technological literacy.

THE CONCEPTS

Technology (or applied science) embraces far too many concepts for coverage in any one book, no matter how brief the treatment of each. Modern technology has three primary parts:

Systems Materials Energy

Any of these three could have been selected; we rather arbitrarily chose the first: systems. In particular, we have focused on a field which is often called *information systems* by scientists and engineers. The title indicates the concern with understanding information, how it is communicated, how it is used to control systems, and how men and machines interact in a system.

Why select systems? There are two reasons. First, the computer seems to be the single, most significant development in modern technology. With the computer (and automation), man can be relieved of the drudgery of routine decision making. Second, the systems approach is increasingly important in economic, political, and social studies.

EDUCATIONAL OBJECTIVES

In the 1970's, any text or course should include a clear statement of educational objectives. What should the reader or student be able to do after the course that he could not do before? How can we measure the success of the program?

While ECCP has prepared specific educational objectives for teachers to evaluate their effectiveness, the true goal of the course is a start toward technological literacy. Several concepts are introduced in the following chapters: decision making, modeling, dynamics, feedback, stability, and logical design. We hope for

Dr. A. Jay Goldstein
Bell Telephone Laboratories

Dr. Newman A. Hall
Commission on Education
National Academy of Engineering

Dr. L. D. Harmon
Bell Telephone Laboratories

Professor W. H. Hayt
Purdue University

Mr. Charles Hellman
Bronx High School of Science, N. Y.

Professor D. A. Huffman
University of California at
Santa Cruz

Professor W. H. Huggins
Johns Hopkins University

Professor C. E. Ingalls
Cornell University

Mr. J. G. Johnson
Sidwell Friends School,
Washington, D. C.

Dean Edward Kopp
Univ. of S. Florida, Tampa

Mr. A. E. Korn
James Caldwell High School, N. J.

Dean William A. Lynch
Polytechnic Institute of Brooklyn

Professor George Maler
University of Colorado

Dr. W. R. Marshall, Jr.
University of Wisconsin

Dr. John R. Pierce
Bell Telephone Laboratories

Mr. Robert Putman
Highline H. S., Seattle, Wash.

Dr. James Reswick
Case Western Reserve University

Mr. Rolla Rissler
Aurora Schools, Colorado

Mr. G. I. Robertson
Bell Telephone Laboratories

Mr. B. A. Sachs
Brooklyn Technical High School
Brooklyn, New York

an understanding of these concepts and then a familiarity with how a computer operates and what it can and cannot do.

For example, queueing is discussed in the first half of Chapter 3. After this study, we hope the reader will recognize queues as he moves through his normal life, will understand various ways the queues can be controlled, and will realize that certain queues are desirable while we should attempt to reduce others by appropriate economic or personal sacrifices. He should better understand the operation of modern urban traffic-control systems, for example.

Thus, two features found in many introductory science courses are relatively unimportant here. This is *not* a course on the scientific method. We are not emphasizing the methodology of the pure scientist. Second, mathematical problem solving is secondary to the process of logical thinking.

As a result, the classes where this course has been taught in the past have been characterized by much more discussion than usual, much less lecturing by the teacher. This book does not present a series of scientific facts which are really irrefutable by the student. Rather, the book presents a series of significant, current problems in which the concepts provide understanding.

ACKNOWLEDGMENTS

Over the five years of *The Man-Made World*, the Project has amassed an awesome list of people and organizations to whom we should express gratitude. Any such attempt at a complete listing would inevitably omit some of the many sources in education, industry, and government, and we can only list here a few.

The National Science Foundation has been the primary financial support for the Project throughout its history.

The National Academy of Engineering's Commission on Education (previously the Commission on Engineering Education) was the originator of the Project.

The Bell Telephone Laboratories and the Polytechnic Institute of Brooklyn have made major contributions in manpower and facilities, with significant contributions by IBM, Johns Hopkins University, M.I.T., Purdue University, American Machine and Foundry Co., Measurement Control Devices, and General Electric.

Finally, during the five years of development, more than 10,000 students have taken the course and, through their teachers, provided feedback which has been the primary guide in the transition from the early notes of 1966 to the present book.

Dr. Samuel Schenberg
Board of Education
New York, N. Y.

Mr. Robert Showers
East H. S., Green Bay, Wisconsin

Professor W. M. Siebert
Massachusetts Institute of
Technology

Mr. M. Simpson
West Essex High School, N. J.

Dr. Jeffrey D. Ullman
Bell Telephone Laboratories

Professor Andires van Dam
Brown University

Mr. R. A. Went
West Essex High School, N. J.

Professor G. Brymer Williams
University of Michigan

Dr. E. E. Zajac
Bell Telephone Laboratories

Universities involved in the preparation of teachers of *The Man-Made World* are:
University of Colorado
Georgia Institute of Technology
University of Houston
University of Illinois (Chicago)
University of Illinois (Urbana)
University of Massachusetts
Memphis State University
Harvey Mudd College
North Carolina State University
Pennsylvania State University
Polytechnic Institute of Brooklyn
University of South Florida
University of Washington
University of Wisconsin

E. E. David, Jr.
E. J. Piel
J. G. Truxal

June 21, 1970

TABLE OF CONTENTS

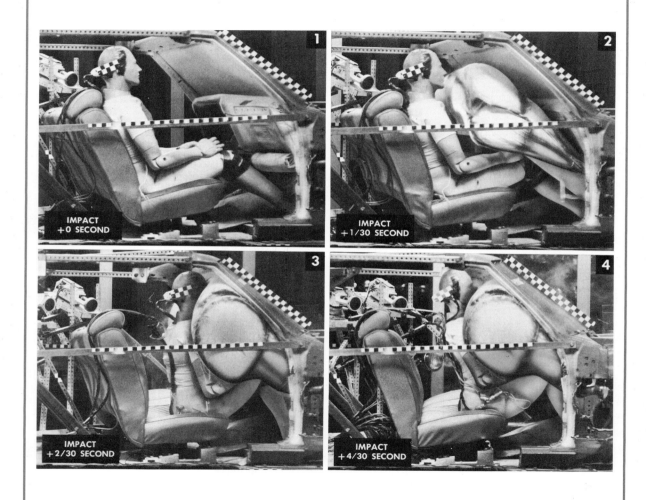

1 IMPACT +0 SECOND

2 IMPACT +1/30 SECOND

3 IMPACT +2/30 SECOND

4 IMPACT +4/30 SECOND

Technology and Man

1

I | THE MAN-MADE WORLD

The man-made world includes all the devices and systems made by man for the use of man.

These are the devices which surround us and touch on every aspect of our lives. Picture yourself in a city apartment house. The four-room apartment contains a kitchen where frozen foods are stored and prepared, a living room with television bringing in entertainers, educators, and politicians, and a bedroom and bath with fresh water, heat, and electricity available in essentially unlimited quantities. Over a wide range of income, man is provided with a span of conveniences unknown even to the very wealthy a few decades ago.

Yet in this same apartment house, incinerators burn the rubbish and garbage from its 1000 inhabitants. If these are average people, 5000 pounds of rubbish appear each day. The smoke and soot from the incinerator add to the pall of smog which so often rests over the city. The rubbish not burned is hauled away to be dumped into the nearby offshore waters, where it gradually changes marine life. Sewage from the building joins that from neighboring houses and flows through the underground pipes to the ocean or nearby lake.

Throughout the building, air conditioners hum 24 hours a day during the hot summer months. The resulting noise adds to the street noise and slowly impairs the hearing of the men and women who live there. Furthermore, the air conditioners demand so much electricity that the local electric company continues to use outdated coal-burning generating equipment. The results are further air pollution, and occasional, localized, electrical blackouts on particularly hot days.

If we glance into one apartment at 8:00 P.M. one summer evening, we find the wife loading the dishwasher and wondering why the oven is not working. In the living room, her husband

This series of photographs shows the operation of the General Motors Experimental Safety Air Cushion during a simulated automobile crash. (Fisher Body Division, General Motors Corporation)

sits on the sofa opposite the color television, watching a baseball game being played a thousand miles away. He glances occasionally at the game as he tries to write a letter to his health insurance company explaining that there was an error in the computer-printed bill he received four months ago. His three earlier letters have all been answered by form responses stating that an investigation has revealed that there has been no error on the company's part. Furthermore, he is now being charged $4.01 for late payment of his bill. As soon as he finishes the letter, he must complete the form to renew his automobile registration. (He recalls with a sigh that last year the job took an hour with the "simple" form. This year the state has decided to increase the information stored in its files.)

The man-made world surrounds us—the comforts and the pleasures, the complexity and the pollution. The rapid increase in productivity has meant more and more food and goods produced by fewer and fewer man-hours of labor. The normal working hours decrease. More and more of the labor force is in the service industries, rather than production. The United States population gradually moves away from the small towns and toward the large metropolitan areas, usually along the oceans or the Great Lakes. These changes accompany our motion into an *age of technology*. The health of our society, its cities, and its social institutions depends on our ability to adapt to modern technology and to control the development of that technology for the benefit of man.

In this book, we obviously cannot discuss in detail any significant fraction of the man-made world. The problems of urban transportation alone could easily occupy the full text. Within this problem, we would have to worry about not only the movement

Fig. 1–1.
A model of the city of 30 years from now. Such models serve as a guide for new product development by industry. They also allow study of city planning (for example, where should schools be located to minimize the need for buses or decrease the number of street crossings for children who walk). Finally, they serve as a partial guide for the many new cities which are now being built in the United States. (General Electric Company)

Fig. 1–2.
High-speed rapid-transit systems help the urban environment by decreasing cities' dependence on combustion-engine vehicles. One set of tracks of the Chicago Transit Authority, for example, can carry 40,000 people per hour versus 2000 for an auto lane.

The train system was built down the middle of the auto expressway purposely with the idea that motorists, caught in rush-hour traffic jams, would see the trains racing by. The next day they would certainly switch to the train.

Unfortunately, the first few weeks there were several train derailments. Motorists, crawling by, watched the stranded passengers. (*General Electric Company*)

of people within a city, but also the movement of goods of all types, both consumer goods (food, newspapers, clothing and garbage) and industrial materials (for construction, manufacturing, and fuel). We would have to look at how all forms of transportation can be developed together for the best total system. Furthermore, the transportation system cannot be considered unless we also study city planning, air pollution, noise pollution, and so forth.

Instead, we shall look at a few of the important concepts which underlie technology. Understanding these concepts will allow us to begin to understand modern technology. How does it work? What can it do? What can it *not* do? What are the potential problems? What are the possible dangers?

The Objectives of the Book

This book has two closely related objectives. From the study of certain concepts of technology, we hope to achieve sufficient understanding to be able to:

1. live with technology and adapt technology to our needs.
2. make intelligent judgments about problems associated with technology.

In more dramatic terms, we want enough understanding to survive in a technological society—to survive as individuals and to survive as members of a society which controls its technology.

What do we mean by "survival as individuals"? We are approaching a time when most of our actions bring us into direct contact with technology. Anyone driving in almost any large city today is part of a large traffic-control system. Increasingly, this total system is automated. In the near future you and your car will be following the detailed instructions of the system. On travel between major cities in the future, you will enter an automated highway, after which your car will be driven automatically at high speed to your desired exit. Communication to you will be from the computer through your car radio.

"Survival as individuals" means having enough understanding to live with such technology—to benefit from its advantages and capabilities, to obtain a fuller life. "Survival as individuals" means learning to cope with the problems created by technology. It means modifying your behavior to adapt to what is possible.

What do we mean by "survival as members of society"? If we are to survive as a group, we must be sure technology develops under our control. We must find a balance between personal privacy or liberty and the welfare of all the people. We must control all parts of the environment: air, water, and land. Technology must be used to improve education and health services for all the people.

In the future, there will be an increasing number of situations in which difficult political and social decisions will be required. In the not-too-distant past, political "bosses" made such decisions when they were necessary for the good of the people. The strength of these political leaders was so great they could disregard objections by special-interest groups.

Today in most United States cities and states, the political leaders are no longer all-powerful bosses, and the voters are much better educated and informed. Difficult decisions are possible only if the people support the action. For example, a city mayor launches a major tow-away program for illegally parked cars in order to improve traffic flow. He is immediately attacked by irate motorists and special-interest groups who are used to parking illegally. Subjected to this roar of protest, often in public, the mayor is likely to retreat unless he is assured of reasonable public support.

The decisions of the future will be much more difficult and also more closely related to technology. How strictly should industrial water pollution be controlled? Are we willing to force a factory out of business if it does not stop discharging waste into a river? Are we willing to use tax money to help the factory? How can competing factories be treated fairly? How can different industries with different waste products be handled fairly?

The voters in any town, state, or nation are being asked to form opinions on many technological questions. Some understanding of the basic ideas of technology is no longer a luxury associated with the "well-educated man." It is a necessity in an age of technology. Survival of our society depends on the ability of the public to react wisely to technological issues, to vote with understanding when difficult political decisions must be made, and to support governmental leaders who suggest appropriate compromises.

2 | MATCHING TECHNOLOGY TO THE HUMAN USER

Technology should serve man. Technology should fill a need; it should make life safer, richer, and more pleasant. If these goals are to be reached, technology must be *matched* to the human user. In other words, we must design the technology by considering the way man behaves.

The need to match technology to the human user is illustrated by the traffic-light problem of this chapter. In this problem, we focus on the place where technology meets the human user. How do we measure the important characteristics of man? How do we then decide how the technology must work? What is the best possible technology? In Section 4 we consider the problem of matching technology to society.

The Traffic-Light Problem

Figure 1–3 shows the problem. A side road, we call it Residential Street, leads from a group of homes onto a main state highway, which we call Main Turnpike. The speed limit on Residential Street is 40 miles per hour, although during the early morning hours men going to work often drive as fast as 60 mph. Likewise, in the evening, after children are no longer playing along the street, cars occasionally move at 60 mph.

The speed limit on Main Turnpike is 50 mph. There is heavy truck traffic on this street during the day. In the late evening, drivers are sometimes observed driving with apparent disregard for safety. Ambulances and fire trucks often use Main Turnpike.

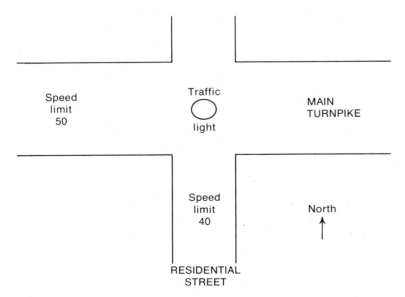

Fig. 1–3.
Intersection of two-lane
Residential Street and four-lane
Main Turnpike. A traffic light
is installed at the intersection.

A series of accidents at the intersection of Residential Street and Main Turnpike has finally convinced officials that a traffic light should be installed at the intersection. A red-yellow-green light is to be purchased. We must decide what kind of cycle the light should have.

1. Should the light have a regular cycle? That is, should the light regularly change? If so, we might call the four periods of time:
   a) Green for Main Turnpike, Red for Residential Street.
   b) Yellow on Main, Red on Residential.
   c) Red on Main, Green on Residential.
   d) Red on Main, Yellow on Residential.
   Instead of this regular cycle, we might have a detector on Residential Street. The light would stay green for Main until a car drove over the detector. Then the light would turn green for Residential for a short time.
2. If a regular cycle is chosen, should it be the same all day long, or should periods *a*, *b*, *c*, and *d* be changed at midday or late at night?
3. If a regular, unchanged cycle is chosen, what should *a*, *b*, *c*, and *d* be?

We should first point out that the problem is not easy. More cars can be moved through the intersection if the lights do not change very often. Every time cars start up after a light turns green, they pick up speed slowly. Therefore, we might like to make *a* three minutes long. That is, the light would stay green on Main for a full three minutes, while cars wait on Residential.

Even though this long green on Main might be a good decision, it won't work with human drivers. The people who live near Residential will learn about the *long* red light as they enter Main. A visitor, however, who drives up Residential just after the light turns red will be convinced the light isn't working. After a minute or two, he will try to drive out on Main even though the light is still red for him. Our light is likely to cause more accidents than it prevents.

Thus, the total cycle cannot be so long that drivers will not accept it. On the other hand, the shorter the cycle, the fewer cars we can move through. If the cycle is very short, we will soon have a long line of waiting cars on Main.

Furthermore, it is difficult to pick the duration of the yellow (parts *b* and *d* of the cycle) in either direction. Suppose that as a car drives up to the intersection on Main, the light turns yellow. The driver has to make a decision *immediately* whether to stop or not. If he is too close, he will decide to go through. We would like the yellow to last long enough for him to make it through the intersection before the cars waiting on Residential are given

Fig. 1–4.
Traffic lights are only one way to signal instructions to the driver. Road signs are even more common, although in the United States there is an unfortunate lack of uniformity.

Throughout almost all of Europe, road signs have been standardized according to recommendation of the United Nations. The signs at the top are easily recognized without written explanations which woud have to depend on the driver's recognition of the language.

The lower part of the figure shows a road sign in Afghanistan. Any visitor driving in Afghanistan would know immediately that he is to drive on the left (unless he is traveling by camel).
(United Nations)

a green light and surge forward. Unfortunately, if we make the two yellows b and d long enough, fewer cars are moved through the intersection and traffic piles up.

In choosing a traffic-light cycle, we must compromise. If drivers obeyed the light, the safest arrangement would be red in all directions at all times. No one would ever be injured, but no car would ever move either. We cannot decide on the basis of maximum safety alone.

Similarly, we cannot decide only on the basis of moving the most cars through the intersection. We have already seen this would mean a very long light cycle and impatient drivers would "run" the light.

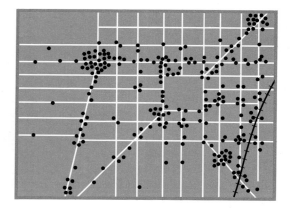

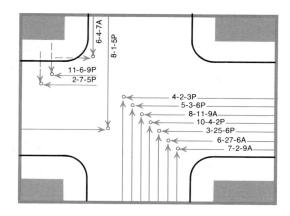

Our decision must be a compromise. We want to move cars as rapidly as possible. We also want to prevent accidents. Finally, we must choose a cycle which is accepted by most drivers. The cycle chosen must be *matched* to the human drivers, their attitudes and their behavior. The technology must be matched to man if it is to serve him well.

Timing the Traffic Light

Finding the best timing cycle for our traffic light is a complex problem. We must learn about driver habits. We also must measure the times cars need to stop at different speeds to decide how long to make the yellow. We need to measure the number of cars passing each minute at various times of the day to be sure that we can handle the expected traffic.

In the next section, we will look further at some of these measurements. Before that, however, we should mention how traffic lights are actually timed in most towns. A light is purchased from a manufacturer. The engineers guess that the traffic on Main is about five times that on Residential. Therefore, it makes sense to set the green on Main five times as long as on Residential. The yellow periods are made about three seconds long. If one full cycle is 60 seconds, this means:

a—Green Main	45 seconds
b—Yellow Main	3 seconds
c—Green Residential	9 seconds
d—Yellow Residential	3 seconds

The light is installed. The first day, a patrolman watches the traffic flow. If it seems to pile up on Residential, *c* is lengthened and *a* is correspondingly decreased. After one or two adjustments of this type, the light is left operating. If these adjustments are made at the wrong time of day, rush hour traffic may be snarled worse than ever.

Fig. 1–5.

Various models or graphs are used to describe the particularly dangerous streets and intersections in a town. The left figure shows where accidents have occurred. From this record, the traffic engineer can see at once where traffic lights should be changed or where additional protection must be provided.

The right diagram is a record of collisions at one intersection. The solid lines are cars, dotted lines pedestrians. The numbers refer to month—day—hour of the collision. (American Automobile Association)

After the light is working, accidents are observed at the intersection. If one or two accidents occur because cars on Residential do not make it through before the light for Main turns green, part *d* of the cycle is lengthened, possibly to five seconds. If accidents continue, *d* is made even longer or the speed limit on Residential is lowered.

Thus, in actual practice the timing cycle is chosen by trial and error. If the engineer has put in traffic lights before, he can often guess a suitable timing. There really is little attempt to find the best cycle. Once a cycle is found that seems to work most of the time, it is left alone.

Furthermore, the cycle is often not changed during the day as traffic patterns change. Equipment to change the timing automatically is expensive and often fails. If a man is assigned to change the timing at certain hours, he may forget or be busy with other work. If the traffic piles up too badly in one direction, a patrol car is sent to the corner and a policeman takes over traffic direction.

When traffic is not too heavy and speeds are low, this sort of trial-and-error approach is all right. As traffic becomes heavier, we may find long waiting lines on Main back up to block other intersections a quarter of a mile away. Traffic blocked in a few intersections can lead to a gigantic tie-up (Fig. 1–6). Also, when speeds become greater, accidents tend to be more serious. The timing of the yellow lights then is particularly important.

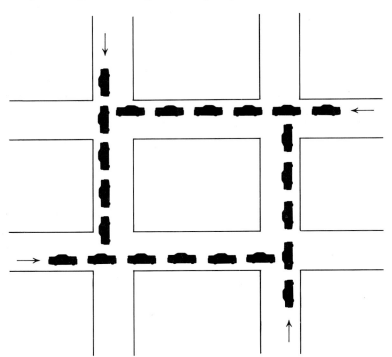

Fig. 1–6
When cars back up from an intersection in a network of city streets, there can be total disaster. In the tie-up shown, motion is impossible. As the cars sit there waiting for something to happen, the lines grow longer and longer. Soon adjacent intersections are affected.

Visions of such problems convince traffic engineers that the only solution for most city problems will come when late one night all streets are made one way out of the city.

Thus, as our traffic system becomes more complex, we soon reach a point at which we must have a more logical approach to decisions. The technology must be more carefully matched to man. Under these circumstances, we must use a *quantitative approach.* That is, we must measure the parts of the problem and make our decisions in such a way that we develop the best possible system.

3 | A QUANTITATIVE LOOK AT THE PROBLEM

The idea of taking a quantitative look at the problem is central to modern technology (and this book). The same idea appears in modern social sciences, economics, and psychology. We attempt to define our problem in terms of numbers, graphs, or mathematical relations. Once we describe the problem in such quantitative terms, we can compare different possible solutions. We can then make a logical decision.

To illustrate this quantitative approach, we return to our problem of timing the traffic light at Main and Residential. We look at only one part of the problem: How long should the light on Residential be yellow? This is part *d* of the total cycle described in the last section.

Fig. 1–7.
Inscription on a tombstone in Munich, Germany. In the fourth line, the date appears M·DCZ4, or 1624. Note the combination of Roman and Arabic numerals. Western Europe gradually accepted the "technological" change from Roman to Arabic numerals. (K. Menninger, Number Words and Number Symbols, M.I.T. Press, Cambridge, Mass., 1969.)

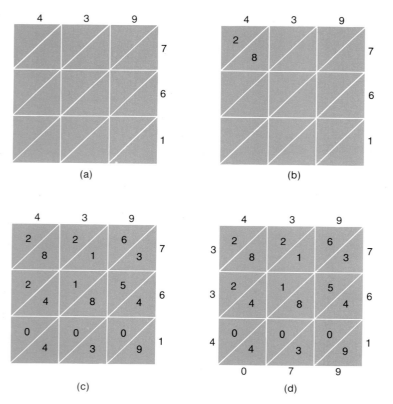

(a)

(b)

(c)

(d)

Fig. 1–8.
When Arabic numerals were first introduced, a major advantage was ease in multiplication. This table illustrates the method often used to teach multiplication with the "new" numbers.

To multiply two 3-digit numbers (439 and 761), we make a 3 × 3 table and place the numbers as shown (a). The diagonal lines are included. In the upper left square, we place the product 4 × 7 (b). Each square is filled with the product of the two integers representing that square (c).

Now the numbers are added along the diagonals and written below and to the left of the table. When there is a carry, this is added to the next diagonal sum (d).

The answer is 334,079.

Important Factors

A driver on Residential sees the light ahead turn from green to yellow. If he is far enough from the intersection, he will apply the brakes and stop. If he is close to the intersection, he will go through, hoping to "beat" the red light. Figure 1–9 shows the GO and NO-GO zones. If he is in the GO zone, he continues. If he is in the NO-GO zone, he stops.

Of course, there is no sharp separation between the GO and NO-GO zones. Different drivers will behave differently. Even a single driver will vary depending on his mood, his speed, the slippery state of the road, whether he has passengers, and so on.

We certainly should keep the light yellow long enough to allow most of the drivers to make it through the intersection. Should this be two seconds, four seconds, six seconds, or what? We also do not want to make the yellow longer than necessary for two reasons:

1. A long yellow means longer times when traffic is not moving through the intersection.
2. Most drivers on Residential drive here frequently. If they learn the yellow is long, they will gradually extend their GO zones.

Thus, again we want a compromise. The yellow should be long enough to allow most drivers who decide to "make a run for it" to get through, but no longer than necessary.

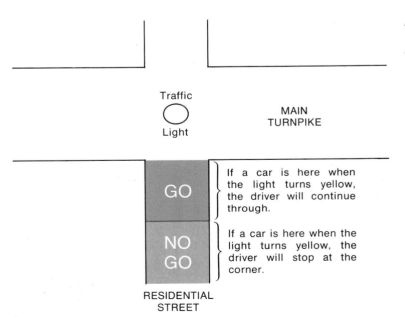

Traffic ○ Light

MAIN TURNPIKE

GO — If a car is here when the light turns yellow, the driver will continue through.

NO GO — If a car is here when the light turns yellow, the driver will stop at the corner.

RESIDENTIAL STREET

Fig. 1–9.
GO and NO-GO zones as driver comes north on Residential toward the intersection with Main Turnpike.

The GO Zone

The experienced driver decides not to stop when he judges a stop would be uncomfortable or dangerous. Obviously, he doesn't want a passenger on the front seat to crash into the windshield. If he stops too suddenly, a car behind him may plow into him. Thus, the driver seeing the light turn yellow judges rapidly whether he can make a convenient stop. If he has driven here often before, he may remember how long the light is yellow and decide whether he can safely make it through. Even if he is unfamiliar with the light, he may guess at the timing. We certainly do not want to make the yellow unusually short.

What determines the comfortable stopping distance? What happens between the driver's first sight of the yellow and the final stop? There is a sequence of events.

1. First, there is a human reaction time. The eye observes the yellow, a message is sent to the brain, a decision is made, a message is sent to the muscles controlling the right foot, and the foot begins to move.
2. Time is required for the foot to move to the brake and then to push down on the brake pedal.
3. After the brake is applied, the car decelerates, or slows down. The greater the deceleration, the more the passengers are thrown forward and the more uncomfortable the stop.

Hence, there are three different times involved in stopping. During each of these times, the car travels a certain distance. The sum of these three distances is the total distance needed for the stop. In order to find the minimum convenient stopping time or distance, we need to estimate each of these three times.

Human Reaction Time

How long is the human reaction time? We can gain some idea of this from a very simple experiment, which is often performed in biology courses.

The person being tested (subject) sits with hand outstretched (Fig. 1–10). The thumb and forefinger are ready to close and grab a ruler. A 12-inch ruler is held by someone else (the tester), with the 12-inch mark at the top, 0 at the bottom. The tester drops the ruler at any instant he chooses. The subject observes the ruler falling and grabs it as soon as possible. We note the distance the ruler has fallen before it is grabbed.

If the ruler merely falls and is not thrown downward, the distance it falls is a direct measure of the time. The relation is shown in Fig. 1–11. A fall of 7.7 inches corresponds to a time of 0.2 second. This time is the human reaction time for this particular experiment. It is the time needed for the eye to observe the ruler

Fig. 1–10.
Testing human reaction time. The tester drops the ruler at some unexpected moment. The subject closes his fingers to grab the falling ruler. The distance fallen is a measure of the reaction time.

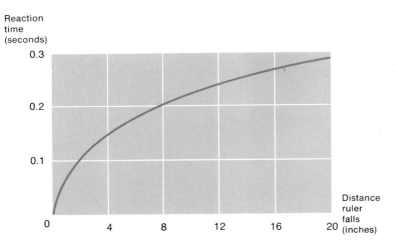

Reaction
time
(seconds)

Distance
ruler
falls
(inches)

*Fig. 1–11.
Reaction
time found
from the distance
the ruler
falls.*

falling, the message to go to the brain, and a message to go to the muscles controlling the thumb and forefinger.

This experiment measures the reaction time in a very simple situation. If we are interested in the reaction time for our driver who spots the light turning yellow, we really should measure the human being performing that task, rather than grabbing a ruler. In the experiment, the decision to act is not difficult, and the signal has to travel only to the hand rather than the foot.

In spite of these shortcomings, the experiment does indicate that the *minimum* human reaction time is 0.2 second, though it varies appreciably from person to person. In the driving task, the actual reaction time seems to be about twice this long, or 0.4 second. In this time interval, the car at a speed of 40 mph travels about 23 feet. In other words, the car moves 23 feet before the driver's foot even starts to move from the gas pedal.

As an aside, the same ruler experiment shows the effect of alcohol on reaction time. If we measure an occasional drinker while he is having a beer or two, we find that his reaction time steadily gets worse. Indeed, very soon he cannot even catch the ruler. The most interesting part of this experiment is that the subject is convinced his reaction time is as good as ever. When he misses the ruler, he is sure that the tester is throwing it down instead of dropping it. Small amounts of alcohol tend to make men much poorer performers in complex tasks such as driving. At the same time, the alcohol relaxes the man and weakens his ability to judge himself.

Time To Start Braking

Once the driver starts to move his foot, there is additional time before his foot hits the brake to start the car slowing down. Here again we need experiments to measure the typical human

performance. When we make such tests, we find that the usual time is slightly more than 0.2 second (the same as the reaction time). These tests are made on subjects who are waiting for a certain signal. They are keyed up for the quickest possible move. Consequently, their performance is much better than the average driver in a car.

While drivers vary over a wide range, the usual time to "hit" the brake is about twice as long as the measured 0.2 second. That is, once the driver's foot starts to move, 0.4 second pass until the brake is hit. Again, the car travels 23 feet at its speed of 40 mph.

In other words, when we combine "driver reaction time" and "time to start braking," 0.8 second passes from the light turning yellow until the brakes are applied. The car travels 46 feet.

Both the reaction time and the time to start braking are characteristics of the human being. Now you see that to understand the timing of a traffic light, we must make measurements on man. In Chapter 9, we will consider further the various things man cannot do and how we can use machines for these tasks. Here we are interested in only one limitation of man: the fact that he just cannot respond immediately.

It is really only in the last few decades that we have recognized that the man-made world can only be studied by also studying

Fig. 1–12.
This experimental television installation in Detroit relays traffic information to a control center, where an operator can regulate lane utilization, vehicle speeds, and the opening and closing of entrances to, and exits from, the freeway. The object is to prevent or reduce traffic congestion and delays, to increase the traffic-carrying capabilities of the freeway, and to make it safer. The circle in the right photograph surrounds one of the cameras used in the installation.

This type of installation may be commonplace in the future and will provide direction and control based on up-to-the-minute knowledge of traffic conditions. (American Automobile Association)

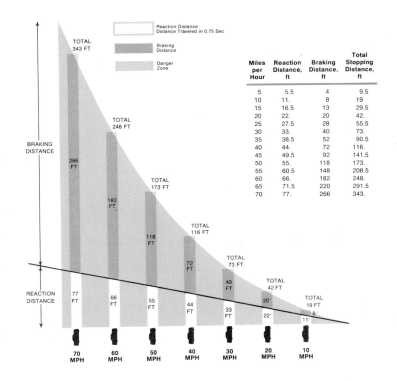

	Reaction Distance Distance Traveled in 0.75 Sec
	Braking Distance
	Danger Zone

Miles per Hour	Reaction Distance, ft	Braking Distance, ft	Total Stopping Distance, ft
5	5.5	4	9.5
10	11.	8	19
15	16.5	13	29.5
20	22.	20	42.
25	27.5	28	55.5
30	33.	40	73.
35	38.5	52	90.5
40	44.	72	116.
45	49.5	92	141.5
50	55.	118	173.
55	60.5	148	208.5
60	66.	182	248.
65	71.5	220	291.5
70	77.	266	343.

BRAKING DISTANCE

REACTION DISTANCE

TOTAL 343 FT — 266 FT — 77 FT — 70 MPH

TOTAL 248 FT — 182 FT — 66 FT — 60 MPH

TOTAL 173 FT — 118 FT — 55 FT — 50 MPH

TOTAL 116 FT — 72 FT — 44 FT — 40 MPH

TOTAL 73 FT — 40 FT — 33 FT — 30 MPH

TOTAL 42 FT — 20' — 22' — 20 MPH

TOTAL 19 FT — 8' — 11' — 10 MPH

Fig. 1–13.
These minimum stopping distances are based on tests made by the Bureau of Public Roads. The chart shows how greatly braking distances and danger zones increase when you increase your speed. The distances are for a stop in minimum time (under emergency conditions). Reaction time is 0.75 second. (American Automobile Association)

man. The part of biology dealing with man is an important part of our understanding of man-made devices and systems.

This appreciation of the importance of man really started in the 1920's. After World War I, the Versailles Treaty did not allow Germany to have aircraft with motors. In order to continue research, the Germans became experts in gliders. They tried to use men to provide some power to propel the gliders and give them greater range and maneuverability. As a result, they became interested in learning how to increase the power a man could provide.

In other words, what was the best way to draw power from a man? Should he wave his arms, pull and push with his arms, use his feet? Which way would provide the most power over a period of time?

The studies showed that the best arrangement is for man to pedal with his feet and legs. A back-and-forth pedaling yields the greatest power. This discovery is interesting if we compare it with crew-racing shells. It means that a boat driven by eight men pumping with their feet could outrace a normal shell with eight oarsmen.

In the 1930's the Germans under Hitler disregarded the Versailles Treaty and turned their attention to regular airplanes. Interest in studying the human being waned. Then during World

War II, anti-aircraft gunners seemed to be extremely unsuccessful in shooting down high-speed planes at close range. A study of the human being showed that his reaction time was causing trouble.

Interest in measuring this reaction time grew when radar was successful in detecting planes. To avoid radar, pilots tried to fly at very low altitudes. When a plane is moving at 400 mph, it travels almost 600 feet every second. If the pilot of the low-flying plane spotted a hill 600 feet ahead, he had only one second to increase the altitude of his plane to avoid crashing into the hill. As the speed of planes has continued to increase during the last 25 years, pilot reaction time has limited the performance. The pilot must be replaced by an automatic pilot which has negligible reaction time.

In the 1950's the rising toll from auto crashes in the United States led to more detailed studies of the human driver. When it was discovered that more than half of the fatal accidents involve a driver under the influence of alcohol, quantitative studies were made of the effects of drinking on human characteristics.

Stopping the Car

Our car is moving at 40 mph. How long does it take us to stop? How far do we travel before stopping? To answer these questions, we would have to know: How hard does the driver hit the brakes? Are the tires and brakes in good condition? Is the road dry?

Traveling at 40 mph, the fastest stop which doesn't throw the passengers into the windshield takes about 2.9 seconds. The car travels 85 feet. Such a stop is decidedly uncomfortable for the driver and passengers.

We are interested in comfortable stops. How long will it take after the driver decides to stop? If he is experienced, he will only try to stop if he knows it can be done comfortably and easily.

To determine what a normal stop from 40 mph means, we can measure with various drivers. As the driver passes a chosen point in the road, he has his foot on the brake and is ready to stop. We find that an easy stop takes about 4.6 seconds and the car covers 135 feet.

Stopping for Yellow

If we combine the three steps after the driver observes the light turn yellow, we see that typical values are:

1. Reaction time is 0.4 second.   Car travels   23 feet.
2. Time to brake is 0.4 second.   Car travels   23 feet.
3. Braking time is 4.6 seconds.   Car travels 135 feet.
Total time to stop is 5.4 seconds.   Car travels 181 feet before stopping.

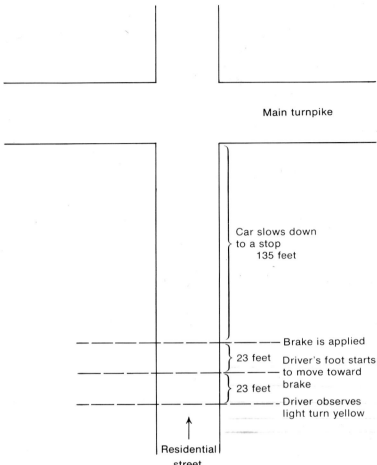

Fig. 1–14.
Three steps in stopping
after the driver sees the
yellow light (40 mph initial
speed; distances are
greater for
higher
speeds).

Main turnpike

Car slows down
to a stop
135 feet

Brake is applied

23 feet — Driver's foot starts
to move toward
brake

23 feet — Driver observes
light turn yellow

Residential
street

If the car speed is greater than 40 mph, the distances are longer. For example, if the car speed is 60 mph, the distances are 35 feet, 35 feet, and 304 feet, a total of 374 feet, or more than twice as far.

Timing of Yellow on Residential Street

Now we are ready to decide logically how long the yellow light should be on for Residential. We have found that a car traveling 40 mph needs 181 feet to stop. The experienced driver knows this. Consequently, he is not likely to stop if he is closer than about 200 feet. Thus, we can guess that the experienced driver will plan to go through on yellow if he is closer than 200 feet.

How long will it take him to make it through from 200 feet? At 40 mph, he is moving at 59 feet/second. To travel 200 feet requires 3.4 seconds. Certainly, we should keep the light yellow for 3.4 seconds. (Typical time seems to be about 2 seconds.)

If we want to be sure cars at 60 mph get through the yellow, we need to provide for about 400 feet of a GO zone. The corresponding time is 4.5 seconds.

The yellow-light example is important for two reasons:

1. It shows that many of today's technological decisions are not made very intelligently. They are really not matched to the human user. The yellow traffic light is often 2 seconds long. This is just not enough time for many of the cars to stop.

   Because of human impatience and our desire to move cars through the intersection, we compromise on the safety side. Certainly a major share of the 55,000 auto deaths each year in the United States results from this compromise with adequate and strict enforcement of safety regulations. Frequently, statewide attempts to be more strict are met by a chorus of public protest.

2. The most important feature of the example is that it shows the importance of a quantitative approach. Once we measure the times and distances in each part of the stopping process, we can make a logical decision about the timing cycle of the light. We can determine the effects of improper timing.

As we move through the later chapters of this book, we will find that a unifying theme is this idea of a quantitative approach. Once we have a numerical or quantitative picture of the problem, we can understand the operation of the device or system. We can see its capabilities and its limitations. We can judge whether the device is matched to the human user.

Consider now the problem of an inflatable bag to protect car passengers in a collision (see Frontispiece). The idea is simple. At the time of impact in a collision, a large nylon bag is quickly inflated between the people in the front seat and the dashboard and windshield. This bag protects the people in the car, because it prevents them from being hurled against or through the windshield and panel.

Is such a scheme feasible? What are the problems? In an average-sized United States car, there is about 0.040 second between the collision and the passenger hitting the inside of the car. To fill the bag in less time, high-pressure gas must be stored and then exploded into the bag. For small cars, there is even less time.

There are some serious problems with such a safety device. We don't want to injure the passengers with the exploding bag. Furthermore, the system must never operate if there is no collision. We can picture a poorly designed system in which the bag mistakenly inflates when we hit a severe bump while driving along at 60 mph.

The Toyota Corporation of Japan has proposed a different system. A small radar set in the front of the car measures approaching objects and a computer predicts a collision. This allows a little more time for the bag to inflate. The danger, of course, is that a car coming toward you may actually miss you, but still cause your bag to inflate. Furthermore, we cannot help but be somewhat suspicious of the reliability of a radar set and computer, even if the entire system were simple enough to cost only $100, as promised.

Thus, while the bag idea seems attractive (particularly since 80% of the drivers refuse to use seat belts), a quantitative study raises serious questions. This quantitative approach is also emphasized in our second example. In the next case, we want to consider the use of technology in tackling an important social problem.

4 | MATCHING TECHNOLOGY TO SOCIETY

Having discussed matching technology to the human user we turn now to the problem of matching technology to society in which we illustrate the use of technology to improve social conditions. We ask: Can technology be used to improve the general health of the people? In particular, can we provide *better health care for more people* if we use man-made devices to help the physician?

Preventive Health Care

If we hope to improve health, one possible way is to measure frequently the health condition of large numbers of people. If disease or a serious condition is found, the individual is referred to a doctor for personal treatment. An example of this type of *mass screening* (testing of many people) is the mobile chest X-ray unit which travels from location to location and takes pictures by the thousands. The same type of mass screening is often found in elementary schools when pupils are tested for hearing, sight, or dental problems.

In the last few years, electronic equipment for testing a man's health condition has become much more elaborate. We can now test for many different diseases and malfunctions. We can often spot trouble before it is apparent to the man. Furthermore, in many cases the early detection makes medical treatment possible and simple.

At the same time, computers are now available which can accept very large amounts of data, remember and "file" these data, compare different test results, and then print out the data with a special mark beside those results which seem abnormal. The computer, in other words, allows the processing of quantities of data which just could not be handled by human beings.

When we consider these two developments (better instruments and computers), the next step is evident. We can build a health testing center: a building where thousands of people each year can receive health examinations. A computer collects all the data and automatically sends to each person's physician a health report on that individual. The building, or center, need have no doctors present. It is merely a testing center.

Such a building is called a *multiphasic health testing* center (abbreviated MHT center). The word *multiphasic* means that many different phases, or aspects, are measured. If we were to take only chest X-rays, the center would not be multiphasic. We usually also measure hearing, sight, blood pressure—often as many as twenty different tests. Thus, the MHT center gives an exam which covers a wide span of health.

MHT centers are now in operation in more than a hundred locations, and hundreds of others are being planned and built. The center tests about 20,000 people per year. If the present growth continues, 6000 centers should be operating by the late 1970's. Consequently, we can expect to be testing over 100 million people annually in the United States within a short time.

The Health Problem

In spite of all the publicity about heart transplants and medical research, there are strong indications that the health condition of the American people is not impressive. For example, life expectancy in the United States has not improved significantly during the past twenty years (Fig. 1–15a). From 1959 to 1966, the United

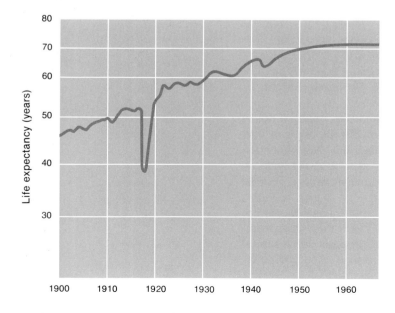

Fig. 1–15 a.
United States life expectancy at birth for females.

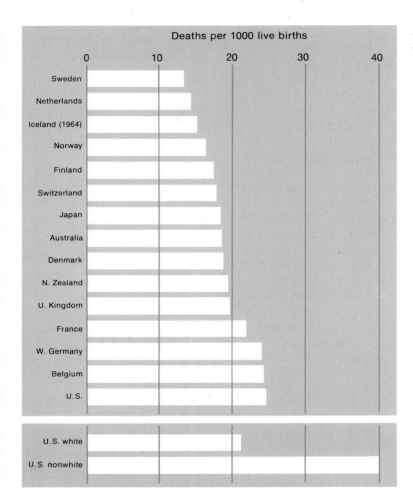

Fig. 1–15 b.
Infant mortality rates in 1965.

States dropped from 13th to 22nd place among the nations of the world in male life expectancy.

Even more dramatic is the low standing of the United States in infant mortality. In this country, this figure is about $\frac{22}{1000}$. In other words, of each 1000 babies born alive, 22 will probably die during their first year. The corresponding figure in Sweden is about one half as large (Fig. 1–15b).

What are the reasons for this standing of the United States? Certainly many factors contribute. In a recent year in New York City, for example, 40% of the women who gave birth had not visited a doctor during their pregnancy. These people are separated from the medical and health-care system. Apparently they do not know how to obtain medical help. Many live a life separated from the familiar social services.

This separation between a large part of the population and the health-care system is emphasized by the infant mortality rates for different groups. In the major cities, for instance, the rate is about $\frac{35}{1000}$ compared to $\frac{22}{1000}$ for the entire United States. There has been a slight drop since the infant-care program of the late 1960's, but city people are still at a severe disadvantage.

The statistics are even more troubling if we look at the non-white population. Infant mortality is over $\frac{40}{1000}$ (almost twice the figure for the entire country). Clearly, this part of the population is separated from the health-care system. The separation is emphasized by the fact that only 2.2% of physicians are black, while 11% of the total population is black.

Thus, certainly a major part of the United States health problem arises from our failure to bring reasonable health care to a large fraction of the people. The American makes 5.5 visits per year to a doctor, compared to 3 visits by the people in Sweden (the usual standard for health care). This seems to indicate that we have an adequate supply of doctors. We simply do not use them properly.

In region upon region of the central city, we find an alarming shortage of physicians (often perhaps one doctor for 25,000 people). The MHT center offers the possibility of bringing this neglected population into the health-care system. Through a network of MHT centers in the cities and mobile centers for rural areas, we can hope to bring reasonable health care to *all* the people.

5 | AN MHT CENTER

To learn what an MHT center is, we project ourselves a few years into the future and travel through a typical center. First, the center is located well away from any hospitals or doctor's offices. It is truly a personal health center, and everything possible is done to make the visit pleasant. Two of us have made appointments together, so that we can talk as we travel from room to room.

As we pass through the front door, we enter a reception room. A pleasant hostess welcomes us and explains the purpose and procedures since we have not been there before. She hands each of us a personal identification card, which we will show in each room to be sure the test results are credited to the right person in the computer. We are then led to chairs before a TV set, where we see a seven-minute program describing what we will be doing during the next two hours. As soon as the program ends, we are asked to go to Room 1.

Fig. 1–16.
Patient flow schedule.
(Searle Medidata Inc., a subsidiary of G. D. Searle & Co.)

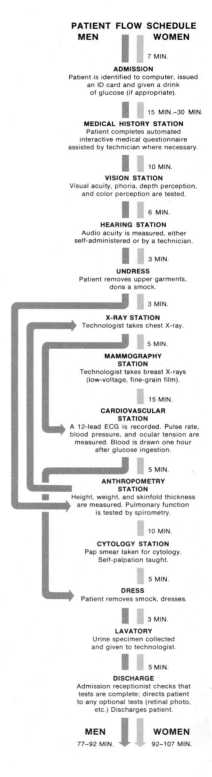

PATIENT FLOW SCHEDULE
MEN   WOMEN

7 MIN.
ADMISSION
Patient is identified to computer, issued an ID card and given a drink of glucose (if appropriate).

15 MIN.–30 MIN.
MEDICAL HISTORY STATION
Patient completes automated interactive medical questionnaire assisted by technician where necessary.

10 MIN.
VISION STATION
Visual acuity, phoria, depth perception, and color perception are tested.

6 MIN.
HEARING STATION
Audio acuity is measured, either self-administered or by a technician.

3 MIN.
UNDRESS
Patient removes upper garments, dons a smock.

3 MIN.
X-RAY STATION
Technologist takes chest X-ray.

5 MIN.
MAMMOGRAPHY STATION
Technologist takes breast X-rays (low-voltage, fine-grain film).

15 MIN.
CARDIOVASCULAR STATION
A 12-lead ECG is recorded. Pulse rate, blood pressure, and ocular tension are measured. Blood is drawn one hour after glucose ingestion.

5 MIN.
ANTHROPOMETRY STATION
Height, weight, and skinfold thickness are measured. Pulmonary function is tested by spirometry.

10 MIN.
CYTOLOGY STATION
Pap smear taken for cytology. Self-palpation taught.

5 MIN.
DRESS
Patient removes smock, dresses.

3 MIN.
LAVATORY
Urine specimen collected and given to technologist.

5 MIN.
DISCHARGE
Admission receptionist checks that tests are complete; directs patient to any optional tests (retinal photo, etc.) Discharges patient.

MEN   WOMEN
77–92 MIN.   92–107 MIN.

In Room 1, an operator measures height, weight, and important body dimensions. All results are fed immediately into the computer, which from now on keeps a record of all test results for each of us. Very often, there is also a measure of obesity. This is done by measuring the thickness of excess fat in the upper arm. If the computer decides we are too fat, we will receive a booklet or a monthly magazine on the danger of obesity and ways to lose weight.

In Room 2, hearing is checked. In Room 3, we have eye tests. We read the usual eye chart. The field of vision is measured automatically—a computer moves a flashing spot around, and we indicate when we can see the flash. Finally, the pressure inside

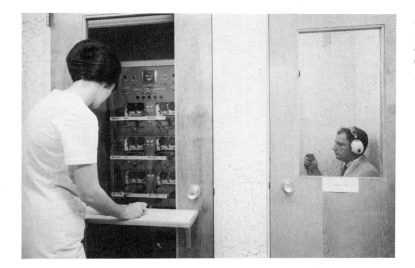

Fig. 1–17.
Hearing is tested with
an automated audiometer.
(*Kaiser Foundation*
Hospitals)

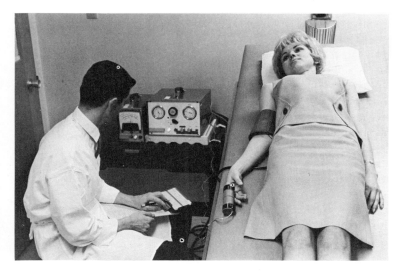

Fig. 1–18.
Pulse rate and blood pressure
are taken and recorded on the
marksense test card. (Kaiser
Foundation Hospitals)

Fig. 1–19.
Vital capacity of lungs is measured by blowing into a respirometer. (Kaiser Foundation Hospitals)

the eye is determined. (Particularly in middle-aged people, an increase in this pressure is a warning of glaucoma, a disease which inevitably causes blindness if it is not treated.)

Another room is devoted to spirometry, the measurement of the amount of air in each breath. We exhale into a balloon. The size of the balloon after one breath is measured. Finally, the data are fed to a computer which then decides whether a person is breathing normally on the basis of age, size, and sex.

As we travel through the series of testing rooms, measurements are made of blood pressure, heart rate, and electrocardiogram (the electrical signals causing the heart to beat). Blood and urine samples are taken and chemical analysis is done by automatic equipment which sends the results directly to the computer. This automatic chemical analysis is very important in keeping the cost of the tests low. It is only in the last decade that we have had such equipment, which can determine the amounts of twenty different chemicals in the blood.

In this way, perhaps twenty different sets of tests are made to check for a variety of problems—heart malfunction, venereal disease, breast cancer in women, and so forth. In each case, test results (in numbers) are sent to the computer, where they are combined into a total picture of the health of the patient. In each room, there are signs and short TV programs explaining what is being done.

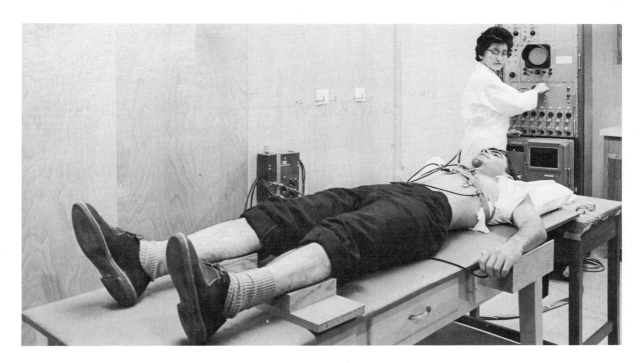

Fig. 1–20.
Electrocardiogram (*EKG*) *of heart is
taken and recorded on patient's
data-processing card.* (*Kaiser
Foundation Hospitals*)

In the last room, we answer questions from the history questionnaire. This is a series of maybe 300 questions regarding family background, habits such as smoking and drinking, a history of diseases and medical problems, and any recent symptoms such as headaches or dizziness for example. The questionnaire takes about as long as all the other tests together. The information from this questionnaire, however, is of vital importance to the physician. In many cases, the patient is able to describe his problems much better than any measurements can reveal.

At the conclusion of the testing procedure, the patient is told his results will be sent directly to his personal physician within a few days. An appointment is made for the patient to visit his physician. If the patient has no regular doctor, he and the receptionist select one from a list of the doctors who call regularly at a clinic associated with the center.

Under no conditions are the test results given directly to the patient. Some of the results may be in error or cause undue concern. Certainly they must be interpreted by a doctor before the patient is alarmed by some result which seems abnormal. If any results do appear dangerous, the doctor will undoubtedly want to have additional tests made before treating the patient.

Thus, the MHT center is a collection of sensitive instruments and a computer. The center is run by nurses, equipment operators, and technicians. These staff members use the equipment to make

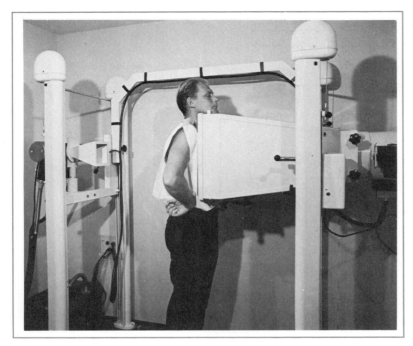

Fig. 1–21.
A 70mm X-ray is
taken for careful study by
a radiologist. (Kaiser
Foundation
Hospitals)

a complete survey of the patient's health for the physician. By using mass testing methods, the examination costs less than half as much as when done on an individual basis. Furthermore, in a large testing program, equipment of the best quality can be used and operators can be especially trained. Most important, the MHT center can bring health care to the people. The center can be located in the middle of the population it serves, it can be staffed by people recruited from this same population, and testing at the center can be a pleasant and attractive experience.

6 | PROBLEMS WITH THE MHT CENTER

There are serious problems in building such an MHT center. These are problems which appear whenever technology is used in a social problem. Following is a discussion of three important problems.

Measuring the Data

The first problem is largely scientific: What data should be measured and which tests should be used?

There are hundreds of tests which indicate a man's health. We might measure the strength with which he can grip a ball. As described in Section 3, we can measure reaction time in various tasks. We can measure a man's body temperature, his heart rate after different amounts of exercise, and the condition of his teeth.

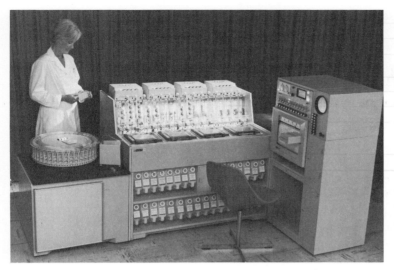

Fig. 1–22.
Automating the analysis
of blood samples and other
body fluids has contributed
enormously to the feasibility of
multiphasic health screening. Once
tediously and expensively done by
hand, analysis of blood constituents
can now be done by machine. Tests on
great numbers of people are thus
economically feasible. The machine's
rate of operation is 60 blood samples
per hr or 720 individual tests. Results
are recorded on a chart in the form
of a profile. Gray regions indicate
the normal range. (Caption from
Machine Design. *Photo from*
Technicon Corporation. SMA
is a registered trademark
of Technicon
Corporation.)

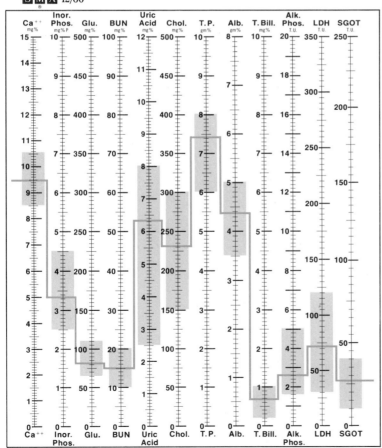

In any system designed to help a social problem area, there is a tendency to measure everything. If we are not to be swamped by data, we must measure only data which meet four conditions:

1. The data must be accurate and dependable, and there must be instruments which will make the measurements.
2. The physician must receive meaningful data. He should be able to diagnose from the data or, at least, decide what extra tests to make.
3. The data must deal with diseases which can be treated.
4. The total data must not make the doctor's job impossible. One center measured 1800 facts about the patient. The typewritten description filled six single-spaced pages. In such a case, the doctor would probably rather give a new health exam than try to figure out what the computer output means.

To meet these conditions, the usual MHT center has at most about twenty different tests. Since each test may give several different numerical results, even this program results in a computer portrait of the patient's health which fills several pages. To help the doctor read this material, the computer "flags," or marks distinctively, those results which seem unusual. Often the first page contains only the abnormal data.

Thus, the design of the MHT center requires careful selection of the tests to be given. The decision must depend on what technology is available, what knowledge is useful medically, and how much data can be profitably used.

Public Acceptance of an MHT Center

The second major problem is: How do we gain acceptance of the technology by the public for whom it is built? In the case of the MHT center, we emphasized in Section 4 that the center is for a part of the population which has been separated from the health-care system. These are people who are probably not highly educated, may be unemployed, and are not familiar with most modern technology. Furthermore, they are not familiar with the present health-care system.

Several principles have emerged from experience with the early MHT centers. While these ideas may seem obvious now, they were often not recognized when the first centers were built.

1. The center should be located in the community it is to serve. Transportation is often very difficult for people in the central city. If the center serves rural people, or groups widely separated, it should be mobile (for example, in a trailer). One of the most successful programs has been the mobile MHT center for the California cannery workers.

2. Going through the center should be a pleasant experience. Enough staff should be present to answer questions and explain the center. Instructions must be clear.

3. Persons employed in the center should be from the same ethnic group as the patients.

4. Signs and questionnaires must be in a language familiar to the customer. If this is Spanish, for example, the correct Spanish dialect should be used.

5. The questionnaire must be at an appropriate literacy level and must not ask for information which the patient cannot be expected to know. One proposed questionnaire contained the question:

    Have you had recurrent abdominal pains? ☐Yes ☐No
    Obviously many people would have no idea what the question meant.

    Another problem with such questions is whether to include a "Don't know" answer in addition to "Yes" and "No." If "Don't know" is included, some people soon learn that a "Don't know" answer is an easy way to move to the next question and finish the work in a hurry. We may find 140 consecutive "Don't know" responses. On the other hand, there are times when the patient really does not know. Many adults do not know whether they ever had mumps. If we have only "Yes" and "No" answers, how are they to respond?

6. The instruments used for testing must either be hidden or be attractive to the patient who is afraid of complex equipment. The success of many tests depends on the patient's cooperation. For example, in measuring sensitivity to pain, a hammer is pressed against the Achilles tendon in the heel to determine the pressure when the patient moves or objects. The measurement is valueless if he decides not to cooperate. Other tests (heart rate, blood pressure) require that the patient be relaxed if the doctor is later to draw any conclusions from the data.

Problems occur again and again when we try to use technology to improve social conditions. The MHT center is a product of modern technology. It is one small step forward in trying to bring benefits to all the population. In this sense, it is typical of the way technology can meet social needs. Yet it is only successful if we make sure the center is built for those people it is supposed to help.

The Effect of the MHT Center on the Total Health System

A third major problem is whether the MHT centers can be placed in the total health-care system. At the moment, the United States has a total system which includes doctors, nurses, hospitals, nursing homes, and so forth. This system is providing excellent health care to a sizable portion of the population.

If, in a very few years, we add several thousand MHT centers, can the other parts of the system adapt? Doctors are already very busy. If we test millions of people who have not been using doctors, can we save enough time from the doctor's time with regular patients to take care of this new group?

If millions of people are tested, we can expect tens of thousands to be referred to ophthalmologists (eye specialists) for eye problems they had not been aware of. Are there enough specialists in each category?

Hospitals are already overcrowded in many cities. What happens if breast operations are needed rapidly for large numbers of women each year?

In other words, new technology can make major changes in a total system. In order to provide the best care for the most people, the other parts of the system must adapt to these changes. Otherwise, we are apt to find that the new technology has unexpected side effects. If we do not plan carefully, the net result may be a final system which is poorer than the one we started with.

The problem is particularly serious because of the speed with which change occurs. The first MHT center (Kaiser) was partially automated in about 1965. By 1969, a half dozen other centers were in operation. By 1970, four hundred additional centers were planned. There is little time in which to build new hospitals or train technicians and operators. To educate a new generation of doctors or even to increase the number of doctors significantly requires additional time.

7 | PROBLEMS WHICH APPEAR IN MOST SOCIAL SYSTEMS

In the last section, we saw that there are three major problems when we try to add MHT centers to the health-care system:

1. The danger of data pollution (so much data we are drowned by it).
2. Matching to the target public.
3. Fitting into the total system.

These three problems arise whenever technology is used in a social situation.

There is always a tendency to measure everything possible. One major city, for example, decided in 1967 to put in a new traffic-control system. The first step was to put two radar sets at every intersection to measure the traffic flow in both directions at each corner. All these data would go to a computer, which then would calculate the best timing of the traffic lights throughout the city.

Even if all the radar sets could be kept operating (a ridiculous hope), there would be no way to decide what to do with all the data. Undoubtedly, the computer designer would first disregard 95% of the data, then use the 5% of the measurements to try to find a reasonably good timing for the traffic lights. Any reasonable system must keep the data to manageable size.

The first problem is really that of *data pollution*. As our society moves into the age of technology, we have to fight continually against overwhelming amounts of data. Complex technical systems always seem to be highly efficient paper producers. For example, one newly developed military weapon has instruction manuals totaling 1.5 million pages. Two hundred and fifty man-years are required just to write this much. We might guess that the greatest part of it will seldom, if ever, be read.

The second problem, matching to the target public, can also be illustrated by recent uses of technology in social areas. Several cities have developed mass rapid-transit systems (subways and high-speed trains) to move large numbers of people rapidly and economically. Too often, these systems end up serving primarily the suburban commuter, rather than the large numbers of city dwellers who cannot find jobs within reasonable travel times. Too often, these modern rapid transits do not even reach into the low-income areas of the city.

Finally, in every new application of technology, the problem of side effects arises. It is often very difficult to predict the effect of a new development on other parts of the total system. In New York City, considerable fanfare accompanied the building of apartments over a new highway approach to the George Washington Bridge to New Jersey. This was the beginning of new, efficient use of air space. After the apartments were occupied, it was found that the carbon monoxide level from automobile exhaust was dangerously high inside the building.

New highway construction in a city almost always has unexpected side effects. There are new points of traffic congestion. Feeder streets become seriously clogged when motorists learn of the new highway. Parking problems are shifted. Patterns of department store patronage change. City government has the extremely difficult task of anticipating changes and making sure the total system adapts to the new technology. If government fails to do this, unemployment may increase, the environment may deteriorate, and the general state of the city will become worse.

8 | FINAL COMMENT

This chapter serves as an introduction to the book. As perhaps you can begin to see, this book is in many ways an unconventional discussion of science. There is a story which is appropriate.

A successful scientist, thirty years out of college, returned to look up a professor he remembered and admired. He sat in one of the professor's classes in a course called "Applications of Science." He was amazed to hear the professor lecture on trolley cars, dirigibles, and cross-country passenger trains. Talking to the students after class, he discovered that the entire course had been on the technology of the 1930's.

Later, he asked the professor, "How can you describe such applications of science when we live surrounded by computers, space travel, artificial hearts, and the information revolution?"

The professor replied thoughtfully, "Yes, I guess I really should do something about the course."

Almost a year later, the scientist received the new catalog from the college. Looking through it, he found the professor's course. Indeed, it had been changed. The new title was "Old Applications of Science."

In contrast, *The Man-Made World* is a book which attempts to present some of the science of the future.

Questions for Study and Discussion

1. Describe three technological advances which improve the living conditions of man and which also contribute to his discomfort.

2. *The Man-Made World* deals with man-machine interaction and society-technology interaction. What is the difference between these two phrases?

3. What is the difference (if any) between *science* and *technology*?

4. Bring to class at least one recent article from a newspaper or magazine which is directly related to the problem of society-technology interaction. Be prepared to discuss the problem and offer a possible solution.

5. What is meant by the phrase "survival as individuals" as used in this text?

6. Curves on some superhighways are banked. What would be the effect on the driver if the straight stretch of highway just 500 feet prior to the curve were banked in the same direction?

7. An analysis of the value of using an inflatable bag to protect car passengers in a collision raised serious questions. Assuming these questions can be resolved, what are the major advantages of this safety device as compared to seat belts? How do these advantages relate to the principle of matching technology to the potential human user?

8. In Section 7, the text outlines three major problems which arise when technology is used in a social situation. Cite an example (which is not in the text) for illustrating each of the problems. Maybe you can find an example such as MHT which illustrates all three problems.

9. In Fig. 1–9, the GO and NO-GO zones are shown. On what do the lengths of these zones depend? Is there always a clear division between the two zones? Explain the reason for your answer.

Problems

1. Using the graph and data table of Fig. 1–13, answer the following questions:

   a) Suppose an automobile is traveling at 30 mph when the driver suddenly sees an obstacle in front of him. How far will he travel before he stops?

   b) Suppose the obstacle referred to in (a) is another automobile traveling in the opposite direction at 30 mph. How far away must it be in order to insure that there is no accident if the drivers see each other at the same time, and have the same reaction time and braking ability as indicated in the text?

   c) Suppose the driver is traveling at 60 mph. How far away must he be from a stationary obstacle when he first sees it in order to insure a safe stop?

   d) (Optional) How do you account for the difference in the answers to (b) and (c)?

2. In Fig. 1–13, the total stopping distance for a car traveling at 40 mph is given as 116 feet and in the text on page 16 the stopping distance for a car traveling at 40 mph is given as 181 feet.

   a) What is the main reason for the difference of 65 feet?

   b) If 116 feet is used as the total stopping distance for a car going 40 mph, what should be the minimum time duration of a yellow light on a highway which has a 40 mph speed limit?

   c) Why does the graph of reaction distance *vs.* speed result in a straight line, while the graph of braking distance *vs.* speed results in a curved line? (See Fig. 1–13.)

3. A one-horsepower electric motor can be operated for about $0.04 per hour. A man can do physical work for an eight-hour day at the rate of about 0.1 horsepower. If the minimum wage for a laboring man is $1.50 per hour, how much would an employer save each day if he could replace the laborer with an equivalent electric motor? What message does this give to the boy who wants to quit school now because he can make $50/week as a laborer?

4. A good football quarterback learns from experience how far to "lead" a pass receiver in order for the receiver and the football to arrive at the same place at the same time. How far does a football player (who can run 100 yards in 10 seconds) travel during the shortest human delay period? What happens to the quarterback's time delay late in the game when he is very tired? How far would a fresh receiver (still able to run 100 yards in 10 seconds) travel during this longer time-delay period?

Laboratory and Projects

I | THE GAME OF FUTURE

At various times during the course, there will be opportunities to participate in a classroom version of a "game" which is played in all seriousness by many engineers and non-engineers around the world.

In this first approach to the game you are given some information about possible future technological developments and their possible consequences. You are then asked (in small groups) to (1) predict the decade during which this event might take place, (2) make judgments on the likelihood that the predicted consequence will occur, and (3) make value judgments on the effect of the consequence on society.

After the small groups have made their decisions on 1, 2, and 3, the entire group will discuss any differences in time or consequence predictions.

Following this discussion, the opinions of "experts" will be given and discussed.

II | MEASURING HUMAN HORSEPOWER

Part A

There has always been interest in the possibility of propelling an airplane by using the muscle power developed by a man as the energy source. Studies of the weight-to-power ratio of animals show that man should be capable of supporting himself in air for short periods of time. The problem is not only of supplying the motive power, but also of obtaining lightweight material for wings of large surface areas.

In this project, we wish to determine the power a man can produce. As mentioned in Chapter 1, this power depends on how man does work (using his arms, legs, and so on). Here we consider only one example: the work done in climbing stairs.

Before determining the horsepower developed by a man, we must define certain terms:

Horsepower: One horsepower equals 550 foot-pounds per second.

Foot-pound: The work accomplished when a one pound force pushes something through a distance of one foot. The displacement must be in the direction in which the force is acting.

If a 120-pound person raises his body from any level to a position two feet above the original level, he will have done 120 pounds $\times$ 2 feet, or 240 foot-pounds of work.

To determine the time rate of work, a measure of the time required for the action is essential. Dividing the work done (foot-pounds) by the time in seconds will give foot-pounds per second.

A horse is considered to be able to work at the rate of 550 foot-pounds per second for an extended period of time. Comparison of your rate of working with this standard will show your horsepower.

$$\text{H.P.} = \frac{\text{Force} \times \text{vertical displacement}}{550 \times \text{seconds}}$$

To determine our horsepower we need to know:

1. The force applied (your weight in pounds).
2. The vertical displacement (distance between floors in feet).
3. The time required for the displacement in seconds. The 550 is a constant or proportionality factor.

Items 1 and 2 will remain fixed during the experiment. However, time will vary; hence a record will have to be kept for each trial.

1. Prepare a table to record your data. After obtaining the data for Items 1 and 2,

   a) Find the time in seconds required to walk at normal speed as you climb a flight of stairs. Record the time required in the table.

   b) Find the time required to climb the stairs taking two stairsteps at each step. Record.

   c) For your final trial, find the time required if you climb the stairs as quickly as possible, supplementing your leg muscles by using your hands to grasp and pull on the handrail. Record.

NEW METHODS OF MODIFYING THE ENVIRONMENT			How likely is it that the result will be a consequence of the development?				What will the effect of the consequence be?				
If these developments were to occur	**During this decade**	**They might result in:**	Virtually certain	Probable	Possible	Almost impossible	Very favorable	Favorable	Little or no importance	Detrimental	Very detrimental
Limited weather control, predictably affecting regional weather at acceptable cost.		**A.** Improvement in agricultural efficiency.									
		B. Disruption in ecological balance.									
		C. Weather being used as military, economic, or political weapon.									
		D. Civil suits alleging damage caused by weather.									
		E. Emergence of a power group of "weather makers."									
		F. Others—specify.									
Widespread use of automobile engines, fuels, or accessories which eliminate harmful exhaust.		**A.** Increased traffic congestion since smog-free automobiles will be allowed to increase in numbers.									
		B. Delay in development of high-speed mass transit systems.									
		C. Continued economic domination by the automobile industry.									
		D. Higher efficiency of engine performance.									
		E. Relaxation of efforts to reduce industrial and municipal air pollution.									
		F. Other.									

NEW METHODS OF MODIFYING THE ENVIRONMENT			How likely is it that the result will be a consequence of the development?				What will the effect of the consequence be?				
If these developments were to occur	**During this decade**	**They might result in:**	Virtually certain	Probable	Possible	Almost impossible	Very favorable	Favorable	Little or no importance	Detrimental	Very detrimental
Widespread use of self-contained dwelling units using systems that recycle water and air to provide independence from the external environment.		**A.** Further fragmentation of society.									
		B. Reduction in degree of dependence of suburban residents on municipal government.									
		C. Further development of units for living in space and underseas.									
		D. Rejection by most people.									
		E. Other.									
Establishment of a central data storage facility with wide public access for general and specialized information in the areas of library, medical, and legal data. *(Adapted from "Forecasts of Some Technological and Scientific Developments and Their Societal Consequences," T. J. Gordon and R. H. Ament, Report R6, The Institute for the Future.)*		**A.** Use of home terminal for education.									
		B. Information storage becoming a salable service resulting in revision of business practices.									
		C. Improvement in social science research.									
		D. Individual citizens becoming "expert" in law and medicine.									
		E. Information overload problems will arise in decision making because people will not be able to handle all the data.									
		F. Invasion of privacy.									
		G. Other.									

■ 2. Calculate the horsepower generated in trials *a*, *b*, and *c*.

■ 3. Obtain statistics which give the weight and maximum power developed by the engine of:

a) a light, private, propellor-driven plane.

b) a light, private jet.

c) the Boeing 747.

d) the Saturn 5.

■ 4. Compare the weight per horsepower for each of the engines for which you obtained information with your own weight per horsepower.

Part B (Demonstration Experiment)

In Part A of this experiment, you found that human beings are quite limited in their ability to generate power. Machines can be used to compensate for man's inability to do work at a rapid rate. Amplifiers are one group of machines which have the capability of using a small amount of power to control a large amount of energy (see Fig. 1).

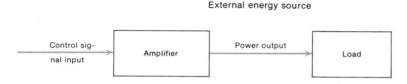

Fig. 1.
Block diagram of an amplifier.

All amplifiers have three properties in common:

1. In addition to the energy of the control signal at the input, they require an external source of energy.
2. They exercise some degree of control over the flow of the external energy.
3. The power delivered to the load is greater than the power of the controlling signal. The ratio of the output power to the input power is called the amplification factor.

In this experiment, you will study the characteristics of a mechanical amplifier (model of a windlass).

1. Observe the operation of the amplifier by comparing the force required to stretch the spring without amplification and the force required when the string is wrapped around the rotating amplifier drum three or four times.

■ 2. Obtain data and plot curves of *Force*$_{in}$ vs. *Force*$_{out}$ over the range of deflection of the spring when there are 0, 1, 2, 3, and 4 turns of string around the drum.

3. From the data of Step 2, make a plot of $Force_{out}/Force_{in}$ (amplification factor) versus number of turns of string on the drum.

4. Is there a gain in power or just in force?

III | COORDINATION OF HUMAN SYSTEMS

The human body is not just one system, it is a rather complex system of systems. Tasks which normally seem quite simple to us actually require a considerable amount of coordination among various systems in the human body.

If you were asked to put a piece of tracing paper over the diagrams in Fig. 2 and trace the figures, you would have very little difficulty in doing it, yet a one-year-old child would have a considerable amount of difficulty in performing the same task. Through training and experience you have learned to coordinate the various systems which involve seeing and the moving of muscles so that you can make an excellent replica of the figures. It is difficult to erase all of this training and experience, but it is possible to confuse the previously learned cooperation by using a mirror.

With the arrangement shown in Fig. 3, and looking in the mirror and not directly at your hand, trace the diagrams of Fig. 2.

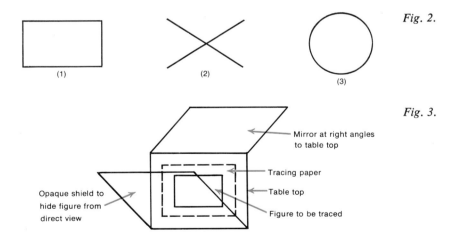

Fig. 2.

(1) (2) (3)

Fig. 3.

Mirror at right angles to table top

Tracing paper

Opaque shield to hide figure from direct view

Table top

Figure to be traced

1. Which of the diagrams in Fig. 2 was easiest to trace? Which was most difficult?

2. Did you improve your ability to trace after practicing each diagram three times? Five times?

3. Does this experiment demonstrate that the human body is made up of cooperating systems? If so, how?

IV | THE YELLOW LIGHT

In Section 3 of this chapter we took a long quantitative look at the problem of the yellow light. While this was quite thorough, it was also theoretical. This laboratory experience is designed to have you look at the real situation.

Part A

In order to make a model of the yellow-light problem at an intersection in your town it is necessary to get some data. If it is not practical for students to get the data as a group, the teacher can make other arrangements.

First it is necessary to have certain measurements of the intersection:

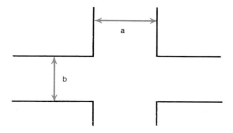

Fig. 4.

The distances *a* and *b* (Fig. 4) should be measured by pacing across the street during the period when crossing is allowed. The time during which the light is yellow for each street should be obtained with a stopwatch. The speed limit for each street should be obtained from signs or by questioning local police.

■ 1. From the graph given in Fig. 5 calculate the distance a car will go during the period of the yellow light if the driver does not try to stop but continues at the speed limit. Starting from the far side of the intersection, measure this distance and mark it GO. This is the farthest a car can be from the intersection and still get through legally while the light is yellow.

■ 2. From the graphs given in Fig. 5 and Fig. 6 calculate the stopping distance in feet for the time the light is yellow. Assume that the first 0.8 second of time is spent in reaction and getting the foot to push on the brake pedal, and the remainder of the time is braking time. For example, if the car is traveling at 30 mph, it will go 35 feet during the time it takes to get the foot on the brake (Fig. 5) and would require at least 3.45 seconds and an additional 75 feet in order to be comfortable during the braking period (Fig. 6). This means that at 30 mph the car would need to be 103

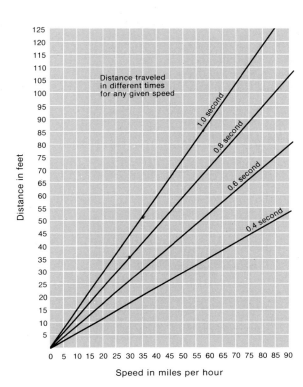

Fig. 5.

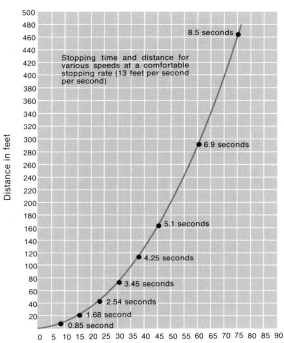

Fig. 6.

feet from the near side of the intersection in order to stop comfortably. Mark this off on your model. It also means that the comfortable stopping time is 4.25 seconds (0.8 second reaction time plus 3.45 seconds comfortable stopping time). If there is a 3-second yellow light and the intersection is 30 feet wide, there will be a dilemma zone of 27 feet in which the car can neither stop comfortably nor pass through legally (and therefore presumably safely). This is pictured in our theoretical model in Fig. 7.

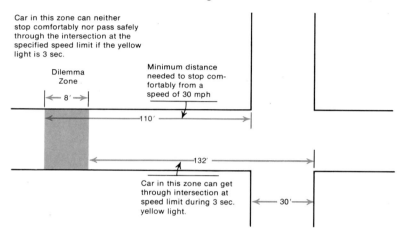

Car in this zone can neither stop comfortably nor pass safely through the intersection at the specified speed limit if the yellow light is 3 sec.

Dilemma Zone

|← 8' →|

Minimum distance needed to stop comfortably from a speed of 30 mph

110'

132'

Car in this zone can get through intersection at speed limit during 3 sec. yellow light.

← 30' →

Fig. 7.

Part B

After you have drawn a diagram such as Fig. 7 for your data, answer the following questions:

1. At the speed limit and the yellow-light time for the street you used what was the length of the GO zone?

2. At the speed limit for the street you used what was the minimum comfortable stopping distance?

3. What was the width of the dilemma zone (if any)?

4. What should be the timing of the yellow light in order to eliminate the dilemma zone (if any) at the intersection you studied?

5. If this time is less than the time obtained by adding 0.8 second to the comfortable braking time obtained from the graph in Fig. 6, how do you account for the difference?

6. If you are beyond the dilemma zone and there is a car in the dilemma zone in front of you when the light turns yellow, and he decides to stop, how does this affect the stopping distance for you? What does this suggest for motorists approaching an intersection even if the light is green?

Part C

1. Suppose the speed limit is 45 mph, the street intersection is 50 feet wide, and the yellow light is on for 3 seconds.

   a) How far from the near side of the intersection is the beginning of the GO section? (Fig. 5)

   b) How far from the near side of the intersection is the beginning of the comfortable stopping section? (Fig. 5 and Fig. 6)

   c) How wide is the dilemma zone ($b - a$)?

   d) What must the yellow-light time be to eliminate the dilemma zone?

   e) Is this practical? What usually happens if the yellow light is extended for too long a period?

2. What are some possible solutions to the dilemma raised in Question e?

3. What will happen if a driver approaches the intersection at 60 mph after the yellow light has been adjusted to meet the requirements of 1d? This is 15 mph above the speed limit. State any important changes in distances and times from those calculated in 1a, b, c, and d. Should situations such as these be considered in setting speed limits and the timing of yellow lights?

STEVENSON

"*Have you gone nuts? One hundred and fifty computer-programmed plays to call in the book, and you say, 'Everybody run downfield and I'll heave one.'*"

Decision Making
2

I | THE ELEMENTS OF DECISION MAKING

You have just been given permission to drive the family car, but your father keeps the key to the locked gas cap. He tells you that he will fill the tank on Friday night and you can have the car all day Saturday, but to drive slowly because the slower you drive, the better your gas mileage will be. You have a single decision to make: Drive slowly and get maximum mileage or drive fast and get minimum mileage. Or is it that simple?

The board of education of a school district asks the principal of the local high school why they should not ban all student cars from the school parking lot. At the same time the student council asks the same principal why he will not allow all licensed students to drive to school. Both groups indicate that the decision is really quite simple.

The Committee for Clean Air of a large city meets with the mayor and demands that all incinerators used for burning garbage within the city limits be shut down immediately to cut down air pollution. The mayor promises to discuss the problem with his commissioners of traffic, sanitation, and rivers and harbors, and report back to the committee. The committee chairman snorts, "More bureaucracy," and marches out of the meeting to a press conference where he reports that the mayor is stalling on a simple question which has a very simple answer. The mayor in return replies, "For every complicated question there is an answer which is forthright, simple, direct, and *wrong!*"

In each of these cases, there are many decisions to be made—some insignificant and others crucial. In this chapter, we look at the nature of decision making and at some of the techniques which can be used to arrive at intelligent decisions.

As an example, let us look at the first "simple" situation described at the beginning of this chapter. This situation contains elements common to all decisions. How fast to drive a car is a simple question. The rule is not, Drive slowly and get maximum mileage, or drive fast and get minimum mileage. We have to look at the problem more carefully.

Figure 2–1 shows how gas mileage depends on speed for a particular car. At a standstill, with the engine running, the effi-

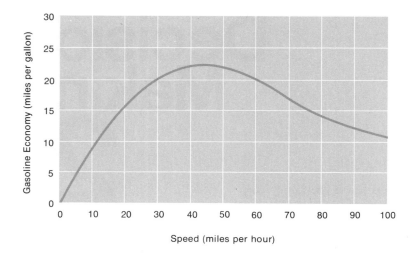

Fig. 2–1.
Gasoline
economy as
a function of
automobile
speed.

ciency is zero—no matter how much fuel we use, the automobile doesn't get anywhere. As the speed increases, the gas economy improves, until at 45 miles per hour, it reaches a maximum of twenty-two miles per gallon. Beyond 45 mph, the economy falls off because of friction in the engine, in the wheel bearings, and in the air through which the car moves, as well as because of a decrease in engine efficiency at high speeds. It is obvious that the most economical operation of this car is realized when we drive at a constant speed of 45 miles per hour.

It is rarely possible, in any realistic situation, to travel at a constant speed, because of traffic conditions, curves, hills, and weather conditions. Fortunately, in this example, small departures from 45 mph do not cause a serious problem. At all speeds between 35 mph and 55 mph, the gasoline economy is within 5% of the maximum. As long as we stay reasonably close to 45 mph, we are operating nearly at an optimum condition.

We base our decision to drive at 45 mph on our desire to maximize the gasoline economy (minimize the cost of gasoline per mile). In the real world, decisions seldom are so easily made. There often are many factors which we must consider. Also, many times we are not sure what is really best (or optimum).

If we were to drive through the center of a city, where the speed limit is set at 25 mph, the police would take a very unsympathetic attitude toward a speed of 45 mph. Our decision making becomes somewhat more complex because of this constraint. Its effect is shown in Fig. 2–2. The legally imposed constraint lets us drive our car at any speed between zero and 25 mph. If we wish to maintain maximum gasoline economy *under this constraint*, we must operate at 25 mph. This is the highest point on the curve of gasoline economy versus speed within the allowed region.

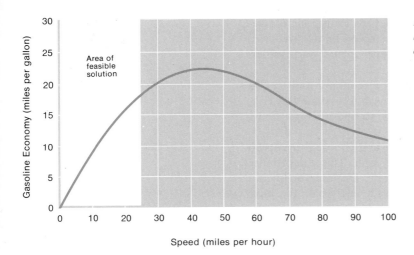

Fig. 2–2.
Speed limit
added.

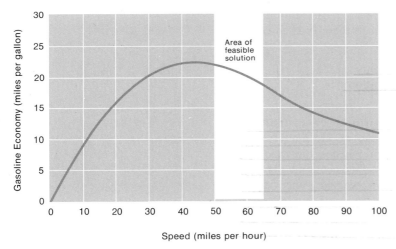

Fig. 2–3.
Case with
minimum and
maximum
speeds.

If, on the other hand, we drive along a limited-access road such as the New York Thruway or the Pennsylvania Turnpike, a different set of constraints may be imposed. One such set is shown in Fig. 2–3. Here, we are not permitted to travel faster than 65 mph, nor slower than 50 mph. We should note here that if our minimum-speed constraint had been 40 mph rather than 50 mph, the decision would have been to drive at 45 mph rather than at either limit. We want the *maximum* possible economy.

Frequently, we are required to make decisions on the basis of conflicting desires. In our limited-access-road example of Fig. 2–3, we may be required to travel between two points in as short a time as possible. Perhaps we have a date with a girl who becomes angry if we arrive late. In addition, we may wish to use as little fuel as possible. We see from Fig. 2–3 that we cannot simultaneously drive at maximum speed (within the speed limit, of course) and maximum economy. The two goals just do not go together. Such conflicts are typical of real decisions.

We may decide to drive at 50 mph for economy and to risk the wrath of our girl friend. If we do not wish to assume this risk, we may drive at 65 mph and arrive on time to find a happy young lady.

As an alternative, we may compromise and choose some speed between 50 mph and 65 mph, perhaps 58 mph. This compromise lets us save some money and not be too late. Deciding on such a compromise really means we have chosen a new goal. We have decided that the few minutes we save are worth the cost of the extra gas.

Unfortunately, most decisions involve such a compromise. We have to decide on the relative importance of the different goals.

The speed example contains all of the parts of a decision-making problem. Let us list these so we can return to them in later examples. There are four elements:

Model   Constraints

Criteria   Optimization

The definitions of these terms are:

1. *Model.* The model is the mathematical or quantitative description of the problem. In our example, we are interested in obtaining the maximum gasoline economy; hence the model is the curve of Fig. 2–1, which shows how gasoline economy depends on the speed. In the following chapters we consider the various forms of models in much greater detail, since the model is the item which changes the problem from one of intuition or common sense into a quantitative problem which we can hope to solve precisely.

2. *Criteria.* The criteria are the goals or objectives of the decision-making problem. In our example, the criteria are to achieve the best possible gasoline economy and to arrive at our destination in as short a time as possible. When there are several criteria which are incompatible, we have to select a criterion which is a compromise among these.

One of the most dramatic decision problems occurs when there is a major catastrophe (fire, explosion, earthquake, or battle) with a very limited number of doctors available. The problem is to decide how to use the doctors in the best way. The injured people are then sorted into three groups: those who cannot be saved, those who can probably be saved by a doctor, and those who will survive without immediate attention by a doctor. The decision problem is then normally solved by the doctors working on only the middle group.

3. *Constraints.* Constraints are added factors which must be considered in the solution of the decision problem. In our speed example, the speed limits represent constraints. Other constraints in our example would be unsafe road conditions and driver fatigue. (Speed may be limited by road conditions, and the time involved in long trips may have to include rest periods.) Thus, the constraints specify the region within which we should look for a decision. We must find the best solution which satisfies the constraints.

4. *Optimization.* Once the problem is formulated (the model), we decide what we really want (the criteria), and we know what is permissible (the constraints), we are ready to try to find the best, or *optimum*, solution. In our speed problem, we found a solution by merely examining the model and considering the constraints. In more complex problems, it may be necessary to find special engineering or mathematical techniques; in many practical cases, we have to use a trial-and-error approach.

In this section we have considered a very simple example of a decision problem (the selection of the optimum speed at which to drive a car). We found all four elements of the typical decision-making problem: model, criteria, constraints, optimization. Next, we want to show a few other decision problems with different forms for the model, criteria, and constraints. We look also at a few different ways to optimize. Finally, we want to suggest the wide variety of decision problems.

2 | TYPES OF DECISIONS

In the preceding section, we found that the speed example of a decision problem has three parts. The model describes the situation. The constraints say what we cannot do. The criterion is the basis for a decision. Once these three parts are known, we optimize. That is, we find the *best* solution.

It is amazing that so many different decision problems can be viewed in this way. In very simple problems, nothing is gained by trying to find the model, constraints, and criteria. A man is quite good at making simple decisions without worrying about how he does it. For example, in a restaurant he orders according to the prices and what he feels like eating. Only a crazy scientist would think that the menu is part of the model. The rest of the model is how well he likes each food. The constraint is the amount of money he is willing to spend. The criterion is to find the best combination of foods which costs less than the limit. Obviously, if we had to think about all this, we would die of starvation. But there are many problems where we can gain much better understanding if we recognize these parts. The two examples which follow should provide some additional insight.

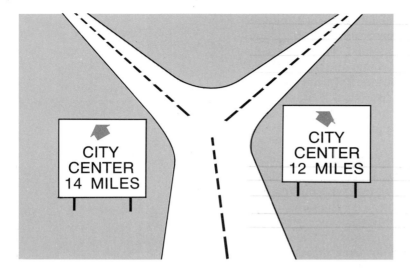

Fig. 2–4.
Which road
to take?

Choosing a Road

We are driving into the center of a city. We reach a "V" in the road. Here we can reach the city by taking either route, and we have to make a decision: right or left. Figure 2–4 shows the problem. Our only criterion is to reach the city center as soon as possible.

If the speed limit is the same on each road and if there is no traffic, the decision is simple. However, we have driven both ways before. We know that there is always traffic, usually heavier on the shorter route. Furthermore, any accident ahead on either road is likely to cause enormous delays. How do we decide?

Obviously in a problem like this, we cannot make an intelligent decision. We just don't know the model. Is there an accident today? If so, where? What is the speed of traffic five miles ahead on each road? How long will it probably take on each road? We cannot make an intelligent decision without knowing the model, at least approximately.

This example emphasizes one of the simple improvements in traffic systems which could be made with relatively little money. A thruway in a major city may cost as much as $28,000,000 per mile to build. For a very small fraction of this amount, we could measure the average traffic speed at several points. This information could be sent back by telephone wires to the driver's decision point. As we approach the "V" of Fig. 2–4, we would see on a large sign the expected time to the city center on each road. Then, we would have a model on which to base a decision.*

Perhaps this example illustrates the value of thinking about the parts of a model. One can argue that all we have done is use common sense. This is true. Indeed, much of modern technology and

*
If such a system were built, we would have to be careful. Otherwise, imagine that at 9 A.M. the signs say 60 minutes by the right road, 18 minutes by the left. Obviously, everybody goes left (except the stubborn driver who simply cannot stand technology). By 9:15 the signs read 30 minutes right, 80 minutes left. To avoid this, we have to calculate the times to take into account the probable behavior of motorists now reaching the "V."

Fig. 2–5.
*An all-too-familiar sight
in an industrial area of a
major city. The reduction of
air pollution by legal action has
inevitable economic and social
consequences. New laws to curb
pollution must consider the
total effect on the city.
Otherwise, the changes
may be more harmful
than the pollution.
One help would be
to have a better idea
of which types of
air pollution are
most injurious to
humans.* (*The
Bettman
Archive*)

mathematics is useful primarily because it guides us into good common sense.

Controlling Air Pollution

Reducing air pollution is a major goal of most city mayors today. The easy solution is obvious. We should first find out where the pollution comes from, then stop it at the source by appropriate laws.

If we attempt to do this, we find much of the pollution comes from cars, incinerators, and industrial stacks. We should ban cars, close incinerators, and stop industrial emissions. Unfortunately, we have used too simple a model. The mayor probably can do none of these.

Banning cars from the downtown area might be political suicide. As Henry Barnes, the late Commissioner of Traffic in New York, enjoyed saying, "Julius Caesar banned chariots in downtown Rome.* You know what happened to him." Even requiring taxis and city cars and buses to be electric means a tremendous expenditure of money.

Closing incinerators is no easier. If apartment houses must not burn rubbish, the city must provide for collection and disposal. Land-fill area in which to dump rubbish is rapidly disappearing, and several cities are already making plans to ship rubbish by train to distant locations. A cooperative effort between society and technology needs to be made to develop an efficient system for recycling trash.

Finally, forcing industry to clean stack emissions raises the cost of operating in the city. Many industries are already considering moving out of the big cities. An important job of the

*
Actually, population density in Rome was over 220,000 per square mile—far larger than today in any United States city, even though there were no high-rise buildings.

mayor is to increase the number of jobs in the city, particularly to take care of many relatively unskilled people moving in from other sections of the country. In many cities, the electric utility is a significant source of air pollution. Forcing the utility to clean its smoke raises the cost of electricity for both homes and industry. Again, industry is encouraged to move out.

Thus, decisions about air pollution require that we have a model which not only shows the sources, but also indicates the effects of legislation. The criterion must be not only clean air, but also a desirable total city. There must be jobs for people who want to work.

Perhaps this example illustrates the complexity of some of the decision problems we face today. The problem is even more complex, since it depends also on what we do in suburbia. If we install our traffic-information system of the first example, perhaps we will encourage more people to drive into the central city. In such problems, we have to determine a suitable model, even if this step requires great work. We cannot make decisions without consideration of the many different effects.

3 | ALGORITHMS

You and one opponent are playing the following game. You are seated across a table from each other. A pile of 11 matches is on the table. You and your opponent alternate turns. At his turn, the player may pick up 1, 2, or 3 matches (never 0 or more than 3). The winner is the one who forces his opponent to take the *last* match.

For example, one playing of the game might see the following moves:

You take 3	8 left
Opponent takes 1	7 left
You take 2	5 left
Opponent takes 1	4 left
You take 3	1 left
Opponent takes 1	0 left

Opponent moved last, you win.

Is there a strategy you can use which will ensure that you always win? In other words, are there rules you should follow? Such a rule might be: Always take 1 if there is an even number left, 2 if there is an odd number left, never 3. (This is not the correct rule.)

Such a set of rules is called an *algorithm*. Before we look at any more decision problems, we need to define this new term, algorithm.* _An algorithm is a sequence or a list of steps that can be taken to solve all problems of a particular type._

---

* The term algorithm is derived from an Uzbek (ŏŏz′bek) mathematician, al-Khwarizmi, who developed such sets of rules in the ninth century. (The Uzbek Soviet Socialist Republic is a portion of the Soviet Union north of Afghanistan.)

We solve most decision problems by using algorithms. Consequently, in this section we want to become familiar with various algorithms and understand how they are used.

An Algorithm for Determining Averages

A very simple algorithm is that for finding the average of a set of measurements. The algorithm is described as follows:

1. Add the given measurements.
2. Count the given measurements.
3. Divide the answer to (1) by the answer to (2).

For example, if the given measurements are -2, 8, and 21, we find the average in three steps:

1. The sum is $-2 + 8 + 21 = 27$.
2. The number of measurements is 3.
3. The average is $\frac{27}{3}$ or 9.

The algorithm works for all possible sets of measurements.

An Algorithm for Number Selection

As we shall see, not all algorithms are so simple. It is not unusual for an algorithm to have hundreds, or even thousands, of steps. Regardless of size, an algorithm is a step-by-step procedure to solve a problem. Thus, an algorithm is just what we need to give instructions to a machine or computer. A computer can only do exactly as instructed by man. Hence, for a computer to solve a problem, we must first put into the computer the algorithm for that class of problems.

People also think in terms of algorithms. For example, the problem is to find the largest of the following positive integers:

13, 6, 47, 21, 5, 3, 7, 33

We first scan the list and eliminate all one-digit numbers, leaving

13, 47, 21, 33

Next we eliminate all two-digit numbers. There is nothing left. Then we back up one step and consider the two-digit numbers. The answer is among these:

13, 47, 21, 33

Now we can proceed in several ways. We might compare the first to the second. 47 is larger than 13, so we throw out 13. This leaves

47, 21, 33

Comparing the first two and throwing out 21 leaves

47, 33

Finally, the answer is 47.

Having worked this example, we can now write an algorithm for solving all such problems described by:

Find the largest of a given set of positive integers.

The algorithm must work for *any* problem of this type. For example, the algorithm must work for: 13, 8, 12, 13.

An interesting feature of the preceding problem is that the algorithm can have several different forms. For example, we might just compare the first two numbers and throw out the smaller. Then we continue this process until only the answer is left.

It is really not surprising that there may be several different algorithms for solving the same problem. Different people certainly solve the same problem in different ways. Some people are very clever at finding algorithms which give an answer quickly. Other people tend to use familiar approaches which may require a much longer time.

An Algorithm for a Decision Problem

The first two examples of algorithms were for number or arithmetic problems. Algorithms are also useful for solving problems which are not easily put into mathematical terms.

Consider the following problem: A town precinct covers a grid of streets (Fig. 2–6). A police car starts at A, where the police station is located. We wish to route this car so that all streets will be patrolled once. The patrol is to be completed in a minimum time so that the next trip can start as soon as possible.

We can cover the precinct in minimum time if we can find a route which covers each block once only. If this is not possible, the car should retrace as few blocks as necessary (or those blocks in which travel time is least). The problem is then: By inspecting Fig. 2–6, find a route which covers each block once only.

Exactly the same problem faces the mailman as he chooses his route, particularly when he travels by car and all mailboxes are at the curb. The newsboy and milkman have the same problem. Before considering a decision problem of such complexity, however, we will look at a much simpler version.

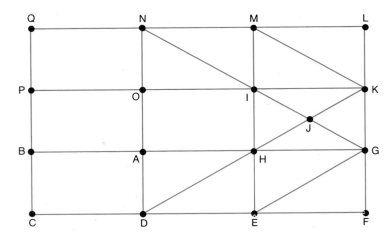

Fig. 2–6.
Precinct streets.

Early in the eighteenth century, the city of Königsberg (now part of the Soviet Union) included seven bridges over the Pregel River (as shown in Fig. 2–7a). A favorite pastime on a pleasant Sunday afternoon was to start at any point in the city and to see if one could walk in such a way as to *cross each bridge just once.* One could start at any point and end at any point.

In 1736, Leonhard Euler* spoiled this Sunday afternoon pastime of the Königsbergers by answering the question mathematically with an algorithm which solved the problem for any number of bridges. In order to understand the solution, we consider the map of Fig. 2–7a. There is an island C, two sides B and D of the main river, and a region A between the two sections of the river after it divides.

The interesting parts of the map are the bridges. Therefore, we simplify the map by showing only the seven bridges. Each bridge connects A to B, or A to C, or B to C, and so on. Let us make A, B, C, and D points and show only the bridges connecting these regions or points. This is done in Fig. 2–7b.

Each line in Fig. 2–7b represents a bridge. Each point, called a *vertex*, represents a land region. The graph has four vertices and seven lines.

Now suppose we can travel over each bridge once and only once. If the total route starts from A, we might go first to B. We would enter B on one line and leave on another line. Thus, *each time we pass through a vertex, there must be exactly two lines connected to that vertex.*

After the total path is drawn, we can say: Every vertex has an even number of lines connected to it. The only possible exceptions are the vertex from which we start and the vertex at which we end (e.g., if we start at A and go to B and so on, we leave A along one line; thereafter, we may pass through A again and two lines to A are added).†

Thus, a closed path covering all bridges is possible only if:

1. Every vertex has an even number of lines (then we start and end at the same point), or
2. Exactly two vertices have an odd number of lines (then we start at one of these and end at the other).

Inspection of Fig. 2–7b reveals that the vertices have the following numbers of lines:

$$\begin{array}{ll} A\ 3 & C\ 5 \\ B\ 3 & D\ 3 \end{array}$$

All four numbers are odd. Hence there is no hope of walking over all seven Königsberg bridges without retracing any path. It makes no difference where we start and end or what route we follow.

Fig. 2–7.
Königsberg-bridge problem.

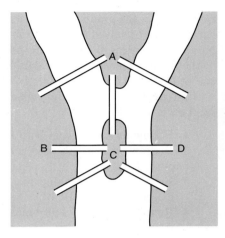

a) Seven bridges of Königsberg.

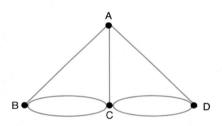

b) Graph of part (a).

---

*
The most famous of Swiss mathematicians who published an enormous number of articles on all phases of mathematics and physics.

†
If we start and end anywhere between vertices, every vertex must possess an even number of lines.

If one additional bridge were built (or we were permitted to swim across one river), the graph would be changed by adding one link or path between two regions. If the bridge were built from A to C, the graph would be as shown in Fig. 2–8. The problem can now be solved (Euler also showed this part of the solution). Now only vertices B and D are odd (as shown by the numbers in the figure). Hence, we can start from B, cross all bridges, and end at D.

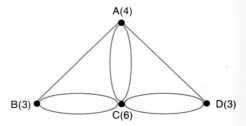

Fig. 2–8.
Königsberg with one additional bridge.

Can we find an algorithm which works in all cases? Euler's solution is as follows:

1. Follow any path from B to D (e.g., BAD).
2. Redraw the graph and omit the path already traced (Fig. 2–9).
3. From any vertex along this omitted (BAD) path, trace a closed loop (e.g., BCB).
4. Redraw the graph and omit this loop which has just been traced (Fig. 2–10).
5. Continue this process until all of the original graph has been covered (e.g., we might next trace the double loop ACDCA).
6. A solution now consists of the following steps. We start at B. We first follow all loops out of B (in our example, BCB only in Step 3). We then proceed to the next vertex on the main path. We continue this process until the entire diagram is covered and we have reached the terminal point. At each stage, we can redraw the graph in order to keep track of the positions which have been traveled. In other words, our solution says we should travel in the sequence:

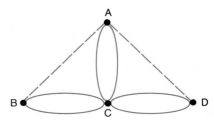

Fig. 2–9.
Graph redrawn with original path omitted.

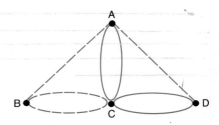

Fig. 2–10.
Graph redrawn with first loop omitted.

$$
\begin{array}{ll}
\text{B C B} & \text{First loop} \\
\text{B to A} & \\
\text{A C D C A} & \text{Second loop} \\
\text{A to D} &
\end{array}
$$

We might show this by indicating the vertices encountered in sequence

B C B A C D C A D

in order to cross all eight bridges with no retracing of steps.

Thus, for this example we have derived two algorithms. The first decides whether a solution is possible and also tells us where we can start and end. The second determines the route to be followed.

Routing Police Cars

We now try to apply the algorithm just discussed to the problem of routing police cars at the beginning of this section. The

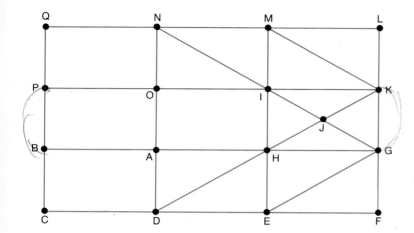

Fig. 2–11. Precinct streets.

precinct includes the grid of streets (Fig. 2–11). A police car starts at A, where the police station is located. We wish to route this car so that all streets will be patrolled once, preferably in a minimum time (so that the second patrolling can be started as soon as possible). We can cover the precinct in minimum time if we can find a route which covers each block once only. If this is not possible (as in this problem), the route must retrace as few blocks as possible (or select those blocks in which travel time is minimum). In order to find a desirable route for the car, we can apply the method of the preceding bridge problem. For this case, every intersection (marked by a letter in Fig. 2–11) is a vertex.

Ideally, we should like to start from A, travel all streets once, and end at A. Inspection of the figure reveals that this is impossible because the following vertices have an odd number of lines:

$$\begin{array}{cc} \text{B } 3 & \text{K } 5 \\ \text{G } 5 & \text{P } 3 \end{array}$$

In order to permit a path from A back to A and through all streets, we must add paths which change these odd numbers to even (and leave the others even). In other words, we can add an extra path from P to B and from G to K. The car will traverse the P-B and K-G blocks twice during the patrol. This change is shown in Fig. 2–12.

Now the route-selection algorithm of the bridge example can be used to find an appropriate route for the patrol car. Two comments are important in this particular example:

1. In a more complex problem, there may be numerous widely separated vertices with an odd number of lines. In such a case, there are many different ways to add extra lines (retracing of certain streets) to make all vertices even. If

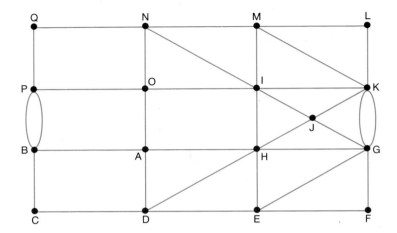

Fig. 2–12.
Street map
with added paths
from P to B, K to G
(corresponding to
two traversals
of these two
blocks).

we wish a patrol requiring minimum time, we must evaluate the extra time required for each possible retracing.

2. We normally wish to vary the patrol pattern of the police car, so that the patrol car does not pass a particular point at regular intervals of time. We can do this if we find different solutions from the route-selection algorithm.

An Algorithm for Games

As discussed at the beginning of Section 3, algorithms can also be used for games. Consider the following game which is more complex than our 11-match game. There are 27 matches on a table and two players, A and B. The players *alternately* pick up and keep 1, 2, 3, or 4 matches. Each player knows how many matches the other has at all times. The winner is the player who has an even number of matches at the end of the game.

In a game such as this, an algorithm for either player A or B consists of a set of strategies for the player to follow to make sure that he will always win or, if this is not possible, that he will maximize his chances of winning. In this particular game, if A goes first, he can always win if he moves or plays according to the following algorithm:

1. On the first move, A picks up 2 matches.
2. At each move thereafter, A proceeds as follows:
    a) If B has an even number: Divide number of matches still on the table by 6 and find the remainder of this division. Take one less than the remainder.
    b) If B has an odd number: Take one more than the remainder unless the remainder is 4, in which case take four (as soon as there are 1 or 3 matches on the table and B has an odd number, A should take them all).

This example of an algorithm demonstrates that there may be alternate paths which we have to follow to find a solution. In this case, the strategy to be used by player A depends on whether player B has an odd or an even number of matches. However, if player A plays according to the algorithm, he will win regardless of the path followed.

This example also illustrates that there may be several different forms in which the algorithm can be stated. In this case we might simply list all possible ways the game may develop (the number of matches remaining on the table after each move is shown in a circle).

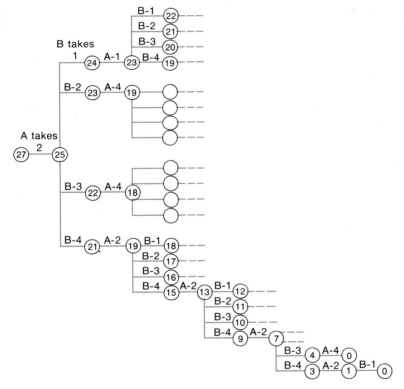

Chart 2–1.

Chart 2–1 is incomplete as drawn; if it were complete, it would indicate all possible ways in which the game might go if A always plays according to the algorithm.

In this section, we have considered several examples of algorithms. The decision-making problems which we look at in the remainder of this chapter and in the next chapter are solved in terms of algorithms. An algorithm is a set of rules for going from the statement of the problem (the model, constraints, and criteria) to the best solution.

4 | CRITERIA

In Section 1, the four elements of the decision-making problem are: model, criteria, constraints, and optimization. In the preceding section, we have discussed the optimization part, particularly when it is possible to find an algorithm or logical, complete procedure to a solution. In this section, we consider the *criteria*, or what it is that we want to optimize.

Probably the most significant decision made by most men throughout their lives is the selection of a wife. In this case, if one were rational and logical, a man might try to look at the problem in the following way:

1. Model: description of the characteristics of different girls. (appearance, beauty, intelligence, personality, character, aesthetic preferences, sense of values, attitude toward marriage, etc.)

2. Criteria: the relative values given to each of the above characteristics by the young man.

3. Constraints: the limited number of girls who would accept a proposal. If this number is precisely one, the problem is easy.

4. Optimization: selection of the girl to maximize the criteria within the constraints.

This particular problem perhaps illustrates the fact that most decision making in life is carried out intuitively and on the basis of poorly defined criteria. It is also clear that the two critical parts of an optimization problem are the model and the criteria— particularly the latter. While two young men might agree fairly well on the model (the different characteristics to be included and the relative ratings of particular girls), they would probably disagree strongly on the relative importance of these various characteristics. The criteria would vary greatly from one young man to the next. This is, of course, most fortunate; otherwise every man would want to marry the same type of girl.

In a simpler example (simpler because we can all agree on the criterion), we can illustrate another aspect of the criterion: the fact that we often have to change the criterion as the problem is being solved. In many cases, we discover that no solution can be found unless the criterion is simplified; in other problems, we can sharpen up the criterion as we move toward a solution. The latter situation is illustrated in the following example.

Figure 2–13 shows a model of a system of corridors in a building. There are three corridors: one from *a* to *b*, one from *a* to *c*, and one from *b* to *c*. These three corridors connect the points *a*, *b*, and *c*. During class changes in a school, the flow of people is from *a* to *b* in that corridor; hence, it is not convenient to try to walk from *b* toward *a*. We show this by placing an arrow from *a* toward *b*.

Similarly, the corridor from *b* to *c* is essentially one-way from *b* to *c*. The corridor from *a* to *c*, however, is wide, and we can move rapidly in either direction. This situation is represented by a double line connecting *a* and *c*, with an arrow in each direction. Thus, Fig. 2–13 represents the convenient pattern of travel throughout the system of corridors.

Now we turn to the problem we wish to solve. Our task is to station one or more men at the intersections *a*, *b*, and *c* in such a way that every intersection is covered by a man at most one "block" away. In other words, we might use three men, one at *a*, one at *b*, and one at *c*. Each man would then be responsible for monitoring his intersection.

Alternatively, we might place a man at *b*. He is able to cover both *b* and *c*, since he can travel to *c* rapidly in case he is needed there (there is a one-way path directly from *b* to *c*). Our man at *b* cannot, however, monitor intersection *a*. To reach *a*, he would have to travel two "blocks," from *b* to *c* and then from *c* to *a*. Hence, if we place one man at *b*, we must place another man at either *a* or *c* to cover intersection *a*.

The problem is where to locate the men in order to minimize the number required. The system of Fig. 2–13 is so simple that we can see that one man stationed at *a* could cover all three intersections (*a*, *b*, and *c*). If our only purpose is to solve this single problem, we are now finished. It is not difficult, however, to imagine a much more complicated pattern of corridors in which the answer is not obvious. In the hope of being able to solve problems of any complexity, we consider this simple example and look for an algorithm which works for all problems of this type.

The Corridor Algorithm

We must station our men to cover, or monitor, each intersection. We first consider intersection *a*. This can be covered by a man stationed at *a* or at *c*. We represent coverage of *a* by the expression

<table>
<tr><td>*To cover*</td><td>*Man at*</td></tr>
<tr><td>*a*</td><td>$(a + c)$</td></tr>
</table>

Here the expression $(a + c)$ means a man at either *a* or *c*; the $+$ sign means the same as "or" in ordinary English. In other words, $a + b + c$ would mean *a or b or c*.*

Similarly, intersection *b* can be covered by a man at *a* or at *b*, represented by

<table>
<tr><td>*To cover*</td><td>*Man at*</td></tr>
<tr><td>*b*</td><td>$(a + b)$</td></tr>
</table>

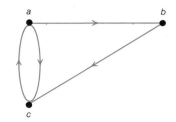

Fig. 2–13. A set of corridors.

*George Boole, 1816–1864, an English mathematician, introduced this idea when he showed how to represent statements in logic by algebra. Boolean algebra is routinely used today in analyzing and designing switching circuits such as those in a computer.

Now to cover both intersection *a* and intersection *b*, we can represent the stationing of men by the product

$$(a + c) \cdot (a + b)$$

Coverage   Coverage
of *a*   of *b*

We are using the product to represent coverage of both intersections. In the algebra Boole developed, multiplication is read as *and*. Thus, *a* · *b* means *a* and *b*.

The above expression thus means we can have coverage of both intersections *a* and *b* by having a man at *a* or *c*, and a man at *a* or *b*.

Once this algebraic expression is written, we can work with it just as in ordinary algebra. If we multiply out, we have

$$(a + c) \cdot (a + b) = a^2 + ab + ca + cb$$

What does each term on the right mean?

The first term, a^2, means that a man at *a* covers intersection *a* (from $a + c$), *and* a man at *a* can cover intersection *b* (from $a + b$). The second term, *ab*, means the man at *a* covers intersection *a* (from $a + c$), *and* a man at *b* covers intersection *b* (from $a + b$). Similarly, *ca* means a man at *c* covers intersection *a* (from $a + c$), while a man at *a* covers intersection *b* (from $a + b$); *cb* means the man at *c* covers *a*, at *b* covers *b*. In other words, the product

$$(a + c) \cdot (a + b)$$

represents all possible ways of covering intersections *a* and *b*.

To cover all three intersections, we have

$$(a + c) \cdot (a + b) \cdot (a + b + c)$$

Coverage   Coverage   Coverage
of *a*   of *b*   of *c*

If we multiply out this product, we find one term, a^3, involving only one letter. In other words, there is just one location *a* at which we can place a man who can cover all three of the intersections, as we saw by simple inspection of the figure.

The problem of Fig. 2–13 is too simple to show the power of this method. Let's turn to Fig. 2–14, where we have six intersections (*a*, *b*, *c*, *d*, *e*, and *f*). This might again be a problem of stationing monitors in school corridors. Exactly the same problem arises if we are stationing policemen on street corners.

In our example, we want to place men as needed to cover the important intersections *a*, *b*, *c*, and *d*. Each of these four intersections must be within one block of a man. Intersections *e* and *f* are less important. (We might have found very little trouble at these

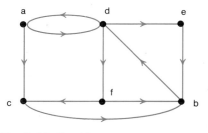

Fig. 2–14. Corridor map.

two locations. Consequently, it is all right if a man has to travel two or more blocks to reach e or f.)

The easiest solution is to put a man at a, another at b, one at c, and one at d. We want to *provide the required coverage with a minimum number of men*.

The statement of the problem is now complete. We must cover each of the four intersections, a, b, c, and d. Looking at Fig. 2–14 we see that intersection a can be covered by a man stationed at intersection a or d, intersection b can be covered by a man stationed at b or c or e or f, and so on. When our short-hand description of the problem covers the four essential intersections, we have the following expressions for the stationing of men:

$$(a + d) \quad (b + c + e + f) \quad (a + c + f) \quad (a + b + d)$$

| To cover | To cover | To cover | To cover |
| a | b | c | d |

By multiplying this out, we obtain 72 terms ($2 \times 4 \times 3 \times 3$), each of which represents a suitable location of men to provide the required coverage. A few of the terms are identical, so there are slightly less than 72 different possibilities.

In this horribly complicated product, typical terms are,

$$dcab \quad a^2bc \quad a^2b^2$$

The term $dcab$ means a man at d, one at c, one at a, and one at b. The term a^2bc means men at a, b, and c with the man at a covering two intersections. Since there are terms with only two letters, we can provide the required coverage with two men only; we can forget about all terms in the product with more than two letters.

The product can then be written*

$$(a + d)(b + c + e + f)(a + c + f)(a + b + d) =$$
$$a^3b + a^3c + a^3f + a^3e + a^2b^2 + c^2d^2 + a^2c^2 + a^2f^2 + d^2f^2$$
$$+ \text{ terms with more than two letters.}$$

We now have nine different ways to station two men to provide the coverage required. The meaning of the terms is illustrated by the following two examples:

a^3b The man at a covers three intersections, that at b covers one;

c^2d^2 The man at c covers two intersections, that at d covers two.

The existence of nine solutions is an unexpected delight. In practice, however, we must choose one of these. Apparently we did not ask for enough in the original choice of a criterion. Instead of just asking for the minimum number of men, we can add additional criteria (i.e., we can sharpen our original criterion).

We might require that each man be responsible for only two intersections. In other words, we ask for as even a division of the work load as possible. In our example, this added criterion rules out the terms $a^3(b + c + f + e)$ and leaves us with only the five possible solutions:

$$a^2b^2 + c^2d^2 + a^2c^2 + a^2f^2 + d^2f^2$$

Even with five solutions, we are unusually wealthy and we can add another criterion. We ask that each man be responsible for the intersection at which he is stationed plus one other (in order to minimize his travel time). This condition rules out a^2f^2 and d^2f^2, since in both cases the man at f is responsible for intersections b and c. We then have three possible solutions:

$$a^2b^2 + c^2d^2 + a^2c^2$$

Finally, we might select one of these three merely on the basis of which location is most important or possibly the personal desires of the men involved. The men we are stationing might prefer not to be placed at a because it is in a drafty location, so we would choose c and d.

General Comments

The interesting part of this problem is the way the final criterion was gradually worked out. We started with the simple criterion of minimizing the number of men. We found that we could add additional parts to the criterion: evening the work load, minimizing required travel, and finally catering to personal preferences.

In most optimization problems the initial criterion should be as simple as possible. Once solutions are found, we can add more criteria to the problem.

The question of what makes a satisfactory criterion underlies a large part of decision making. The continuing debate in the United States on the need for new weapon systems (for example, anti-missile missiles) stems from different criteria. Many of the people who want to build anti-missile missiles are worried about the defense of the country. They feel that, if we are strong defensively, we can prevent war. Opponents, on the other hand, argue that our economic and social needs must be faced. We cannot afford to meet these needs and engage in an armament race too. The difficulty of decision arises because of the constraints imposed by the need to limit government spending to control the national economic picture. When radically different criteria are used in decision making, it is not surprising that different decisions are reached.

In Section 4 we emphasize the importance of the criteria in decision problems. Of our four elements—model, constraints, criteria, optimization—the model represents the system with which

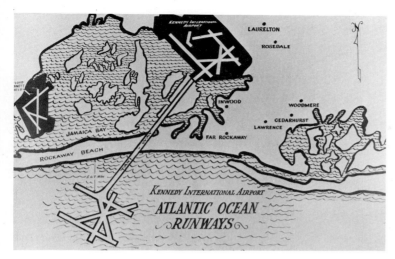

KENNEDY INTERNATIONAL AIRPORT
ATLANTIC OCEAN RUNWAYS

Fig. 2–15.
The choice
of an airport
location is one of
the most difficult decision
problems faced today. No
homeowner wants a jet airport
near his home. The noise is
continually annoying and may actually
lead to physical and mental illness. An
agency planning a new airport wants to
place it as close to the city center
as possible.
　The figure illustrates one possible
solution for airports near an ocean
or lake. The runways for landing
and take-off are located well
out in the water, with a
taxi-way to the terminals.
The only significant
problem with such a
solution is the cost.
("Toward a Quieter
City," report of
the Mayor's
Task Force on
Noise Control,
New York
City,
1970.)

we are working, the constraints represent the limitations imposed on the system, the criteria represent the objectives of the system, and the optimization is the best solution to the problem. In applied science, as well as in personal decisions, the criterion is that part of the problem which is most difficult to describe precisely.

5 | OPTIMIZATION WITH FEW ALTERNATIVES

When optimization involves only one choice from a few alternatives, the selection can often be made by a simple comparison. In this section we consider two examples of such an approach. The first involves only two alternatives and the second requires a selection from among a small set of alternatives.

Choice from Two Alternatives

If there are only two feasible designs or plans, selection of the better of the two is usually simple if the people in charge of the design can agree on the criterion (that is, the basis on which the two possibilities are to be compared). For example, if there are just two possible choices for the location of a new airport, the only differences between the two might be as shown in this summary:

Location	Total cost	Time from center of town
East of town	$3,000,000	45 minutes
West of town	$5,000,000	25 minutes

The choice between the two locations depends on the importance of each of the factors. If we decide that the location nearer town will result in increased use of the airport and hence a much greater

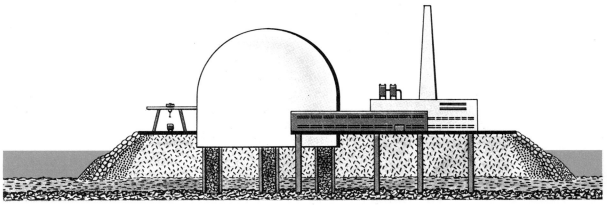

(*Max Gschwind for* Fortune *magazine*)

*Fig. 2–16a.
The most
common method
of building in
water is to use land
fill. When the water
is not too deep, dirt or
solid material is used to
make an artificial island. Stone
is used around the sides
to prevent washing
away the fill. The lower
figure shows a
16-acre island
in Lake Huron
built for an
electric
generating
station.
(Photo
The Detroit
Edison Company)*

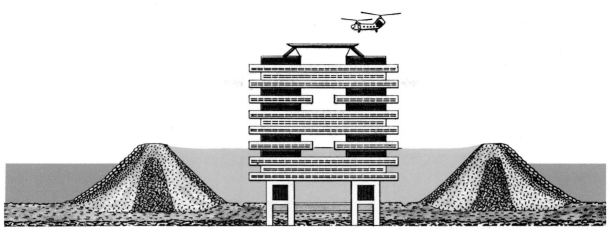

(*Max Gschwind for* Fortune *magazine*)

Fig. 2–16b.
Dry land can
be created by building
dikes to surround the area
to be made dry. Seepage through
the dikes is pumped out. The lower
picture is a scene from the
Netherlands, where more
than 1,600,000 acres
have been claimed
from the sea. (Photo
Netherlands
Information
Service)

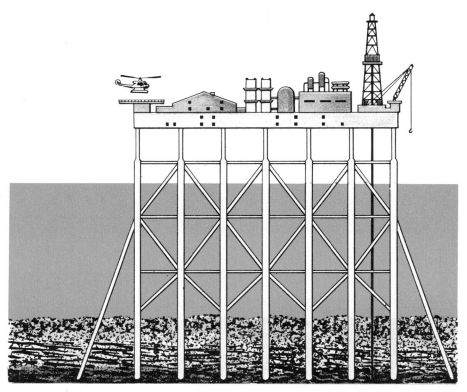

(*Max Gschwind for* Fortune *magazine*)

*Fig. 2–16c.
Piles can be
used to support
buildings. Much of Venice
rests on such columns. The
deepest structure is an oil platform
in 340 feet of water in the
Gulf of Mexico. Shown
here is a sulphur mine
seven miles off the
Louisiana shore.
(Photo Freeport
Sulphur Co.)*

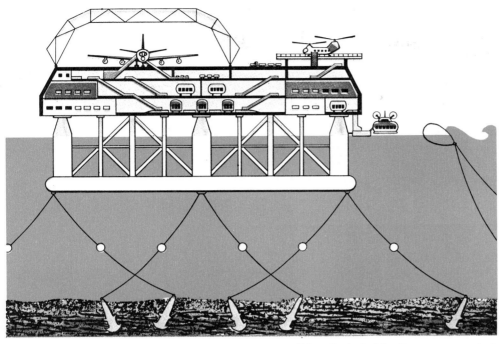

(*Max Gschwind for* Fortune *magazine*)

Fig. 2–16d.
Floating structures
support airports, oil rigs,
and housing projects (as shown in
the model). The structure
must be even stronger
than a ship, since it
must stay steady
during waves. (Photo
R. Buckminster
Fuller)

tax and revenue return to the town, we may agree on the west location.

This class of problems in which there are only two alternatives is illustrated by a simple route-planning problem.

Let us look at Fig. 2–17. Bill is late for a date with Jean. He is anxious to drive from his house to Jean's house in the minimum time. By the rural road, which is six miles long, he can drive directly to Jean's house averaging 30 miles per hour. Driving over a winding feeder road, he can reach a parkway on which he is permitted to drive at the rate of 60 miles per hour. He then takes another feeder road to Jean's house. He can drive at 30 mph on the feeder roads which are each one mile long. The parkway section is six miles long, so the total route is 6 miles + 2 miles = 8 miles, as compared to the six miles for the rural road. Which road should Bill take?

The rural route can be driven in $\frac{6}{30} = \frac{1}{5}$ hour, or 12 minutes. On the other route, the two-mile trip along the feeder road can be completed in $\frac{2}{30} = \frac{1}{15}$ hour, or 4 minutes, and the 6 miles of parkway require $\frac{6}{60} = \frac{1}{10}$ hour, or 6 minutes: a total of 10 minutes. Thus the longer route is two minutes shorter in time. Bill should take the parkway route because he is late and must use the faster route. If he were not late, he might choose the rural road because it requires less gasoline, or because he enjoys the scenery. Clearly the best route depends on the criterion that is used.

This problem may seem rather unexciting: one can hardly care very much about a time difference of a few minutes. If we were dispatching fire equipment from a fire station to a fire, however, the saving of a few minutes by optimum routing could be of critical importance.

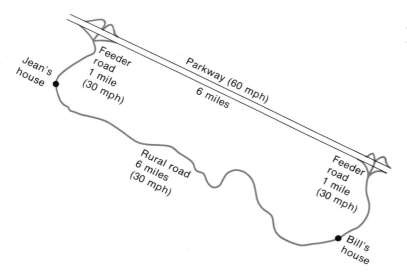

Fig. 2–17.
A route-planning program.

Choice from a Few Alternatives

If a small number of designs or plans is available, it is possible to study each plan carefully and to select that plan which is best. This is sometimes referred to as the "brute force" approach. To illustrate this method, we consider a very common *replacement problem:* When should one "trade in" his car to keep the average yearly cost as small as possible?

The problem is simpler if we consider trade-ins only after 1, 2, 3, 4, 5, . . . years of use. We must estimate the cost of operation of the car for each year of use. This cost must include the cost of fuel, oil, tires, batteries, maintenance, insurance, registration fees, taxes, and repairs. The cost of operation depends on the number of miles the car is to be driven each year, since this use determines depreciation, repairs, and so forth.

In Table 2–1 we list the data (columns 2 and 3) and the computations (columns 4 to 7) needed to answer our question. In column 1 we list the *age* of the car, and in column 2 the estimate of the *sale value* of the car. The car is assumed to depreciate $600 during the first year. The actual value depends on the make of the car, the market for second-hand cars, the mileage driven, and the quality of the care given to the car.

Column 3 lists estimates of the *operating expense* for each year. The operating costs listed are low and imply that the car is driven only about 10,000 miles per year. Operating costs do increase slightly as the years go by because of more repairs and poorer performance.

Column 4 is the *depreciation*. This is just the original cost ($3000) minus the sale value in column 2. Column 5 is the *cumulative operating cost*, which is the total of the operating costs in

Table 2–1.

Year (1)	Sale value (2)	Oper. cost (3)	Depreci-ation (4)	Cumu. oper. cost (5)	Total cost (6)	Av. cost per year (7)	
0	$3000	—	—	—	—	—	
1	2400	$ 800	$ 600	$ 800	$ 1400	$1400	
2	1920	850	1080	1650	2730	1365	
3	1540	900	1460	2550	4010	1337	
4	1230	950	1770	3500	5270	1318	
5	980	1000	2020	4500	6520	1304	
6	740	1050	2260	5550	7810	1302	←— Sell
7	520	1100	2480	6650	9130	1304	
8	320	1150	2680	7800	10480	1310	
9	150	1200	2850	9000	11850	1317	
10	50	1250	2950	10250	13200	1320	

column 3 up to and including the present year. Column 6 is the *total cost* which is the sum of the depreciation and the cumulative operating cost (sum of columns 4 and 5).

The last column is the *average cost per year*. This is the total cost divided by the number of years. This column gives the cost per year if we trade in after 1, 2, 3, . . . years.

Thus, we want to find the minimum in the last column. The best operating cost per year is achieved if we trade in after six years. It is important to note that this minimum is very "flat"; it costs only $35 per year more (about 3% more) to trade in the car after the third year. If the reliability of operation or the status which comes from driving a late model car are considered, the car should be traded before the sixth year.

The decision is, of course, no better than the estimates that have been made of the sale value and the operating costs. The collection of the necessary data, upon which realistic estimates can be made, is the major difficulty in solving this problem. Furthermore, the solution of the problem does not take into account the probable future changes in the price of cars and in the cost of operation or the changing value of the dollar. As an extreme illustration of this latter factor, if severe runaway inflation should occur (as in Germany in the 1920's), money becomes worthless, costs rise at uncontrolled rates, and any economic analysis is useless days after it is made. Even a slight inflation, however, can change the best trade-in time.

The route-planning and trade-in examples described are very simple cases of optimization. In both cases, once a model is constructed (the data are collected), determining the optimum solution is straightforward. In the next section, we consider a more complex problem in which there are so many possibilities they cannot all be listed and compared one by one.

6 | DYNAMIC PROGRAMMING

In many problems, a large number of possible designs must be analyzed to find the optimum solution. A direct approach takes much too long. In such a case, we look for an algorithm which reduces the number of computations. We saw this approach in the Königsberg-bridge problem where tedious trial and error was eliminated by the use of Euler's algorithm.

Now let us consider an idea called *dynamic programming* which lets us find a simple way to solve a complex problem. Dynamic programming is an approach to those decision-making problems in which a sequence of decisions is required. Dynamic programming says: Once we know how to proceed in the best way from a particular point to the end, we follow this path regardless of how we reached the particular point. Consider the following example:

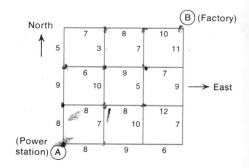

Fig. 2–18.
Costs per block
for new feeder line.

Fig. 2–19.
The street was built for three
lanes of traffic. Parked trucks
loading and unloading block both
curb lanes. The middle lane is very
close to blocked because of trucks
looking for a place to park. One block like
this obviously can result in a ten-minute
delay for an ambulance which happens
to pick this route. There just is no place
for the cars and trucks to go to leave
a clear path for the ambulance.

Angry citizens often urge a ban
on truck deliveries during the day.
This is really no solution. If trucks
have to deliver at night, labor
costs will increase. We can
expect higher costs of food,
clothing, and other necessities.
(Vincent J. Lopez for
Department of Water
Resources. "Toward a
Quieter City," report
of Mayor's Task Force
on Noise Control,
New York City, 1970.)

A power company, located at A in Fig. 2–18, plans to install a feeder line to a new factory, located at B. Streets have to be torn up and trenches dug for the new line. Some streets are paved with asphalt, others with concrete; the number of buried gas and telephone lines varies from street to street. On streets with heavy traffic, extra guards must be hired to direct traffic. The costs for the installation of the feeder line *in any particular block* are given by the numbers of Fig. 2–18. These range from a low of $3,000 to a high of $12,000. (Costs in the figure are in thousands of dollars.) How can we find which route from the power station at A to the factory at B will cost the least?*

This problem can be solved by calculating the cost for each possible route from A to B. We then can select the route with the minimum cost. This procedure is not particularly attractive, especially if we have a somewhat more complicated problem. For

*
In the text, the problem is to find the minimum-cost path from A to B in Fig. 2–18. The cost along each block is given.

The solution we find to this problem is equally useful for the problem of routing an ambulance from A to an accident at B in minimum time. Then the numbers in Fig. 2–18 represent the time to travel each block. In the same way we could talk about moving a fire truck from the station at A to a fire at B.

In a major city today, ambulances or fire trucks are often routed in a very illogical manner. The driver is given his destination and he selects the route he guesses will be fastest. He receives no information about traffic congestion along the various streets. It would be relatively inexpensive to measure traffic speeds along city streets at all times, feed these data to a computer, and have the computer use our algorithm to determine the best route. The decision would then be sent by radio to the driver.

74

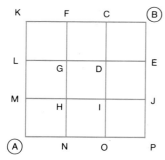

Fig. 2–20.
Grid of streets
with intersections labeled.

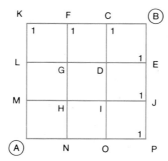

Fig. 2–21.
Number of routes
from intersections K, F, C,
E, J, P indicated.

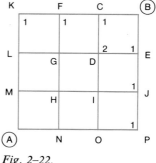

Fig. 2–22.
Two routes
from D to B.

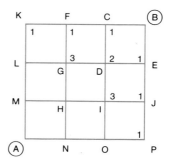

Fig. 2–23.
Three routes from G or I.

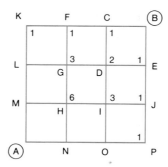

Fig. 2–24.
Six routes from H.

example, a grid eight blocks in each direction would have 12,870 different possible paths; calculating the cost for each would be a discouraging task.

Before we solve this problem by using a simple algorithm, let us digress briefly to see how the number of possible routes for any grid can be calculated. How did we obtain the number 12,870 of the last paragraph, or how many different routes are there for our 3 × 3 grid of Fig. 2–18?

Number of Alternate Routes

We assume that any route from A goes either to the north or to the east. In other words, no route moves south or west, and we always move *toward* the destination at B. There is a simple method for computing the number of possible routes.

The easiest method is to work backward from destination B. In Fig. 2–20 we have labeled each intersection with a letter. First consider intersections K, F, and C along the top. From each of these, there is only *one* route to B—east. Likewise, from each of E, J, and P there is only *one* route to B—north. Hence, we place a small one beside each of these intersections (Fig. 2–21).

Next we consider intersection D. From here we can go to E or to C; from either E or C there is only one route. Hence, from D there are two routes to B (Fig. 2–22).

If we continue to move away from B, we next consider G. Here if we go to D there are two routes, to F there is one. Hence, there are three routes from G to B. Similarly, we find three routes from I (Fig. 2–23). From H, there are three routes provided we go first to I, three if we go first to G—or six in all (Fig. 2–24).

In this manner, we find the total diagram of Fig. 2–25. There are twenty possible routes from A to B—twenty different costs to be evaluated and compared.* We have answered the question *by working from B back toward A*.

*
If the routes did not always move from A to B to the north or to the east, there would be any number of possible paths, since we could loop around a square block any number of times.

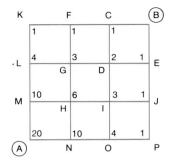

Fig. 2–25.
Number of routes
from each intersection
to B.

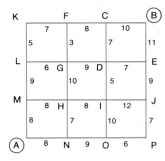

Fig. 2–26.
The original problem
with each intersection labeled
and costs per block shown.

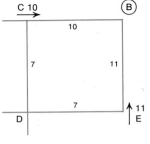

Fig. 2–27.
Best paths from C and E.

If the original problem had more blocks, we would find the following data:

No. blocks on a side	3	4	5	6	20
No. possible routes	20	70	252	924	137,846,528,820

The power company would be faced with a gigantic task if it had to figure out the costs for all possible routes from A to B for a grid twenty blocks in each direction. Consider the time it would take to figure out all the above possibilities. Some method of reducing the enormous number of possibilities is needed.

The Solution

The solution to the decision problem can be found if we work from the destination B backward, rather than starting at A and listing all paths. In other words, in Fig. 2–26, we first ask: If we are laying the pipeline and reach C, how do we proceed from here? We then repeat the question for E, which is also one block from B.

The answers to both of these questions are obvious. From C we must go east and the cost is 10; from E, north is the only allowable path and the cost is 11. We indicate these two "best solutions" on Fig. 2–27.

Next, how do we move if we are at intersection D? Here we can go north to C; from C we know east is the optimum. Hence north from D brings us to B with a total cost of 7 + 10, or 17. East from D takes us to E at a cost of 7, then on to B at a cost of 11, a total cost of 18. Hence, if we are at D we should move north and the cost to B is 17 (Fig. 2–28).

We turn to the next set of intersections away from B: the intersections at F, G, H, I, and J. F and J are simple, since there is no choice (Fig. 2–29). G is calculated next. Going north yields a total cost of 3 + 18 or 21; heading east yields 9 + 17, or 26 (from

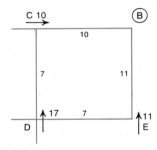

Fig. 2–28.
Best paths from D, E, C.

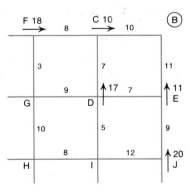

Fig. 2–29.
Optimum paths from F,
J, C, D, E.

D we already know the minimum cost is 17). Hence the desired path from G is north at a cost of 21.

In the same way the path from I is calculated as north at a cost of 22. Once I and G are fixed, H is calculated and we obtain the data of Fig. 2–30.

We next consider the intersections K, P, L, O, M, N, and A. This process finally leads to the minimum-cost values for every corner, shown in Fig. 2–31. Arrowheads at each corner show whether to travel north or east.

Thus, Fig. 2–31 is the solution of the original problem and gives the minimum-cost path from A to B. The total cost of this best path is $44,000.

The total grid has twenty possible paths from A to B. Instead of evaluating the twenty costs, one by one, we look at the problem as a set of binary (two-choice) decisions. There are nine of these in the total grid—at the intersections D, G, H, I, L, O, M, N, and A in Fig. 2–31. The solution of the problem requires the comparison of only nine pairs of numbers. These are, at each of these intersections, the cost if we go north or east.

To see how great this algorithm really is, let us look at the problem with twenty blocks each way. There are 137,846,528,820 possible paths. With the algorithm, we need compare only 20 × 20, or 400 pairs of numbers. If we can carry out four comparisons per minute, the algorithm solution takes less than two hours. If four path costs could be evaluated per minute (a rather high rate in a 20 × 20 problem), looking at all possible paths would take over 65,000 years of work. The algorithm turns an unsolvable problem into a solvable one.

Because a variety of optimization problems can be solved in the same way as the above example, it is useful to review the steps involved. We are faced with a problem in which there is a

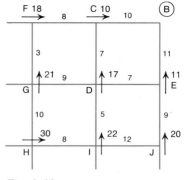

Fig. 2–30.
Optimum paths from G, I, H *added*.

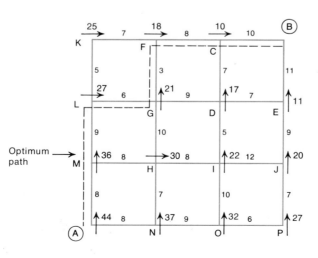

Fig. 2–31.
Minimum-cost values for the entire problem grid.

large number of alternatives. We try to find a way of looking at the problem as a *sequence of simple decisions*, rather than one complicated decision. We can find this sequence by working from the end B back toward the beginning A.

As an example, consider intersection H in Fig. 2–31. Before considering H, we determined the minimum-cost paths from G to B and from I to B (Fig. 2–32). If we are at H, we have only two choices, north to G or east to I. If we go north to G, we thereafter follow the already-determined, minimum-cost path to B (cost 21). Thus, from H the cost is 31 if we head north, 30 if we head east. We obviously should move east.

The key to success here is that, once we determine the minimum-cost path from G to B, it is the path to follow regardless of how we reach G. Recognizing this is what simplifies the decision at H.

There are many different kinds of optimization problems that can be solved with dynamic programming. Airlines take advantage of the wind direction on the different legs of the transatlantic routes to find minimum-time or minimum-fuel routes. Problems dealing with inventory control in warehouses, in programming of expansions of telephone switching centers, and in automatic control of industrial factories and chemical plants are other examples.

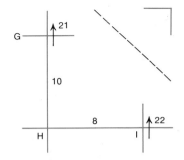

Fig. 2–32.
Decision at H,

7 | PROBLEMS WITHOUT SOLUTIONS

In a book, we of course look at problems which have solutions. Unfortunately, many real-life decision problems are not so simple. When we find the model, constraints, and criterion, we find that we have an impossible situation. In this section, we describe one important problem of this type.

Assume you are the owner of an apartment house in the center of a city. You are concerned about your property. The money you are receiving does not seem to be enough, and you must make a decision on what to do.

First, we describe the model. The building has five stories (no elevator) with eighty rooms. It is divided into twenty apartments, each with four rooms. There are 54 people living in the building. You bought the building in 1964 at a price of $80,000, which is its value. At that time, you invested $10,000 of your own money and the bank gave you two mortgages totalling $70,000. The annual interest on these mortgages is $10,500.

The city rent-control law allows you to charge $18 for each room each month. When all apartments are filled and everyone pays his rent, you receive $17,280 each year. To make repairs and keep the building in modest shape, you have to hire a full-time superintendent, someone part-time to collect the rents, and a lawyer part-time to help you meet all city laws. Also, you have to pay for repairs to the heating system and building. The cost of this maintenance comes to $15,000 per year.

We can summarize all the above in a table:

Investment	Income	Costs		Net loss each year
10,000	Rent 17,280	Mortgage payments	10,500	
		Maintenance	15,000	
		Total	$25,500	$8,220

Clearly, you are in trouble. You have invested $10,000 of your own money for the opportunity to lose $8,220 each year. A few more investments like this and you will be eligible for welfare payments.

Obviously you should raise the rent or cut down on maintenance costs. Here the constraints appear. The city rent-control law limits what you can do. If you want to raise the rents, you are told by your lawyer that the law has the following conditions:

> You are allowed to make 8% profit on your $80,000 building (or $6,400 each year). But you cannot count mortgage costs as part of your expenses. You cannot ask for a rent increase unless the rent is not enough to cover maintenance (yours is). Also, you must have lost money (not counting mortgage payments) two years in a row before you ask for higher rents.

Since your rent income is greater than maintenance costs, the law does not allow you to raise the rent.

Your other possibility to stop losing money is to cut maintenance costs. You hire the superintendent for only half time and try to cut repairs. The tenants complain, the city building inspector is called, he finds violations and recommends to the tenants that they pay only $1 per month until the repairs are made. Suddenly your rent income drops from $17,280 per year to $240 per year. Now you are really losing money!

You are hopelessly trapped. There just is no decision you can make to change your annual loss into a profit. You have only one possible decision: abandon the building.

Unfortunately, all the preceding numbers are not imaginary. They are typical 1970 data for the housing problem in New York City. They illustrate the economic basis of the problems which promise to create a national housing crisis in our cities in the 1970's.

As a result of this situation, 30,000 apartments are being abandoned each year in New York City and the number is growing. In the last three years of the 1960's more buildings were abandoned than could be torn down in twenty years of slum clearance. The city is filling up with abandoned buildings. With practically no new apartment construction, the 1970's are likely to see 25,000 families a year in New York City with no place to live.

Fig. 2–33.
A scene in East New York
is an example of the abandoning of
housing in the city.

They will have to live in armories or move out of the city, unless the model and constraints described can be changed in a major way.

The discussion above is presented in terms of the landlord. We might also have viewed this as a tenant. Faced with poor maintenance leading to inadequate heat, decay of the building, and dangers for his children, he has no recourse but to appeal to the city officials for help and to refuse to pay rent—steps which lead so often to abandoning of the building and his eviction. Likewise, we might take the view of the city government and consider the problems of writing a successful rent-control law.

The tendency to blame only the tenant, only the landlord, or only the city government clearly simplifies the problem. The model and constraints in this case are such that there is no optimization possible. To improve the situation, we must start anew with changes in the model and the constraints.

8 | FINAL COMMENT

Our goal in this chapter was to introduce the four elements of decision making. The three parts of the problem are:

MODEL   CONSTRAINTS   CRITERIA

Once we know these three parts, we can proceed to the

OPTIMIZATION

Several simple examples are used to show what each of these parts means.

From this chapter, two features of decision problems should also be clear:

First, decision problems are everywhere. The high school senior must decide whether to go on in school, what career to follow, what school to choose. The college senior again faces similar decisions. Marriage, choosing a job, picking a home, buying goods, deciding on a vacation—every major action we take involves a decision. Businesses and government also face decisions continually, often decisions which have a great effect on the future.

Obviously, as we think about a particular decision, we don't usually ask ourselves: What is the model? What are the constraints? What criteria should we use? Why then should we worry about the ideas in this chapter?

Some of the decisions we make are major factors in determining our future happiness, the relationship we have to other people, and our role in society. Because these decisions are of such importance, we must attempt to bring to them as much intelligence, wisdom, and logical thought as possible. Understanding the elements of a decision problem often helps in making a better decision.

Furthermore, government, business, and social organizations are more and more being run by men who try to use *logical* decision techniques. It is now common for a town to have a master plan for the future. Deciding on a plan requires a model, constraints, and a criterion. Even though estimates of the future are certain to be wrong, in order to make our decisions as logical as possible we must use a quantitative model. That is, we must describe our model in numbers. If we as voters and citizens are to understand government action, we must be able to see how the model, constraints, and criteria are measured.

Decision methods are often criticized because of this dependence on numbers. People ask, "Since the decision depends on predictions or guesses and since the numbers may be wrong, why is the decision any better than a wild guess or a hunch?" The basic idea of all modern system studies is that a model, even if imperfect, improves the decision. The more nearly correct the model, the better the decision. Hopefully, the decision will be good even if the model changes slightly.

The second interesting feature of decision problems is that there is no simple method for finding the answer. We looked at a few algorithms in this chapter. These work only for special kinds of problems. When we face an entirely new problem, we have to use our imagination and ingenuity to try to find an answer.

This situation is quite different from the usual problems considered in science and math texts. Often there we can learn a method for solving a whole category of problems (the quadratic formula in math). When we consider difficult problems from the real world, there often is no straightforward way to the answer. We have to try different attacks. We can only hope for success. Decision problems are challenging because they have this feature.

Questions for Study and Discussion

1. Why do decision makers bother with a model? Why don't they just go ahead and decide?

2. Test the steps of the algorithm which can be used to solve any problem of the Königsberg-bridge type. We can assume we start with a graph as a model for the system; the solution desired is either a statement that the problem is not solvable or a step-by-step procedure for finding a path which travels on every line once and only once.

3. Numbers may be numerical, of course, as in arithmetic, or literal as in algebra. Would it ever be possible to construct an algorithm in a situation which involved no numbers, as defined above?

4. The girl at the check-out counter in a supermarket has to make change. The usual method can be described by an algorithm, or by a set of algorithms for use in different situations. Work out algorithms for these cases:
   a) A dollar bill is given for a 22¢ purchase.
   b) A five-dollar bill is paid for goods costing $2.78.

5. What criteria would you use in selecting a piece of sporting equipment for yourself if money were not a constraint? Select any kind of equipment that you feel you are a good judge of, for example, skis, ski boots, baseball glove (for which position?), surfboard, tennis racket, 4-iron.

6. Suppose in the previous case that you are able to spend only 75% of the cost of the "best" item. Assume that you don't have to shop around, because the store carries every conceivable item in stock—if you can imagine such riches. What happens to your criteria? That is, do you relax all your criteria about equally, or do you eliminate or greatly weaken some requirements in order to retain others unchanged? (Be as specific as you can and see if you can produce "logical" reasons.)

7. For the system of Fig. 2–12 with the extra lines from P-B and K-G added, determine an optimum path on the basis of the route-selection algorithm. Start with the A-to-A path made up of

   ABCDEFGHIJKLMNOA

   and then add loops from the diagram remaining after this path is deleted. If each block requires one minute for traversal, what is the minimum time for a complete patrol? How much time is saved compared to the common back-and-forth patrol represented by the path:

   AONQPOABPBCDAHEDHIONMINMLKMI
   KJIHJKGJHGEFGHA

   (For evaluation of the time required here, the original street diagram of Fig. 2–11 should be used since this drawing represents the actual problem.)

8. In a small city in the Midwest, there is a three-mile stretch of one street on which the traffic lights are timed so that a motorist who drives at 30 mph can expect to be confronted by a red light only once or twice during his trip west on this street. Depending on the traffic density (number of cars per mile), the timing is changed so that 28 mph, 23 mph, or some other speed becomes the optimum speed. What constraints might affect such a system?

9. The following problem can be solved by the Königsberg-bridge algorithm. Explain what you would do to fit the algorithm to this problem. Problem: Draw a continuous line that passes through each edge of every rectangle in the diagram once and only once.

10. Apply the method you developed in Question 9 to find out whether the problem can be solved if the diagram changes to this:

11. Various performance criteria are appropriate in evaluating a record player. Discuss those which would be used:

    a) by a committee of students buying a record player for a school recreation room.

    b) by a restaurant owner planning to pipe background music into his various dining rooms.

c) by a lover of contemporary "serious" music (the orchestra is likely to have a large percussion section).

12. Could a walker cross all the bridges of Fig. 2–8 (starting at B and crossing each bridge once and only once) by any other path than that given in the text? If so, describe one such path.

13. A small town has a grid of streets like the corridor map of Fig. 2–14. The town government decides to install street lights, but wishes to minimize cost. Intersections *a*, *b*, *c*, and *d* should be illuminated (more or less) in each case by a light not more than a block away, but *e* and *f* are considered less important. Give an expression which describes the possible placement of street lights. Do not expand the expression.

14. Figure 2–5 has a caption which says that the reduction of air pollution by legal action has inevitable economic and social consequences which may be more harmful than the pollution in the short run. Describe a pollution situation in which immediate solution would cause serious economic and social consequences. Describe a pollution situation in which the failure to solve it immediately would cause serious economic and social consequences.

15. You wish to run a wire from your record-player amplifier at A to a speaker at G in the room pictured. You must run the wire along the edges of the room. List all the possible ways to install the wires following the constraint given. What is the optimum path to use to insure that future shifts in furniture (other than the record player or speaker) would not require rewiring?

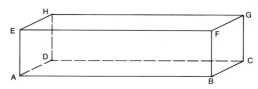

16. In the last two decades, astronomers have found several new ways of learning more about the stars. Radio astronomy, for example, occupies the attention of many, and quite recently the study of X-ray astronomy has begun. X-rays do not penetrate the atmosphere very far, and X-rays in space were first discovered by carrying detectors on rockets. An X-ray astronomer must make a decision as to which of various available rockets to use in his work. How would you describe his model? What criteria can you think of? What constraints occur to you?

17. Sketch a map of the footpaths in one of the parks of your town, and apply the Königsberg-bridge algorithm to work out a minimum route for a snowplow to follow in clearing the paths. (If you live where it doesn't snow, assume that paper and leaves are on the paths and a sweeper is assigned to the park clean-up.)

18. A young married man with one child, working for a "national company," is sent to a new town where he must buy or rent a house or apartment. What are likely to be (a) the constraints and (b) the criteria which will govern his decision?

19. Refer to Question 18. Would you expect either of these to change if the man were an executive officer of the company, aged about 55, with no children at home, but grandchildren not many miles away? If so, how?

Problems

1. The following information represents the grades (to nearest 10%) of 300 students on a test.

   a) Make a graphical model of the data below.

   b) If the performance criterion is that 80% of the students should pass, what would be the passing grade for this test?

   c) If we are using this test to help us select students for an honor class, and school policy (a constraint) allows a maximum of only 10% of the students to be admitted to honor classes, what is the (optimum) grade for honor students? Based on your answer, what recommendations would you make to the school policy makers?

Number of students	Test grade
10	20%
10	30
20	40
20	50
30	60
90	70
90	80
30	90

2. One of the newer members of the United Nations has decided to change its system of money. It intends to use coins and paper money instead of beads. The chancellor of the exchequer has decided to use dollar bills and higher order denominations of paper money, but he cannot decide what to manufacture in the way of coins.

He notices that the United States uses five coins worth 50, 25, 10, 5, and 1 cents. He, too, has recognized that the decimal system (100 cents to the dollar) is desirable and he wants to make only five different coins. He is disturbed by the large number of coins required to make change in the United States. To give 94¢ change, for example, Americans have to use eight coins. This use of many coins is particularly annoying in his tropical country, since the men normally wear a minimum of clothing and pockets are correspondingly small.

Faced with this problem, the chancellor decides he will choose values for his five coins such that the smallest is worth one cent, and such that no more than six coins are ever required to make change.

a) What values should he give to each of his five coins?

b) If every value from 1 to 99 cents is equally probable, on the average how many coins are required in each transaction in the United States?

c) Repeat part (b) for your choice of coins in (a).

3. The curve of Fig. 2–1 for gasoline economy versus speed is assumed valid for this problem. We have rented a car from the No. 1 rental agency and will have to pay a fixed 9¢ per mile for the distance traveled. We wish to drive from the airport to a meeting 80 miles distant. Half of the trip distance is on an expressway with a minimum speed of 40 mph and a maximum of 70 mph; the other 40 miles of the trip are on a country road with a maximum speed of 45 mph and a minimum of 20 mph. Since we are stockholders of the No. 1 rental

agency, we are anxious to minimize our fuel consumption during the trip. How should we drive?

a) How many fewer gallons of gas shall we use by following this optimum schedule, rather than following the schedule which would bring us to our destination with maximum fuel consumption?

b) How many fewer gallons do we use than if we drove at maximum speed at all times?

c) How much longer is required for our "optimum" trip than for the fastest trip? (In other words, how much time do we have to waste to do minimum damage to rental agency No. 1?)

4. The process for finding the square root of a number can be simplified by using an algorithm. We wish to find the square root of 46 to three significant digits. We take half of 46 as a trial root (23); add this to the fraction of the original (46) over the trial root (23) and divide by 2. $(23 + \frac{46}{23}) \div 2 = 12.5$. This number squared is 156.25 (too large). We repeat the process $(12.5 + \frac{46}{12.5}) \div 2 = 8.09$. This number squared is 65.4 (closer). Repeat the process $(8.09 + \frac{46}{8.09}) \div 2 = 6.89$. 6.89 square = 47.5 (closer). Repeat the process until you get the correct answer. Could you have started with 6, or 7, or 10 instead of 23? Using this algorithm, find the following: (a) $\sqrt{72}$ (b) $\sqrt{6}$ (c) $\sqrt{948}$.

5. As an example of an algorithm, we consider the following game. There are 20 matches on a table and there are two players, A and B. The players alternately pick up and retain 1 or 2 matches. The winner is the player who has the twentieth match. Devise an algorithm for A to always win.

6. A simple "model" of the car trade-in problem (p. 00) is to assume that the sale value of the car at the nth year is a fraction of the sale value at the $(n-1)$ year, for example:

$$v_n = a v_{n-1} \text{ where } 0 < a < 1, \quad (a = \text{constant})$$

and to assume that the operating cost for the nth year is increased by a fixed amount over the operating cost for the $(n-1)$ year; i.e.

$$c_n = c_{n-1} + b \quad (b = \text{constant})$$

Use this "model" for the case where

$v = \$3000$ (initial sale value)
$a = 0.80$ (depreciation factor)
$c = \$800$ (operating cost for the first year)
$b = \$50$ (increase in operating cost per year)

and determine the year to trade-in in order to minimize average operating cost per year.

7. Let us reconsider the question of when to turn in a car, assuming steady inflation at the rate of 5% a year. Since money is worth less, after one year it takes (on the average) $1.05 to buy what $1 would the year before, after two years it takes $1.10, and so on. However, the automobile dealers might stick by their usual formula in making up the "little red book," because they would profit by pricing used cars at the old, uninflated rate ($2400 by the table on p. 00 ought to become 2520 inflated dollars). The operating cost, on the other hand, would be measured in terms of money that buys ever less, so the first two entries in column 3 must change to $840 (from $800 + \frac{5}{100} \cdot 800$) and $935 (from $850 + \frac{2 \cdot 5}{100} \cdot 850$). The last four columns are a bit sticky, to say the least, because we must add dollars whose values keep changing. We simplify the problem by using the numbers given in columns 2 and 3 at their face value.

Making these assumptions, should you still trade in after six years?

8. In the four-by-four grid shown below, find the minimum-time path from A to B going only north and east. There are 70 possible routes, but you need to calculate only 24 numbers. Block times are shown.

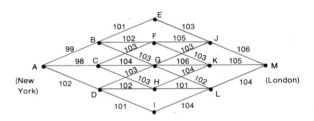

9. An unscrupulous cabdriver who is paid by time rather than distance wants to have the *maximum*-time route from A to B in the previous problem, still traveling only north or east (to avoid suspicion). Can you find it for him?

10. Many of the airlines flying the North Atlantic now use computers to find minimum-time jet flight paths taking into account the strong winds that usually occur at jet cruising altitudes, and the restrictions on possible routes imposed by air traffic control requirements. Savings on the order of 15 minutes on the nominal seven-hour flight are obtained this way. A grid of "checkpoints" is se-

lected and all routes must consist of generally east-west straight-line segments between checkpoints. A simplified version of such a grid is shown. Imagine that checkpoint A is New York and M is London (or Paris if you prefer). Checkpoints B through L are points in the ocean (located by giving their longitude and latitude). Using wind data from "weather ships" on the North Atlantic and characteristics of the particular jet airplane the airline uses, the flight planner computes the time to fly each segment (results for east-to-west flights differ, of course, from west-to-east flights). He

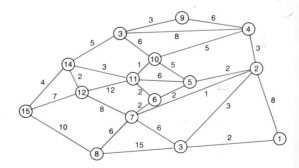

then has a "maze problem" very similar to the one treated in Section 6 to find the route that requires minimum time. In practice there are many more checkpoints than are shown in the figure so that a computer solution is essential. For the data given in the figure (time in minutes), use the dynamic programming algorithm to find the minimum-time route. (There are 19 possible routes.)

11. The text examples of a shortest-route problem used a very special street pattern, namely a square grid of streets. The technique developed for solving that problem can also be used on other street patterns. In the illustration shown next we have a pattern of streets, and the time in minutes to travel each block is shown. We wish to find the shortest route from corner 15 to corner 1 where on any block we may travel only toward the right; e.g., we may travel from corner 7 to 6 but not from 6 to

7. Use exactly the same algorithm to solve this problem as was used on the square grid pattern.

12. Notice that in Problem 11 the route would have been 3 minutes shorter if the last step (2, 1) had been replaced by (2, 3, 1). However, this route traverses a block in a direction not permitted by the stated conditions. Suppose this restriction is relaxed: how can we find the best route when we may traverse a block in any direction? A method called *relaxation* can be used. We must have, to begin with, a time estimate at each junction that is certainly long enough (there must be room for optimization). At each intersection in turn, beginning with those nearest the destination, see if this time estimate can be improved by taking any other, previously forbidden, route, and if so enter in a chart the new estimate and the new direction. When the whole pattern has been surveyed, transfer the new times and directions to the map. For example, suppose we start with the final time estimates obtained in Problem 11. The correction chart for the first few entries looks as follows:

Junction	Old estimate	Arrow to	1st new estimate	Arrow to
2	8	1	5	3
3	2	1	2	1
4	11	2	11	2
5	10	2	10	2

After making this improvement successively at all the junctions, we repeat the process again and again until no improvements can be made at any junction. We then have at each junction the minimum time to junction 1 and the arrows indicate the route. In this example, on the second time around we find that 4 and 5 have been revised to:

Junction	1st new estimate	Arrow to	2nd new estimate	Arrow to
4	11	2	8	2
5	10	2	7	2

Complete this problem.

Laboratory and Projects

I | MINIMUM FENCE FOR MAXIMUM AREA

The conclusions you draw from this experiment and the algorithm which you derive can be useful in many areas. This could be a problem faced by a farmer who wishes to fence in a rectangular field along an existing fence. It could be a problem faced by an urban housing development where we wish to fence in a playground area alongside an existing building. It could also be a problem at a beach resort where we wish to fence in a "safe" area in the water, using the shoreline as one side of the fence.

Part A

Suppose we have 1000 feet of fence and we wish to obtain the greatest possible area within the boundaries of this fence. We shall assume an unlimited length of existing boundary on one side of the area to be fenced as in the diagram on the right.

The solution to the problem is one of optimization and must contain the four fundamental elements of a decision-making problem: model, criteria, constraints, and optimization.

■ 1. What are the criteria in this problem?

■ 2. What are the constraints in the problem?

Procedure

Using several sheets of cross-sectioned graph paper, some wire, and a ruler (for scaling), design a satisfactory scale model for this problem. Search for an optimum solution by experimenting with different fence designs on your model. Remember always to keep within the constraints.

▪ 3. If the fence were to be built according to your conclusions, how much area would the fence enclose?

Use your model to determine the optimum fence design assuming a curved border. Include some irregularly shaped borders in your trials.

▪ 4. What would be the maximum area if you used a curved fence?

Part B

You have been using a scale model and the technique of trial and error to solve this optimization problem. This method is widely used by engineers to solve very complex problems which cannot be readily solved by other means. For example, after the shape of a new airplane is devised, a small model is tested in a wind tunnel. It is changed many times before an actual airplane is built. We have used this technique because it is a quick and fairly accurate way of determining the maximum area.

If we limit our considerations to rectangular fields, another approach is possible. Let us now try to develop a mathematical model. We were able to use the wire scale model because it related the length of the fence to the area enclosed by the fence. Once we had this information for several different fence designs, we were able to choose the optimum design.

The mathematical model should be set up to give us the same information, that is, a relation between the perimeter of the field and the enclosed area.

A rectangular field has the length of its sides represented by a and b. The fenced perimeter can be represented by the equation:

$$P = 2a + b \tag{1}$$

The area of the field is

$$A = a\,b \tag{2}$$

Solve the perimeter equation for b:

$$b = P - 2a$$

and substitute b into equation (2):

$$A = a\,(P - 2a)$$
$$A = Pa - 2\,a^2$$

This last equation allows us to calculate the area of the rectangle when we know the fenced perimeter of the rectangle and the length of one of the sides. It is therefore a good mathematical model for our optimization problem. To obtain a solution, we simply calculate the area for different values of a and then select the value of a which gives the maximum area.

Procedure

Calculate the area for the values of *a* indicated in the table below:

Length *a* (ft.)	Area (ft.²)
0	
50	
100	
150	
200	
250	
300	
350	
400	
450	
500	

■ 1. Plot a graph of Area versus Length *a*.

■ 2. Estimate the value of *a* which results in a maximum area. What is the corresponding value of *b*? Do these results agree with your findings from the scale model?

Suppose that the beach situation described in the introduction of this problem had the additional constraint that the maximum distance of beach front which could be fenced was 350 feet and the minimum was 300 feet, but the shape of the fenced-in area was optional.

■ 3. What shape fenced area would you use to get the maximum area?

■ 4. What would be the dimensions of the fenced area?

II | OPTIMIZATION AND THE PAPER AIRPLANE

Recently the *Scientific American* magazine sponsored a contest to design paper airplanes. Many of the country's top aeronautical engineers and physicists entered the contest. (The book *How To Make and Fly Paper Airplanes* has been written by the winner, Ralph S. Barnaby—Bantam Books.) While you may not be an aeronautical engineer, you have probably at some time built a paper airplane. Using the criteria and constraints listed below try to design a paper airplane.

Part A

The major objective (criterion) is to build a paper airplane which will land the greatest distance from the launch point upon being launched from a height of six feet above the landing level.

Constraints

1. Paper must be 8½″ × 11″ and of the same weight for all.

2. Masking tape 2″ × ½″ must be used in construction.

3. Two paper clips must be used in some way.

4. Airplane must be constructed during laboratory period and may undergo changes in design or construction following a maximum of two test launchings.

Part B

Original criterion may be replaced with others such as:

1. Fly the straightest path.

2. Stay in the air the longest.

3. Return closest to launch point after traveling at least ten feet from launch point.

III | GAMES AND ALGORITHMS

Algorithms have been defined many ways and one definition is: A system for winning every time even without cheating.

Part A

One of the many games which depend not on the outcome of chance events but on the ingenuity of the players is the game Eleven Matches. Eleven objects, such as matches, are on a table. Each player in turn may pick up 1, 2, or 3 matches, as he thinks best. The players keep taking turns until there are no more matches. The player who is forced to pick up the last match is the loser. What is the algorithm, if any, by which the player who has the first turn can always win?

Part B

One popular course for adults (business executives, engineers, and so forth) is problem solving and creative thinking. The class is given a problem. Free discussion is allowed under a group leader. The group leader is familiar with at least one possible solution. The following problem is the type which is usually presented to the group: You are facing three men (one to your left, one in the center, and one to your right). One of these men always tells the truth, one always lies, and one randomly decides whether to tell the truth or lie. Each man knows who is the truthteller, the liar, and who varies his answers. You are allowed three yes-no questions. Each question can be addressed to any one man. After hearing the three answers, you must decide which is which. What three questions do you use?

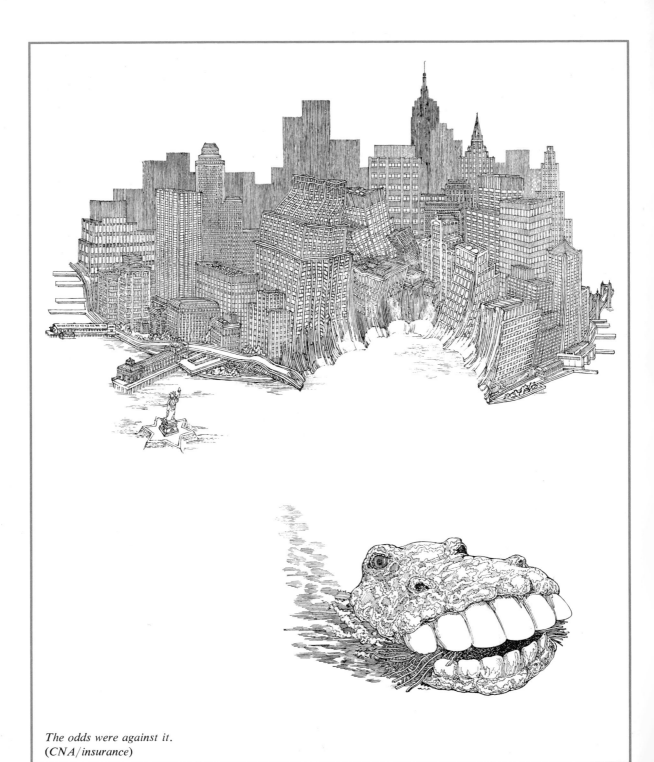

The odds were against it.
(CNA/insurance)

Optimization

3

1 | INTRODUCTION

In this chapter, we will consider problems of three types. Type 1 is a *queueing* problem. Type 2 represents a group of problems described by the name *games*. Type 3 comes under the general heading of *linear programming*. While we will not go into the detail required to solve the problems described, we will look at methods of solution which can be used in such problems. All of these are decision problems. They have the elements described in the preceding chapter. All are solved by optimization. These decision problems are illustrated by two examples.

Problem 1. By 1980, 30 million people a year will be using Chicago's three airports. Three quarters of these people will go to the airport by car or taxi. How many lanes should we build for the main highways entering the airports in order to prevent excessive traffic jams except on very rare occasions? The answer turns out to be six. Once this answer is obtained, the airport designer next has to worry about parking places for all the cars.

Problem 2. We have limited total resources, too little to do everything we would like to do. How do we allocate these resources in an optimum decision? For example, a sausage-maker has a recipe for frankfurters which includes a number of ingredients for which the prices fluctuate widely from week to week. He wants a computer program to determine the proportions each week to maximize the profit. Of course, there are constraints to limit the variations of flavor so the customers do not object.

2 | QUEUEING PROBLEMS

A queue is a waiting line. An example of a queue is found at a theater box office just before a show starts. Some of the worst queues exist at a typical college when new freshmen arrive to register or at an army induction center.

For our discussion, we must recognize that queues are likely to form when customers arrive at a service facility. In order to be general, we talk about *customers* arriving. In actual examples, these customers may be people (arriving at a supermarket checkout counter) or machines (airplanes arriving at an airport or subway trains waiting to enter a station).

These customers are to be *serviced*. The service facility processes the customers, in many cases one at a time. Once the customer is serviced, he leaves the system.

A Supermarket Queueing Problem

A small supermarket is built with eight check-out counters. The manager is ready to open for business and is hiring people. He needs to know how many check-out girls to hire. How many of the eight counters should he plan to open?

If too few counters are open, long queues are likely to form. Customers who have to wait fifteen minutes to check out are not likely to come back to this store. On the other hand, if he opens too many counters, he will be paying the salaries of people who are working only a fraction of the time. Since the profit of a supermarket is not large under the best conditions, he cannot afford very many unproductive people.

This problem is interesting because it is complicated by changes which occur randomly, or without a definite pattern, during the day. In a supermarket, there are times when, for no apparent reason, the check-out system is very heavily loaded. At other times, hardly any customers are checking out. These changes are partly predictable (Friday evening is a busy time). They are also partly unpredictable. The strategy the manager chooses must take care of the unexpected load.

How does the manager make decisions? As we look at queueing problems in the next few sections, we will find some of the techniques which can be used to make the system adapt to occasional heavy loads. Three steps are obvious in a supermarket.

1. An express counter is set up for customers who do not require very long to service. This removes these customers from the queues at the slower service channels.
2. If the system becomes overloaded, we try to reduce the service time for each customer. In a supermarket, the manager rapidly switches some men from arranging merchandise on the floor to helping the check-out clerk by putting the items in bags. Once the queues are reduced, the men are returned to the other tasks.
3. Finally, the manager or other supervising personnel can man unused check-out counters.

As we discuss queueing, we will call these three strategies:

1. Changing the priority system (that is, which customers are served first).
2. Reducing the service time.
3. Increasing the number of service channels.

These are three basic techniques for keeping queue lengths small. In the supermarket case, they are easily carried out. In other problems, they may not be possible, especially if customers insist on being treated on a first-come, first-served basis.

Fig. 3–1.
A queueing
problem. (Drawing
by Stevenson;
© 1969
The New Yorker
Magazine,
Inc.)

"What did I say about leaving on a Friday afternoon!"

Examples of Other Queueing Problems

Queueing problems arise in such diverse situations as:

1. Tollbooths on a parkway, where long lines of cars may build up.
2. A barbershop or hairdresser operating without appointments. Even though the barber may be idle a significant part of the time, there are other times when a queue of three or four customers forms. The queue seldom becomes longer because customers turn away rather than wait.
3. An airport in which only one runway can be used for take-offs and landings. Because of traffic demands generated by men anxious to return home for the weekend, many major airports in the United States have very large queues of planes circling and of aircraft waiting to take-off every Friday afternoon around 5 P.M.
4. A telephone exchange. On Christmas and Mother's Day, long-distance toll lines are frequently saturated.*
5. A production facility dependent on a number (perhaps ten) of machines which must be maintained by two mechanics. Each machine breaks down occasionally and can be repaired by a mechanic. The amount of time required for servicing varies according to the nature of the difficulty. In this case, the queue consists of inoperative machines being repaired or waiting to be repaired. The customer for the service in this case is the machine.

*
The telephone-system example is particularly appropriate, since the basic work on queueing problems was originally developed for guiding the design of telephone systems. A. K. Erlang of the Copenhagen (Denmark) Telephone Company was a key worker as far back as 1907.

Importance of Random Behavior

Queueing problems are interesting because both the customer arrivals and the servicing times are random. They vary in an unpredictable way.

If the customers arrive at regular intervals and each is serviced in a specific time, the queueing problem is trivial. For example, in the case of a single service channel (Fig. 3–2), if customers arrive to be serviced every 5 minutes and servicing requires 4 minutes, we can state:

> The service channel is busy $\frac{4}{5}$ of the time. The maximum queue length is one (the person being serviced). The average queue length is $\frac{4}{5}$ because $\frac{4}{5}$ of the time it is one, and $\frac{1}{5}$ of the time zero.

There is really no queueing problem. If the service time were greater than 5 minutes, the queue would steadily grow larger. More and more people would be waiting longer and longer.

If random behavior is considered, the situation is quite different. For example, we can imagine that customers arrive completely at random with an average of one every 5 minutes. In other words, in 500 minutes we expect about 100 arrivals. These come, however, with equal probability in every short interval of time.

Furthermore, we can assume that *on the average* 4 minutes are required per customer. Any one customer may require more or less time. In other words, we assume that the servicing times vary at random in the same way that the times between customer arrivals vary.

If these two types of random behavior occur, the queueing problem is much more difficult. Now it is quite common for a group of customers to arrive very close to one another. The second, third, and fourth customers may arrive before the first is served. Indeed, if we study such situations, we find that over a

Fig. 3–2. Regular arrivals, single channel.

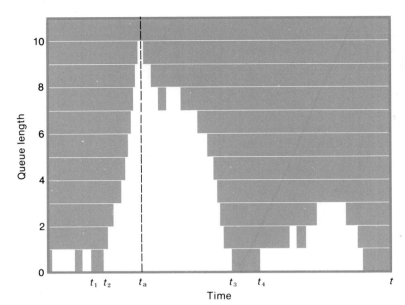

Fig. 3–3. Possible plot of queue length for random arrivals and service times.

long period of time, the average queue length is four. On the average, there are three people waiting and one being served, even though the service channel is still busy only $\frac{4}{5}$ of the time.

In this case, we can run an experiment and observe the way the queue length q changes with time. Figure 3–3 shows one such graph obtained from a computer study. From t_1 to t_2 and t_3 to t_4, there is no queue at all. The service channel sits idle. Right after t_2, customers arrive rapidly. By t_a, the queue length is ten.

Even though the service channel does not seem overloaded (it is idle $\frac{1}{5}$ of the time), the customers who arrived around t_a are obviously not well treated. The random feature creates the problem.

Because the randomness is so important, we can only study queueing problems experimentally with a few basic ideas of probability. These are presented in the next section. We will consider queueing again in Section 4.

3 | PROBABILITY

One of the striking features of the past twenty years is the growing awareness that many things can only be described in terms of probability. Even weather reports now often contain the statement, "The probability of rain tomorrow is 20%." The ideas of probability, which were first developed in the eighteenth century to analyze gambling games, are essential today for understanding many important systems, from traffic flow on highways and through tunnels to queues. In this section, we will discuss a few aspects of probability.

If we thoroughly shuffle a deck of 52 playing cards and draw one card at random from that deck, the probability that it is the 7 of spades is $\frac{1}{52}$. If we repeated the experiment 52,000 times, we would expect that the card selected would be the 7 of spades about 1000 times.

Can we describe this experiment more carefully? In the experiment, there are 52 *equally likely* results. The phrase "equally likely" is very important here: Each of the 52 cards is equally likely to be the one selected. We can list these equally likely results:

Spade A, 2, 3, . . . , 10, J, Q, K;　Heart A, 2, 3, . . . , K, . . .

Before the experiment is performed, we decide which results we will call favorable: In our example, we selected only the spade 7. Hence, of the 52 equally likely outcomes, only one is desired. Then the probability of a favorable result is $\frac{1}{52}$.

As long as the possible results are equally likely, we can find the probability of any group, rather than a single outcome. For example, the probability the card drawn is a 7 is $\frac{4}{52}$. (There are 4 sevens among the 52 possible outcomes.) The probability of a spade is $\frac{13}{52}$, of a face card (jack, queen, or king) is $\frac{12}{52}$.

Listing Equally Probable Outcomes

Thus, to determine the probability of a favorable result, we first list the equally likely outcomes. We must be careful that the items of the list are indeed equally likely. As an example, consider the problem of tossing two coins. There are now three possible results:

2 heads　　　1 head and 1 tail　　　2 tails

These three outcomes are not equally probable, however. If one coin is a penny P and the other a nickle N, the possible results are

P	*N*	*Overall*
H	H	2 heads
H	T	1 head and 1 tail
T	H	1 head and 1 tail
T	T	2 tails

Thus, the probability of two heads is $\frac{1}{4}$; the probability of a match is $\frac{2}{4}$, of a mismatch also $\frac{2}{4}$.

One way to calculate the probability is, therefore, to list all equally likely outcomes of the experiment, count these (call the number N), count the favorable results (call this n), and finally calculate the probability as

$$\frac{n}{N} \text{ or } \frac{\text{favorable results}}{\text{total, equally likely results}}$$

This method works only when the outcomes are equally probable.

Meaning of a Probability

What a certain probability means cannot be measured easily. It depends on the importance to you of the various outcomes of the experiment.

For example, you are entering the hospital for an operation. Statistics indicate the probability of full recovery is some number y, where y might be $\frac{8}{10}$, or $\frac{9}{10}$, or $\frac{999}{1000}$. Above what value of y do you stop worrying about the operation? Obviously there is some value above which you are safer in the hospital than going to work. Yet most of us would worry even if the probability of full recovery were one. The significance of probability depends strongly on the person's individual response to the various outcomes and on the value he places on each possible outcome.

As another example, a probability of 20% for rain tomorrow seems like an optimistic forecast if you are planning nothing unusual. It is worrisome if you have tickets for a World Series game and, if there is a postponement, you cannot go the following day.

Intuition and Probability

Intuition is not always much help in probability problems. Simple mathematical "illusions" are as common as optical illusions. In the most familiar example of such a mathematical illusion, an honest coin is tossed repeatedly with a sequence of heads and tails. As soon as four heads in a row come up, the tossing pauses while bets are placed on the single next toss. Most people would prefer to bet on a tail, and occasionally we can even find an "educated" individual who will give us favorable odds if we allow him to bet on a tail coming up next. Actually, the probability of a head coming up on the next toss is exactly $\frac{1}{2}$, since preceding tosses are irrelevant past history.

The examples of such breakdowns of common sense are numerous. As another example, consider a system in which a series of men check their hats as they enter a restaurant. Unfortunately, the checkroom attendant is hopelessly unintelligent and simply gives each man, as he leaves, a hat selected at random from her collection. What is the probability that no man will receive his own hat? In other words, how often does it happen that everyone receives the wrong hat?

We do not wish to solve this problem completely because the solution would take us into an area of mathematics which is not really important to the following chapters. The answer does depend, however, on the number of men who have checked hats. If only one man has checked his hat, there is no chance he will get the wrong hat. With two men, there is a 50% chance they both will; with three men, there is a 33% chance all three men will receive wrong hats; and we can continue enumerating possibilities.*

*
To obtain these results, it is easiest to count all the ways in which the hats might be distributed. We construct three rows: the first is the hat man A receives, the second B, and the third C. Then we call man A's hat a, man B's hat b, and man C's hat c. The hats can be distributed as follows:

```
To A—a a b b c c
To B—b c a c a b
To C—c b c a b a
```

The underline denotes the wrong hat. There are only two columns (fourth and fifth) with all entries underlined — two of the six possibilities (33%) in which *every* man has a wrong hat.

If the table is worked out for four men, it shows 24 different ways to return the hats. In 9 of them (38%) every hat ends on a wrong man.

As the number of men becomes very large, the answer approaches 37%.

The important feature of the problem is not the particular answer, but two aspects of the answer:

1. The probability of no one receiving his own hat is essentially the same if there are 8, 20, 800, or 4000 men. Once the number exceeds 8, the actual number is largely immaterial (37% of the time this experiment is tried, no one receives his own hat).

2. The probability is higher for an even number than for either adjacent odd number. In other words, the event is more likely to occur with four men than with either three or five, more likely with six than with five or seven.

Both these results are rather startling and counter to the intuition of most people.

Probability of Several Events

We need one more characteristic of probability. If two *independent* (unrelated) experiments are performed, the probability of success in both is the product of the individual probabilities.

As an example, in a football betting pool, you are asked to pick the winners of four different games played the same day. Each of the games is an even bet, that is each team is equally likely to win. In an actual pool, points are added to the score of the weaker team to make the bet even. The probability of your selecting a winner on each game is $\frac{1}{2}$; the chance of picking all four winners is

$$\frac{1}{2} \times \frac{1}{2} \times \frac{1}{2} \times \frac{1}{2} \text{ or } \frac{1}{16}$$

If you bet $1, you should receive $16 if you are successful. Typically, the man operating the pool pays off $10, so his profit averages $6 on each $16 bet. Actually, his profit may be even greater if a tie game results and all bettors lose.

We can also use this idea of independent experiments to calculate the probability of drawing the spade 7 from the shuffled deck of 52 cards. The suit drawn is independent of the number of the card. Hence, there is a probability of $\frac{1}{4}$ of drawing a spade, a probability of $\frac{1}{13}$ of drawing a 7. Success in both directions has a probability of

$$\frac{1}{4} \times \frac{1}{13} \text{ or } \frac{1}{52}$$

As another example, a high school senior applies for admission to four colleges. He knows the colleges act independently on admission decisions. He estimates his chances of admission at each college are

College A	$\frac{1}{5}$	College C	$\frac{1}{2}$
College B	$\frac{1}{2}$	College D	$\frac{5}{6}$

What chance does he have of being admitted to all four?

Since the four "experiments" are independent, the probability of success in all four is just the product of the individual probabilities:

$$\frac{1}{5} \times \frac{1}{2} \times \frac{1}{2} \times \frac{5}{6} \text{ or } \frac{1}{24}$$

He has 1 chance in 24, which is a little over 4%, of receiving four acceptances.

What is the probability he will receive *no* acceptances? Now to "succeed" in this experiment, he needs to be rejected by all four colleges. At each, the rejection probability is*

We assume at each college he is either admitted or rejected. Hence, the probability of rejection is one minus the probability of acceptance: if he has $\frac{5}{6}$ chance of being accepted, he has $\frac{1}{6}$ chance of being rejected.

College A	$\frac{4}{5}$	College C	$\frac{1}{2}$
College B	$\frac{1}{2}$	College D	$\frac{1}{6}$

The probability of four rejections is

$$\frac{4}{5} \times \frac{1}{2} \times \frac{1}{2} \times \frac{1}{6} \text{ or } \frac{1}{30}$$

He has a probability of $\frac{29}{30}$ (about 97%) of receiving at least one acceptance. Again, what does this mean? Should he worry until he hears from the schools? Obviously, whether 97% success is favorable odds or not depends on the individual interpretation of the outcomes.

The college-admission example above illustrates a common class of problems: We are given the probabilities of success in various, independent experiments; we want the probability of success in *at least one* experiment. To solve this problem, we find the probabilities of failure in each, then the probability P of failure in all; finally $1 - P$ is the probability of succeeding in at least one experiment.

This same approach is used to solve a classic probability problem. In a room of N people, what is the probability that at least two have the same day of the year as birthdays? To solve this, we start around the room and ask each person in turn what his birthday is. The answer of the first man A is immaterial. The second man B has a probability

$$\frac{364}{365}$$

of having a different birthday; the third man C has a probability

$$\frac{363}{365}$$

of having a birthday different from *both* A and B. The probability that three men (A, B, C) have different birthdays is then

$$\frac{364}{365} \times \frac{363}{365}$$

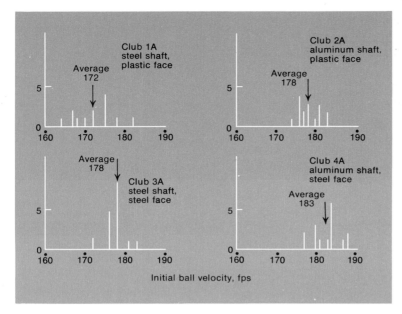

Fig. 3–4.
Results
from a study
by freshman
engineering students
of a comparison of four
types of golf clubs (drivers).
For each club, the horizontal
axis shows the ball speed just
after it was hit. The vertical axis
indicates the number of times the
result was obtained. Thus, for Club 4A,
a speed of 180 feet/sec was measured
three times.

The data illustrate the probability
characteristic found in most
experiments, particularly when
human beings are part of the
system. The human golfer just
cannot hit the ball identically
every time. In such a case,
we can use the average
(shown above). If we
are golfers, we might
also be interested in
using the club with the
greatest consistency.
(Technology
Review of M.I.T.)

What is the probability that five people in a room all have different birthdays? What is the probability, with five people in the room, that two or more have the same birthdays?

If one continues this calculation for larger and larger groups of people, we find that with only 22 people in the room, the probability that they all have different birthdays is less than one half. Hence, if we have a room with 22 or more people, the chances are that at least two have the same birthday. We assume, of course, there are not identical twins in the room. Furthermore, we neglect the possibility of a February 29 birthday.

A similar technique can be used to solve the following problem. Although a complete analysis is much too tedious, the problem does illustrate the difficulty of guessing answers to probability problems. Often our guesses are not even close to the correct answer. Here is the problem.

A gum manufacturer puts a baseball card in with each package of gum. There are 100 different cards available, placed at random in the packages. Johnny decides to collect cards and asks his father, "Tell me, before I start buying gum, how many packs will I probably have to buy before I have the complete set of 100?"

Johnny's father answers, "If you do no trading with your friends, you can expect to have to buy _____." What number should be inserted in the blank? Make your own estimate before turning to the answer in the footnote at the end of this section. This is an interesting example, since scientists and engineers often guess far from the correct answer.

Fig. 3–5.
*Picture of a golf club
and ball in contact as the
ball is being driven. As the ball
moves off the tee, it closes an
electrical switch at the right to
start a series of high-speed
photographs. The ball
velocity is determined
from successive pictures
taken a known time
apart. Of incidental
interest is the
astonishing extent
to which the ball
is compressed.
(B. L. Averbach,
M.I.T. 13—5082,
Cambridge,
Mass. 02139)*

One final example illustrates the difficulties we can get into when we talk about situations governed by probabilities. Consider this problem. Three men stand 50 yards apart from one another in a field; each has a gun and bullets. We call the men:

S   for sharpshooter
M   for marksman
D   for duffer

Each shoots one bullet at a time in rotation: first D, then M, then S, then D again, and so forth. Each man can see whether the other two are still alive and standing, each can hear the others' shots, but they cannot see who the others are aiming at as they shoot.

They have quite different abilities. S always hits what he aims at (the probability of a hit is 1), M hits 80% of the time, and D hits 60% of the time. If D shoots at S and misses, S will hear the shot and know D has shot at him.

D shoots first. Where should he aim?

Then M shoots if he is still alive. Where should he aim?

Then S shoots if he is alive. Where should he aim?

In other words, you should take the part of each of the men in turn and decide on the best strategy.

The answer for D is easy. He certainly does not want to shoot at M; if he should hit M, S would certainly shoot him, and the three-way duel would end with S the sole survivor. Likewise, D probably should not shoot at S. If he misses and M does not hit S, S will be apt to aim at D in revenge. If D happens to shoot S,

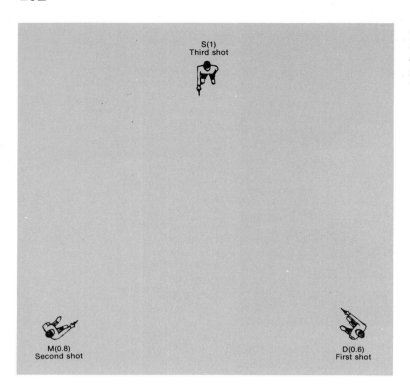

Fig. 3–6. Position of the men in three-way "duel."

M will certainly shoot at D. Thus, the best strategy for D is to aim well away from both M and S.

Now it is M's turn. He certainly won't aim at D; he might succeed and S would then shoot him. And if M shoots at S, he seems to gain little: if he shoots S, D will certainly shoot at him; if he misses, S is likely to shoot at him with catastrophic results. Hence, one can argue M should also shoot well away from S and D.

Now S prepares to shoot. So far he knows D and M have both shot, but obviously not at him (he's heard no bullet whizzing by). He probably is in no danger from another round, since presumably D and M will make the same decisions again. On the other hand, if he shoots M, D will certainly shoot at him. On the basis of this reasoning, S fires far from M and D, and we start the cycle all over again.

If the preceding argument is correct, the shooting continues forever, with no one ever being hit. Of course, S must anticipate when M or D will tire of the suspense and decide to start aiming more purposefully; just before this time, S should shoot at M and take his chances with D. The selection of a desirable strategy (the optimization problem) depends not only on a study of the probabilities, but also on a guess as to the likely behavior of the other people in the contest.

This particular example has been used as a very elementary model for business competition among three companies. Then instead of bullets we can discuss price and product competition. In such a situation, it may be advantageous to everyone to have all three companies survive.

In this section, we have introduced some basic ideas associated with probability. We now wish to return to queues and see how these ideas are useful in doing experiments to measure the properties of queues.*

4 | QUEUEING STUDIES

In the study of queues, we might ask a variety of questions:

1. With a given level of business (average number of customers) and specified service facilities (e.g., number of toll booths), what is the average size of the queue?
2. How often will the queue be longer than a given length? How often does the barber lose possible customers who look in the window and then leave if there are more than two men waiting?
3. What is gained by adding another service station such as an additional toll booth or an extra barber and chair?
4. What can be gained by reducing the time required to service customers?

We would like to be able to answer such questions mathematically (or by computer studies), because in many cases we just cannot experiment on the actual system. We can hardly expect political approval for constructing another runway at an airport if we merely express the desire to experiment. We must be able to show that another runway will be helpful. Thus, queueing theory is concerned with seeking solutions to questions on the basis of studies of the model.

In this section, we will introduce the subject. In particular, we look at simple queueing problems. In more complex problems, we have to use the digital computer to simulate the system. That is, the computer operates as a model of the actual system. The following paragraphs consider primarily the question: What is the queue length in a very simple class of problems?

Arrival of Customers

In our attempts to simplify the problem, we first consider a very special way in which customers arrive at the service facility. We assume that they arrive completely at random and independent of one another. In other words, if on the average a customer arrives every 7 minutes, in any short interval T, the probability of a customer arriving is $\frac{T}{7}$. This is assumed true even if we look

*Fig. 3–7.
Lengthy queues build
up during a peak traffic
period at the tollbooths
serving the Triborough Bridge
in New York City.
(Triborough Bridge
and Tunnel
Authority)*

at an interval just after the arrival of a customer. All we are saying here is that the customers arrive at a certain average rate. When they arrive is pure chance. Occasionally several customers arrive in close succession; on other occasions, there are long intervals between arrivals.

Now we can do an experiment to simulate a possible arrival of customers. We are interested in representing the arrival of customers at a store after it opens at 9 A.M. For the next hour, the customers arrive at an average rate of one every 4 minutes. We use the following steps:

1. We first divide the average 4 minutes into a number of smaller intervals. We might select one-minute intervals. The shorter the interval, the more accuracy we obtain, but also the more work we have to do. We should use an interval $\frac{1}{4}$ or less of our average time.

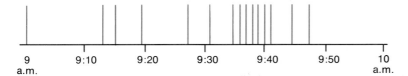

9 a.m.   9:10   9:20   9:30   9:40   9:50   10 a.m.

Fig. 3–8.
Possible arrivals
(average spacing
is four minutes).

2. We now look at each one-minute interval in turn. The probability of an arrival in this interval is $\frac{1}{4}$ because the interval is $\frac{1}{4}$ of the average arrival time. We first want to determine whether there is an arrival in the first interval, then the second interval (9:01 to 9:02), and so on for the hour.

3. We look now for some random experiment where the favorable outcomes are $\frac{1}{4}$ of all equally probable outcomes. If we have a deck of cards, we shuffle thoroughly and pick one card. We decide beforehand that if it is a spade (probability $\frac{1}{4}$), there is an arrival. If it is any other card, no arrival.

   Alternatively, we can roll two dice. The probability of rolling a 3 or a 4 or a 5 is $\frac{1}{4}$.* Hence, we roll the dice. If a 3, 4, or 5 comes up, there is an arrival; if 2, 6, 7, . . . , or 12 comes up, no arrival.

4. We repeat the experiment for each of the 60 intervals during the hour in which we are interested.

When we perform this experiment, we might find an arrival pattern as in Fig. 3–8.† Each vertical line represents the arrival of one customer. It is apparent that there are long periods when nobody arrives, for example, from 9:01 to 9:13 and from 9:47 to 10 A.M. Likewise, there are periods when customers are arriving one right after the other, particularly from 9:35 to 9:41.

Actually our way of determining the arrivals is not quite correct, since we do not allow more than one arrival in any one minute (and we could never have more than 60 arrivals). The graph does, however, indicate the idea of *random* arrivals.

This idea of random arrivals is often used in queueing problems because the problem is thereby greatly simplified. Actually, however, there are many cases in which the assumption is very close to the actual situation. Telephone calls during a busy period may be placed at random; customers do often arrive at a barbershop nearly at random. When highway traffic density is not too great, cars may arrive at a toll station according to our assumption.

Servicing Time

We now have the picture of customers arriving at random at our service facility. Each of these customers must be waited on,

*
When two dice are rolled, the probabilities of the various sums are:

Sum	Probability	Sum	Probability
2	$\frac{1}{36}$	8	$\frac{5}{36}$
3	$\frac{2}{36}$	9	$\frac{4}{36}$
4	$\frac{3}{36}$	10	$\frac{3}{36}$
5	$\frac{4}{36}$	11	$\frac{2}{36}$
6	$\frac{5}{36}$	12	$\frac{1}{36}$
7	$\frac{6}{36}$		

These numbers can be found easily. If one die were white and the other red, (so we can recognize which is which), there are 36 equally likely outcomes:

White 1	and	Red 1
White 2	and	Red 1
	Etc.	

Of these results, three give a sum of four (White 1, Red 3; White 2, Red 2; and White 3, Red 1). Therefore, three of the 36 equally likely outcomes are favorable if we want a four. The probability of a four is $\frac{3}{36}$.

The probability of a 3 *or* 4 *or* 5 is just the sum of the three separate probabilities, or $\frac{9}{36}$.

†
The data in Fig. 3–8 were determined by the author by rolling two dice. Each time a 3 or 4 or 5 appeared, it was interpreted as the arrival of a customer. Every other number rolled corresponded to the passage of a one-minute interval with no customer arriving. Merely by accident there were 15 arrivals, exactly the average per hour.

Dice are easier to use than cards. The trouble with cards is that the complete deck should be thoroughly shuffled before each draw of a card. Not only is this tiresome, but it is impossible for a human being to shuffle a deck thoroughly. There always is a tendency for groups of cards to stay together.

or *serviced*. Next, we assume that there is only one service facility: one runway, one barber, or one toll collector. Furthermore, we assume that the times required to service the various customers are either:

a) all equal, or
b) vary in exactly the same way as the inter-arrival times of the customers.

Assumption (a) means every customer takes the same time. This is approximately true for airplanes landing or haircuts.

In (b), the service times change from customer to customer. One man is serviced rapidly, another takes a long time. In a queue waiting at a bank teller's window, one customer may just be cashing a check. Another customer is bringing the day's receipts from a store and a bagful of coins must be counted.

Actually, we could complete this section by considering only the case of equal service times. We add the random variation to show the effect of different customer service times, an effect with which we are all familiar as a result of waiting in line behind the sweet little old lady who has a thousand questions she must ask.

Utilization

Before considering the problem of the queue length, we need to ask one question: Can the service station do the job of handling the arriving customers? In other words, is the service facility, with its specified servicing time, able to process the customers with their average arrival rate? Or, does the system break down completely, with the number of waiting customers tending to grow and grow?

The answer to this question depends only on a ratio which we call β:

$$\beta = \frac{\text{average servicing time}}{\text{average time between arrivals}}$$

If β is less than one, the service facility can process the customers. A β greater than one means the system cannot work and the concept of a queue is meaningless. In other words, on the average the service facility must be able to process customers faster than they arrive on the average.

This result is hardly startling. If airplanes arrive at an airport at an average rate of two per minute and they can land on the single available runway at a maximum rate of one per minute (corresponding to a servicing time of one minute per customer), it is clear that the system cannot operate successfully. The sky will gradually fill with aircraft.

The quantity β, which is called the *utilization factor* in books on queueing, is a measure of the fraction of time the service facility is used. Thus, if $\beta = 0.7$, the service personnel and equipment are being utilized 70% of the time. In an eight-hour day, the barber, for example, would be giving haircuts 5.6 of the hours, on the average.

Length of the Queue

We are now ready to discuss the queueing problem, and in particular the average length of the queue. We have considered the arrival of the customers, the servicing time, and the utilization factor β. What would be nice now would be a simple mathematical formula for the length of the queue. Indeed, there are such formulas. We are assuming the arrivals are random and the servicing times either fixed or random. There is a formula for the average queue length in each case:

Fixed service times: $\dfrac{\beta}{1 - \beta}\,(1 - \tfrac{1}{2}\beta)$

Random service times: $\dfrac{\beta}{1 - \beta}$

Unfortunately, these formulas aren't of much help except as very general guides. They are formulas which describe the average queue length after very long periods of time have elapsed. For example, we may require several thousand arrivals before we begin to approach the average given by the formula. In a barbershop, supermarket, or airport, we essentially start the problem over again every morning and arrival rates change during the day. Formulas which work only after hundreds or thousands of arrivals are not much help.

If mathematics does not help us, we use experiments to find the typical way the queue changes with time. Consequently, we return to our store example, where we had found arrivals at

① 9:01	④ 9:19	⑦ 9:35	⑩ 9:38	⑬ 9:41
② 9:13	⑤ 9:27	⑧ 9:36	⑪ 9:39	⑭ 9:44
③ 9:15	⑥ 9:31	⑨ 9:37	⑫ 9:40	⑮ 9:47

Let us assume that each customer requires 3 minutes for servicing. We can plot the queue length during the hour as shown in Fig. 3–9.

This graph is derived as follows. At 9:01 a customer arrives and is serviced until 9:04. At 9:13 another arrives who is serviced until

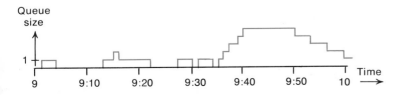

Queue size

Fig. 3–9. Queue length in the store example.

9:16. At 9:15 another customer arrives and has to wait until 9:16 before his 3-minute service period starts. At 9:19 he is finished, but customer 4 arrives and is serviced until 9:22.

At 9:35 the seventh customer arrives. Thereafter, it is easier to calculate who is being serviced.

9:35— 9:38	Customer 7
9:38— 9:41	Customer 8
9:41— 9:44	Customer 9
9:44— 9:47	Customer 10
9:47— 9:50	Customer 11
9:50— 9:53	Customer 12
9:53— 9:56	Customer 13
9:56— 9:59	Customer 14
9:59—10:00	Customer 15

Customers who arrive before they are serviced must wait in the queue (the store services on a first-come, first-served basis). Consequently we can finish the graph by determining just after each minute who is still waiting or being serviced. If we do this, we obtain Fig. 3–9.

Several features are interesting. First, there is nobody waiting or being serviced 17 minutes of the hour. In other words, the service facility is being used only $\frac{43}{60}$ or 72% of the time. This number would, over a long time interval, be very close to 75% or β. Thus the utilization factor β indicates what fraction of the time the service facility is busy.

Second, even though the service facility is busy less than 75% of the time, there is another long period during which *five* customers are in the queue—one being served and the other four waiting. We might even calculate how long each customer has to wait by simply comparing his arrival time and the time when he starts being serviced.

Customer ① —0 ④ —0 ⑦ —0 ⑩ —6 ⑬ —12
         ② —0 ⑤ —0 ⑧ —2 ⑪ —8 ⑭ —12
         ③ —1 ⑥ —0 ⑨ —4 ⑫ —10 ⑮ —12

These times or the long queues may be intolerable. We may lose customers to a competitor.

Finally, we can calculate the average queue length during this hour. The graph on page 107 shows that, if we count a customer being serviced as a queue of one, the 60 minutes is divided as follows:

Queue of zero	17 minutes
Queue of one	19
Queue of two	5
Queue of three	5
Queue of four	4
Queue of five	10

The average queue length during this period is then

$$q = \frac{17 \times 0 + 19 \times 1 + 5 \times 2 + 5 \times 3 + 4 \times 4 + 10 \times 5}{60}$$

or

$$q = 1.8$$

On the average, we find 1.8 customers in the store during this period.

Comments

In the preceding example we studied the queues in a service facility which opens at 9 A.M. Customers arrive on the average every 4 minutes. Every customer is serviced in exactly 3 minutes. You remember that by an experiment such as throwing dice or drawing cards from a shuffled deck we could simulate the system. From this experiment, we determined the queue length during the hour from 9 A.M. to 10 A.M. Several comments should be made.

1. To study the store operation, we should repeat the experiment a number of times. Any one "day" may not be typical at all.
2. We can easily make the service times random instead of constant. Then for each customer we find experimentally the time he requires for service. The procedure is exactly the same as the way we found the times between arrivals. We find that this added randomness increases sharply the queue length.
3. We can study queueing problems with more than one service channel. We need only decide how the customer selects the channel he enters. He might, for example, always go to the channel with the shortest queue.
4. We can study experimentally the effect of reducing service time. In our store experiment, for instance, a cut in the service time to 2 minutes would cut the average time in the queue from 6.6 to 4 minutes and the average queue length from 1.8 to 1. These are the numbers which result when we use the same times as above.
5. We can study the advantages of an express check-out counter or an exact-change gate on a parkway toll plaza. In this case, we find the service times for all customers. We take those with short service times and process them through a separate channel.

Queueing problems are interesting for several reasons. They confront us continually in life. They show so dramatically the way randomness can affect system behavior. Finally, they can be studied experimentally and, indeed, experiments are often more revealing than advanced mathematical analysis. In complex studies, we of course have to use computers to perform the calculations. For ex-

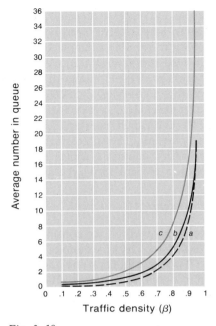

Fig. 3–10.
If the queues are too long, there are several ways to improve the system. The three curves show the mathematical formulas for queue length versus β in different cases: (a) Service time is halved, (b) Two servers work from one queue, and (c) Two queues are used, with customers sent randomly to one or the other.

The curves show it is best to reduce service time. Using two servers and one queue is almost as good. (Adapted from "Queues," Martin A. Leibowitz. Copyright © 1968 by Scientfic American, Inc. All rights reserved.)

ample, the study of airplane landing and take-off queues at a busy airport may involve several customers a minute.

An impressive example of queueing is the design of an international terminal at a modern airport. The facilities are built to process passengers in less than one hour, including (for departing passengers) baggage checking (perhaps 35 seconds/passenger) and ticket checking (maybe 65 seconds/passenger). Arriving passengers must go through passport clearance, health check, baggage pickup, and customs. The successful operation of the terminal depends on providing enough parallel channels and holding down service times to prevent very large queues.

5 | GAMES

In this section, we consider games as a new class of problems. We first look at an example.

Two friends, Dick and Joe, decide to play a game. At an agreed upon signal, they each push forward one or two fingers. To make the game interesting, they agree ahead of time that there will be a payoff after each play. The following rules determine the payoff:

If they match (both one or both two), Dick wins 10 cents.
If they don't match, Dick loses 10 cents.

These rules can be summarized in a table as shown below. Along the left, we list Dick's two possible strategies (One or Two).

		JOE	
		One	Two
	One	10	−10
DICK			
	Two	−10	10

Along the top are listed Joe's strategies (One or Two). There are now four boxes. In these we place the *payoff to Dick* from each combination of strategies by Dick and Joe. Notice we arbitrarily decided to focus attention on Dick, and we list the payoff to him. What Dick wins, Joe loses, so the payoffs to Joe would be just the negatives of those shown.

The table above is called the *payoff matrix* of our game. It includes the two players, the strategies for each, and the payoff from each possible combination of strategies.

Dick and Joe play for awhile, with each selecting one or two fingers to show. Neither follows a regular pattern, so neither can predict his opponent's moves with any success. After playing all morning, they find their fortunes have varied. For awhile Dick was ahead, then Joe. At lunch, Joe owes Dick 20 cents.

Walking back from lunch, Dick comments, "The game seems a little boring. Let's try to introduce some excitement. Suppose we change the payoffs. If we match, I'll still win 10 cents. If I have

One and you Two, I'll pay you 20 cents. If I have Two and you One, no one will pay at all."

Joe agrees something must be done to make the game more interesting. As he understands, the payoff matrix now has the form:

		JOE	
		One	Two
DICK	One	10	−20
	Two	0	10

Joe thinks the change is trivial and actually wishes they would try a new game. But he agrees and they start playing again.

Dick asks if Joe would mind if he rolled a pair of dice before each play, just for diversion. Joe is now convinced Dick is foolish, but decides to humor him. They spend the afternoon playing.

By quitting time, they have played 800 times during the afternoon. Joe owes Dick $20.40. As Joe leaves for home, he wonders whether he has just had a streak of bad luck, or is there something about the game he doesn't understand?

We can "solve" this game. That is, we can find the best strategy for each player and also the average payoff expected each time the game is played. As you may already have noticed the game is really loaded in Dick's favor.

Are there other similar games that can be solved? For what class of problems can we find a solution?

The problems we are considering here make up the subject of *game theory*. A more accurate description is the theory of *two-person, zero-sum, finite* games. What does each of these italicized terms mean?

The games are *two-person*. That is, there are two players competing with each other. Furthermore, both players are intelligent. Both have the same picture of the payoff matrix. And each knows the other is intelligent.

The games are *zero-sum*. What one player wins, the other loses. There is no chance for the two players to cooperate so they both win. We have strictly a competitive situation.

The games are *finite*. Each of the two players has only a certain number of possible strategies. For example, we might have a 3×4 game: one player X has three strategies, the other player Y has four. In this case, the payoff matrix might look like

		Y			
		y_1	y_2	y_3	y_4
X	x_1	3	6	−2	9
	x_2	4	−7	−3	2
	x_3	−5	−9	8	1

If player X chooses strategy x_2 and Y chooses y_3, the payoff to X is -3. X loses 3, Y wins 3. The entries in the payoff matrix are the payoffs to the player listed at the left.

Solution of Simple Games

When the game is described by only two possible strategies for each player, there is a simple algorithm for the solution.

We first look at the game to see if it can be simplified. Is there one strategy for either player which is obviously better than the other? For example, in the game

		Y	
		y_1	y_2
	x_1	4	7
X			
	x_2	3	2

X obviously should always play x_1. No matter what Y does, X will win more than if he played x_2.

Now that we know X will always play x_1, let's see what Y should do. Y is intelligent and knows X is going to play x_1. Then Y certainly should play y_1. Thus, if both players are intelligent, each time the game is played X will win 4. We call the average amount X wins from each play the *value* of the game. Here the value is 4.

Similarly, in the game

		Y	
		y_1	y_2
	x_1	6	2
X			
	x_2	5	3

Y should always play y_2. Then X should play x_2. The value of the game is 3.

If neither X nor Y has a single best strategy, we can find the way each should mix his strategies. First consider X in the game

		Y	
		y_1	y_2
	x_1	4	7
X			
	x_2	6	2

1. Subtract the second column from the first

$$4 - 7 = -3$$
$$6 - 2 = 4$$

We have the two numbers

$$-3$$
$$4$$

2. Reverse these numbers and throw away minus signs. We have

$$4$$
$$3$$

3. These represent the relative number of times X should play x_1 and x_2, respectively. In other words, $\frac{4}{7}$ of the time he should play x_1. In a single play, he selects x_1 or x_2 at random with probabilities $\frac{4}{7}$ and $\frac{3}{7}$.

We find Y's strategy by the same procedure. Subtracting the second row from the first gives

$$-2 \qquad 5$$

Interchanging and discarding the minus sign gives

$$5 \qquad 2$$

Y should play y_1 $\frac{5}{7}$ of the time, y_2 $\frac{2}{7}$ of the time.

The value of the game can be found by considering X's best strategy against either move by Y. If we pick y_1, X can expect to win 4 in $\frac{4}{7}$ of the time, 6 in $\frac{3}{7}$ of the time. The expected winning for X is then

$$4 \times \frac{4}{7} + 6 \times \frac{3}{7}$$

The value of the game is $\frac{34}{7}$.

The preceding description of the algorithm is not very satisfactory. We have not explained where the algorithm comes from. Indeed, the algorithm only works for two-strategy games. If each player has three or more possible strategies, we have to expand the algorithm with several additional algebraic steps. A more complete discussion would lead us into mathematics which is not of importance in the remainder of the text.

Thus, the algorithm for the 2×2 game is included here merely to describe the sort of solutions which we can obtain for games.

With the algorithm, we can solve our original game. Dick and Joe started by playing the game:

		JOE	
		One	Two
	One	10	−10
DICK			
	Two	−10	10

Application of the algorithm shows each player should select One or Two at random, each half the time. The value of the game is zero. If they play long enough, the average (per play) payoff should approach zero.

After lunch they switched to the game:

		JOE	
		One	Two
	One	10	−20
DICK			
	Two	0	10

Now the best strategy for Dick is to play One $\frac{1}{4}$ of the time, Two $\frac{3}{4}$ of the time. On any single play, he throws the dice. If a 3 or 4 or 5 comes up, he plays One. If any other sum appears on the dice, he plays Two. (The dice are used merely to assure a random selection. He wants to avoid following a pattern which Joe might detect.) If Dick follows this strategy and Joe plays One, Dick will win ten $\frac{1}{4}$ of the time, zero $\frac{3}{4}$ of the time. The value of the game is 2.5 cents. If they play 800 times, Dick can expect to win about 800 × 2.5 cents, or $20.

Comment

The examples of this section serve as only an introduction to games. Game theory started during World War II with studies of different tactical choices. During the last two decades, many economic and social situations have been studied. Current emphasis in research is on trying to find ways to solve games in which both players can profit if they adopt cooperative strategies. The hope is to apply game theory to social science problems in which we are looking for a strategy which, for example, will benefit both the power company seeking to locate new generators and the groups interested in improving the environment.

6 | LINEAR-PROGRAMMING PROBLEMS

An ice cream plant can make two flavors, vanilla and chocolate. The plant capacity is 1000 quarts per day, and the sales department says that it can sell any amount of vanilla up to 800 quarts and any amount of chocolate up to 600 quarts. If the profit per quart is 10¢ for vanilla and 13¢ for chocolate, what is the most profitable daily production?

Since the profit per quart is larger for the chocolate than for the vanilla, we should produce as much chocolate as we can sell. This would set the production of the plant at 600 quarts of chocolate. All excess production could then be applied to the vanilla ice cream, to add to the profit. The daily production of the plant is thus 600 quarts of chocolate and 400 quarts of vanilla, not a very difficult problem. We now solve this problem in a systematic manner because we want a technique which can work for more complicated problems.

The type of problems in the remainder of this chapter are called *linear-programming* problems. One of the earliest famous problems was a study in 1945 by George Stigler of Columbia University. He was looking for the cheapest diet which would provide the daily needs of calories, proteins, calcium, iron, vitamin A, thiamine, riboflavin, niacin, and ascorbic acid. He considered 70 possible foods.

The lowest-cost diet was a combination of wheat flour, cabbage, and hog liver. By mixing appropriate amounts of each of these, a man could live in good health for $59.88 a year. Costs today are higher, so it might require 30¢ a day.

This original solution was limited because computers were not available. Too many constraints could not be handled. If we were to solve the same problem today, we could add constraints which would ensure that on occasional days the diet was a little more appetizing.

Inequalities

In order to find a general method of solution, we need to use mathematical statements called *inequalities*. The inequalities are:

$x > 20$	meaning	x is greater than 20
$x \geq 20$	meaning	x is greater than or equal to 20
$x \leq 20$	meaning	x is less than or equal to 20

In other words, an inequality describes a range of permissible values. For example, in the first example, the inequality is just a simple way of saying x can have any value larger than 20.

We often combine two inequalities into a single expression. For example,

$$0 \leq x \leq 30$$

means that 0 is less than or equal to x, and x is less than or equal to 30. In other words, x can have any value from 0 to 30. As an example, if we have $30 and we want to buy a radio, the money spent on the radio R can be in the range

$$0 \leq R \leq 30$$

If we are also spending T dollars to buy tickets for a movie,

$$0 \leq R + T \leq 30$$

We can now describe the ice cream problem in terms of inequalities. We let V be the number of quarts of vanilla to be produced in one day and C be the number of quarts of chocolate to be produced in one day. One thousand is the total number of quarts produced by the plant for one day.

The number of quarts of vanilla to be produced may not exceed 800, but may be less than that quantity. We express this by the double inequality:

$$800 \geq V \geq 0$$

The vanilla ice cream produced may be any amount between 0 and 800 quarts. Similarly the quantity of chocolate ice cream to be produced can be represented by:

$$600 \geq C \geq 0$$

The number of unknowns can be reduced from two to one by remembering that the total production must be 1000 quarts of ice cream. With this limitation, if V quarts of vanilla are produced, there can only be $(1000 - V)$ quarts of chocolate ice cream.

Our statements are thus:

$$0 \leq V \leq 800$$
$$0 \leq (1000 - V) \leq 600$$

At this point, we have a model for the decision problem. The model states that there are several constraints on the quantity of vanilla, V, produced each day:

$V \geq 0$ (The amount of vanilla must be positive or zero, we cannot have a negative amount of vanilla.)

$V \leq 800$ (We cannot sell more than 800 quarts of vanilla.)

$1000 - V \geq 0$ (The amount of chocolate must be positive or zero.)

$1000 - V \leq 600$ (We cannot sell more than 600 quarts of chocolate.)

The last two inequalities are hard to interpret. They can be simplified if we manipulate them so that V appears alone on one side (as in the first two). To do this, we need some rules on how to change inequalities.

We can add any quantity (positive or negative) to both sides of an inequality. For example, the inequality is

$$x + 10 \geq 2$$

We can subtract ten from both sides to obtain

$$x \geq -8$$

We can multiply both sides of an inequality by any positive number. If

$$3x \geq 24$$

then multiplying by $\frac{1}{3}$ gives

$$x \geq 8$$

If we multiply both sides by a negative number, we must change the direction of the inequality. That is, "less than" becomes "greater than."

With these rules, we can simplify the four inequalities for the ice cream problem. We started with:

$$\begin{cases} V \geq 0 \\ V \leq 800 \\ 1000 - V \geq 0 \\ 1000 - V \leq 600 \end{cases}$$

In the third, we add V to both sides. In the fourth, we subtract 600 and add V. Then we have:

$$\left. \begin{array}{l} V \geq 0 \\ V \leq 800 \\ V \leq 1000 \\ V \geq 400 \end{array} \right\} \text{ Model}$$

The second of these is clearly stronger than the third, the fourth stronger than the first. That is, if the second is satisfied, the third is automatically o.k. Hence,

$$\boxed{\begin{array}{l} V \geq 400 \\ V \leq 800 \end{array}} \text{ Final model}$$

Now that the model is determined, we can turn our attention to the *criterion:* what is to be optimized? In this problem, we wish to maximize the profit. The problem specifies that the profit is 10¢ for each quart of vanilla and 13¢ per quart of chocolate. Thus, if P is used to represent profit (in cents), we can write

$$P = 10V + 13C$$

Once again, C is equal to $1000 - V$. The criterion (profit) equation becomes

$$P = 10V + 13(1000 - V)$$

$$\boxed{P = 13000 - 3V} \quad \text{*Criterion* (to be maximized)}$$

Finally, we must consider the *constraints* before we start the optimization. In this particular problem, there is no constraint specified. A constraint might be that the owner never permits production of less than 500 quarts of vanilla because of his personal desire to cater to his vanilla customers.*

Solution

Now that the problem statement is complete we turn to the optimization of the solution. We must select V to satisfy the two inequalities of the model, and at the same time to maximize P. If we look at P, we see that the profit is reduced as V is increased. Therefore, we should pick V as small as possible.

* One might argue, for example, that the factory limit of 1000 quarts/day of ice cream is a constraint rather than a part of the model. It makes no difference whether a given part of the problem is considered part of the model or a constraint. In this example, we put everything into the model.

The model shows that V must be 400. The solution is then to produce 400 quarts of vanilla and 600 quarts of chocolate. The resulting profit, found from the relation for P, is 11,800 cents.

7 | GRAPHING INEQUALITIES

In retrospect, we can only marvel at the last section. There we took a ridiculously simple problem which can be solved by inspection. We managed to make a complicated problem which involved the definition of a model and criterion and the manipulation of a set of inequalities. Of course we did this because we need the more complicated approach when we come to more difficult problems. In the next section, we will look at a problem which cannot be solved by inspection. Then we will need to find a model in terms of a set of inequalities.

Before we consider this more complicated problem, we have to look at graphs for inequalities. We will need to portray graphically what we mean by an inequality with two unknowns. For example, we might have

$$2x + y \geq 6$$

How do we show on a graph what this inequality means?

The two unknown quantities are x and y. Therefore, we try to show this graph in the $x - y$ plane (Fig. 3–11). It makes no difference which we choose to plot horizontally, which vertically. Here we show y vertically.

The inequality divides the plane in two parts. In one part, the inequality is true. In the other part, it is not. The dividing line is found by plotting the *equation* we have when the inequality sign is replaced by an equal sign.

In our example, the equation is

$$2x + y = 6$$

This equation is plotted as a straight line. On one side of this line, the inequality is satisfied; on the other side, it is not.

The easiest way to plot this equation is to find the two intercepts. When y is zero, the equation states x is 3. When x is zero, $y = 6$. Figure 3–12 shows the straight line through these two points. (Any two points can be used, or we can find one point and the slope.)

Once the line is drawn, the inequality is satisfied on one side of the line. Which side? We can take any point on either side and try these values of x and y in the inequality. For example, the origin (x and y both zero) does *not* satisfy the inequality. Then, we can add the words on Fig. 3–13. If we want to be sure of the above calculation, we can pick a second point on the other side of the line and see whether the inequality is satisfied or not. In our

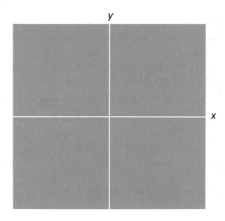

Fig. 3–11. The plane where we will show $2x + y \geq 6$.

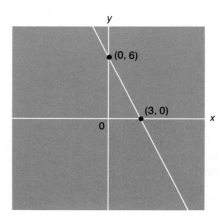

Fig. 3–12. Plot of $2x + y = 6$.

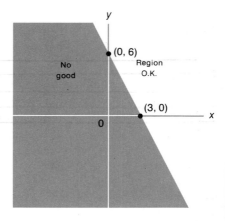

Fig. 3–13. Region for $2x + y \geq 6$.

example, we might pick an x of 3, a y of 6. We are above the line and the inequality *is* satisfied.

Actually, we usually use different terms than those of Fig. 3–13. Figure 3–14 shows the form. The region in which the inequality is satisfied is called the *feasible region.* We use shading on the *other* side of the line to show that this area is forbidden. The inequality is *not* good in the shaded region.

When we have several inequalities, each defines a line in the x-y plane, to one side of which the inequality is met. The total *feasible region* is then that area where *all* the inequalities are met.

An example is provided by the three inequalities:

$$2x + 4y \geq 12$$
$$x \leq 5$$
$$y \leq 3$$

The first inequality gives the allowable region shown in Fig. 3–15. The straight line passes through (6,0) and (0,3). The point (0,0) is not in the feasible region.

The second inequality allows anything to the left of $x = 5$. The third inequality admits all below $y = 3$. When these two lines are added, we obtain Fig. 3–16. The feasible region is the triangle.

With this background, we can now turn to a more interesting linear-programming problem.

8 | A TRANSPORTATION-PLANNING PROBLEM

The ice cream problem involves a search for a "best" solution to a problem for which more than one solution is possible. The "best" solution is one which achieves either a maximum or a minimum under the constraints that are imposed. These constraints are expressed mathematically as either *linear equations* or *linear inequalities.* Such problems have been named linear-programming problems.

By a linear equation here, we mean each variable, x or y, appears only to the first power. For example, $3x + 4y = 2$ is a linear equation. The term linear also refers to the fact that the plot is a straight line. In a similar fashion, a linear inequality is an inequality which becomes a linear equation when the inequality sign is replaced by an equal sign. Our examples of linear-programming problems in this chapter are purposely simple, that is, the number of variables has been limited. Practical problems frequently involve a large number of variables, often hundreds. Then high-speed computers must be used to find the solution.

It is fascinating to discover the variety of design and planning problems that involve linear inequalities, and once we learn to use linear inequalities, we can apply them to many different types of problems. In this section we treat a *transportation-planning problem.*

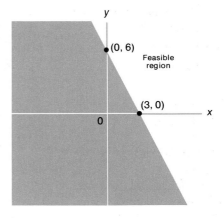

Fig. 3–14. Feasible region for $2x + y \geq 6$.

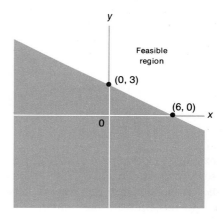

Fig. 3–15. Feasible region for first of three inequalities.

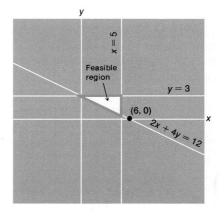

Fig. 3–16. Feasible region defined by all three inequalities.

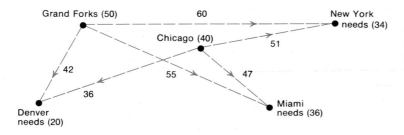

Fig. 3–17.
Sources and needs of wheat
and shipping costs (numbers in
carloads and in dollars/carload).

Statement of the Problem*

A grain dealer owns 50 carloads of wheat in Grand Forks, North Dakota, and 40 carloads in Chicago. He has sold 20 carloads to a customer in Denver, 36 carloads to a customer in Miami, and the remaining 34 carloads to a customer in New York. He wishes to find the minimum-cost shipping schedule, on the basis of the freight rates shown in Fig. 3–17 and Table 3–1. For

*
Example suggested by "Operations Research for Students of Business," notes by James R. Jackson, published by University of California at Los Angeles.

From \ To	Denver	Miami	New York
Grand Forks	42	55	60
Chicago	36	47	51

Table 3–1.
Wheat freight
rates in dollars per
carload.

example, it cost $42 per carload for shipments from Grand Forks to Denver. Different modes of shipment involve rates which are *not* proportional to the distance between the cities.

In this problem the quantity of wheat which has been sold is just equal to the quantity of wheat in storage. This need not necessarily be the case in practice, but we have designed this problem to illustrate the general method with the use of the simplest mathematics.

We can present the shipment data in a table, which indicates quantities as shown in Table 3–2. The figure in each box will be the quantity to be sent from the source at the left to the destination at the top.

↓ Quantity in storage	Quantity needed → / From \ To	20 / Denver	36 / Miami	34 / New York
50	Grand Forks			
40	Chicago			

Table 3–2.
Data
arrangement for solving
grain problem.

Our problem is to find the quantity of wheat to fit into each of the six empty boxes so that three conditions are met. These are:

1. The amounts in the first row add up to 50 and the amounts in the second row add up to 40 (the total amounts to be shipped from Grand Forks and Chicago, respectively).
2. The amounts in the first, second, and third columns add up to 20, 36, and 34, respectively (the amounts to be delivered to Denver, Miami, and New York, respectively).
3. The total freight cost is a minimum. This cost for each shipment is obtained by multiplying the amount of wheat in each box by the rate for that trip. The six products are then added to give total cost.

It is not enough to pick the numbers in each box so that the sums of the rows and columns are correct. This would be a feasible solution, but we must also minimize the total cost. With "cut and try" methods, we might find such a solution. However, a systematic approach usually takes less time. For problems with more shipping points and with more destinations, a systematic approach (an algorithm) and a computer are essential.

If we designate the quantity shipped from Grand Forks to Denver as x units, then $20 - x$ units must be shipped from Chicago to Denver to complete the order. Similarly, if we designate the amount shipped from Grand Forks to Miami as y, the amount from Chicago to Miami must be $36 - y$. Now the amount from Grand Forks to New York must be $50 - x - y$ if the total out of Grand Forks is to be 50. The amount from Chicago to New York must then be $34 - (50 - x - y) = x + y - 16$. We automatically satisfy the requirement that the total shipped from Chicago is 40, since the total amount sold equals the amount ordered.

Table 3–3 now represents the quantities of wheat to be shipped from each storage house (Grand Forks and Chicago) to each destination (Denver, Miami, and New York). Because of the way the six entries are selected, each row and column adds up properly. Our task in optimization is now to select x and y and, hence, each of these six entries:

$$x, y, 50 - x - y, 20 - x, 36 - y, x + y - 16$$

↓ Quantity in storage	Quantity needed → / From \ To	20 Denver	36 Miami	34 New York
50	Grand Forks	x	y	$50 - x - y$
40	Chicago	$20 - x$	$36 - y$	$x + y - 16$

Table 3–3.
Table 3–2 with unknown quantities
x *and* y *defined.*

There is one additional part of the model (or the constraints). Each of the entries in Table 3–3 must be *positive* or *zero*. We cannot ship negative quantities of wheat. Therefore, we must add the six inequalities:

$$x \geq 0$$
$$y \geq 0$$
$$50 - x - y \geq 0$$
$$20 - x \geq 0$$
$$36 - y \geq 0$$
$$x + y - 16 \geq 0$$

These inequalities can be rewritten in the following form, if we rearrange terms:

Derived by adding terms to both sides of the inequalities
$$\begin{cases} x \geq 0 \\ y \geq 0 \\ y \leq -x + 50 \\ x \leq 20 \\ y \leq 36 \\ y \geq -x + 16 \end{cases}$$

Determination of the Feasible Region

We now have the model represented by Table 3–3 and the six inequalities. Before considering the criterion function (the cost to be minimized), we can represent our six inequalities on a graph of y versus x. Each inequality states that we must work on one side of a straight line in this plane.

For example, the first inequality

$$x \geq 0$$

requires that we operate to the right of the y axis. The fourth

$$x \leq 20$$

places us to the left of the vertical line $x = 20$. The effect of these two inequalities is to restrict our choice of x and y to the region shown in Fig. 3–18.

Figure 3–19 shows the six straight lines and the corresponding permissible regions. The six inequalities require that we operate (select x and y) either within the feasible region indicated in Fig. 3–19 or *on the boundary of this region*.

Criterion Function

Figure 3–19 and Table 3–3, together, now represent the model and the constraints of the total problem. Before we can start the optimization, we need to determine an equation for the criterion, the cost. Optimization then consists of looking for a point (x and y) within the feasible region at which the cost is minimized.

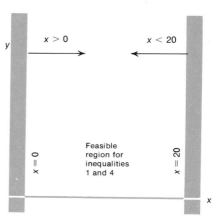

Fig. 3–18. Plot of the vertical lines $x = 0$ *and* $x = 20$.

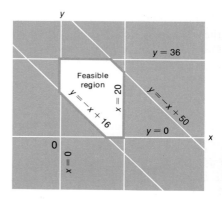

Fig. 3–19.
Graphical representation of the six inequalities of the model.

The cost of each shipment in each of the boxes of Table 3–3 can be expressed by multiplying the number of carloads shipped times the cost of shipping for each carload. The sum of all these individual shipping charges then gives an expression which represents the total cost of the entire shipment. Thus, the shipment from Grand Forks to Denver is x carloads. Each costs \$42. Then the cost for this shipment is $42x$. In a similar fashion, we can compute the cost of shipment between any two cities for each of the squares. The total cost of shipping is:

$$C = 42x + 55y + 60(50 - x - y) + 36(20 - x) + 47(36 - y) \\ + 51(x + y - 16)$$

If this expression is simplified,

$$C = 4596 - 3x - y$$

This C is the total cost of all shipments. It gives us the cost for any values of x and y we may choose.

Thus, the problem we started with now can be stated as follows. How do we pick values for x and y which

1. are in the feasible region of **Fig.** 3–19 or on the boundary, and
2. minimize C as given above?

Solution

After all this preparation, we are finally ready to solve the problem. There are two ways we can find a solution.
Trial and error. We can guess at x and y in the feasible region. We can find C. We then try different x and y values and again find C. We continue this until we have the least possible value of C.

This trial and error would be hopeless if we had to try every pair of x and y values in the feasible region. Fortunately, the solution always comes out at a vertex or corner of the feasible region. Figure 3–20 shows there are only six vertices. We only need to try these six pairs of x and y values.

Vertex	x	y	$C = 4596 - 3x - y$
A	0	36	4560
B	14	36	4518
C	20	30	4506
D	20	0	4536
E	16	0	4548
F	0	16	4580

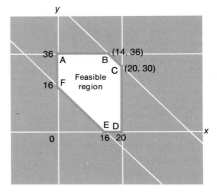

Fig. 3–20.
Feasible region with vertices shown.

The minimum cost occurs at C. The problem solution is

$$x = 20 \qquad y = 30 \qquad C = 4506$$

Finding the x and y values at each vertex and calculating C is not difficult. Even in quite complex problems, a computer can be used to search for the best vertex. In very complex problems with hundreds of constraints, even computers take too long. Then there is a search procedure called the simplex method which allows you to program the computer so that we first guess one vertex, then always move to a vertex where C is less. We don't bother to look at vertices where C is larger.

Graphical. We can also solve our problem by a graphical approach. This procedure shows why the solution always is at a vertex. Let's return to our problem statement which has two parts:

The feasible region of Fig. 3–19
The C equation, $C = 4596 - 3x - y$

We can rearrange terms in the cost equation in the form

$$y = -3x + (4596 - C)$$

This is the equation of a straight line. The slope is -3. The intercept on the y axis is $(4596 - C)$. Each different value of C gives a different line. For example, if C is 4596, the line is $y = -3x$ (Fig. 3–21). We change C to 4580, the line moves as shown in Fig. 3–22.

If we look at Fig. 3–22, we see that as C decreases, the cost line moves parallel to itself upward to the right. The minimum C must occur when the line is in the position shown in Fig. 3–23. The best C occurs when the line passes through

$$x = 20 \qquad y = 30$$

There the equation for C tells us that

$$C = 4596 - 3(20) - (30) = 4506$$

The total solution for least cost is shown in Table 3–4.

To From	Denver	Miami	New York
Grand Forks	20	30	0
Chicago	0	6	34

It is interesting that *no* wheat should be shipped from Chicago to Denver even though this is the cheapest single route.

Figure 3–23 shows that the largest C occurs when the cost line is downward to the left as far as possible. Then C is 4580.

If this is a real problem, the differences between maximum and minimum cost do not seem very impressive.

Least cost: $4506   Greatest cost: $4580

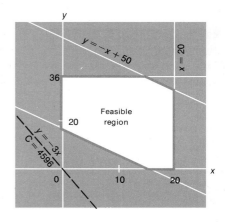

Fig. 3–21. Cost line (C = 4596) added.

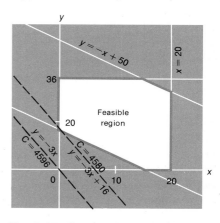

Fig. 3–22. Cost line (C = 4580) added.

*Table 3–4.
Minimum-cost
solution for wheat
problem.*

All this work only saves $74. If the profit we can expect in running this business is only 3%, however, the profit expected from this sale is about $136. By using optimization, we have added $74 to this figure. This is hardly a trivial gain.

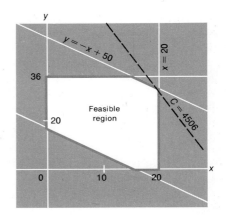

Fig. 3–23. Feasible region with maximum-cost line added.

9 | FINAL COMMENT

In this chapter, we have continued the discussion of decision making. We have considered three kinds of decision problems: queues, games, and linear programming. In each case, our purpose is to introduce the reader to a class of decision problems. We try to see what are typical elements of the problems. There is no attempt to go deeply enough to permit the reader to solve any more than the simplest problems.

Thus, our major objective is to show that various decision problems can be approached intelligently and logically. Decision making need not be entirely a guess and a prayer. While no one should run a business strictly on the basis of these mathematical viewpoints, knowledge of problem-solving techniques can help the decision maker to gain a better understanding of his problem.

Let us consider queueing as an example. We hope that the reader will gain some appreciation of what a queue is, why queues form, and how queues can be avoided. Once we understand what a queue is, certain facts are important:

1. The system must be built so average service time is less than the average time between arrivals.
2. The randomness of arrivals and service times is really what causes queues.
3. Even though the average queue length may be small, there are frequent periods when the queue length is very large (possibly greater than 10).

Once we understand the general behavior of queues, we can see obvious actions that can be taken to reduce the queueing problem:

a) The arrivals can be made more regular. (Drive-in theaters have a playground to attract some customers early in an attempt to reduce queues just before show time.)
b) Service times can be made less random. Buses that require exact change need much less time to load passengers, and have the side advantage of discouraging robbery.
c) Service times can be reduced. Banks require the customer to fill out the deposit or withdrawal slips before he enters the line at the cashier's window.
d) Priority rules can be established. Airline passengers who already have tickets need not check in at the counter, but can go directly to the gate.

Fig. 3–24.
Traffic congestion. (City of New York
Department of Traffic
photo by C. DeLuise)

Fig. 3–25.
Congested highway
with minimum speed
optimistically
indicated.
(Photo by
Richard O'Leary)

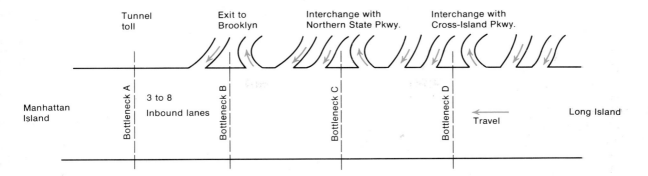

Fig. 3–26.
A few
bottlenecks on a
main highway.

e) Additional service channels can be added. When bank queues become so long customers are irked, the alert manager quickly opens other cashier windows.

Finally, the chapter should point out the complexity of common problems. Often we find a system in which there are a sequence of service channels or bottlenecks. For example, the Long Island Expressway is the main highway from Long Island into Manhattan. The normal traffic congestion has given the 50-mile road the affectionate title of the "world's longest parking lot."

On a typical morning rush hour, there are several points at which severe bottlenecks occur (Fig. 3–26). Often enormous queues form at each of these points. Frequently, one queue backs up all the way to the preceding bottleneck, and traffic almost comes to a halt.

It is clear that reducing the queues at Bottleneck D may really not help the system at all. All this does is make the situation that much worse at C, B, and A. Indeed, a severe bottleneck at D (perhaps caused by a stalled car in the middle lane) may be just what the system needs. Drivers coming from east of D may have an unusually long wait as they approach D, but from there on in to Manhattan traffic flow may be very smooth.

In other words, bottlenecks may actually be desirable. Perhaps an inexpensive method of control for our highway would be to arrange for a stalled car or two each morning at appropriate locations.

In a variety of more significant problems, we sense that bottlenecks really may be advantageous. For example, the rapid development of computers during the past two decades led to the use of more than 75,000 computers by 1970. The resulting problems in business management, city operation, and many other fields can only suggest that we should not complain about the various bottlenecks which have slowed down this progress. This slowdown may have been essential because our social institutions change slowly.

Questions for Study and Discussion

1. "Most engineering design activity currently relies on engineering judgment." Mention some of the kinds of judgment on which people rely when they buy manufactured goods, for example, canned or frozen vegetables, electrical appliances, automobiles.

2. Independent testing laboratories have been formed to help consumers by checking on various products. If you were to start a small testing laboratory in your community how would you go about rating canned tomatoes? Hamburgers? Hot dogs?

3. More than half the people in the United States are under thirty, and the average age of all Americans is going down slightly each year. Suppose you inherit $20,000 and decide to start your own store. How would you do a market survey to decide what kind of store to open in your community? State your major criterion and the various constraints which you see.

4. Is $xy = 5$ a linear equation? Why, or why not?

5. What is meant by the term "feasible region" in the solution by graphical means of a linear-programming problem?

6. In developing the problem on shipping wheat (Section 8), the text shows the inequality $50 - x - y \geq 0$ rewritten in the form $y \leq -x + 50$. How would the graph of the feasible region look if the original inequality had been changed to $x \leq 50 - y$?

7. a) Graph the equation $3x = y$. What is the slope of this line?

b) Graph the equation $x = 3y$. What is the slope of this line?

c) Graph the equation $x + y = 3$. What is the slope of this line?

d) Graph the equation $x - y = 3$. What is the slope of this line?

e) What is meant by the "slope of a line"?

8. In taking a 25-question True-False test you pick T as the answer to the first three questions. If the correct responses are truly randomly distributed throughout the test what are the chances that the correct response to the fourth question is T?

9. What examples of queue formation in your school can you think of? (Consider, for example, classrooms just before or after class, science labs, cafeteria, gym, principal's office, library, bulletin boards.) In each case, is it a single-service queue or a multi-service queue?

10. a) If customers arrive at the teller's window in a bank at random, but at an average rate of one every 30 seconds, what is the average time between arrivals in terms of seconds? Of minutes? Of hours?

b) If the teller always takes 20 seconds to service each customer, what is β?

c) How long is the average queue length? (It is easier here to use common fractions rather than decimals.)

d) Suppose the teller still averages 20 seconds per customer, but the lengths of individual transactions are random, does this change β?

Problems

1. A man is faced with the decision of driving his car to work or taking the bus. He assumes that the cost of driving the car is $0.75 and the possible outcomes are:

   that he will arrive safely, probably—0.85
   that he will be late due to traffic—0.145
   that he will have an accident—0.005

He associates the following values with the above possible outcomes of the decision which he must make:

   arriving at work on time without incident—0
   arriving late due to traffic—$1.00
   having an accident—$50.00
      (he had $50.00 deductible)

The cost of taking the bus is $0.30. The probabilities are:

arriving on time—0.10
being late due to bad connections—0.90

The values are rated as with the car except that he assumes no accident if he uses the bus.

In order to make a realistic decision he must know:

a) Which outcome is most likely?

b) Which action could lead to the most favorable outcome?

c) What is the least favorable possible outcome?

d) Which decision has the largest expected value?

2. The 747 airliner can carry 360 passengers.

a) Assuming that it takes 30 seconds for one gate attendant to service a customer at the gate (check baggage and verify ticket), what will be the minimum time to process the passengers if there are four gate attendants?

b) What is the minimum number of gate attendants needed to process the passengers if no passenger is required to arrive at the gate sooner than 30 minutes prior to departure time?

c) Calculate β for the situations stated in (a) and (b).

3. Assume that the airline uses five gate attendants, and that the passengers begin to arrive at the gate 45 minutes prior to departure time of the 747 airliner described in Problem 2.

a) What is the average inter-arrival time in each queue if all queues stay the same length? For simplicity, assume arriving passengers go one-by-one to each queue in turn. (Hint: divide the number of passengers by 5 and assume one queue.)

b) If the service time is constant at 30 seconds, what is the average queue length?

c) If the service time is random but averages 30 seconds, what is the average queue length according to the formula in the text?

d) If the airline passengers complain about arriving 45 minutes prior to flight time and 90% start arriving no sooner than 30 minutes prior to scheduled departure, what is the utilization factor?

e) What is the average queue length if service time is constant?

f) What is the average queue length if service time is random?

g) If the airline wants to maintain the same average queue length as you obtained in (c), what are its alternatives?

4. In a barbershop the service time is 15 minutes per customer. The cost for a haircut is $2.00.

a) If the average inter-arrival time is 20 minutes, find the percent of time the barber will be working.

b) Assuming that we are talking about *constant* service time, what is the average queue length?

c) What is the gross income per day (neglecting tips), assuming an 8-hour day?

d) In order to stay in business the barber must gross at least $40/day. If he does not wish to speed up his service time, how much must his business increase before he can afford to hire another man, costing $20/day?

e) If he had not hired a new man when business increased by the amount in part (d), what would be the average queue length (still assuming constant service time)?

5. In the previous problem we assumed constant service time.

a) Use the same figures and calculate the queue lengths of (b) and (e) assuming random service time.

b) Which is a more realistic model for serving time, constant or random?

c) How could the barber decrease his queue length without getting more help?

6. In the queueing problem described in the text (arrivals from 9 to 10 A.M.), calculate the average length of time a customer has to wait before he is serviced.

7. The queueing problem described in the text (arrivals from 9 to 10 A.M.) describes a system in which the customers are served on a first-come, first-served basis. This situation describes the usual barber shop, store, movie box office, or highway tollgate. We might alternately work on a last-come, first-served basis. In other words, when the service man is free, he turns next to the last arrival among

the customers. Would this change the average queue length? The average waiting time of the customers?

8. a) Given the inequality $y \geq 0$, show on regular x and y axes how this inequality divides the plane into two regions. What is the coordinate of y when $x = 0$ (i.e., what is the y-intercept for the equation of the dividing line)? Find the point 5 units higher than this y-intercept. If the coordinates of this point are substituted into the inequality, is the inequality satisfied? Is the point in the feasible region?

b) Now divide the plane (using the same axes) by the inequality $2y - x \leq 10$. What is the coordinate of y when $x = 0$ (i.e., what is the y-intercept for the equation of the dividing line)? Find the point 5 units higher than this y-intercept. If the coordinates of this point are substituted into the inequality, is the inequality satisfied? Is the point in the feasible region?

c) Lastly, divide the plane by the inequality $2y + x \geq 10$. What is the y-intercept of the equation of the dividing line? Find the coordinates of the point 5 units higher than the y-intercept. Is the inequality satisfied by this point? Is it in the feasible region?

d) State as a general rule the method used here to find which part of a plane is "feasible" when an inequality is plotted.

9.

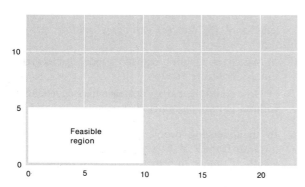

a) What are the inequalities that give this feasible region?

b) The cost equation is $C = 20 - x - 2y$. Rearrange this in a form similar to $y = mx + b$: the "slope, intercept" form of the equation. What is the slope? To minimize the cost, should y be as small or as large as possible?

c) Through which corner of the feasible region must the graph of the cost equation run to minimize cost?

d) If the cost equation were $C = 20 + x - 2y$, answer the questions asked in parts (b) and (c).

10. Work out the ice cream production problem again in terms of C, the number of quarts of chocolate produced, rather than V.

11. Suppose that a radio manufacturer turns out only two types of radio: a standard model, selling at a profit of $20 each, and a luxury model, selling at a profit of $30 each. The factory has an assembly line for each model, but their capacity is limited. It is possible to produce at most 8 standard radios per day on the first line, or 5 luxury radios per day on the second. The manufacturer is faced with another constraint: owing to limited supply of skilled labor he has only 12 employees, so the available labor amounts to 12 man-days per day. To assemble a standard radio requires one man-day, but it takes two man-days to make a luxury radio.

a) How many radios of each type should he make to maximize his profit?

b) What will this maximum profit be?

12. In the transportation problem of the text, suppose that the amount of wheat at Grand Forks is 30 carloads and at Chicago is 60 carloads. Find the minimum-cost shipping plan.

13. An oil company has 200 thousand barrels of oil stored in Kuwait (on the Persian Gulf), 150 thousand barrels stored in Galveston, Texas, and 100 thousand barrels stored in Caracas, Venezuela. A customer in New York would like 250 thousand barrels and a customer in London would like the remaining 200 thousand barrels. The shipping costs in cents per barrels are shown below. Find the minimum-cost shipment schedule.

	Kuwait	Galveston	Caracas
New York	38	10	18
London	34	22	25

14. An interesting variation on the grain dealer's problem occurs when the dealer has three ware-

houses from which to furnish two cities. The problem is described in the table below:

To	Denver		Miami	
Available | From		42		55
50 | Grand Forks	x		$50 - x$	
		36		47
40 | Chicago	y		$40 - y$	
		28		60
20 | Omaha	$60 - x - y$		$x + y - 40$	
Needed	60		50	

a) What is the optimum amount to ship from each warehouse to each buyer?

b) What is the minimum cost?

15. The machinery in a certain chemical plant can be adjusted so that in every 1000 lb of output, at least 800 lb are product A. The remainder, if any, of the output will be composed of product B anywhere between 0 and 100 lb, and product C.

a) Let x = no. lb of A produced, and y = no. lb of B produced. Then $1000 - x - y$ = no. lb of C produced, or $x + y \leq 1000$. Construct a graph showing the feasible area.

b) Product B has long been sold at a profit of 12¢/lb, and C at a profit of 10¢/lb. A, however, is liable to variations in price. Let us therefore say that it brings a profit of T¢/lb. Write an equation showing the profit to be expected from 1000 lb of output, in terms of x, y, and T.

c) What is the best mixture of A, B, and C, and what is the resulting profit, if $T = 9$¢/lb?

d) What if $T = 11$¢/lb?

16. A student in *The Man-Made World* course gets a job in General Custards Ice Cream and Soda Stand. Each day he has 80 units of milk, 70 units of ice cream, and 30 units of syrup. The specialty of the house is the "Last Stand" which consists of 2 units of milk, 3 units of ice cream, and 1 unit of syrup. "Vanilla Plains" contains 2 units of milk and 1 unit of each of the other basic ingredients. The "Last Stand" sells for 80¢ and the "Vanilla Plains" sells for 50¢. What is the most profitable division of the available ingredients to make these two types of soda? Assuming that the basic ingredients cost General Custards $20, how much profit is made if all of the sodas are sold?

17. Suppose a mother is shopping for food for her family and she wants to know how much steak and potatoes to buy. Being concerned that the family receive proper nutrition, she refers to a chart which says that for each day a normal person should have:

8 units of carbohydrates
19 units of vitamins
7 units of proteins

She checks the food value table to find:

	Units of carbohydrates	Units of vitamins	Units of proteins
1 unit of potatoes has	3	4	1
1 unit of steak has	1	3	3

She checks the prices of these two foods and calculates that:

1 unit of potatoes costs 30¢
1 unit of steak costs 80¢

She wishes to get the most nutritional value for the money spent. Having taken a course called *The Man-Made World* back in 1969, she decides to solve her problem by linear programming. If she does it correctly, how many units of potatoes and steak will she buy for each member of the family?

18. A typical automobile assembly plant has a maximum production capacity of 1200 cars per day. The sales force predicts that it can sell 600 Cloudstars per day and 800 Dachshunds (the long wheelbase jobs). The profit for a Dachshund is $500 and the profit for a Cloudstar is $650. Let C = the number of Cloudstars, and D = the number of Dachshunds produced each day.

a) Write five inequalities that satisfy the conditions of the problem.

b) Graph the inequalities.

c) Write an equation to represent the profit in the example.

d) Plot the line for the equation and show the feasible region.

e) What is the most profitable daily production and what is the maximum daily profit from this assembly plant?

Laboratory and Projects

I | QUEUEING

This experiment is designed to permit you to observe the development of a queue, to change the factors that affect the queue, and to observe the influence of these factors on the length of the queue. Section 4 of this Chapter is background for this experiment.

Part A

A Barber Shop: The customer arrival times are random, but the service time is constant. We assume that 16 minutes are necessary for servicing each customer. The customer who wants "just a little off the sides" requires the same time as the nearly bald gentleman for whom the barber has to hunt and search for hair to trim. There is only one barber, and no prospective customers depart before being serviced.

Furthermore, we will make some simplifying assumptions. An arriving customer will enter the shop no matter how many people are waiting. Likewise, the barber does not speed up his service when people are waiting. Finally, no one has an appointment or any special priority. Use a single die (one of a pair of dice) to estimate customer arrival time. Consider a customer to have arrived each time a "two" is thrown. Also assume that the die is rolled at three-minute intervals.

■ 1. Prepare a customer-arrival chart similar to Fig. 3–8, but based on your data. Use the period from 9:00 A.M. to 12:00 noon.

■ 2. From the customer-arrival chart, prepare a graph of queue length versus time.

3. Calculate

   a) The average inter-arrival time $\dfrac{\text{number of arrivals}}{\text{total time}}$

   b) Utilization factor: $\beta = \dfrac{\text{average service time}}{\text{average inter-arrival time}}$

   c) The average queue length: $q = \dfrac{\beta}{1 - \beta}\left(1 - \dfrac{\beta}{2}\right)$

4. Does the calculated value of queue length compare reasonably with your observed value?

Part B

Laboratory experiments are interesting and usually illustrate the points made in the text, but the real situation does not always follow the laboratory situation. It is therefore important for your understanding that you measure some real queues. One of the best places to measure queues is the local supermarket during a busy period. The queueing situation which changes most rapidly is the express lane.

After checking with your teacher and the supermarket manager, go to the supermarket during a peak period and take data for one hour in the following categories:

   a) Length of queue at the start.
   b) Length of queue at two-minute intervals.
   c) Actual arrival times of customers.
   d) Actual service time for each customer.

From these data, calculate the following:

   e) Average queue length.
   f) Average inter-arrival time.
   g) Average service time.

Apply the data from f and g to the equation for average queue length and compare this figure with the average queue length obtained in e.

1. How does the average queue length from actual measurement compare with the average queue length obtained by calculation using $q = \dfrac{\beta}{1 - \beta}$. Was there a significant difference? If so, explain what you think the reason was for this difference and give evidence which will support your contention.

2. Did the customers ever leave the express check-out line to get into another line? If so, what was the situation which caused this?

■ 3. What action, if any, did the store manager take during the hour you were there that affected the length of the queue?

■ 4. What action, if any, did the store manager take while you were there to affect the queues at any of the other check-out counters?

II | GAMES

There are many mathematical games which illustrate the use of strategies in competitive situations. As we make games more complex, the problems rapidly become very difficult. Such games are important because they involve some of the same reasoning as in much more important situations (e.g. international competition).

Part A

This game called Hit-and-Run was invented by Jurg Nievergelt of the University of Illinois and was reported on in the July 1969 issue of *Scientific American*. One player has a new book of red-tipped matches (20), the other has a book of blue-tipped matches. Start the game on the matrix (Fig. 1). Players place matches alternately. The object of the game is to construct a path connecting the two "red" or "blue" sides of the board. Each line segment is one match long.

After you have played this game once or twice and developed a strategy for winning, go on to Part B.

Part B

With the same matches and rules, play the game on a matrix which is three segments long on each side (Fig. 2).

There is a winning strategy for this game also, but it is not obvious. Once you have solved it go on to Part C.

Part C

With the same matches and rules, play the game on a matrix which is four segments long on each side (Fig. 3).

From the problems which you have in devising a strategy for this game, what do you think of the value of taking the strategy used for winning simple games and applying it to more complex versions of the same game?

What other competitive situations are simulated by this game?

III | DESIGN OF A REMOTE ELECTRIC HEATER

The task in this experiment is to design an electric heater which is located in a shed a long distance away from a source of electric power. All electric heaters contain an element called a resistor, which heats up when it is connected to a source of electric power.

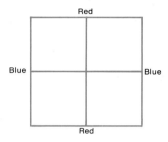

Fig. 1.

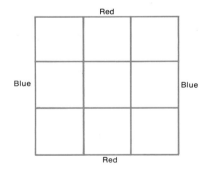

Fig. 2.

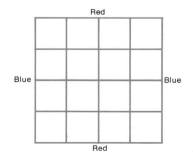

Fig. 3.

In the electric toaster or broiler, the heating resistor is a wire which gets red hot when the power is turned on. In our design we have the additional problem that a very long electric cable is needed to connect the heater to the power supply and the cable itself has its own electric resistance. We construct a crude but very useful model of this system by using small cylindrical carbon resistors.

The model heater is set up as shown in Fig. 4. The electrical power for the heater may be obtained from a low voltage source, either A.C. or D.C., which supplies approximately 10-15 volts. The large cylindrical resistor is used to simulate the resistance of the long cable and in this experiment it has a resistance value of 82 ohms. (A unit of electrical resistance is called an ohm.)

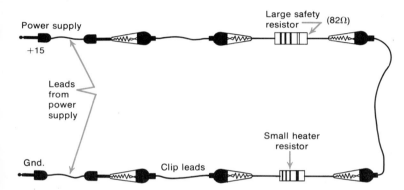

Fig. 4.
Model of an electric heater.

Our problem is to determine which one of the small resistors will make the best heater. (Each resistor has a different resistance, which can be identified by the colored bands on the resistor. Table 1 lists the resistances and the respective color code.) The best heater will be the one which heats up the shed the fastest.

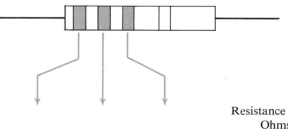

Table 1. Values of heater resistances.

			Resistance Value Ohms
Red	Black	Black	20
Orange	White	Black	39
Gray	Red	Black	82
Brown	Red	Brown	120
Orange	Black	Brown	300

We use a small block of paraffin to simulate the shed. The resistor which sinks farthest into the paraffin in a given time is the best heater in this system.

Part A

Procedure

With the power off, connect the resistors as in Fig. 4. Since the weight of the small heater resistor on the paraffin block will affect the speed with which it melts through the block, it is important that the clips be the same distance from the resistor body. This will insure uniform melting and will eliminate the weight of the clips as a variable in the experiment.

With the resistor not in contact with the wax, turn the power on and let the resistor heat up for one minute to allow it to reach a stable temperature. Carefully place the resistor on the wax block as in Fig. 5.

It may take some practice to develop your technique so that each resistor exerts the same force on the block. Keep the resistor on the block for one minute. As you remove the resistor quickly tip the wax block so that the melted wax runs out of the notch.

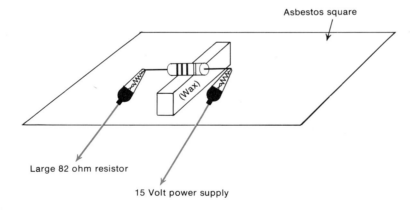

Fig. 5.
Measurement of heating.

Repeat for the other four small resistors.

1. Record the data and plot the resistance *vs.* depth of the notch in millimeters.

2. What is the optimum value of heater resistance?

3. Using this method is it necessary to understand the physics of electrical power generation to pick the optimum heater?

If you are interested in pursuing the physics principles which explain this situation, your teacher will provide you with the appropriate references.

Part B

Another way of doing this experiment is to measure the time necessary to melt the same amount of wax using the five small resistors.

Procedure

Place the 300 ohm resistor on a square of asbestos (to prevent damage to the table) and allow it to heat for one minute (Fig. 4). Place the block of paraffin on the resistor as in Fig. 6 and continue the heating until the paraffin block has melted a u-shaped slot as deep as the diameter of the resistor (until the block has settled to a horizontal position). Record the time. Repeat with the remaining resistors.

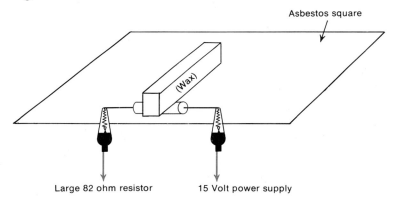

Asbestos square

Fig. 6.

Large 82 ohm resistor   15 Volt power supply

■ 1. Record the data and plot the resistance *vs.* time.

■ 2. What is the optimum value of heater resistance?

■ 3. Does this agree with the results of Part A?

■ 4. Discuss the relative merits of these two procedures.

A   Power generation

B   Refuse disposal

Autos   C

Heating   D

E   Industry

Modeling 4

I | THE NATURE OF MODELS

According to an old story, six blind men who had never seen an elephant tried to decide what it was.* The first man, feeling the elephant's flat, vertical side, concluded that the beast was similar to a wall. The second man touched a round, smooth, sharp tusk and decided that an elephant is similar to a spear. Grasping the squirming trunk, the third blind man said that the animal resembled a snake. The fourth man, who touched a knee, observed that elephants resembled trees. From an exploration of the ear of the elephant, the fifth man was convinced that the animal had the shape of a fan, while an examination of the tail convinced the sixth blind man that an elephant was similar to a rope.

Each, of course, was partially correct, but insofar as a complete representation of the elephant was concerned, all were wrong. Each man, after observing the "real world," formulated a description, or *model*. But since the observations were incomplete, the models were incorrect.

Every time we describe an object, we are really making a model. We use our senses to find information about the real world. From this information, we decide what the object is. Then we pick out the *important* features. These make up the model.

The model is an efficient way of viewing things. A good model includes only those parts which are useful. But, by restricting our thoughts to a few features, we are able to understand the object or system. We can anticipate the effects of actions we might take. On this basis, we can select the best action. Thus, man's ability to control his environment and to build useful systems depends directly on his capacity to find models.

Models Are Usually Quantitative

Models begin as conceptions; they are ideas about the structure and nature of something. Before we can go very far with the model, we usually have to develop a *quantitative model*, one which tells how much, where, and when in terms of numbers. In other words, we use the language of mathematics to describe our situation.

Five sources of air pollution. (State of New York Department of Environmental Conservation. N.Y.S. Health photos B, C, and D by M. Dixson)

* The Blind Men and the Elephant (J. G. Saxe, 1816–1887). The poem is given in Question 15 at the end of the chapter.

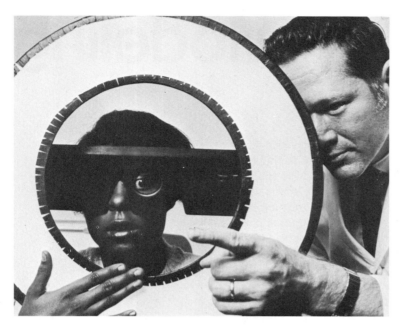

Fig. 4–1.
The girl is testing
her left eye. Every few
seconds, a point of light
appears on the screen. She
moves a lever in the direction
where she sees the light. From
this test, we can make a map of
the field of vision (the regions
where she sees the light).

This map is an example of
a model. Here we model the
seeing ability of the patient.
From a study of this model,
the physician measures
the condition of the
eyes and can detect
the beginning of
certain disease.
(IBM
Corporation)

The importance of a quantitative model can be illustrated by the task of the man who is in charge of scheduling elevators in a twenty-story office building from 8:45 to 9 A.M. every weekday morning as people arrive for work. There are six elevators, each able to stop at every floor.

When the building is first opened, the elevator supervisor simply loads each elevator in turn. As soon as it is loaded, it departs. Our supervisor notices that service is extremely slow. Occasionally, an elevator is loaded with one passenger for each of 15 different floors. The poor passenger who works on the highest floor has a ten-minute ride. (Especially since the passengers at the back of the elevator as it starts up always seem to be the ones who want to get off first.)

As the complaints about service become more and more bitter, our supervisor wonders if service might be improved by using a better plan. He might, for example, use two elevators for floors 2-8 only, two others for floors 9-15, and the other two for 16-20. He might stop only at even-numbered floors and force employees to walk down one flight of stairs to reach the odd-numbered floors. Obviously, there are many possible strategies which could be tried to improve service.

The choice of a desirable strategy depends on the way employees arrive in the morning for each of the floors. For example, the fifteenth floor holds the executive suite for the top officers of the company. If any of them should arrive during this busy period, they must be delivered as rapidly as possible. This priority for

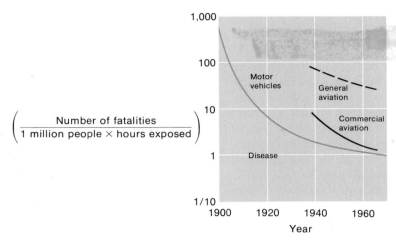

$$\left(\frac{\text{Number of fatalities}}{\text{1 million people} \times \text{hours exposed}}\right)$$

Fig. 4–2.
A different model which shows the willingness of people to accept risk. The vertical axis is a measure of the risk involved in different activities (the number of deaths in the U.S. divided by the total number of hours a million people are exposed to the risk). For example, 10 means that there are ten deaths every hour a million people engage in this activity.

The interesting part of this model is that just living (with possible disease) has about the same risk as driving a car or flying a commercial airplane. (Only about 5% of the people ever fly; 85% ride in a car.) People seem willing to accept this risk of 1. Cigarette smoking has a risk of 3, skiing about 1, and normal railroad travel 1/10. This model gives us some idea how safe we must make new forms of transportation (moving sidewalks or 300-mph trains).

The model is discouraging if we are interested in cutting down traffic accidents. It suggests that the average person may be willing to accept today's death rate. (Chauncey Starr, "Social benefit versus technological risk," UCLA report, 1969.)

Floor 15 means that at least one elevator should go directly to this floor. Similarly, certain floors may have many employees and should be given preference. In other words, to derive a sensible plan or strategy our supervisor needs at least a rough quantitative picture of the flow of people. This is the model.

The model may be very approximate and rough, guessing only that the number of employees is the same for each floor. Or it may be very detailed with the number for each floor, their probable arrival times, and their relative importance to the company. If the model is very simple, the supervisor may be able to decide intuitively on a strategy; if the model is very detailed, a computer may be required to evaluate and compare the possible strategies.

The important feature of this example is that no intelligent decision is possible unless a model is used, whether or not the supervisor calls his picture a model.

2 | THE GRAPH AS A DESCRIPTIVE MODEL

To form a model, we need to collect data about some aspect of the real world. We might wish to determine whether a simple relationship exists between the heights and the weights of twenty-year-old men. After making several experimental measurements, we may secure a set of related numbers such as 5'6", 130 lb; 6'1", 180 lb; 5'7", 155 lb. But it is difficult to discover any systematic relationship in this way. Even though we may have a reasonable expectation that as height increases, weight will increase, this verbal model is vague and imprecise.

In order to present the data in a way which we can interpret more easily, we make a graphical plot, as in Fig. 4–3. Each point represents the height-weight data for one man. We notice now that the points are not scattered, but seem to be closely grouped. What can we say about the relationship between height and weight?

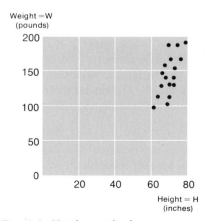

Fig. 4–3. Height-weight data for 20-year-old men.

A straight line can be drawn through the points to represent averages for these data (Fig. 4–4). This picture of the data is a *graphical model* of the relationship between height and weight.

This graphical model presents a clear but simplified description of the real world. It can be used as the basis for some reasonable predictions. From the straight-line average we can estimate the probable weight of a twenty-year-old man even though we know only his height. Let us suppose that we wish to estimate the weight of someone who is 73″ tall. The corresponding weight for this height can be immediately obtained from our graph. The graphical model permits us to estimate the weight of such an individual, even if our original data did not include any individual of this height. Thus, we may predict from our graphical model that an individual who is 73″ tall probably weighs about 180 pounds.

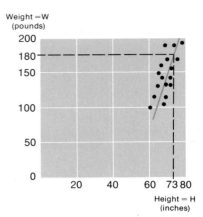

Fig. 4–4. Weight expected from a man 73″ tall.

An Equation As a Model

While the graph of Fig. 4–4 is an appropriate model for our weight-height relationship, it is also often convenient to have a mathematical equation. The straight-line graph is exactly equivalent to an equation, which in this case happens to be

$$W = 8H - 407$$

(We derive this in the next paragraph.) Even though the graph and the equation say exactly the same thing, it is often a nuisance to have to redraw the graph every time we want to tell someone what the model is. Furthermore, it is frequently easier to work with the equation. For example, what is the height expected for a 150-pound man? From the equation,

$$150 = 8H - 407 \text{ or } H = 70''$$

Since our graph is a straight line, we know from algebra that the equation has the general form $y = mx + b$, or in this case,

$$W = mH + b$$

where W is the weight, H is the height, m is the slope, b is the vertical axis intercept. To complete the equation, we must therefore determine the values of the constants m and b. Since b represents the W, or vertical intercept of the line, Fig. 4–5 reveals that the line cuts through the W-axis at a value of -407. Hence,

$$W = mH - 407$$

We can measure the slope m directly. When H increases by 10 inches (for example, from 60 to 70), W increases by 80 pounds. The slope is thus $\frac{80}{10}$ or 8, and the equation is

$$W = 8H - 407$$

Once the equation is determined, the weight W can be found for any given height, or the height for any given weight.

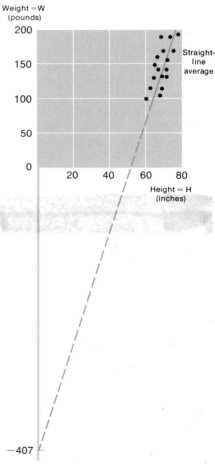

Fig. 4–5. Straight line extended until it hits the W axis.

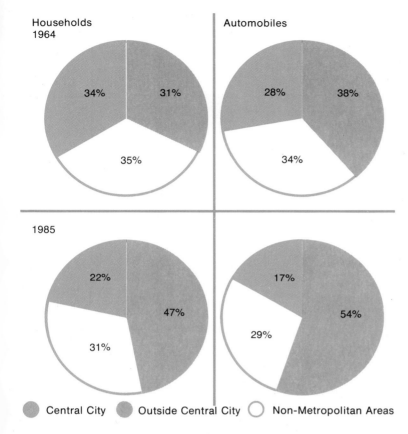

Households
1964

Automobiles

34% 31%

35%

28% 38%

34%

1985

22% 47%

31%

17% 54%

29%

● Central City ● Outside Central City ○ Non-Metropolitan Areas

*Fig. 4–6.
A completely
different form
of graphical
model. In this case,
we are interested
in showing the changes
expected from 1964 to
1985. The "pie" charts show
how the suburbs will grow,
primarily as the central city
becomes less important (if
present trends continue).
Thus, a model can be
portrayed in many
different forms.*
(Technology
Review *of
M.I.T.*)

We can use this model to predict points which were not orig-inally in our data sample. But these predictions must be carefully examined. For instance, the line we have drawn tells us that an individual who is 54″ tall will weigh about 25 lb; worse yet, a 24″ person can be expected to tip the scale at an impressive *nega-tive* 215 lb! Surely these are curious figures for twenty-year-old men. What is wrong with our model?

The straight-line average that we drew was extended so that the W intercept (-407) could be found. But we are not entitled to say that all points on the entire line must represent real situations. Actually, the only use we can make of this model is to predict within our data field. We know that anywhere inside of the cluster of measured points we are in the neighborhood of a real-world possibility. However, *we run into the danger of unrealistic prediction if we apply the model beyond the region that has been measured.* This danger applies to either the graph or the equation.

Our model, therefore, has its limitations. It must not be used to predict beyond the region of results experimentally obtained unless there are very good reasons to believe that real-world laws are not being violated. To test this model requires that we obtain a fresh sample of twenty-year-old men, either the data from the same school for other years or from other schools or from the army. Then we either enter them on the plot or compare them with predictions made from the algebraic model.

The equation is a mathematical model which says exactly the same things as the graphical model. Both of these models are more useful than the verbal model with which we started.

One of the most interesting aspects of this model is that it turns out to be so simple. This is quite unexpected. If we think of the people we meet while walking down a busy city street, we know that all sizes and weights are combined. It is true that the sample studied was passed through two strainers (age and sex) to make it more manageable.

We should give one final word of caution. We must remember where our original data came from. If we had measured the heights and weights of the members of the Kansas City Chiefs professional football team, we probably could not expect to use the model to predict the weight of a six-foot, 25-year-old starving artist. At least, we should be suspicious of the prediction. The model is only useful as long as we stay with the part of the world from which the model was derived.

Fig. 4–7.
City traffic congestion is not a new phenomenon. This picture was taken in 1905 at the corner of Dearborn and Randolph Streets in Chicago. (The Bettmann Archive)

3 | A DESCRIPTIVE MODEL OF TRAFFIC FLOW

A key urban problem is the question of how to handle motor traffic in the streets. Some cities have gone so far as to ban all automobiles from a few streets. This does not so much solve the problem as eliminate it—at least from those streets. Furthermore, it sometimes substitutes other problems, for example for the elderly and infirm, or for the shopkeeper trying to attract customers. It also increases the cost of goods since delivery expenses rise.

If a traffic engineer is to improve present conditions he must be able to predict the results of changes. To do this he must construct a model of the traffic flow in and around the city. Since such a model is too complicated, we use a study of the simpler circumstances within a school; even this we limit to what goes on at a single corridor intersection, such as that shown in Fig. 4–8. The limited model we derive for a single intersection could be extended to a whole building. The resulting larger model can be of practical use to school administrators, schedule-makers, and architects.

In order to construct our model, we must determine what affects the behavior of the system. We are primarily interested in the rate at which people (including teachers and custodians) pass from one corridor to another, in other words, in the density of traffic as measured in people per minute. The measurement will be made by *sensors*, devices which respond in some way whenever a person passes. An example would be the device often called an electric eye, but for short-term service a much more practical sensor is a person stationed at the proper spot.

Figure 4–9 shows the intersection with measurement points identified. Traffic problems usually arise at intersections rather than in the main corridors.* What we need for the present is a count, minute by minute during the time when classes are being changed, and for a minute (or more) before and after.

Table 4–1 shows a possible result of such a set of measurements. The actual number of people who go through the intersection during the counting period is about half that shown because each person was counted twice, once when he entered and once when he left (if we forget people like Joe and Susan).

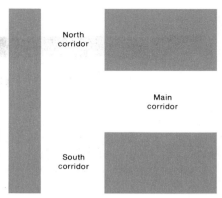

Fig. 4–8. A school corridor intersection.

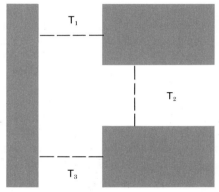

Fig. 4–9. Where traffic density is measured.

* As Joe passes position T_2 going east, he sees that Susan is going south past T_1. He reverses his field and meets her in the middle of the intersection. Therefore, they both hold up traffic and he is counted three times at T_2; he might even stroll a few yards down the south corridor with Sue and add two more tallies to his total.

Table 4–1.
Detailed traffic count at one intersection, end of first period.

Minute number	Counting station			Total counts
	T_1	T_2	T_3	
1	26 counts	50 counts	30 counts	106
2	42	63	47	152
3	61	102	55	218
4	112	184	73	369
5	38	42	28	108
6	22	17	9	48
			Total for period	1001

The data can be shown more effectively if we present them in a plot, rather than a column of numbers. One form of plot which is often used in displaying a count of events is called a histogram. It has the form shown in Fig. 4–10 for the total number of individuals passing through this intersection.

The height of each vertical column is proportional to half the total count for the minute indicated below it, half the total for the reason just explained. It is evident that most of the traffic, but not all, occurred during the interval between classes. The school administrator confronted with such a bar graph might well be suspicious of the large traffic during the minute before the bell rang for class changing: Are some teachers dismissing their classes early? He might also be disturbed by evidence of a good deal of tardiness.

Further information, useful to the scheduling officer, can be obtained if another bar graph is made, this time of total counts at the end of each period during the day (Fig. 4–11). Here the peaks shown for traffic at the ends of the first and fourth periods might suggest altering the room assignments in such a way as to lessen the traffic through this intersection at these times. For example, more students might be scheduled for successive classrooms in the same corridor to avoid the intersection.

A full study of a school's traffic pattern requires that data be obtained at every intersection for every class-changing period throughout the day or even the entire week. If there are ten important intersections, the total data can be portrayed in fifty histo-

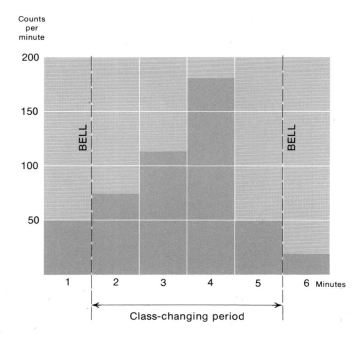

Fig. 4–10.
Bar graph of traffic at one intersection at the end of the first period.

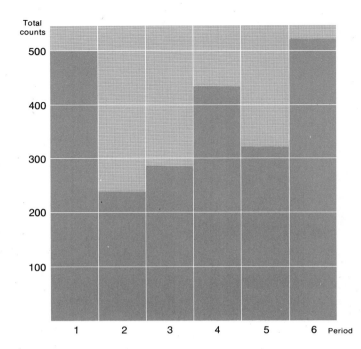

Fig. 4–11.
Bar graph
of total daily traffic
at one intersection.

grams similar to Fig. 4–11 (ten for each of five days) or similar curves on a minute-by-minute basis. Inspection of these data then reveals the times and locations of major congestion.

In processing data of this type, it is common practice to indicate those particular data points which are either unusually large or notably small. We might show in red all parts of the histograms which are larger than a predetermined amount (corresponding to troublesome congestion). We may choose to make up a small separate list showing critical data only. Such selective presentation of critical data is called "flagging"; we flag data which are especially important.

This type of traffic-flow study is closely related to the queueing problem described in Chapter 3. Queues occur when traffic density increases. In the city traffic problem, the results are obvious. Excessively long travel time ruins emergency services (ambulances, fire trucks, and police cars), lowers the income of taxi drivers, raises the cost of goods, and causes lost time for workers. One of the reasons some small businesses are moving out of the cities to the suburbs is the consistent tardiness of workers because of transportation problems. In one recent study, a manufacturing business employing 700 workers measured tardiness in the morning, lost time at the lunch hour, and early departures in the afternoon. The model showed that if the company moved to the suburbs, production would increase 12% with the same size labor force.

Fig. 4–12.
Broadway as it
approaches Times Square
in New York City. The movie title
is an appropriate commentary.
(*The Bettmann Archive*)

Even if all employees were given free lunches, profits would increase 4% just from the greater production.

Studies based on such models reveal the importance of adequate transportation in the life of any town or city. They also are a commentary on the way business is carried out in a city. When commuting trains, buses, and subways so often bring the businessmen to work an hour late, one has to wonder whether most organizations don't employ more men than they really need.

4 | MODELS FOR RESOURCE MANAGEMENT

Management of resources is a fertile area for the use of models and decision making. In this section, we illustrate this by looking at the history of the buffalo in the western United States. Then we show how this resource could have been managed with some simple models and logical decisions.

The ideas which we present are currently being used in many different ways. In the northwestern United States, the salmon population is being modeled in detail in order to determine how to regulate salmon fishing and changes in water to ensure that the salmon supply will continue in the future. France has several programs for sea "gardening," growing fish and shellfish for food. The invasion of the Great Lakes by lampreys (eels) and the resulting changes have been under study for more than two decades, with a variety of decisions made to restore the fishing industry and sport.

The buffalo provide an interesting example of what is apt to happen with no national policy for resource management—no model, no logical decision.

A Model for 1830

In 1830, there were 40 million buffalo roaming the western United States. In terms of weight (at 1000 pounds each), there were 40 billion pounds of buffalo compared to only 24 billion pounds of human beings today in the entire United States. The buffalo dominated the western plains to an extent unknown for any other animal in history, including human beings.

In 1830, the railroad arrived and the rapid westward expansion of the United States began. By 1887, there were only 200 buffalo left. In this slaughter, animals were often killed for only the tongues and hides. An average of only 20 pounds of meat (of a possible 500) per buffalo was eaten. The peak was reached in 1872 when national heroes like "Buffalo Bill" Cody led the killing of more than seven million.

In less than sixty years, the lack of any sort of policy led to the destruction of what could have been a major source of meat for today's population of this country. A single buffalo could provide the entire meat supply for at least five people for a year.

Had the nation been aware of the potential problem in 1830, a model could have been constructed. In order to decide how many buffalo can be killed each year, we first must know how the population rises or falls from natural causes. Recent studies of buffalo have indicated the following facts:

1. Buffalo reach maturity at age 2.
2. 90% of the females age 2 or older have one calf a year.
3. 53% of the calves are male, 47% female.
4. 30% of the calves live for two years, or to maturity (infant mortality is high for most animals).
5. 10% of the mature beasts die each year (from tuberculosis, drowning, predation, and so on).

Fig. 4–13.
"Buffalo Bill"
Cody. The picture is
from a poster announcing
the appearance of
"Buffalo Bill"
sometime during
the 1890's. (The
Bettmann
Archive)

From these observations, we can construct a model in mathematical form to show the number of mature females alive at the beginning of each year.* For example, let's look first at year seven. How many females are there at the beginning of year seven? Of course, this can be any year picked at random.

First, the answer depends on the number there were at the beginning of year six. Ten percent of those have died. Some other fraction have been *harvested* (killed for meat and hides). Let us call this fraction k. Then, of those who started year six, $(0.9 - k)$ as many start year seven.

This is not quite the whole story. During year six, some female buffalo have reached maturity (two years old). For every 100 mature females at the beginning of year five, 90 calves are born; 47% of these, or 42.3, are female; 30% of these, or 12.69, live to be two years old. Thus, 0.1269 of the female population at the start of year five become mature females at the start of year seven.

---

* The same general sort of model is used for males. The total population is, of course, the sum. To simplify the discussion, we consider only the females. (The male calves born depend on the number of females, so it is easiest to find the number of females first.)

Therefore, the number at year seven is $(0.9 - k)$ times the number at year six plus 0.1269 times the number at year five. If we call F_7 the number of females at the start of year seven, we can summarize this model very simply by an equation

$$F_7 = (0.9 - k) F_6 + 0.1269 F_5$$

We can change the 7, 6, 5 to any three consecutive years. All this equation says is that the mature population depends on the size a year earlier and also two years earlier. The first term arises from the total death rate, the second term from the births two years ago.

A Decision Policy

Once the model is known, we can choose k to control the population. We might want the population to increase, stay constant, or decrease.

As an example, if the decision in 1830 had been to keep the population constant at 20 million mature females, how should k be chosen? Now in the preceding equation, we want F_7, F_6, and F_5 to be the same (20 million). Then we can cancel these F's or divide each term by F. This leaves an equation

$$1 = (0.9 - k) + 0.1269$$

which states that k should be 0.0269. This means we can harvest 2.69% of the mature females each year. The population will not

*Fig. 4–15.
Seal of the U.S.
Department of the
Interior. It is ironic that
the animal we slaughtered
so enthusiastically is the
symbol of a government
department and even
appears on our coins.*

change. Since 0.0269 times 20 million is 538,000, we can harvest for food and hides 538,000 mature females each year. (Even more males can be harvested, since more males are born.)

If both males and females are counted, we could have provided enough meat for 6 million people a year since 1830 and still had a buffalo population of 40 million. All that was missing was an intelligent decision policy.

If we were actually managing the buffalo population, we would have to measure the death and birth rates each year. If there were an epidemic, we might want to decrease the harvest until things were back to normal.

We could also construct a more detailed model. Birth rates depend on the age of the mature females, the weather, and other factors. Infant mortality varies with location (the number of predators, the food supply, and so forth).

The interesting feature of the example is the obvious benefit from a very simple model. Even if the numbers in the model are slightly in error, the decision (or policy) ensures that the population changes will be slow. If after several years we find the population is decreasing, we can reduce the harvest slightly. Over a period of years, we can reach a policy which gives a stable population. We can continually improve our model and our policy with experience.

5 | A POPULATION MODEL

While we worry about saving the buffalo or the falcon from extinction, there is much greater concern today about trends in the population of human beings. A model of the world's population is interesting because it represents a dynamic situation (one in which events change with time). Furthermore, the derivation of the model illustrates how we often need to refine or improve our model after it is initially determined.

It has been estimated that since the appearance of man on the earth, a total of 15 billion human beings have existed. With a world population of nearly 3 billion today, 20% of all the people who have ever lived are alive today. Our population is growing at an explosive rate.

Demography, or the study of population, is of increasing concern to economists, ecologists, sociologists, political scientists, engineers, and many others who must understand the present and plan for the future. Models of population change are exceedingly important to such study. They make possible analysis and prediction which can lead to more effective planning for the many goods and services that people need.

We wish to obtain a simple model which would let us estimate the world population at some future date. The present average rate of population increase for the entire world is estimated to be close to 2% per year, and we assume in this section that this rate of increase does not change. At the start of 1960, population was approximately 3 billion (that is, 3 followed by nine zeros, or 3×10^9).

If the rate of increase is 2% per year, by the start of 1961 the increase is 0.02(3,000,000,000), or 60,000,000 people, to make a total of 3,060,000,000. We can then calculate the increase for the next year and for all succeeding years. The results are shown in Table 4–2.

Year	Population at start of year	Increase	Population at end of year
1960	3,000,000,000	60,000,000	3,060,000,000
1961	3,060,000,000	61,200,000	3,121,200,000
1962	3,121,200,000	62,424,000	3,183,624,000
1963	3,183,624,000	63,672,480	3,247,296,480
1964	3,247,296,480	64,945,930	3,312,242,410
1965	3,312,242,410	66,244,848	3,378,487,258
1966	3,378,487,258	67,569,745	3,446,057,003
1967	3,446,057,003	68,921,140	3,514,978,143
1968	3,514,978,143	70,299,562	3,585,277,705
1969	3,585,277,705	71,705,554	3,656,983,259

Table 4–2.
Estimated world population, 1960 to 1969. These data can be compared with more recent estimates: In July 1967, the Population Reference Bureau used United Nations and other statistics to estimate that, in the summer of 1966, world population was 3.34 billion, an increase in one year of 65 million.

The table shows that the increase is *greater* each year. The growth is always 2% of the population at the beginning of the year. As the population rises, the growth also increases. In fact, if we continued Table 4–2, we would find a population of 6 billion by 1995. That is, the population will double in about 35 years.

Carrying the calculation further, we would notice that this doubling occurs *every* 35 years. This would be true for any population number we start with, as long as the rate of increase is 2% per year. If the rate of increase were 3%, the doubling would occur in 23.5 years.*

Are these numbers in our table really accurate? From the entry in Table 4–2 for the population at the beginning of 1966, the model predicts exactly 3,378,487,258 people, a precise value. We have, however, ignored the fact that the numerical values with which we started were only approximations: The 3 billion initial population was a rounded number, and the 2% was an estimated average growth rate. If the initial population was *exactly* 3,000,-000,000 and the rate of increase was exactly 2.000,000,000%, then we should obtain ten meaningful digits in our answer. But since precision was lacking in our measurement of both the starting population and the rate of increase, the results can have only a limited number of significant figures. We must, therefore, be content to use rounded numbers. The extent of precision when two

*
This is the same growth as compound interest gives. A rough rule is that the time to double is 72 divided by the rate of increase or interest in %. Thus, 5% interest doubles the money in $\frac{72}{5}$ or a little over 14 years.

numbers are multiplied is restricted by the number with the smaller precision. If in this case it is the 2% figure, and we assume that we are certain of its value to three significant figures (2.00%), then the rounded number having acceptable accuracy is not 3.378487258 billion, but 3.38 billion.

If we continue our example with a 2% rate of increase, we find that in the year 2060, just one hundred years from our starting date, the population will be nearly 22 billion. By the year 2160, it will reach an enormous 157 billion! With a doubling of population in 35 years, the growth after two centuries results in a population which is more than fifty times the original population.

Plots of Population Growth

We have already seen that a plot or graph is much easier to understand than a table of numbers. We now construct such a plot. For this, we use the predicted population at the beginning of each decade from 1961 to 2060 (Table 4–3).

Decade	Population (in billions)
1961–1970	3.06
1971–1980	3.72
1981–1990	4.55
1991–2000	5.55
2001–2010	6.77
2011–2020	8.20
2021–2030	10.06
2031–2040	12.2
2041–2050	15.0
2051–2060	18.3

Table 4–3.
Estimated
world population at start of
each decade, 1961–2060.

As a matter of convenience, we have used the population for the first year of the decade, although the actual number continually grows. In the United States, where a census is made every tenth year, the count obtained is often considered to be the legal population until the next census is completed, even though the Census Bureau issues an annual estimate of the current number of our people. These values are plotted as a bar graph in Fig. 4–16. The height of each vertical column is proportional to the population at the start of that decade as given by the table. Not only do the heights of the bars go up in each ten-year period, but the steps become increasingly larger.

The bar graph is one way of plotting the population growth. At the beginning of each decade (the start of 1961, 1971, and so forth), the figure is an accurate estimate or prediction. Then for the next ten years (Fig. 4–16), we show the population as constant

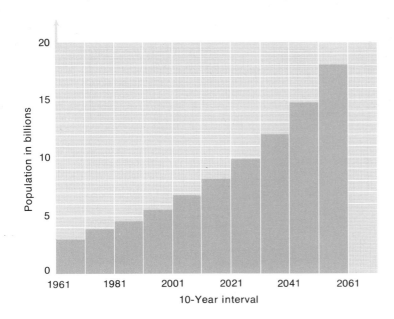

*Fig. 4–16.
Estimated
growth of world
population.*

until the start of the next decade. Actually, we know that the population tends to increase fairly smoothly year by year, month by month, and day by day.

This steady growth in population can be portrayed if we

1. Plot the points corresponding to the start of each decade (the points shown in Fig. 4–17).
2. Draw a smooth curve through these points, as shown in the figure. This smooth curve gives only an approximate model of the real situation, since the population growth does fluctuate from year to year due to famine, epidemics, wars, and so forth. The curve does represent a prediction of world population if we assume:

    a starting figure of 3,060,000,000 in 1961 (or 3.06 billion at the start of 1961) and
    a growth of 2% per year.

Now the extremely fast growth of population is clear. Even though the percentage increase remains constant at 2% per year, the larger increases each year produce a curve which becomes steeper and steeper. This curve is different from those we found in our previous models. The previous plots were linear. The population curve of Fig. 4–17, however, is not a straight line.

Furthermore, this is a particular kind of curve. Each new value of the variable is obtained by adding a constant percentage of the previous value to that value. We have a growth that is proportional to the accumulated size: "the bigger it gets, the faster it grows."

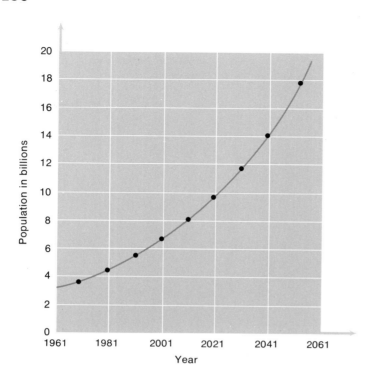

Fig. 4–17.
Fitting a smooth
average line to the population
growth graph.

A snowballing relationship of this type is called *exponential*. The curve of Fig. 4–17 is known as an exponential curve. This very important curve represents a model which is encountered frequently, and we find other examples later in this chapter.

Population Plots Over Longer Time Scales

We now consider another plot of population increase, from 1700 to 2165. This is shown in Fig. 4–18. The eye tends to follow the curve upward to the right; but it is also important to note that the graph drops as we look to the left, or as we go backward in time. In fact, prior to 1800 the height of the curve on the scale of this graph is so small that it is difficult to measure its value. This reflects the fact that a "population explosion" has occurred: The population of the earth in the past was extremely small compared to the present population. The curve makes more reasonable the earlier statement that approximately 20% of all the people who have ever lived are alive today. From a larger scale copy of the curve we could find the even more striking fact that the population *increase* from 1940 to 1963 (just 23 years) was greater than the total population of the world in 1800!

Since we have extended our look backward in time, it is appropriate also to look further into the future. We might attempt to look ahead to the year 2700, a period only slightly more than 700 years

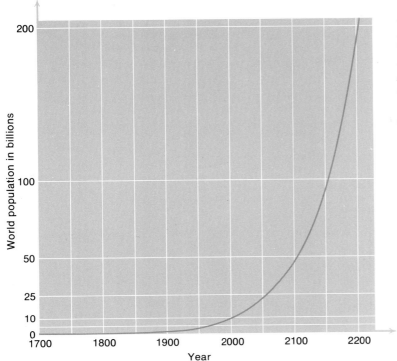

Fig. 4–18.
The growth of world population from 1700 to 2165. Prior to the present time the curve is based on historical fact. Later values are predicted from a model.

from now. This represents about the same time difference as that between the present and the time of Marco Polo. The graphical results of the computations are shown in Fig. 4–19.

Can this really be expected? The curve shoots up at a fantastic rate. The vertical scale on the left is much larger than in the preceding figure—so much so that the steeply rising curve to the year 2165 (Fig. 4–18) cannot even be seen. Our new exponential curve has reached such proportions by the year 2700 that, if we tried to plot it on the scale of Fig. 4–18, we would need a sheet of paper 27 thousand times as high, or 11 thousand feet (more than two miles) high instead of five inches.

What does this curve of Fig. 4–19 tell us? By the year 2510 we should expect to have a world population of nearly 200,000 billion people, by 2635 about 1,800,000 billion people. Thirty-five years after that it will have doubled to approximately 3,600,000 billion. In the year 2692, the model predicts a 5,450,000 billion population.

How large is 5,450,000 billion? We can express it in many ways. The number when written completely would appear as:

$$5,450,000,000,000,000,000$$

It may be written as 5.45×10^{15}, but this form does not give one a good "feeling" for the enormous size.

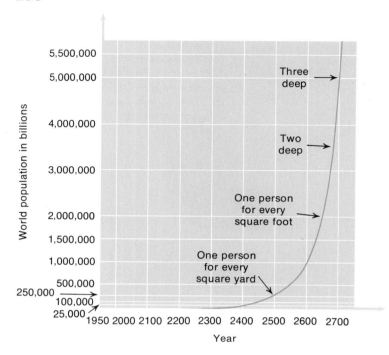

Fig. 4–19.
Modeling prediction
of world population to the
year 2700.

Here is one picture that helps to visualize the magnitude of the number. There are roughly 31 million seconds in each year. If we counted one thousand persons per *second*, it would take us 176,000 years to complete the census.

There is yet another way to grasp the significance of this estimate of 5,450,000 billion population. Let us ask where these people will be: How much room will they have? The surface of the earth contains approximately 1,860,000 billion square feet. About 80% of this area is covered by water, but let us suppose that all of the surface were land. We can calculate that in the year 2510 when the population is 200,000 billion, there will be 9.3 square feet per person, or about one person per square yard *all over the earth*. Worse yet, in 2635 each person will only have one square foot on which to stand, and in 2670 if they insist on retaining that much real estate, they will be standing on each other's shoulders *two deep*. And only 22 years later they will be *three deep*. Now, if we do not assume that these people can tread water but instead must occupy the land area only ($\frac{1}{5}$ of the total area), in 2692 we should expect to see totem poles 15 persons high on every square foot!

6 | EXPONENTIAL GROWTH

The model of the last section obviously is not useful for predicting population too far into the future. The model is based on a 2% increase every year. Once the land becomes too crowded, this growth rate certainly will fall. The population will level off. In the next section, we will consider how to change our model to include this fact.

The model is useful, however, to predict what will happen in the next thirty years. If the present growth rate is 2% per year, we can guess that this will not change very much in the near future. Regardless of the international efforts to control births, any real change requires educating masses of people. Even if reasonable laws were passed to limit the number of children a family could have, years would pass before the laws could be enforced effectively.

Furthermore, there is a time lag involved in any program of population control. This is a factor in the model which is often overlooked when people write or speak about the population problem. To reduce the number of babies born each year, we need to reduce the number of women of child-bearing age. But the number of women 20 to 40 years old depends on the number of babies born more than twenty years ago.

Consequently, even if the number of babies born this year were reduced, it would be more than twenty years before the effects were really noticeable. In other words, the population growth rate is fairly well determined already for the period from now until the year 2000. The real effects of any population-control program would not show up until about the year 2000.

Such a time lag is found in problems where we are trying to control the environment. For example, even if all new cars built from today on were made to have no lead compounds in the exhaust gases, it would be several years before there was a noticeable change in the lead in the air simply because of all the cars already on the road. Ten years would pass before almost all of the cars on the road were "clean."

Because of this time lag, the model based on 2% growth per year gives a reasonable prediction of population for the next thirty years. (We assume there will be no world war, great epidemic, or mass starvation.) In this section, we want to look in more detail at models which have such a constant growth rate.

A Plot with Constant Growth Rate

A signal (or a quantity such as population) with a constant growth rate is called an *exponential*. This simply means that the *percentage change* in each year (or hour or second) is the same. Each year the population increases by 2%.

When we try to plot population, we run into the problem shown in Fig. 4–20. A plot which shows what is beginning around the year 2000 is useless before 1800 (where the curve is almost zero on our vertical scale). Also, the curve starts to rise so rapidly after 2100 that it quickly goes off the paper. The trouble is clear. Our vertical scale only allows us to plot values from perhaps 2 billion

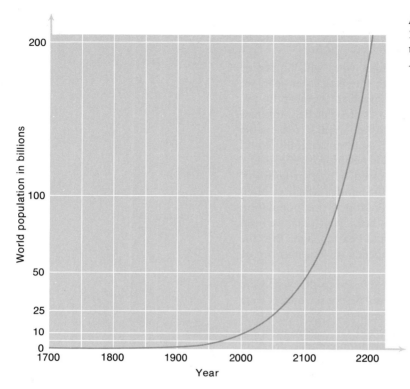

Fig. 4–20.
The growth of
world population from
1700 to 2165.

to 200 billion. Over the years of interest, the population varies over a much wider range than this.

When we have a constant growth rate, we can simplify the plotting if we use a different vertical scale. The population is 3 billion in 1960 and we are interested in the years from 1760 to 2160. How do we plot?

We first take a sheet of paper and mark the years off in the regular way. 1960 is in the middle, and every 50 years are shown from 1760 to 2160. We know the population in 1960 is 3 billion, so the middle of the vertical axis is set at 3.

In the last section we saw that the population doubled every 35 years (with a 2% increase per year). Hence, we have the points

Year	Population
1995	6
2030	12
2065	24
2100	48
2135	96

Along the vertical scale, we show a doubling every equal space (Fig. 4–22). In other words, the distance from 3 to 6 is the same as from 6 to 12. Once the vertical axis is labeled in this "odd" way, we can mark the above points.

The usefulness of this plot is now obvious. *The points are on a straight line.* The population doubles every 35 years. If the vertical axis is marked to show a doubling corresponding to a constant distance up, the curve moves up this same amount every time it moves across by 35 years. This is just the property of a straight line. If we draw the rest of the curve back to 1760, we obtain the complete graph of Fig. 4–23.

Once the horizontal and vertical scales are labeled, we need only two points to determine the straight line. For example, the 1960 population is 3, and the 1995 is 6. We can just draw a straight line through these two points.

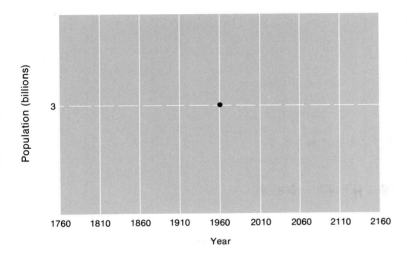

Fig. 4–21.
Years marked off for the population plot.

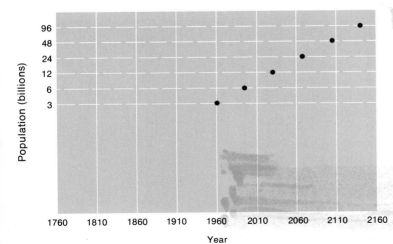

Fig. 4–22.
Points into the future shown with an unusual vertical scale.

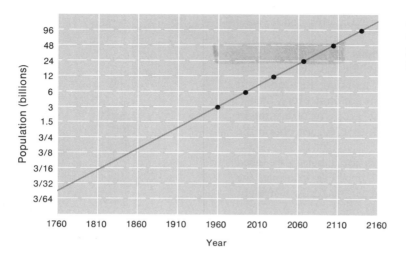

Fig. 4–23.
Plot of population
growth from 1760 to 2160 with
increase of 2% each year.

An Economic Model

The plot of Fig. 4–23 is useful whenever there is a constant rate of growth. The economic plots of the United States gross national product, for instance, show an average increase of 5% each year. If the total is 1 trillion dollars in 1970, the future can be predicted from Fig. 4–24. We construct this plot as follows.

The years of interest are 1960 to 1990, so we label the horizontal axis. The 5% rise each year means we double every $\frac{72}{5}$ or about every 14 years. We now have two points (1 trillion in 1970, 2 trillion in 1984). The straight line can then be drawn.

The graph shows that in 1990 the gross national product will be almost 3 trillion dollars. Since the United States population in 1990 is expected to be about 300 million, the average per person income should be 3 trillion/300 million or $10,000. (In 1970 it was about $5,000.)

One word of caution should be given here. To make Fig. 4–24, we have used an odd vertical scale (Fig. 4–25). The distance from 1 to 2 is the same as from 2 to 4. Each time we move this distance, the quantity being measured doubles. When we asked in Fig. 4–24 what the gross national product would be in 1990, we had to estimate a value between 2 and 4 (point A in Fig. 4–24).

To make this estimate, we must recognize in Fig. 4–25 that if a is halfway from 1 to 2, a does not correspond to 1.5 trillion. If it did, going from 1 to a would mean multiplying by 1.5. Going from a to 2 would mean multiplying by $\frac{2}{1.5}$ or 1.33. But we said equal distances on the vertical scale correspond to equal multiplying factors. If this is to be true, the point a corresponds to $\sqrt{2}$ trillions, or 1.4 trillions; b midway between 2 and 4 is $2\sqrt{2}$ trillions or 2.8 (not 3).

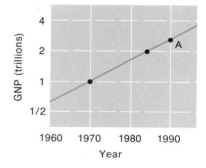

Fig. 4–24. Expected gross national product of the United States.

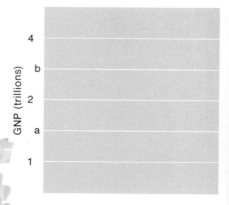

Fig. 4–25. Vertical scale for an exponential plot (a is 1.4, b is 2.8).

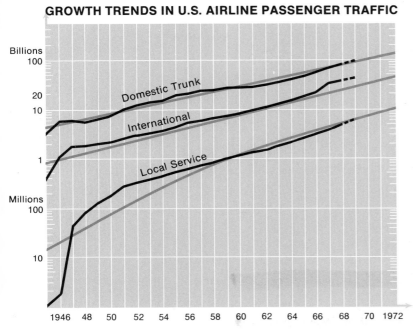

GROWTH TRENDS IN U.S. AIRLINE PASSENGER TRAFFIC

Fig. 4–26.
Another example of
a plot of exponential
growth. The way in which
the actual data follow the
straight lines promises that the
changes in the next few years can
be predicted with reasonable confidence.

These particular curves are used by
airport planners, airlines, travel agencies,
government agencies, and companies
building equipment for the air-travel
industry.

The straight line for international
travel shows that the travel doubles
every 5 years (1962 to 1967).
Thus, the annual increase is
72/5 or 14%. (Aviation
Week and Space
Technology)

The Solid-Waste Problem

If the United States is considered as one vast system (containing as elements the manufacturing plants, the transportation vehicles, the people, and the natural and man-made devices), one of the important signals which can be measured is the solid waste. Such solid-waste material includes all the material which we throw away: the 6 million cars which are scrapped every year, the appliances discarded, the refuse from construction and demolition of buildings, and the garbage. Our highly advanced technology and the associated high standard of living lead to a national problem of increasingly serious magnitude. How can we dispose of this solid waste economically and without dangerously fouling the environment? If we take a longer range viewpoint, how can we recycle this trash? In other words, how can we conserve our natural resources by using again the materials we are throwing away?

The magnitude of the problem is vividly portrayed in Fig. 4–27,* which shows the quantity of solid waste produced per year in the United States since 1920. The significance of this particular signal is perhaps clearer if we note that in 1965 nearly five pounds were produced each day for each person in the country. Furthermore, the rate of increase is appreciably greater than the rate of population increase. (Indeed, the United States is by far the most efficient producer of rubbish in the history of civilization. With less than 10% of the world's population, we generate well over half of its rubbish.)

*Figures 4–27 and 4–28 are taken from the report, "A Strategy for a Livable Environment," published for the U.S. Department of Health, Education, and Welfare, June, 1967.

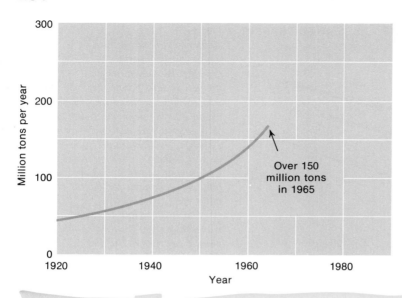

Fig. 4–27.
United States
solid waste.

The importance and urgency of the problem derive from two principal factors:

1. In most cities, available land for dumps is being rapidly exhausted. At the same time, the nature of the solid waste is changing. A few decades ago, the rubbish was primarily garbage and ashes. Today it includes vast quantities of metals, plastics (e.g., non-returnable containers), and other new products, many of which cannot be economically burned without contributing to air pollution.

2. We know very little about the effects of environmental pollution on our physical and mental health. To what extent are polluted air or mounds of junked automobiles responsible for the increases observed in mental illness, in urban unrest, and in such physical illnesses as lung cancer? Even data such as shown in Fig. 4–28 are not easily interpreted. Individuals living in the city may smoke more heavily, may lead lives under greater nervous tension, and so forth. The problem of evaluating the importance of environmental pollution is further complicated by the realization that major effects on the balances of nature and the characteristics of man are unlikely to become evident for a generation or more, when it may well be too late to reverse the established trends. (The problem of the time lag enters again.)

Whether we are worried about a single city or a whole region, the intelligent planning of new methods for disposing of waste requires that we predict how much waste will be produced in the future. For example, if we decide to build a new incinerator to

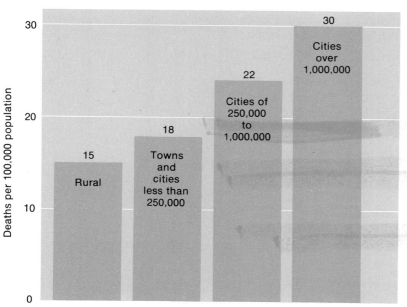

Fig. 4–28.
Annual deaths from lung
cancer per 100,000 population
as a function of the size of the
community in which the people live.

Fig. 4–29.
In a large city-apartment
complex, trash disposal is a
major problem. Incinerators
pollute the air. To remove pollution
economically, the trash must be burned
in an expensive central incinerator
(equipped with stack emission controls)
or collected and hauled away.

The picture shows a pneumatic
tube system for collecting trash.
Tenants place their trash in a
chute. It is drawn to a central
point by an air stream moving
at 60 miles an hour. The
system is designed to
put an end to odors,
littered basements,
and cans of exposed
garbage.
(Aerojet-
General
Corporation)

burn the rubbish, six years may be required until the incinerator is operating. Thus, if a decision is made in 1970, the incinerator which is built should be designed for the quantity and type of rubbish which will appear after 1976.

In this discussion, we consider only the quantity of solid waste produced. Figure 4–27 shows the past history of the system over the years 1920 to 1965. We wish to use the data to predict the signal at least a few years in advance. This need for prediction arises for two reasons:

1. Data are usually available only some time after they are valid (in problems of this broad a nature, a year or two may be required). Thus, the curve of Fig. 4–27 runs only to 1965, even though it was published nearly two years later.
2. Design and construction of the facility require several years, so that the system is truly being built for the future.

Before we try to use the data of Fig. 4–27 to predict solid waste in the future, we notice that the quantity is increasing at a constant rate. Every 12 years, the amount increases by 50%. This is exactly the property of an exponential. Consequently, we can change the plot of Fig. 4–27 to the exponential form.

In 1950, there are 100 million tons generated. In 1962, the figure is 150 million tons. We label the time axis from 1920 to 2000. The vertical axis includes 100, and then successive lines 1.5 times as much (Fig. 4–30).

Once the straight line is drawn, we can predict the quantity probably generated at any future date. For example, by the year 2000 we can expect almost 500 million tons per year. With 300 million people, this is about 1.6 tons or 3200 pounds per person per year. (Almost 10 pounds a day!)

One proposal has been to build an artificial mountain of rubbish from New Jersey to California. When it is completed, it could be covered with grass and landscaped so it would provide a recreation area for future generations. The proposal does not sug-

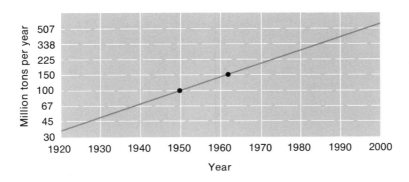

Fig. 4–30.
Solid waste generated
with vertical scale chosen to
show 50% every 12 years.

*Fig. 4–31.
One of today's
pressing problems is
abandoned cars. The number
of abandoned cars grows
exponentially. Attempts
to deal with the
problem have been
essentially
unsuccessful.
(Photo by Richard O'Leary,
St. Louis, Mo.)*

gest how to compensate the people who happen to live nearby and have to watch it grow.

More attractive proposals are to raise the cost of goods so that the manufacturer has to reclaim them. Automobiles are a prime target. In 1969, more than 50,000 cars were simply abandoned on the streets of New York City. Any person driving through the United States soon encounters automobile junk yards steadily growing in size and ugliness.

The Exponential As a Planning Device

In this section, we have seen that exponentials can be plotted as straight lines if the vertical scale is marked as follows. Each unit distance up corresponds to multiplying the quantity by the same factor.

Because exponentials occur so often, this form of plot is particularly important. In using such plots for prediction of the future, it is important to recognize that we usually do not need great accuracy. Whether the solid waste generated in the United States in the year 2000 is 500 million tons or 450 million tons is not very important. Regardless of what the number will actually be, we need to start planning immediately if we are not to be swamped by rubbish. Either we have to stop its growth or we must find better ways to recycle it or dispose of it.

168

> **Box 4–1**
>
> EQUATIONS FOR EXPONENTIALS
>
> For those who like to live by mathematics, we can write an equation for an exponential growth curve. The solid-waste generation data are used as an example.
>
> The general equation for an exponential is
>
> $$y = A\, r^{t/T}$$
>
> Here y is the quantity we are measuring (the solid waste generated per year in millions of tons). r is the factor by which y is multiplied in T years. In our case, r is 1.5 and T is 12 years. t is the time (in years) measured from any reference point. If we select the 100 value in 1950 as the reference, t is measured from 1950. (Thus, a t of 7 represents 1957, a t of -21 represents 1929.) Finally, A is the quantity at the time t is zero. (In our case, A is 100 since that is the solid waste in 1950.)
>
> Thus, the equation for United States solid-waste generation is
>
> $$y = 100\,(1.5)^{t/12}$$
>
> where t is measured from 1950. In 1998, t is 48 and
>
> $$y = 100\,(1.5)^4 = 506$$
>
> The equation predicts 506 million tons generated in 1998.
>
> The equation gives exactly the same information as our straight-line graph.

This type of prediction is called *extrapolation*. We extrapolate (or extend) what has happened in the past into the future. We assume that the growth rate will not change. We should recognize that such extrapolation is not necessarily valid. New changes may occur in our way of living which may increase or decrease the rate of population growth or solid-waste generation. (See Box 4–1.)

7 | AN IMPROVED POPULATION MODEL

Obviously the population model with 2% growth is incorrect, at least after some time in the future. Quite clearly there are some limiting factors which prevent such a population increase. Actually our model is too simple because we did not take into account several important factors that tend to limit our predictions.

To learn more about these factors, it is helpful to examine functional models of the world population. Such models are easy to find in a biology laboratory. Any small organism that reproduces rapidly will do. Fruit flies, yeast, and bacteria are commonly used examples. Here we describe a population model using yeast.

First the experimenter must prepare a food supply, a "nutrient medium." For many yeast species, this may be simply a weak sugar syrup slightly modified by addition of other substances. Then there is need for a jar in which to keep the yeast as they gorge themselves, and for Adam and Eve, so to speak: the syrup must be inoculated with a few yeast cells to start with. The temperature should be kept constant, and the medium should be gently but constantly stirred.

It is hardly possible to take a census of yeast cells as one does of people. Instead, a sampling technique is used. Knowing the starting volume of his experiment, the investigator can withdraw a definite, very small percentage of it and count the yeast cells in that. Since the solution of food has been stirred, he can safely assume that his sample is typical, and that he can simply multiply by the proper factor to learn the total population. Such models as this one are particularly convenient because they take up very little space; moreover, it is easy to try different circumstances ("to vary the parameters" as the professional puts it). It becomes possible to answer such questions as these: What is the rate of increase of population when the experiment begins? Does this rate remain constant as the population becomes larger? Does it matter whether the available space for the organisms remains constant or is made to increase as the population grows? Yeast cells produce an alcohol (there are many kinds of alcohol) from the sugar they consume; what is the effect of leaving the alcohol to accumulate in the nutrient medium? Of removing all but a constant fraction? Of removing all of it as it forms? (Removal can be rather easily accomplished by continually pumping fresh nutrient medium in and at the same time allowing the used medium to trickle out through a filter.)

Which of these possible experiments cast light on our graphical model of world population? First, we know that the entire land surface of the earth is not inhabited but that it is not unlimited. (There is room for population to increase but the space will be used up someday.) This is modeled in the yeast case by using bigger jars (and more medium) up to a certain point, but then no more. Second, we know that food production can be increased for human beings but not without limit. We can supply more sugar to the yeast on a schedule that we think is comparable to the future history of the world; even better, we can try many schedules. In short, we can test our model and thus refine it, by comparing it with what we already know about the course of development of the human population.

Now it turns out that such experiments as those described are practically always alike in one feature. Growth is roughly exponential at first; the *rate* of increase is not necessarily constant, but the population curve is closely similar to that of Fig. 4–32. If the

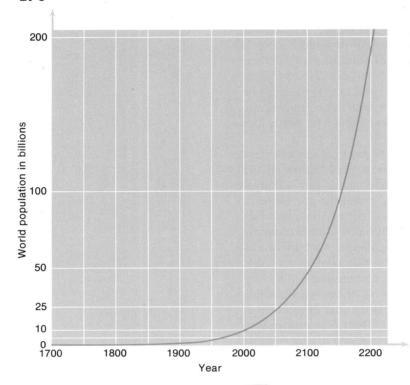

Fig. 4–32.
Exponential
growth curve.

experiment lasts long enough, however, the rate of growth sooner or later begins to decrease and in time reaches zero. The curve stops its exponential growth and tends to level out, as suggested in Fig. 4–33. Because this curve has a kind of S-shape, it is known as a *sigmoid* (from the Greek word *sigma* for the letter S).

The basic reason for the change of shape shown is overcrowding. Without unlimited space in which to grow and unlimited food to support life, the individual yeast cell has neither room nor food to allow it to reach normal size. No doubt there are other reasons, but they are less important.

In the case of small animals and insects tested in a laboratory or carefully controlled environment, the same leveling off of population has been observed. The rate of growth tends to decrease as the animals become badly overcrowded because of both physical and mental deterioration resulting from inadequate supplies of food, air, and water, excessive nervous tension associated with inability to move freely, and so forth. In actual experiments with rats, once severe overcrowding occurs, mothers reject their young, adults kill females, and reproduction falls rapidly. Whether such experiments have any meaning for human beings is not at all clear.

Certainly nature ultimately limits population growth, although typically with a serious deterioration of the species as suggested in the preceding paragraph. One would assume that man will limit his population growth significantly before serious damage has been done to his general physical and mental condition. Indeed, the focus of the recent worldwide emphasis on population control is to limit the population to levels at which adequate food resources and land are available to ensure happy and healthy individuals.

Perhaps the first man to recognize the sigmoid character of population growth (although he did not express it in this way) was Thomas Robert Malthus (1766–1834). He was an Englishman who wrote a gloomy essay pointing out that a time must come when population will outrun food. Then the growth of population would be stopped, he believed, by widespread epidemics, or starvation, or war, or some combination of these. Instead of a truly sigmoid curve, however, his curve would probably actually turn downward. He predicted that the end of the growth period would come during the nineteenth century. That it did not was the result of the discovery of chemical fertilizers. With these an acre of ground brings forth several times as much food as was possible in Malthus' day. We can see, however, that some limit, at some time, must be reached in the number of pounds of food that can be won from an acre of ground or of sea. The supplies of potash and phosphate easily recovered must someday disappear, so the cost of fertilizers must rise. The third major fertilizer element, nitrogen, is available without foreseeable limit from the air. The current research in development of food from the sea promises to provide the possibility of feeding adequately a great number of people, but not an unlimited number.

A Model with Decreasing Growth

Even if we agree that the population model of the preceding section (with 2% annual growth rate) is impossible, we can still use this exponential growth curve to predict moderately into the future. The sigmoid characteristic of Fig. 4–33 does behave as an exponential in the early portion. Recent population data indicate that we are indeed still in this part of the curve.

If we desire to peer further into the future, we need a model which shows the eventual slowing of growth. To obtain such a model, we need to assume that the growth rate will eventually decrease as the population increases.

In the population model, we worked with a growth rate of 2% a year. The population in any year is 1.02 times the population a year earlier. In a stable situation, with the population constant, the value in any year is just 1.00 times the value the preceding year.

Thus, to obtain a sigmoid curve, we need to reduce the 1.02 to 1.00 as the population increases. Exactly how do we put in this change? When does the factor change from 1.02 to 1.01, for example? Then when from 1.01 to 1.00? Where will the population start to level off?

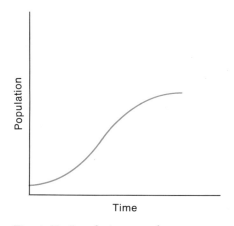

Fig. 4–33. Population growth in laboratory studies.

Box 4–2

MATHEMATICS OF A SIGMOID MODEL

The man who relishes mathematical language can derive a model for the sigmoid characteristic as follows.

Exponential growth In the exponential case, the population in any year is a constant (1.02 in our case) times the value a year earlier. If P_{n+1} is the population in the year $(n+1)$, P_n in the year (n), then

$$P_{n+1} = r\, P_n$$

Thus, if P_{1960} is 3 billion and r is 1.02,

$$P_{1961} = 1.02 \times 3 = 3.06 \text{ billion}$$

Sigmoid behavior To obtain the leveling off, the factor r must decrease as population increases. This change occurs if we choose

$$r = r_0 - c\, P_n$$

r_0 is 1.02. This value is reduced by $c\, P_n$ where c is a constant. As the population P_n increases, the r decreases. Then

$$P_{n+1} = (r_0 - c\, P_n)P_n$$

Example If r_0 is 1.029 and c is 0.003, and we measure population in billions, we obtain

$$r_0 - c\, P_n = 1.029 - 0.003 \times 3 = 1.02$$

when the population is 3 billion. That is, in 1960 the growth rate is 2%.

As P_n increases, the factor $(r_0 - c\, P_n)$ decreases steadily. This factor becomes one when

$$1.029 - 0.003 \times P_n = 1$$

or

$$P_n = 9.7 \text{ billion}$$

According to this model the population levels off at 9.7 billion. This value depends critically on what we choose for the constant c, which represents the rate at which the growth decreases. Since we do not know what factors will limit population, we really can only guess at c.

Obviously, we can only guess. Today we are still in the region of exponential increase. This growth will eventually be slowed, but we do not know what factors will cause the leveling off.

The particular population model of this section is not especially important; indeed, major political or social changes may make our predictions of future world population look ridiculous to the historians of the year 2000. Our purpose in this section, however, is to introduce the idea of model refinement. We discovered that our 2% model is ridiculous if we try to predict hundreds of years into the future. On the basis of our rather superficial understanding of population growth, we then found an alternative model which at best has the property of leading to a limited ultimate population. (See Box 4–2.) Since we know we cannot predict population accurately too far into the future, our simple models are probably as good as much more complicated ones.

8 | USES OF MODELS

Predictive population models are often used with great success in governmental planning at all levels—town, state, and federal. For example, the design of a transportation system for a region requires that we have reasonably reliable predictions of population distribution in order to assess future transportation needs (for transporting people and the materials which people require).

In such a problem, the complete system model includes population models for hundreds or thousands of separate towns. The complete model is often a mathematical model composed of many equations. Some of these are similar to the equations we have used and some are more complicated. Not only must birth and death rates (net growth) be considered, but the relationships among other factors must be included. There are also influences which make the populations of towns interdependent. If one town becomes unduly crowded, there is a strong tendency for neighboring towns to grow more rapidly. Immigration and emigration rates thus are important considerations for the development of an accurate dynamic model: a model in which the interrelationship of factors changes with time.

Many Models for One System

It is possible for one system to be represented by a number of different models. As in the case of the blind men and the elephant, no one model describes the real thing completely, but separate models of sub-systems are often necessary and useful.

An air-conditioner provides an example of this characteristic. One model can be developed which is based on heat flow: how heat is extracted from a room, how the fluid in the unit changes its temperature as it absorbs heat, and how this heat is then transferred outside of the room. This model must include such factors as expected temperature ranges, characteristics of the refrigeration unit and blower, and intake and outlet duct air flows.

Another model to describe the same system might be a control model which includes the thermostat, the various relays and contacts, and the electrical network which links the electrical parts of the system.

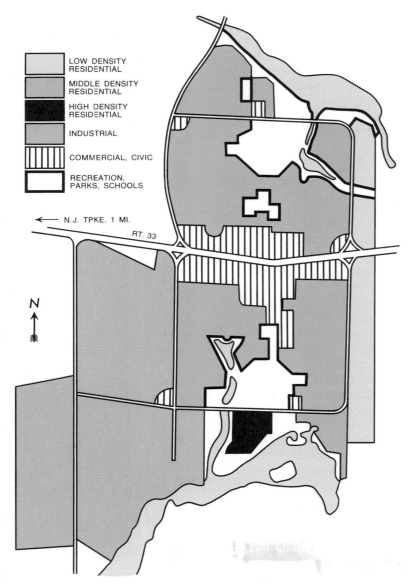

LOW DENSITY
RESIDENTIAL

MIDDLE DENSITY
RESIDENTIAL

HIGH DENSITY
RESIDENTIAL

INDUSTRIAL

COMMERCIAL, CIVIC

RECREATION,
PARKS, SCHOOLS

← N.J. TPKE. 1 MI.

RT. 33

N

*Fig. 4–34.
An initial model for
the planning of a new
town—in this case Twin
Rivers, New Jersey, with a
planned population of 10,000.*

*This first model shows the
general planning to include the land
uses and the highways and lakes. (The
two irregular areas are lakes.) Once this
overall model is adopted, each section
can be planned in detail. When the
town is designed in this way, schools,
parks, and shopping areas can be
located within walking distances
of most residential areas. Streets
can be planned to avoid too many
pedestrian crossings. The goal is
to create a town which avoids
many of the annoyances and
difficulties of existing towns
which have grown without
overall planning. (© 1970
by The New York Times
Company.
Reprinted by
Permission)*

Yet another model of an air-conditioner could be developed
for a study of its mechanical behavior. For example, we may wish
to know how much noise and vibration the equipment will produce
and how to design the air-conditioner to minimize the noise and
vibration. For this objective the model would include a number of
factors such as the characteristics of the moving parts, their
mountings, the location of shock absorbers, and the geometrical
arrangements of the openings, the absorbent surfaces, and the
baffles.

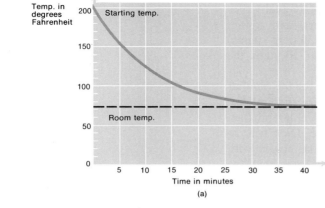

Fig. 4–35.
Three
examples of exponential
systems.

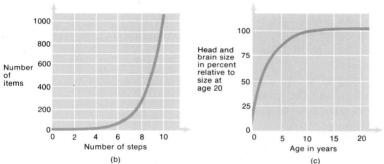

One Model for Many Systems

What is there in common among the way in which a cup of coffee cools, the way in which the numbers of chain letters increase, and the way in which a human head grows? Just as it is possible for one system to be described by several different models, so one model frequently is applicable to many kinds of systems. In Fig. 4–35 there are models of three different processes which one would not ordinarily think of as being similar. In (a) is shown how the temperature of a cup of coffee drops as the coffee cools to room temperature. The initial temperature is just below the boiling point. It drops rapidly at the start, then more and more slowly. After a half hour, the temperature of the coffee has dropped to within a few degrees of room temperature.

The illustration in (b) describes a system in which the production of an item doubles at each step. During the early part of the curve the values are not readily observed because of the scale of the graph axis, but as the number of steps goes from 1 to 2 to 3 to 4, the number of items increases from 2 to 4 to 8 to 16. The increases become larger with each step; for instance in going from 9 to 10 steps the number of items doubles from 512 to 1024. This is a model for the chain-letter process, where an individual writes to two people, each of these writes to two others, and so on. After 20 steps in such a process the number of letters (items) being written is more than one million, and after 30 steps the number becomes greater than one billion.

This rapid growth of an exponential is the basis for the familiar story of the golf game between Mike and Dick. On the first tee, Mike suggested they play for a penny on the first hole and then double the bet each hole. By the eighteenth hole, Dick was so nervous he missed his drive completely, while Mike played with total relaxation. It was only after the match was over that the mathematically slow Mike realized that the bet on the last hole was more than $1300.

Figure 4–35c shows how the size of a human head grows from birth to age twenty. At birth it is a little less than $\frac{1}{4}$ its full size, and it is growing very rapidly. At the age of five the growth begins to slow down appreciably, and at the age of fifteen the head is within a few percent of its ultimate size.

We can see what is common to the cooling of a cup of coffee, the rate of increase in a chain-letter situation, and the growth of the human head. Each displays an exponential rate of change. In each of the processes, as in the population expansion, growth either increases or decreases exponentially.

Our exponential curve thus fits many systems. Such things as the rate at which an automobile coasts to a stop, the growth of plants, and the accumulation of bank interest can also be represented approximately by an exponential curve. All these examples show that one model may represent many different systems.

Another remarkable aspect of modeling is how often we can find mathematical relationships among important quantities. As a final example, violent storms are of four general types: tornadoes, thunderstorms, hurricanes, and cyclones. There is a relationship between the size of the storm and its duration: the larger the storm, the longer it lasts. Indeed, if we call D the diameter of the storm in miles and T the duration in hours, the model is given by the equation:

$$D^3 = 216\ T^2$$

Once we have found such a model from observation of many storms, we can answer such questions as:

How long does a cyclone with a diameter of 600 miles last?
How large is a thunderstorm which lasts one hour?
If a hurricane has a diameter of 100 miles, what would be the anticipated duration in days?
A tornado lasts 11.5 minutes; what is the expected size?

The answers are 1000 hours, 6 miles, 2.8 days, and 2 miles.

Such a model is only approximate. Any given storm may deviate from our equation just as any particular seventeen-year-old boy may be 5 feet 2 inches tall and weigh 300 pounds. In spite of such occasional deviations, however, the model yields a picture of typical average relationships among physical quantities.

Determination of Models

Once an idea of the structure and nature of a thing is conceived, it may be expressed in many different ways. We may have different models. Some, as we have seen, are verbal models. A map is a model, a graphical model. Other models are mathematical in which quantitative expressions are used to describe relationships in a precise way. Some models are developed using computers.

An aircraft represents so complicated an aerodynamic problem that a complete mathematical description may be impossible. Therefore, it is usually modeled by constructing a small-scale version of metal or wood for testing in a wind tunnel.

Models are used, not only to describe a set of ideas, but also to evaluate and to predict the behavior of systems before they are built. This procedure can save enormous amounts of time and money. It can avoid expensive failures and permit the best design to be found without the need for construction of many versions of the real thing. Models evolve, and it is customary to go through a process of making successive refinements to find a more suitable model.

For example, in the development of a model for a nerve cell, there is need for successive refinement. A preliminary model is designed, it is tested against the real nerve cell, then the model is modified so that it becomes more realistic in its behavior. In the process of model construction it is essential to alternate back and forth between the real world and the model.

The essential parts of the model-making process are shown in Fig. 4–36. Measurements or observations of the real world are used to develop a model. After a preliminary model is made, measurements made using the model are compared to the behavior

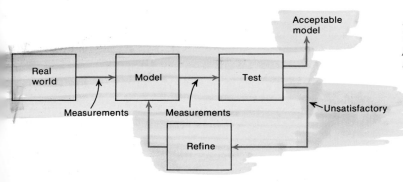

Fig. 4–36.
The model-making process, shown as a block diagram.

of the real world. In most cases these tests show that the model is not completely satisfactory, so that it must be refined. This process is repeated until the model is acceptable.

In the modeling of a nerve cell, or the modeling of the growth of a population of people, the real-world measurements are made on a system which already exists. In this case our model-making process is intended to produce a model which accurately matches the real world. In the case where scale-model airplanes or spacecraft are modeled, the real-world object may not yet exist, and the box marked "Real World" in Fig. 4–36 theoretically contains the real object which we imagine and wish to achieve, as well as all pertinent facts about the real world (such as the properties of air, characteristics of flight systems which have already been built, and the characteristics of various materials and fuels). The model-building procedures are no different from those already discussed.

Models can be *descriptive*, as in verbal, graphical, or mathematical representations. They can also be *functional* (they "really work"), as in scaled-down airplanes for use in wind tunnels or working replicas of nerve cells. The model includes only those parts of the system which are important for our purposes. The model guides our thinking and suggests how we can improve the system.

Questions for Study and Discussion

1. "No model is ever complete." Would it be helpful if one could in fact construct a complete model? Explain briefly.

2. Discuss the differences between

   a) functional and descriptive models.

   b) dynamic and static models.

   Give an example of each.

3. Suggest two reasons why a mathematical model may be desirable.

4. When a model is first designed, what is the next step which should be taken with it?

5. Figure 4–3 shows the height-weight data for twenty-year-old men as a somewhat scattered cloud of points. Explain why it is reasonable and useful to draw a *particular* straight line through these points.

6. The greatest height shown by the graph line of Fig. 4–4 is about 6′5″. Would you be justified in using the graph line to predict the weight of a candidate for center on the basketball team if he is 6′10″ tall? Why, or why not?

7. From the equation for the relationship between weight and height of twenty-year-old men, find the expected height of a young man who weighs 145 lbs.

8. A forester might use a method similar to that of Section 2 to study the relationship between the height of a tree and the diameter of its trunk. It was pointed out that the height-weight data had been "passed through two strainers (age and sex)." What "strainers" would the forester have to use?

9. How do highway engineers take a traffic count?

10. To use an electric eye as a sensor in a corridor, a beam of light is arranged to shine from one wall (where the light source is) to the other wall (where the electric eye is). Whenever the light beam is interrupted by a person passing, the circuit of the

electric eye operates a counter. Why would such a device give an incorrect count in the situation shown in Fig. 4–8? Can you think of a way to make it count correctly?

11. The text discussion on the uncontrolled harvesting of buffalo is but one example of how man has willfully brought a species of animal to the point of extinction. Describe five other examples of how man has either willfully or unwittingly produced this effect.

12. The movements of the elk pictured in Fig. 4–14 are monitored from a satellite. Describe how this is done and explain what elements of the life habits of an animal can be learned in this way.

13. In the text, several factors which affect the growth rate of a town are listed: birth rate, death rate, nearness of other crowded towns, transportation system. Suggest five other factors which might influence the growth rate of a town. In each case, explain briefly why the factor would be likely to increase or to decrease the growth rate.

14. We have seen that it is possible for one system to have a number of different models which apply to it. Many times an engineer finds it necessary to have models of sub-systems. Suggest three models which could be used to describe a submarine.

15. The poem on the blind men is:

THE BLIND MEN AND THE ELEPHANT

It was six men of Indostan
    To learning much inclined,
Who went to see the Elephant
    (Though all of them were blind),
That each by observation
    Might satisfy his mind.

The First approached the Elephant,
    And happening to fall
Against his broad and sturdy side,
    At once began to bawl:
"God bless me! but the Elephant
    Is very like a wall!"

The Second, feeling of the tusk,
    Cried, "Ho! what have we here
So very round and smooth and sharp?
    To me 'tis mighty clear
This wonder of an Elephant
    Is very like a spear!"

The Third approached the animal,
    And happening to take
The squirming trunk within his hands,
    Thus boldly up and spake:
"I see," quoth he, "the Elephant
    Is very like a snake!"

The Fourth reached out an eager hand,
    And felt about the knee.
"What most this wondrous beast is like
    Is mighty plain," quoth he;
" 'Tis clear enough the Elephant
    Is very like a tree!"

The Fifth who chanced to touch the ear,
    Said: "E'en the blindest man
Can tell what this resembles most;
    Deny the fact who can,
This marvel of an Elephant
    Is very like a fan!"

The Sixth no sooner had begun
    About the beast to grope,
Than, seizing on the swinging tail
    That fell within his scope,
"I see," quoth he, "the Elephant
    Is very like a rope!"

And so these men of Indostan
    Disputed loud and long,
Each in his own opinion
    Exceeding stiff and strong.
Though each was partly in the right
    And all were in the wrong!

John Godfrey Saxe
American Poet 1816–1887

A modern version of the six blind men and the elephant is suggested by the following problem. A printed capital letter of the English alphabet is scanned photoelectrically and the resultant signal is read into a digital computer. Seven subroutines in the digital computer inspect it. The first states that the letter is like a U because it has at least one pocket to hold rain coming from above; the second shows that it is like a K because it has at least one pocket to hold rain from below; the third and fourth find that it is like an A because it has no pockets on right or left; the fifth shows that it is like a V because it has two ends; the sixth shows that it is like an S because it has no junctions; the seventh shows that it is like a D because it has two corners. Combining these models of the letter, determine what it is.

Problems

1. The diaphragm control on a camera is often marked with the following numbers (called stops): 11, 8, 5.6, 4, 2.8. In going from any stop to the one with the next *smaller* number, the amount of light admitted to the film at a given shutter speed doubles.

   a) If the light admitted at stop 2.8 is called "*L*," how much light is admitted at stop 11, shutter speed remaining the same?

   b) If the proper exposure for certain conditions is $\frac{1}{25}$ sec at stop 11, what would it be at each of the other stops?

   c) Do the answers to (b) form a linear or a non-linear relation? (*Linear* means graphed by a straight line.)

2. Let us approximate a human body by a cylinder. Since the proportions of the body stay relatively constant as it grows, a tall cylinder will have a larger diameter than a short one. We assume that the height of the cylinder is always 7 times the diameter. Thus, the cylindrical approximation of a 6-foot man will have a diameter of $\frac{6}{7}$ foot and a volume of $\pi r^2 h$, or about 3.5 cubic feet. The human body is about 60% water and weighs about the same as an equal volume of water would. Water weighs 62.4 pounds per cubic foot, so the 6-foot equivalent will weigh about 216 pounds.

   a) Compute the weights for equivalent cylinders whose heights are 2, 3, 4, and 5 feet. Plot the results, including 6 feet, on a graph showing height versus weight.

   b) What kind of curve is this? How does it compare to that of the straight-line-average fit of Fig. 4–4? Discuss any discrepancies and the validity of the earlier model in light of the new one.

3. A paintbrush has just been used and the owner wishes to clean it. After the brush has been scraped against the side of the paint can, it still contains 4 fluid ounces of paint. The owner dips it into a quart (32 fluid ounces) of clean solvent and stirs well until the diluted paint solution is uniform. After draining, the brush still holds 4 fluid ounces, part of which is paint and part solvent, since the diluted solution is uniform. The process is repeated with a fresh quart of solvent.

   a) How much paint is left in the brush after 5 solvent baths?

   b) Prepare a table and plot a curve of the amount of paint remaining after each rinse. What kind of curve is this? Will the paintbrush ever get completely clean? Why or why not?

4. A man receives his weekly salary of $150 every Friday and in paying his various obligations spends half of the amount he has in his pocket each day.

   a) How much money will he have left on the following Friday?

   b) Sketch a graph of his current funds versus the day of the week.

   (c) If he received $300 every other Friday, would he be in better or worse shape on the next payday, assuming that his spending habits remain the same?

5. You are served a hot cup of coffee at 200°F and a cold container of cream at 40°F, and you do not intend to drink the coffee for 10 minutes. You wish it to be as hot as possible at that time. Assume that the coffee cools as shown in Fig. 4–35a and that the cream container stays at the same temperature.

   a) Determine the temperature of the coffee at $t = 5$ and $t = 10$ minutes.

   b) If a volume V_1 of coffee at temperature T_1 is mixed with a volume V_2 of cream at temperature T_2, assume that the temperature of the mixture is: $(T_1 V_1 + T_2 V_2)/(V_1 + V_2)$. What will the temperature of the mixture be if 1 fluid ounce of cream is added to 6 fluid ounces of coffee at $t = 10$ minutes?

   c) Now assume that the cream is mixed with the coffee at $t = 0$. What is temperature T_0 of the mixture at $t = 0$?

   d) The cooling curve for the mixture is similar to that of Fig. 4–35a except that it begins at the new temperature T_0 as calculated in part c and it always lies $(T_0 - 75)/(200 - 75)$ of the distance to the given curve from the straight line showing room temperature (75°). What will be the temperature of this mixture at $t = 10$ minutes?

   e) Will a hotter cup of coffee result from adding the cream first or later?

6. The half-life of radioactive decay is the time in which the amount of the given radioactive material decreases by a factor of two. Radioactive carbon-14 has a half-life of 5700 years, but let us assume that it is 5000 years in this problem to allow simpler calculations. Carbon-14 is created by the action of cosmic rays on the carbon dioxide in the atmosphere, and the amount remains constant with time. Growing plants, and the animals that eat the plants, absorb carbon-14 during their lives, but the process stops when the plant or animal dies. Radioactive decay then causes the relative amount of carbon-14 to decrease. Measurement of the radioactivity of fossils permits an estimate to be made of the time at which they died.

a) What fraction of carbon-14 will remain in a sample after 50,000 years?

b) Approximately how old is a fossil bone in which the amount of carbon-14 is 1.0% of its initial value?

c) Sketch a curve showing the fraction of carbon-14 left in a sample as a function of time.

7. Experimental data on the growth of a population of yeast cells are given in the accompanying table.

Time (hours)	Number of cells
0	6
2	10
4	48
6	117
8	234
10	342
12	397
14	428
16	438
18	442

a) Plot a graph of the number of cells versus time in hours. What is the population at 9 hours?

b) The shape of the curve is exponential at first as the cells multiply, but it soon levels off as the supply of food becomes limited. The curve is called a *sigmoid*. What would you estimate the population to be at 30 hours?

c) Although your estimate may be an accurate one, based on the tabular model above and its graph, it is probably not correct in the real life of a yeast colony. If the table were continued, it would show that the population decreases somewhat as the environment becomes poisoned. During what time intervals is the rate of growth a maximum? A minimum?

8. The current rate of population increase is 2% per year. If this rate had existed since the time of Christ until now, and the world population had been 400 million at the time, what would the present world population be? The land area of the earth is approximately 58,400 sq. miles. How many people would now occupy each square mile? What are three factors which limit population growth?

9. The elevator supervisor (Section 1) decides that one way to solve his problem may be to limit each elevator to certain floors only—but which? He talks this over at home, where his son (a student using *The Man-Made World*) offers to help. The boy makes a series of measurements which show the following facts:

a) When an elevator starts at one floor and stops at the next, the running time is 8 seconds (whether rising or descending).

b) When an elevator has reached full speed either way, the running time between consecutive floors is 3 seconds.

c) Starting at rest and going to full speed, the running time to the next floor from the starting one is 5 seconds.

d) For the reverse situation, slowing down from full speed to a stop, the running time is 6 seconds over the last story before stopping.

e) The waiting period at each stop, while of random length, averages 10 seconds.

1. If you were the son, show how you would display these facts in a convenient form to help your father reach a decision. In other words, how would you model the elevator system?

2. Make these simplifying assumptions: the average traffic to each floor is the same; there is no "executive" floor; the operation is always in a steady state. In the light of your display, what strategy of operation would *you* recommend? Why?

10. In some countries the present rate of increase of population is about 4% per year. The doubling time at this rate of increase is less than 18 years.

If the population of the world were increasing at this rate from 3 billion in 1960, calculate the expected population for each year until 1970 (keep only 3 significant figures).

11. Suppose you have a cube of wood, L units on a side. Now cut the cube into smaller cubes, each $\frac{1}{2}L$ units on a side. Cut these in turn into cubes each $\frac{1}{4}L$ units on a side, and so on.

a) What is the total surface area of the original cube?

b) What is the total surface area of the 8 cubes which result from the first cut?

c) What is the total surface area of the cubes resulting from the second cut?

d) Is the increase in area linear or non-linear?

12. Popville, Nebraska, had a population of 10,000 people in the year 1966. A study of the population trend for Popville shows that the town has been losing people at a rate of 1% per year. Predict the town's population in 1968, 1970, 1972, and 1974, if you assume no change in this rate.

13. Below is a graphical model of the height-age growth function of a group of girls:

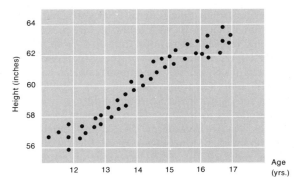

a) Place a straight line in the grid above that most accurately predicts average for all the girls.

b) What is the average growth in inches per year for this group of girls between ages 14 and 16?

c) If Sue is 12 years old and 56″ tall, what statements can be justifiably drawn from this graph concerning her height when she is 17?

Laboratory and Projects

I | IS IT AN ELEPHANT?

Chapter 4 begins with the story of the six blind men who tried to identify an elephant. Since you are not blind, it might be difficult to appreciate their plight. However, in this experiment you will be faced with a similar predicament. You will be compelled to predict without complete information.

Objective of the Experiment

The purpose of this experiment is to try to predict the contents of a container without any access to the interior.

Procedure

Take the container furnished by your teacher, and with all the facilities (touch, weight, sound, motion) at your command (without actually looking inside the container), try to determine the contents of the container and answer these questions:

How many objects are in the container?

What are the shapes of the objects?

What is the material of which they are made?

Additional Comments

After each student in the class has had the opportunity to make his predictions, the teacher will open the container to let the class see the contents. To those whose models were closest to the "real world," we offer congratulations.

Discussion

Try to answer some of the following questions after you have discussed them with your friends:

1. How do you predict the contents of a forthcoming examination in order to prepare properly for it?

2. Can you predict your grade for the term at the end of the first third? The first half? The last third? What factors do you use for making these predictions? Are the same factors useful in all your subjects?

3. Do College Board examination scores predict scholastic accomplishments at college? Why? What factors should be used by a college admissions officer in making a decision? Why?

II | TRAFFIC FLOW

Introduction

Traffic flow problems exist all around us. The problem of air traffic flow is an extremely complicated one—planes coming into airports from different directions at varying speeds and altitudes each containing passengers who want to be first to land. Yet there is only one available runway at most airports. The problem of rail traffic into and out of large cities is similar to the air traffic problem, because of the limited number of available tracks. Auto traffic flow is a very complicated problem with varying times at which peak traffic flows past a given point. Added problems arise from changes in weather, road conditions, and accidents.

Object of the Experiment

If a traffic engineer is to predict future situations from present situations, he must construct a model of the traffic flow in and around the airport, railroad yard, or city. The following experiment is designed to give you some experience with this type of modeling for prediction.

Part A

You are to measure the traffic density at various places in the corridors of a school during a particular passing time. From the data construct a model of the traffic flow. From these and other available data you will be asked to predict the traffic flow for various times during the day.

The area of traffic flow which you will model will be that area of the school which will be assigned to you by a teacher or an administrator. The following procedure will be adopted.

Procedure

1. Obtain or draw a floor plan of the area of the school which your group will study.

2. Station students at the various intersections, as in Fig. 1, for the entire interclass period just prior to your *Man-Made World* class. Students at positions 1, 3, 5, and 7 will count the number of students who enter the intersection each minute from their corridor. Students at positions 2, 4, 6, and 8 will count the number of students who leave the intersection each minute and move past them.

3. Upon returning to the laboratory each person will construct a graph of the number of students counted as a function of time. Such a graph may be similar to the following:

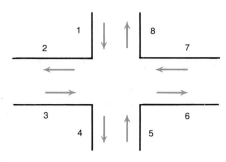

Fig. 1.

Fig. 2.
Graph for position number 1 between periods 2 and 3.

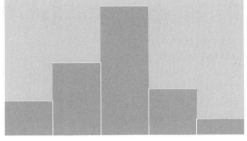

Number of students

Time

4. On a single floor plan (using arrows of different lengths to represent different numbers of students, and different colors to represent different times of observation) display the traffic flow pattern for the interclass movement.

5. Using the master schedule of your school and the traffic flow charts which you have developed, predict the traffic problems which may develop during different interclass periods when the number of students entering the corridors from various rooms is markedly different from what it was during the time when you made the traffic study. Check your prediction.

What are some of the factors which could affect the model you used for your prediction?

Part B

There are many other areas in which you can make the same type of study and prediction. A group of students might very well study traffic flow during a fire drill, or city street traffic at various times during the day (during a vacation period).

The text suggests that the school scheduler might examine the model and make changes in the schedules of some students to alleviate problems. Obviously this would best be done before school opens.

■ 1. Suppose your school scheduler asked you to develop a model which could be programmed into a computer to predict where bottlenecks might occur. He wants to run this simulated school traffic flow before any students are actually in school so that he can make adjustments prior to the opening of school.

What data would you need from the scheduler before you could develop the model?

Be on the lookout for articles in the press regarding traffic problems to see if they have resulted from a model of traffic flow based on real data or on the imagination of the writer using very little data.

III | INTRODUCTION TO THE ANALOG COMPUTER

The analog computer is a device which can perform certain mathematical operations on signals which are fed into it. Large sophisticated computers can add, subtract, scale (multiply a signal by a constant coefficient), integrate, multiply two signals, square, take the square root, divide, and perform certain trigonometric operations. The analog computer used in this course can perform only the operations of addition, subtraction, scaling, and integration. This is adequate for us to solve a wide variety of engineering problems.

In most analog computers today the mathematical operations are performed by circuits of electric and electronic components. Each circuit has an input and an output terminal, and when an electric signal is connected to the input terminal, the circuit generates a signal at its output terminal which is the result of its having performed a mathematical operation on the input signal. If an electric signal is connected to the input terminal of a scalor circuit, the signal at the output terminal will have the same shape as the input signal, but its magnitude will have been multiplied by some constant factor. An adder circuit has necessarily more than one input terminal, and the signal at its output terminal is always the sum of the signals which are connected to its input terminals. In an integrator circuit the output signal represents the area under a plot of the input signal versus time.

The output signal from one circuit can also be connected to the input of another circuit so that a sequence of several mathematical operations can be performed. This feature enables us to solve mathematical equations on the analog computer.

In order to get some insight into how the various parts of the analog computer (Fig. 3) work together to model a simple system, let us see how the flow of water into a tank can be modeled (Fig. 4).

First we need something to represent the source of water and its control valve C_1. These can be represented by the electrical power supply and the constant knob next to it (Fig. 5).

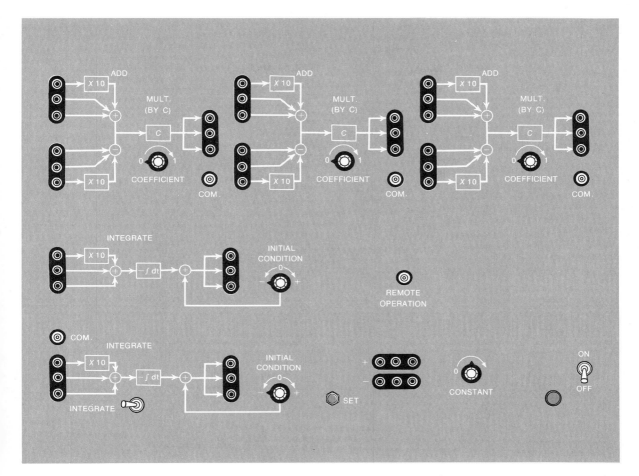

Fig. 3.

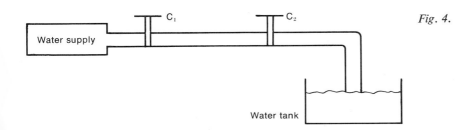

Fig. 4.

Fig. 5.

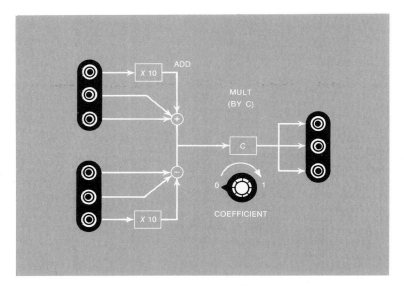

Fig. 6.

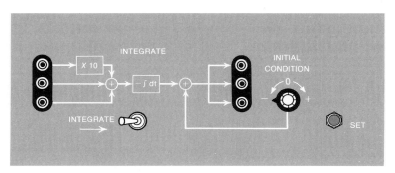

Fig. 7.

When the water gets to the tank, we also want to have a valve there to control the rate of flow of water into the tank. One of the scaling units can be used to represent the tank valve C_2. (See Fig. 6.)

As the water accumulates in the tank, we need some device to measure this accumulation. The integrator is our accumulation-measuring device (Fig. 7).

We are now ready to model the flow of water into the tank. For example,

$$\text{let} \quad 1 \text{ volt} = 1 \text{ gal/sec}$$

Now if we assume that 10 gal/sec of water is flowing out of the water supply, then C_1 must be set so that an output of 10 volts is obtained. This 10 volts then becomes the input to C_2 (scaling unit) which varies from 0 to 1. If we want 2 gal/sec to flow into the tank, then C_2 must be set to 0.2: $(.2(10) = 2)$. Let us try this much of our model on the analog computer.

We are now ready to monitor the flow of water into the tank. This is done by connecting the output of C_2 (2 volts) to the input of the integrator (accumulation-measuring device). We will assume that there is no water in the tank to begin with. This is done by pushing down the SET button (red button) and turning the initial condition knob until the meter reads zero. Next we have to set the timing knob (notice that there are four possible settings on models A, B, & C). For this experiment, the most convenient setting is 1 second. This 1-second setting means that every time we push the integrator lever to the right, the integrator will accumulate for 1 second. For example, if we are interested in finding out the amount of accumulation for 5 seconds, we merely push the lever to the right five times. (For model D, no time setting is necessary, just push the lever to the side marked 1 second and the accumulation is done automatically.)

Let us modify our analog model by assuming that there are two gallons of water in the tank at the beginning. In order to represent this new condition, the initial condition of the integrator must be set to 2 volts (push SET button and turn initial condition knob). Whether it is $+2v$ or $-2v$ depends on the sign of the input to the integrator. If the input rate of flow of water is $+$, then the output accumulation is $-$ (characteristic of this integrator) and therefore the initial condition must also be $-$. You are now ready to measure the accumulation of water when there is an initial amount of water in the tank.

IV | ANALOG COMPUTER MODEL FOR BUFFALO HARVESTING

The equation for buffalo harvesting in the text is reasonably straightforward, especially if the number of buffalo is to be kept constant. The purpose of this experiment is to demonstrate the use of the analog computer in adding and multiplying by a constant as well as to show how its use can aid in the simulation of real situations.

Part A

The equation for buffalo harvesting is

$$F_Y = (0.9 - k) F_{Y-1} + 0.1269 F_{Y-2}$$

in which F_Y is the number of females alive January 1st in any one year.

F_{Y-1} is the number of females alive January 1st the previous year.

F_{Y-2} is the number of females alive January 1st two years prior to F_Y.

If the population of female buffalo is 15 million in 1850 and 16 million in 1851, what should the harvesting factor k be during that year to insure that there will be 17 million buffalo available the next January 1st?

The flow diagram (Fig. 8) indicates the same situation as that of the mathematical model $F_Y = (0.9 - k) F_{Y-1} + 0.13 F_{Y-2}$. If we use Y as 1852, $Y - 1$ is 1851, and $Y - 2$ is 1850, the diagram becomes that of Fig. 9.

The summing-scalor of the AMF model D analog computer is pictured next:

Fig. 8.

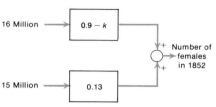

Fig. 9.

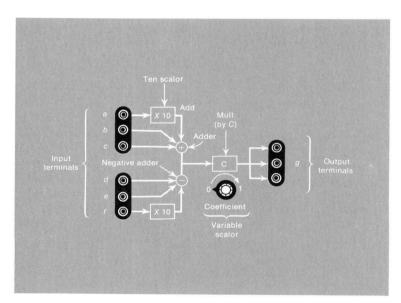

Fig. 10.
Summing-Scalor.

The section of the flow chart which indicates how we use this summing-scalor is shown in Fig. 11.

The black lines on the diagram below indicate how the summing-scalor is wired.

Fig. 11.

Fig. 12.

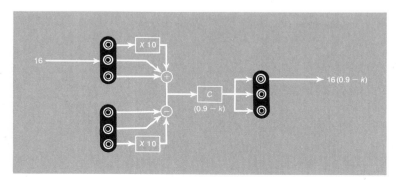

The original signal of 16 can come either from one of the (initial condition) outlets or from the main power supply as demonstrated by your teacher. Next this signal is added to the second scalor. This is obtained by feeding a signal of 15 into the second adder and setting the coefficient of 0.13, as indicated below.

Fig. 13.

Fig. 14.

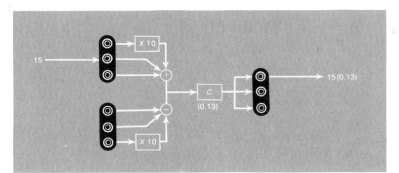

The two outputs are then fed into the third scalor as in Fig. 15.

Fig. 15.

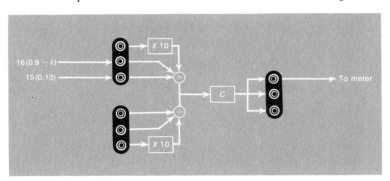

Since the only variable is the value k, turn the potentiometer knob on the first scalor unit until the meter reads 17 (on the same scale as the original 16 and 15 readings).

The complete wiring diagram follows.

Fig. 16.
Addition of two signals.

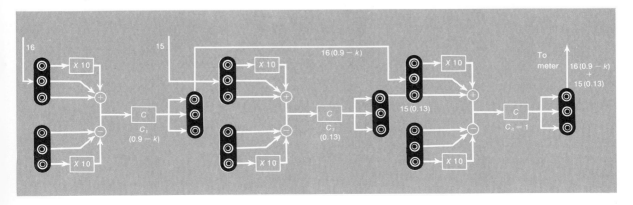

192

The coefficient value of C_1 is $(0.9 - k)$. Check this value by feeding a known signal into C (1.0 volt) and measure the output. This output equals $0.9 - k$.

$$-k = \text{output} - 0.9$$
$$k = 0.9 - \text{output}$$

1. What is the measured value of k using the analog computer?

2. Calculate k from the equation below

$$17 = (0.9 - k)\, 16 + 0.13 \times 15.$$

3. Are the observed and calculated values of k different? Why?

Part B

The number of female buffalo January 1, 1852, is 17 million and 18 million on January 1, 1853, and no buffalo are harvested in 1853.

1. What was k for 1853?

2. How many buffalo are there on January 1, 1854,

    a) by analog computer model?

    b) by paper and pencil computation?

V | MODELS OF EXPONENTIAL GROWTH

Population, money, and solid waste have one thing in common. Under certain circumstances they will grow at an exponential rate. The purpose of this experiment is to learn how to use the analog computer for modeling these dynamic growth situations. Given the percentage growth of any system, the amount of growth can be obtained manually or by using the integrator portion of the analog computer (Fig. 17).

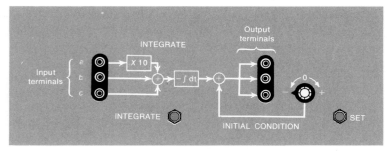

Fig. 17. Summing-Integrator.

The integrator acts as an accumulation device which sums up the amount of anything (people, money, solid waste) at the output, if you input a rate of change. Let us use the now familiar world population situation to illustrate the use of the analog computer for modeling exponential growth. As discussed in Section 5, the rate of growth of the world's population per year is about 2 percent of the existing population and at the start of 1960 the world's population was approximately 3 billion. This information can be depicted by the following flow chart (Fig. 18).

In order to model the world's population growth on the analog computer, we first set the initial condition of one of the integrators to 0.3 volt (3 billion people in 1960) and one of the constant knobs to .02 (2 percent). The time scale of one second on the analog computer is, in this case, one year. We are now ready to use Fig. 18 to wire up the analog computer.

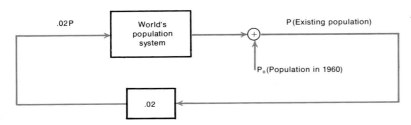

Fig. 18.

Obtain data for 20 years (20 sec) in 1 year (1 sec) intervals and compare it to the information in Table 4–2 of Section 5.

1. Try to predict what the population growth would be if the growth rate is 3 percent and 5 percent. Now obtain data from the analog-computer model by changing the constant knob to 0.03 and 0.05.

    One of the more interesting features of this analog-computer model is that it can also be a model for computing the growth of money in a savings bank, the accumulation of solid waste, and other growth situations. This is but one example of one model being able to represent many systems. For example, if money is placed in a bank and a 5 percent compound interest per year is offered, the amount of money can be computed by the analog-computer model by merely changing the constant knob to 0.05 (5 percent). The initial condition knob was originally set at 0.3 volt (we could let it represent $300) or change it to represent whatever initial amount of money we decide to put into the bank.

2. Use the analog computer to calculate the total amount of money (principal plus interest) in the account after ten years.

3. Many banks now compound their interest quarterly. Using the same principal and interest, modify the analog-computer simulation to find the total amount after ten years.

4. In Section 6, a model of solid waste accumulation was discussed. Modify the analog-exponential model to represent the United States solid-waste problem.

Systems
5

I | INTRODUCTION

One day we read the following news item in the newspapers: "The U.S. Air Force has launched a revolutionary Spy-in-the-Sky which can sweep over military bases in any foreign country, take pictures, and return packages of film on command from ground controllers. It may even operate a television system which could be monitored from the ground!"

On another day we read: "Senator Warren G. Magnuson called today for an international agreement to protect the world's fish supply. He said he will propose a conference for such an agreement when he talks to Russian officials in Moscow later this week."

In a Memorandum of Decision of the Superior Court of the State of California, we read the following: "This is an action for wrongful death. The action arises out of an automobile accident which occurred on May 16, 1960, on a highway known as the Carmel-Pacific Grove Cutoff in Monterey County, California. On the date in question, Don Wells Lyford, a young man sixteen years of age, was driving a 1960 Corvair automobile on this highway and proceeding in the direction of Carmel. The highway involved was a two-lane highway through a wooded section with a number of curves. At the exit of a right hand curve and the beginning of a curve to the left, the Corvair automobile went across the center line and into the opposing lane of traffic and collided with a Plymouth automobile which was proceeding toward Pacific Grove. Don was not thrown from the Corvair but, as a result of the collision, he received certain injuries from which he died before reaching the hospital. Don, the only occupant of the Corvair automobile, made no statement before his death relative to the cause of the Corvair going out of control."*

What do these three stories have in common? They are all news; they are all interesting; they are all unfinished stories; they may have some immediate or long range effect on our lives. But what is important in these stories is that they all involve modeling and they also involve dynamics. We do not attempt to predict the action of a satellite, or the future of the world's fish supply, or the action of a 1960 car, without building a model first.

*Habitat, Expo '67, architect
Moshe Safdie. (Photo by CMHC,
Ottawa, Canada)*

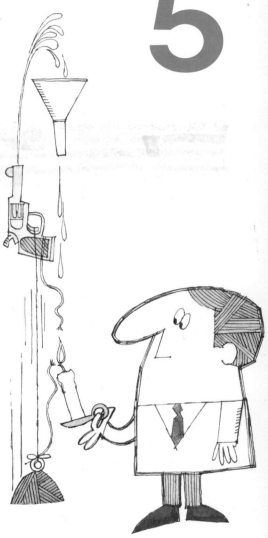

* Memorandum of Decision #771-098, July 29, 1966. The question the Court had to decide was whether the accident resulted from the behavior of the car. As the car switched from the right curve to the left curve, was it impossible to steer properly?

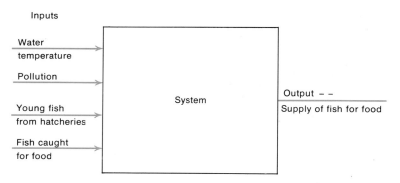

Inputs

Water temperature

Pollution

Young fish from hatcheries

Fish caught for food

System

Output – –
Supply of fish for food

Fig. 5–1. The fish system.

Furthermore, all three stories refer to *systems*. A system is a connection of many different parts which interact, often in a complicated way. In a system, several causes combine to give an end result. To understand the system, we have to find the different causes, and find how each influences the total effect.

The world's fish supply, for example, is a system. The number of fish available for food in the future depends on the extent to which we pollute the waters, the number of fish caught each year, the program to grow young fish under protected conditions, and the future changes in water temperature. In other words, we are concerned about the food supply available in the future. We call this end result, or effect, the output signal. There are also many different input signals, or causes, which can influence this output (Fig. 5–1). We can change some of these inputs easily. Others are largely beyond our control.

This example demonstrates why we want to study systems. We try to understand this system—that is, how the output is related to each of the inputs. Once we have such understanding, we can decide how to change inputs in order to give a desired output. What laws should be passed to regulate fishing? What must be done to control pollution? How should we stock the lakes and oceans with young fish?

In this chapter, we will look at various systems. Our immediate goal is to become familiar with a few of the methods which have proven useful in gaining a better understanding of such problems. Our more basic goal is to begin to understand the "systems approach." Recently, economic, social, and political problems have been attacked with systems ideas or from the systems viewpoint.

2 | INPUT-OUTPUT IDEAS

Systems can be discussed in terms of *inputs* and *outputs*. Figure 5–2 shows the input and output for a familiar system. The system is an automobile. The input is the depressing of the gas pedal. The output is the speed of the car.

Certainly we have over-simplified the system if we describe it by the two curves of Fig. 5–2. The speed of the car depends on many signals other than the position of the gas pedal. (For example, if the road is icy, the rear wheels may spin and the speed of the car remain at zero, no matter how the gas pedal is worked.) The two plots of Fig. 5–2 are, however, useful for describing the system under certain conditions. They might describe how a particular car starts up from a stop on a normal concrete road. We might even use such curves to compare different cars (Fig. 5–3).

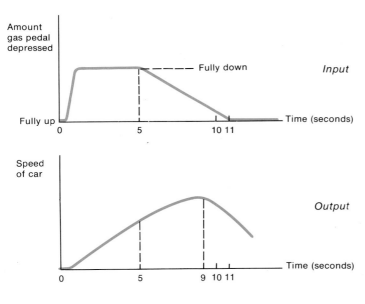

Fig. 5–2.
Signals for system controlling automobile speed.

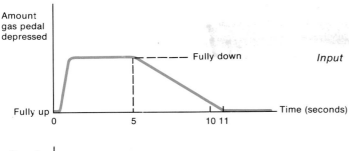

Fig. 5–3.
Comparison of cars a, b, and c. Car a has the most "pick-up."

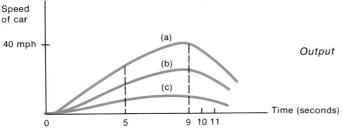

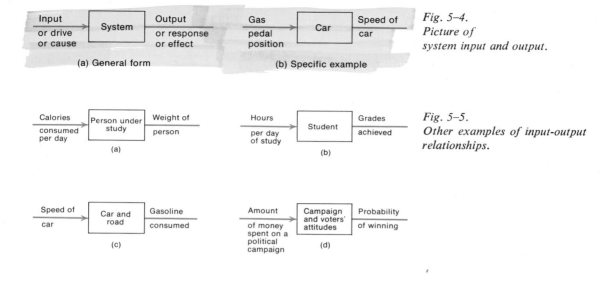

Fig. 5–4.
*Picture of
system input and output.*

Fig. 5–5.
*Other examples of input-output
relationships.*

Thus, in our example, the system is the automobile and the road. The position of the gas pedal is the input signal. This is the signal which moves the system into action. The speed of the car is the output signal, the response of the system to the input, or drive. In other words, when the gas pedal is pressed down, the speed changes, when the system is driven, the system responds.

Rather than write all these words every time we want to describe a new system, we use the simplified notation shown in Fig. 5–4. Here a block represents the system. The input is shown by an arrow pointing into the block. The output is a line coming from the block on the right. This sort of picture is called a *block diagram*. It shows at once the signals to be considered as the input and as the output.

This block diagram is really a picture of the cause-and-effect relationship in a system. The input is the cause, and the output is the effect. In this sense, we can use a block diagram to represent a wide variety of different types of systems. A few of these are shown in Fig. 5–5. We could extend this list to include any cause-and-effect relationship which might exist in nature or among the world of man-made devices.

All the block diagram shows is that there is some relationship between the output and the input. This relationship may not be simple or obvious. For example, in the system of Fig. 5–5d, we sense that the probability of a political candidate winning depends in some way on how much he spends on his campaign (on speeches, billboard signs, television commercials, and the like). It may well be that his chances of winning rise for awhile as his expenditure increases, then fall as the voters become irritated by his over-exposure.

Thus, the block diagram portrays simply what signal we are considering as the input and which signal is the output—the cause and the effect, respectively.

The Block Diagram and the Model

The block diagram defines the system we wish to model. In other words, the block diagram defines the input and output signals. The model must go farther. It must describe exactly how the input determines the output.

The model is *quantitative.* It describes in numbers how the output is determined by the input. Once we have a quantitative model, we can find the output corresponding to a given input. Figure 5–3, for example, is a quantitative model: it shows the way the car speed depends on the position of the gas pedal.

Figure 5–6 illustrates the distinction between a block diagram and a model. The figure is for a very famous example, the model showing the way in which the time required to complete a particular task changes with the number of men assigned to a job. The *complete* figure constitutes the model. Part (a) is the block diagram defining the signals, and (b) shows the *quantitative* relationship between the number of men (N) and the required time (T).

The shape of the N-T curve is familiar to work planners. In building a house, for example, we can reduce the total time required by putting more men on the job. This speeding-up process only works for a while, however. Eventually we reach a point (N_0, T_0 on the plot) after which the addition of more men actually slows down the work.

This slowdown occurs for many reasons. The men may be in each other's way. An excessive effort is wasted in organizing the work and planning each man's activities. The men begin to spend large amounts of time in talking together and enjoying one another's company. Possibly the job is so divided into small parts that the men spend large amounts of time standing around waiting to do their particular parts.

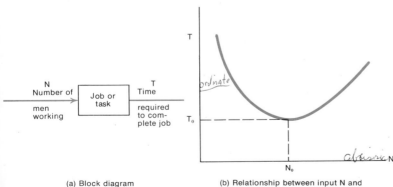

(a) Block diagram

(b) Relationship between input N and output T

Fig. 5–6.
Model describing effect of changing the manpower assigned to a particular task.

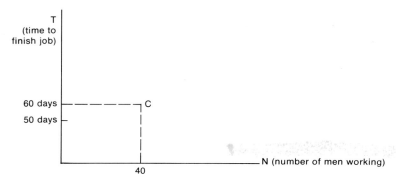

Fig. 5–7.
Part way through the job, we estimate we are at point C. There are 40 men working. Looking ahead, we guess it will take 60 days to finish the bridge. Our contract requires that we be done in 50 days. After that, we have to pay a heavy penalty for each day late.

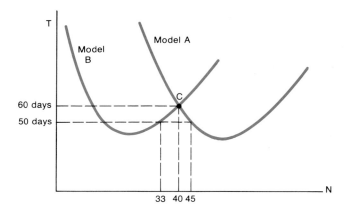

Fig. 5–8.
We are operating at point C (40 men, 60 days). We don't know which model (A or B) is correct. If Model A is the one, we should add 5 men. If Model B is correct, we should lay off 7 men.

The model of Fig. 5–6 is important in managing a business or a project. It often happens, for instance, that part way through a project we find that the work will not be completed in time. Our bridge may not be completed by the scheduled opening of the entire road. In Fig. 5–7, we estimate the time required will be 60 days, instead of the 50 days which are necessary. As manager of the job what problem do we have? We estimate we will finish in 60 days. We must be done in 50 days. Should we add more men?

Before we can answer this question, we have to know the model. There are two possible models (A and B in Fig. 5–8). If Model A describes our job, we should add 5 more men. If Model B is correct, however, adding more men will require more time. We should reduce the work crew by 7 men.

Unfortunately, what sometimes happens is that more men are added. The project is delayed even more. The manager desperately adds still more men. Finally, the work falls hopelessly far behind schedule.

If the manager understands the model and the two possibilities of Fig. 5–8, he can try adding more men. If it appears that the job is delayed even more, he then can consider reducing the work force. An understanding of the model is important if he is to manage intelligently and successfully. Indeed, he must measure his model and try to find the curve of Fig. 5–6 with the numbers for his particular job. Then he can decide how to finish the job on time and at the least cost.

The block diagram is just a simple way of showing cause and effect. An effect may be the cause for a different system. In other words, the output of one block may be the input to another.

Figure 5–9 shows a simple extension of the earlier example of the gas-pedal car-speed system. Here the total system consists of three parts.

1. The driver. The input is the driver's observation of the time and his location. As these signals change, he decides how fast he must go to reach his destination in time. On this basis, he moves the gas-pedal position (the output for the first block).
2. The car. The input here is the gas-pedal position and the output is the speed.
3. The distance traveled. The speed is the input signal which determines the final output signal: the location, or position, of the car.

Thus, we can put together a number of blocks in a block diagram. The output of one block serves also as the input of the next in a sequence of cause-effect relationships. Indeed, many of the systems which we wish to consider in subsequent chapters consist of a combination of several simple elements, each represented by a block. Figure 5–10 shows another example of a system involving more than one block.

Once we have a block diagram, we can make measurements to find a quantitative model. Thus, the block diagram is just a picture of the different inputs and outputs.

Furthermore, the block diagram shows the train of cause-effect relations. In the thermostat system of Fig. 5–10, the operating

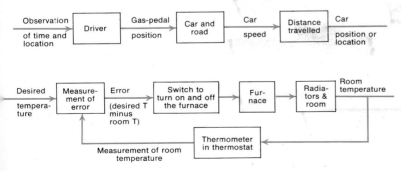

Fig. 5–9.
A more complex block diagram.

Fig. 5–10.
Parts of a home-heating system. When the room temperature drops, the furnace is turned on.

furnace delivers heat. The radiators convert this to a change in room temperature. The input for the total system is the desired temperature (where the thermostat is set). The total system output is the room temperature. In this case, the ideal system would be one which gave an output exactly equal to the input at all times.

If we want to study such a home-heating system, we need to make measurements so we can describe each block in Fig. 5–10 in terms of numbers. How much heat comes from the furnace? What is the resulting temperature change in each room? When does the thermostat go on? How long afterward does the furnace start? It is only after we have such information that we can describe in detail how the system works. The block diagram is just the first step in an understanding of the system.

3 │ FINDING THE PARTS OF A SYSTEM

As we shall see at the end of this chapter, the noise environment in which we live is a complex system. There are various inputs, things we want to do which cause noise. For example, trucks carrying food to supermarkets are often noisy. The system output is the noise which surrounds us and which may damage our health.

This noise system is very complicated. The complexity arises because the system has grown without much conscious control by man. Many parts of the system grew without any worry about noise. Only in recent years have we realized that we must control noise, particularly in the cities.

Because of the complexity of the noise system, we will need to look at some simpler systems first. The study of a simple system will suggest how to tackle more difficult problems such as reducing noise in our cities. Now, let's consider the system mentioned earlier, the heating of a home.

What are our goals in this study?

1. We need to understand how the system works in general. How do we heat a house? How do we measure how well it is heated? What causes changes in the living-room temperature?
2. We want to understand what the system is. What are the important inputs and outputs?
3. How are the outputs determined by the inputs?
4. Once (3) is answered, we can ask: How can we improve the performance of the system? Usually, there are two ways to obtain a better output:

   a) We can change the relations between inputs and outputs.

   b) We can control some of the input signals.

Thus, our goal is really to understand the system enough to make decisions about how to improve performance. There are constraints.

Changes may cost money, and we may be limited in the money we are willing to spend. Then our decision problem is to find the best, or optimum, system within the constraints.

With these general ideas as background, let us begin our study of the home-heating system.

How the System Works

The home-heating system is supposed to keep the inside temperature comfortable even when conditions outside vary greatly. In order to have a single example, let's assume that the furnace burns oil. The burning oil heats water to a high temperature (perhaps 190°F). This hot water is then pumped through the house. In each room there is a radiator. The hot water in this radiator heats the surrounding air. The warm air then rises and cooler air moves toward the radiator. The positions of the radiators and thermostat are shown in Fig. 5–11.

One very important part of the system is the thermostat. In our house, there is a single thermostat located on the living-room wall. We are using an inexpensive thermostat which works as follows. The dial is set manually at 72°F. The thermostat includes an element made of two different metals. As the temperature at the thermostat changes, the metals expand different amounts and the element bends. When the room temperature drops below 72°F, the bending is such that a switch is closed and electricity is sent to the furnace and the furnace starts.

Once the furnace starts, the water is heated. As soon as the water temperature rises, the pump starts and heat is sent through the house. In our system, when the temperature reaches 75°F, the

Fig. 5–11.
Floor plan of a
house showing the position of
radiators and thermostat.

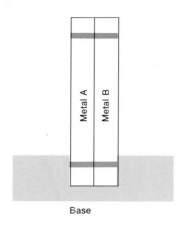

(a) Two different metals are in the shape of strips fastened together at both ends. At a temperature of 70°, both strips are the same length as shown here.

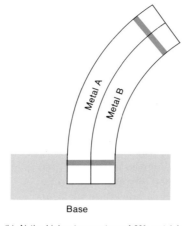

(b) At the higher temperature of 80°, metal A has expanded more than B. Strip A is longer than B. Since the two strips are still together at both ends, the combination must be bent to the right above.

Fig. 5–12.
One possible way a thermostat can operate. The position of the top of the two strips changes with the temperature. Setting the thermostat at 70° means that, as we move from b to a, the switch is closed as we reach position a and electricity is sent to the furnace. There are other possibilities. We can use one metal in the form of a coiled spring. As the temperature rises, the metal lengthens and the spring uncoils.

thermostat turns off the furnace. There is still hot water in the radiators, however, and the room temperature continues to rise until this hot water cools. We find that the heating of the room actually doesn't stop until the room temperature reaches about 77°F.

Thus, on a moderately cold winter day, the living-room temperature changes as shown in Fig. 5–13. With our inexpensive thermostat, the temperature changes over a wide range. At 71° the ladies are putting on their sweaters; at 77° the men are removing their jackets.

Inputs and Outputs of the System

Now that we have a general idea of how the system works, we want to move toward a more refined picture in order to learn what strategies we might use to improve performance. What are the system inputs and outputs? What are the different parts of the system? In other words, what is the block diagram describing the system?

First, what are the *outputs?* What signals measure system performance? Since this is a home-heating system, the output is the temperature inside the house. The system is supposed to keep this temperature constant.

The temperature is not the same throughout the house. The temperature in a bedroom may be very different from that in the kitchen. Which temperatures should we use as system outputs? The answer depends on what we want to know about the system. If we want an overall picture of the house heating, we might choose three temperatures: living room (at the thermostat), kitchen, and one bedroom.

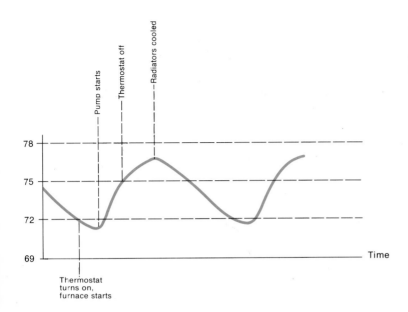

Fig. 5–13.
Temperature at the
thermostat during the day.
(Thermostat is set at 72°.)

Now what are the *inputs?* What causes the three temperatures to change? Here we can list several signals:

1. Outside temperature.
2. Outside wind velocity.
3. Distance the bedroom window is opened.
4. Time the front door is open.
5. Heat from the kitchen appliances (refrigerator, range, and so on).
6. Temperature we set on the thermostat.

At this point, we have to decide what signals to include. There are dozens of different things which affect the inside temperature. We have chosen only six. We have not included the number of people in each room (people give off heat), the windows in other rooms, things causing the air inside the house to move, and additional heat sources such as electric heaters.

In any model, we have to simplify as much as is reasonably possible. We must neglect items which are secondary. After we have finished our study, we can reconsider some of these neglected items. We may find that we have neglected something important. In such a case, we have to start the problem all over again. If we try to include everything, however, the system usually becomes so complicated that there is no hope of ever gaining any understanding.

Once the inputs and outputs are chosen, we can draw a simple block diagram (Fig. 5–14). Figure 5–14 is just a pictorial statement of what we have decided to consider.

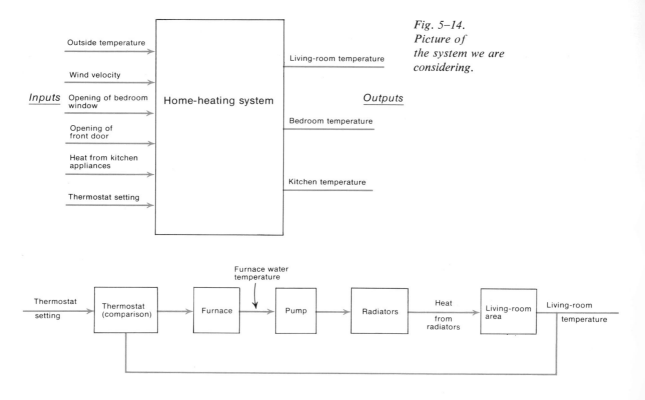

Fig. 5–14.
Picture of
the system we are
considering.

Fig. 5–15.
Block diagram
of the basic heating
system.

Inputs Affect Outputs

Figure 5–14 defines the problem. Now we would like to break this single block down into simpler parts. Our goal is to see how each of the inputs (causes) affects each output (effect).

In order to start, we consider a part of the system. Here we should choose the most important part which is the thermostat setting as input and the living-room temperature as output. The furnace, radiators, and so on are built to allow the thermostat setting to control the living-room temperature.

If we look at just this part of the system, we have the block diagram shown in Fig. 5–15. There are two inputs to the thermostat. We need to compare the setting with the actual living-room temperature. If the comparison indicates the living-room temperature is lower than the thermostat setting, electric power is supplied to the furnace. The pump circulates hot water to the radiators. The heat from the radiators changes the living-room temperature. The temperature change resulting from radiator heat depends on the size of the living room, how much heat leaks out the windows, and so on.

Figure 5–15 is the basic system. Now let's continue to study only the one output (living-room temperature), but let us try to

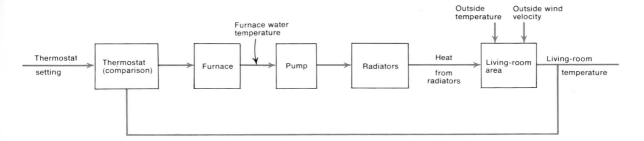

understand what the other inputs do in the system. How do the other inputs (Fig. 5–14) affect the living-room temperature? Can we show these effects by adding to Fig. 5–15?

The outside temperature and wind velocity both affect the system in the same general way. When the outside temperature is very low, large amounts of heat leak out of the living room continually. The same thing happens when a cold wind blows directly on the living-room walls facing the outdoors. Thus, these two inputs represent heat lost from the living-room area. Figure 5–16 shows the new block diagram.

The opening of the bedroom window and the use of kitchen appliances have no effect on living-room temperature. (The kitchen appliances might, but our appliances are fairly far from the thermostat.) The opening of the front door does drop the living-room temperature. Thus, the entire block diagram for determining the living-room temperature is shown in Fig. 5–17.

Finally, we want to show the other two outputs in our block diagram. Each of the other rooms is heated by radiators fed hot water by the same pump system. If we add these parts of the total system, we obtain the block diagram of Fig. 5–18. We now have a complete picture of our home-heating system.

Changing System Performance

Figure 5–18 is a model of the system. Each block shows a cause-effect relation which is a part of the system.

Fig. 5–16.
Block diagram
with two more inputs
added.

Fig. 5–17.
Complete block
diagram for finding living-room
temperature.

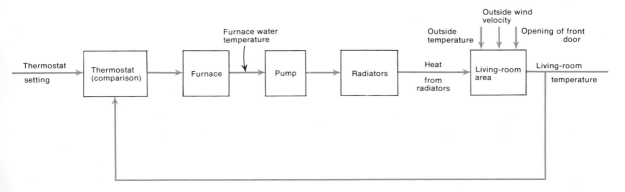

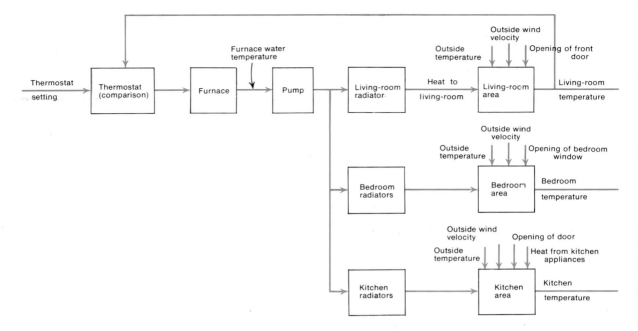

Fig. 5–18.
Complete block diagram for six inputs and three outputs.

Such a system understanding is valuable if we want to change system performance. For example, perhaps we bought the house last summer and it is now mid-winter. We find that the heating system is inadequate. On cold, windy days, the living-room temperature drops to 60°F regardless of the thermostat setting. What can be done? Figure 5–18 suggests we consider the following alternatives:

1. We might try to change the inputs. Four inputs affect living-room temperature: thermostat setting, outside temperature, outside wind velocity, and opening of the front door. We can't do anything about the first two. We might cut the effects of wind by planting dense, high shrubs across the front of the house. We can decide to use only the side door during cold days, so the front door is kept closed.

2. We might try to change the characteristics of individual blocks in Fig. 5–18. Some possible steps are:

   a) Improved radiators. Here we can be sure furniture doesn't block air flow to radiators, and that dust is removed from the radiators. As a last resort, we may decide to add radiators.

   b) Raise the water temperature at the furnace before the pump starts.

   c) Decrease the effects of outside temperature and wind by using storm windows, insulating the walls, and adding heavy drapes across the wall facing the wind.

In the same way, the block diagram indicates steps we might take to change the bedroom or kitchen temperature. For example, if the kitchen is too warm (because of the appliances), we might shut down kitchen radiators. A much more elaborate and expensive solution would be to add a separate thermostat and pump in each of the rooms.

The home-heating system of this section is meant to serve only as an example of the use of block diagrams. When we are trying to understand an unfamiliar system, we first find a block diagram which shows the ways in which the inputs control the outputs. Once this block diagram is drawn, we can see various ways in which we can change system performance.

In a complete study, we would need to take measurements on our system. From these tests, we would find exactly how each input affects each output. We could then predict what would happen to the living-room temperature under specified conditions. We would decide how completely insulated the living room should be to ensure that the system could maintain a desired temperature even with the coldest winter weather.

Thus, the *systems approach* is a way of looking at complex problems or systems. Using the systems approach we try to pick out those features which are important. Then, an understanding of the system suggests what to change to improve the system.

4 | RATE INPUTS

In many cases, we can find the input-output relation in precise terms. For example, we can say the output of a block is four times the input. If we can describe each block in this way, we can find the total system output from any given inputs. Then we can understand in detail how the system operates and how the system can be controlled.

In the examples we will consider, the blocks can be described by three different relations (Fig. 5–19). The first two of these are familiar. The first is called a *scalor*. Here the output is just a constant times the input. In the diagram the constant is called k. This is merely a number (positive or negative). For example, if k is 2, the output at all times is twice the input.

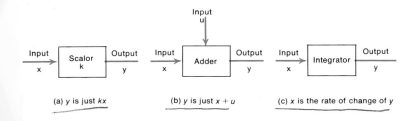

(a) y is just kx (b) y is just $x + u$ (c) x is the rate of change of y

Fig. 5–19. Three types of blocks for simple systems.

210

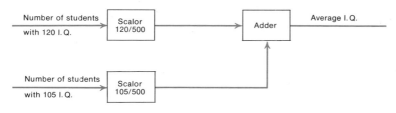

The second type of block is an _adder_. The output is the sum of the two inputs.

These two blocks (scalor and adder) are enough to describe some very simple systems. For example, suppose an entering college class is always 500 students. Only two types of students are admitted: those with an I.Q. of approximately 120 and those with an I.Q. of about 105. The average I.Q. of the class is found as in Fig. 5–20. Two scalors and one adder are used.

The third type of block is called an _integrator_. Actually this is just _rate control_. The input is the rate at which the output is changing. Since this block is so important when we model systems, we will consider two examples of the integrator in the rest of this section.

In Chapter 4 we saw that the world population is increasing 2% per year. The population was 3 billion in 1960. Each year thereafter the population was 2% larger than the year before. The resulting population grows exponentially (and doubles about every 35 years).

We can represent this growth in a block diagram as shown in Fig. 5–21. There have to be two inputs to determine the population output. First, we need the population at the beginning. In our case, this is 1960 and the population (p_0) is 3 billion. Then, each year the population grows 2%. The fraction by which it increases each year is $\frac{2}{100}$. This rate is the main input to the integrator. The output is the starting value plus the sum of all the changes since 1960. Thus, integration is just a way of showing that the output depends on the starting value and on the rate at which the output is changing.

Why do we bother to discuss this type of block called an integrator? First, the integrator is important because we can represent many systems by combining integrators with the adders and scalors of Fig. 5–19. Second, the three blocks of Fig. 5–19 are the basic

Fig. 5–20. System for finding average I.Q. This block diagram is just another way of showing the formula

$$\frac{H \times 120 + L \times 105}{500}$$

where H is the number of students with an I.Q. of 120, and L is the number of students with an I.Q. of 105.

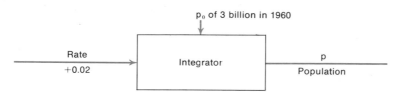

Fig. 5–21. Block diagram for population growth.

elements of analog computers which are used so often to model and study complex systems.

The integrator takes us from the rate of change of any signal to that signal itself. For example, in a chemical plant where hydrochloric acid is made, the chemical engineer must keep account of the *total* amount of acid in the plant's storage tank. The storage tank is shown in Fig. 5–22. In this figure, the acid flows from the filler pipe into the tank *at a rate of q* gallons per hour. The variation of q with time for a typical morning is shown in Fig. 5–23. (The dip at 10:15 A.M. results from a malfunction of equipment, and the noontime dip is caused by the lunch hour.) At 8:00 A.M. on this day, the volume of acid stored in the tank (the residue from the preceding day) is 1700 gallons. In other words, the initial volume (or initial condition) for the day is V_0 or 1700 gallons.

This system is described by the block diagram of Fig. 5–24. The primary input is q, the rate at which acid is flowing into the tank. The output is the amount of acid in the tank at any time.

In contrast to the population example, the rate q is a signal which changes with time. The amount of acid in the tank at any time is the starting amount plus that which has entered since 8 A.M.

From the curve of Fig. 5–23, we can calculate the total quantity which enters the tank in any period of time. Figure 5–25 shows a one-hour period during which q is constant at 2000 gallons/hour.

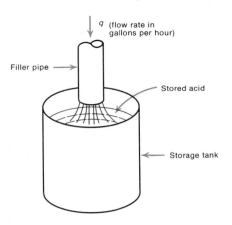

Fig. 5–22.
Hydrochloric-acid
storage tank.

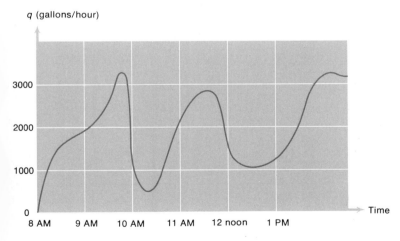

Fig. 5–23.
Rate of flow of acid
into tank of Fig. 5–22
on a typical day.

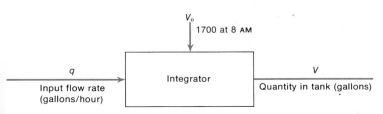

Fig. 5–24.
Block diagram
for hydrochloric-acid
tank.

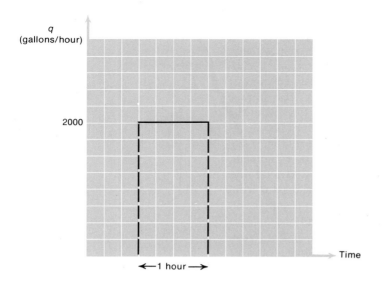

Fig. 5–25.
Plot if q *is*
constant at 2000 gallons/hour
for a one-hour period.

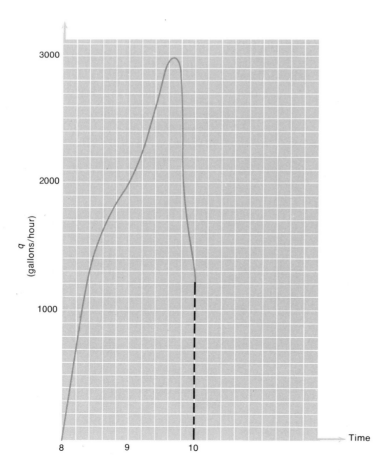

Fig. 5–26.
q *from*
8 to 10 A.M.

During this period, the volume in the tank increases by 2000 gallons. This volume is just the area under the q curve for this period of time.

In Fig. 5–25, this area can be measured by counting the number of squares of the graph paper. There are 32 squares under this plot of q. Hence, 32 squares correspond to a volume change of 2000 gallons (or every 16 squares represents 1000 gallons).

How much acid enters the tank from 8 to 10 A.M.? Now we plot the curve of q versus time (Fig. 5–26). We want the area under this curve from 8 to 10 A.M. If we use the same scales as in Fig. 5–25, every 16 squares corresponds to 1000 gallons. If we count the squares in Fig. 5–26, there are about 120. Hence from 8 to 10 A.M., $\frac{120}{16}$ thousand gallons, or 7500 gallons, enter the tank.

In other words, *the integrator gives an output which is the starting value plus the change. The change is the area under the rate curve from the starting time to the time the output is measured.*

Figure 5–23 shows the *rate* at which acid enters the tank from 8 A.M. on. To find the volume in the tank at 1 P.M., we add 1700 gallons (present at 8 A.M.) to the area under the q curve from 8 A.M. to 1 P.M. (Fig. 5–27).

The preceding example shows that an integrator is essentially an *area-measuring device*. The output at any time (t_1) is the output at the start (t_0) plus the area under the rate curve from t_0 to t_1.

Counting squares on graph paper is a tedious job. Fortunately, there are devices which integrate for us. The simplest is the planimeter (Fig. 5–28). Sometimes the input (rate) signal is an electrical voltage. Then we can use an electronic integrator, a connection of transistors and electrical elements which act as an integrator. We can also integrate with a digital computer. These are all devices which simplify the job of finding areas (or counting squares), and hence permit us to integrate.

Fig. 5–27.
The area of the shaded portion represents the volume of acid added to the tank from 8 A.M. to 1 P.M.

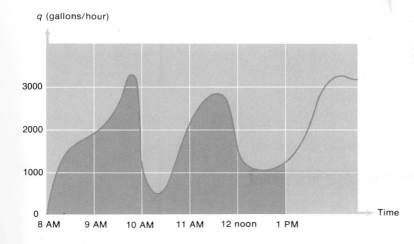

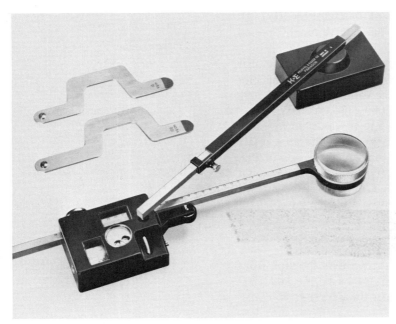

*Fig. 5–28.
A planimeter
is a mechanical
device for measuring
the area under a curve.
When the movable
pointer is carried
all around the
boundary of the
region, the counter
reads the area
within the
region. (Keuffel
& Esser
Company)*

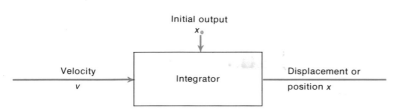

*Fig. 5–29.
Relation
between velocity and
displacement.*

Velocity Input

We can easily measure the velocity of an object. This velocity is merely the rate of change of the position. We might describe the velocity of a car as 30 miles/hour on a road due north.*

From the velocity, we can find the position of the car as shown in Fig. 5–29. The present position (usually called displacement) is the initial value plus the area under the velocity curve from the starting time to the present.

Thus, the speedometer and odometer readings of a car are related by a mechanical integrator. The odometer reading is the output of the integrator, and the speedometer is the main input. The change in the odometer reading is the area under the speedometer curve.

In a similar fashion, acceleration is defined as the rate of change of velocity. In other words, velocity is the output of an integrator which has acceleration as the main input. If an object falls freely from a balloon at $t = 0$ with no significant air resistance, the

* In science, velocity means speed plus the direction of motion.

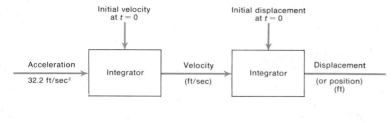

Fig. 5–30.
Two integrators allow
us to find position from a
known acceleration.

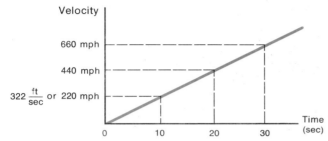

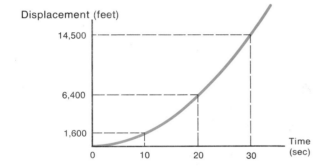

Fig. 5–31.
Velocity and
displacement (distance
fallen) when an object falls
from a balloon. We
assume acceleration
is constant.

acceleration is constant, that is, the velocity is changing by 32 ft/sec every second, written as 32 ft/sec² and called the acceleration of gravity. The way the velocity and position of the object change with time are found from the system of Fig. 5–30. The acceleration is constant in our example. The corresponding velocity and displacement are shown in Fig. 5–31.

When a man jumps or falls from a balloon, Fig. 5–31 is not correct. His velocity does not continue to increase forever. If it did, by 34 seconds he would be passing through the speed of sound. We could expect a sonic boom.

What happens to cause the velocity to level off (Fig. 5–33)? As his velocity increases, there is greater and greater air resistance. There is a force tending to oppose this motion. The actual amount of air resistance depends on the shape of the man (there is less when he is curled up). This air resistance grows until it just offsets the gravity, and then the velocity doesn't change. This is called the "terminal velocity."*

*The air resistance also causes heating. This is the reason for the extremely high temperature of nose cones re-entering the atmosphere from space. In that case, the velocity is already great before the cone enters the air.

*Fig. 5–32.
Free fall in
skydiving. Like a
flock of big dark birds,
skydivers coast earthward
in the French Frog position.
This position is probably
the most popular among
skydiving experts. (Photo
by Bob Cowan,
St. Louis, Mo.)*

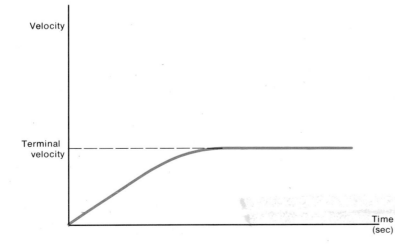

*Fig. 5–33.
Actual velocity
of a man in free fall
from a balloon.*

Thus, acceleration, velocity, and displacement are related signals. Velocity is the rate of change of displacement; acceleration is the rate of change of velocity. Figure 5–30 displays how the integrator block shows the relations among these three signals and allows us to calculate from a measured acceleration.

5 | POPULATION MODEL FOR A TOWN

Our three building blocks of the last section were a scalor, adder, and integrator. To illustrate how these blocks are used to represent a real system, we now consider the modeling of the population of a town.

The population of a town changes for four reasons:

1. Births
2. Deaths
3. Immigration (people moving into the town)
4. Emigration (people moving away)

We want our model to show each of these factors.

Births

Usually we specify the births per year by giving the birth rate (the number of live babies born per 1000 population). For example, we might find from town records that the birth rate is 15/1000. If there are 3000 people in the town, we can expect 45 births. In general terms, if we use p for the population and b for the birth rate, the number of new births per year is bp (Fig. 5–34). This bp is the rate at which babies are produced.

Deaths

In a similar way, deaths subtract from the population. The population change is the output of an integrator (Fig. 5–35). The input is the death rate d times the population. This dp is the rate at which deaths occur.

If there were no immigration or emigration, the population p would just be the starting value (p_0) plus the new births minus the deaths. In other words, the system could be modeled by the block diagram of Fig. 5–36. Notice that a scalor block of -1 follows the death-rate integrator so that deaths are subtracted.

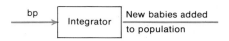

Fig. 5–34.
b *is the birth rate per 1000 population and* p *the population in thousands.*

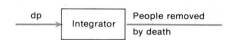

Fig. 5–35.
Population changes from death.

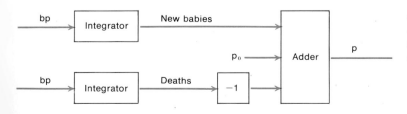

Fig. 5–36.
Effects of births and deaths.

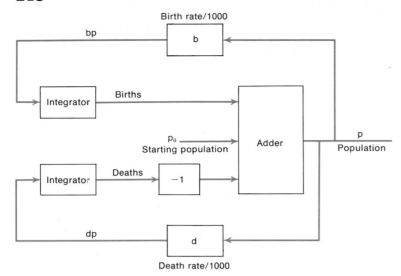

Fig. 5–37.
Complete block diagram
for the town population as it changes
because of births and deaths.

Looking at Fig. 5–36, we notice that the two signals required for the integrator inputs are *bp* and *dp*—constants times *p*. Hence, we can obtain these signals from the output *p* (Fig. 5–37). We merely add two scalors with gains *b* and *d*.

Immigration and Emigration

Finally, we want to include immigration and emigration in our model. If *i* is the number of people per year moving into the town and *e* is the number per year leaving, there is a net flow into the town of (*i − e*). Hence, this (*i − e*) also represents a rate of change of population. If we put this signal into an integrator, the output is the number of new people each year from migration. Then the total model is shown in Fig. 5–38.

What is the value of a block diagram such as Fig. 5–38? Why is this model any better than a discussion in words of the system? First of all, the model is an exact description of the system. The model shows exactly how the population *p* is determined from migration, birth, and death rates.

The model is particularly useful also if we wish to study the way *p* varies with time for several different values of *b*, *d*, *i*, and *e*. We might program a digital computer to represent the model. We could also use an analog computer described by the same set of equations.

Perhaps, the importance of the model can be best emphasized by considering the problem of a town planning new schools, new police and fire facilities, or new hospitals. Intelligent planning in any of these areas requires prediction of town population several years into the future.

The model indicates that the future population is determined by the present population, p_0, and by the four quantities *b*, *d*, *i*,

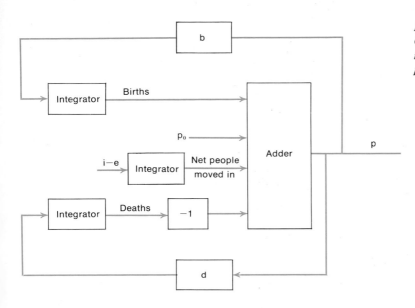

Fig. 5–38.
Complete
model for town
population.

p	population (thousands) at any time
p_0	population at start in thousands
i	immigration rate (thousands of people/year moving in)
e	emigration rate (thousands of people/year moving out)
b	births per 1000 population per year
d	deaths per 1000 population per year

and e. In order to predict p, we must select appropriate values for b, d, i, and e into the future:

b depends on the number of women of child-bearing age and on the economic and sociological characteristics of the community.

d depends on the age distribution of the population, and on the characteristics of public health (epidemics, adequate supply of doctors, accident rate, etc.).

i and e depend on the land available for new housing, the tax policies of the town, the nearby availability of jobs, the quality of the schools, and the general trends in the region (e.g., the extent to which people are moving from a neighboring city to the suburbs).

How do we determine these values? The representation of the model on a computer lets us rapidly find the future changes in p with many different values for b, d, i, and e. Study of these different predictions then allows us to decide on likely future changes in p, as well as the range over which the future population might vary.

Furthermore, the model helps us in setting the tax rate and zoning policies of the town so that in the future there will be enough schools and municipal services for the expanded community. In other words, work with a model of this type allows a much more intelligent approach to city planning and government. The model provides a basis for selection among alternative, possible courses of action.

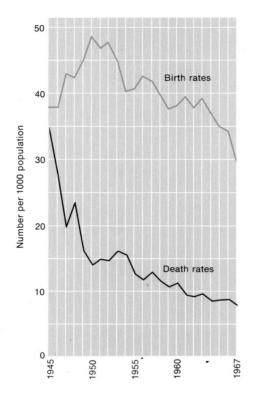

Fig. 5–39
Finding an accurate model
for a town or country requires
that we insert accurate birth and
death rates. For example, a model for
the island of Mauritius in the Indian
Ocean might be based on the data shown.

The population was known to be
700,000 in 1962. If we assume that the
death rate stays constant at 10 per 1000
population per year and the birth rate
levels off at 35 per 1000 population
per year, the net change is +25/1000
population per year (or 2.5% per year).
The population will double in 28 years.

A more accurate model would
insert signals for the birth and death
rates (b and d) into Fig. 5–38—
signals which were the best possible
predictions of the future. For
example, the high birth rate
around 1950 should result
in another high value
around 1975, when
girls born in 1950 are
having children.
Thus, to predict
with confidence we
need to know the
population in each
age category.
(Demographic
Yearbook,
United
Nations,
N.Y.,
1954–67.)

6 | THE NOISE-ENVIRONMENT SYSTEM

For the last fifteen years, the noise in our major cities has been increasing at a constant rate. If this growth continues, every middle-aged resident of New York City could be totally deaf by the year 2000.

If such a statement is true, or even partially "true," we can be certain that New Yorkers will suffer in less obvious ways long before the year 2000. Probably city dwellers are already showing some bad effects of the sound environment. By the time the effects become so obvious that the public demands action, there is likely to be permanent damage which cannot be corrected.

In this section, we want to look at this *noise system*. What are the inputs—the sources of noise? How might these inputs be reduced or controlled? Which are the inputs which are especially harmful to human beings? How can we, as citizens and voters, decide what new laws should be approved to "clean up" the noise environment? How can we understand the problem better?

Where does this example fit into this chapter? We have talked about block diagrams for systems. In both the home-heating system and the town population example, we drew a block diagram. This model showed the cause-effect relations in the system. From the block diagram, we could see which inputs affect which outputs.

Then we could see how to change these inputs to obtain the end result we wanted.

Unfortunately, many interesting systems are very complex. Often there are many inputs and outputs. Even if we know what we would like the outputs to be, we may not be able to decide what to make the inputs.

This complexity often appears in *social systems* (systems which describe problems of society). The system is like a giant balloon. Trying to improve the situation, you change one input. This is like pushing in the balloon at one point. The balloon simply bulges out somewhere else or the pressure inside increases.

Similarly in social systems, an action by city government may improve some part of the system, but actually make things worse in other ways. For example, a city starts a very liberal welfare program to help the people who are unable to work. As a result there is a heavy migration of unemployable people to the city. The generous welfare program becomes much more expensive than planned. Taxes are raised, businesses move out of the city, the number of jobs decreases, and the welfare problem grows steadily worse.

Thus, in social systems, we often cannot draw a neat block diagram. Our system ideas can still be useful. We first study the system. As we gain understanding, we see how to simplify the system. Certain inputs and outputs can be neglected. While quantitative relations between inputs and outputs may be impossible to find, we can try to guess intelligently about which inputs have what effects. From this study, we can begin to understand how to improve the system. The more we understand the system, the more likely we are to be able to make intelligent decisions.

In order to show some of these problems with social systems, we consider the noise environment of a major city. Because it is the largest city, New York happens to be a few years ahead of other cities in most problems. Therefore, some of the following data are for New York, but the problem is very similar even in cities of a few hundred thousand people.

Measuring Noise

If we talk about the noise environment, we must have a way of measuring the amount of noise. (By *noise*, we mean any unpleasant or unwanted sound.) How bad is the noise? How much worse is it now than five years ago? How much worse is it in New York City than in Norwich, Vermont? How bad is it at various times of the day?

These are questions which really cannot be answered easily with numbers. People react differently to noises. This is partly because people have different hearing abilities. But your reaction to noise also depends on your nervousness, your ability to shut out disturbances, and so forth. What we need are some measurable features of noise which indicate its effects on an average person.

Three aspects of noise are important in determining the degree to which the noise disturbs or injures people:

1. *Duration.* The longer the noise lasts, the worse it is.
2. *Loudness.* This is measured in units called *dB* (the abbreviation for decibels, named after Alexander Graham Bell). The dB indicate how loud a sound seems to the human being. A level of 0 dB is the weakest sound an average person can hear. As shown in Table 5–1, normal conversation corresponds to a level of 60 dB. Standing three feet from a riveter results in a noise level of 125 dB.

Conversation	60 dB
Traffic	80 dB
Factory machinery	80 dB
Garbage disposal unit	80 dB
Heavy truck at 25′	90 dB
Food blender	93 dB
Subway train at 20′	95 dB
Power lawnmower	96 dB
Loud outdoor motor	102 dB
Loud motorcycle	110 dB
Hammers, compressors at 10′	110 dB
Large pneumatic riveter at 3′	125 dB

Table 5–1. Decibel levels of common noisemakers.

3. *Frequency.* In Chapter 6, we will see that any signal can be viewed as a sum of sinusoids. The sound signal of noise is no exception. Each noise signal has its own frequencies. Certain frequencies are more annoying than others to human beings. Also, the annoyance depends on the combinations of frequencies. Figure 5–40 shows the frequencies and loudness levels for a few familiar sounds. We see that a boat whistle has frequencies around 200 cycles/second, and the level is about 100 dB. (This is for people standing by the whistle.)

Thus, to describe noise, we have to say how long it lasts, how loud it is, and what frequencies are present.

The effect of noise on a person depends on the duration, loudness, and frequency. The effect depends on the frequency most of all because the human being hears different frequencies differently. Figure 5–41 shows the actual size of a pure note (a sinusoid) at different frequencies when the average person thinks the loudness is the same. In other words, a note at 63 cycles/second must be 55 dB to seem to be as loud as a note of 4000 cycles/sec at 3 dB. (The human ear is very sensitive to frequencies from 2000 to 5000 cycles/sec.)

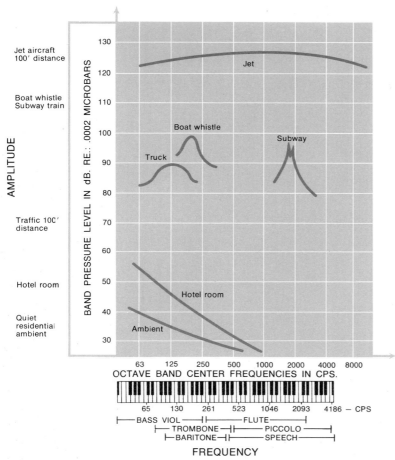

Fig. 5–40.
Noise spectra—
levels vs. frequency.
("Toward a Quieter City,"
Report of the Mayor's Task Force
on Noise Control,
New York, N. Y.)

Thus, a 40 dB noise at 125 cycles/sec would just barely be heard by an average person. A 40 dB noise at 3000 cycles/sec would sound loud. If we worry about the noise environment, our main concern is with noise which has frequencies from perhaps 1000 to 5000 cycles/sec.

To describe the noise environment, we would like a single number. Stating the loudness and frequency of all different parts of the noise is a nuisance. Our single number should measure the noise loudness in each frequency region, and weight this according to how sensitive the ear is for those frequencies. Several measurements of this type are used by engineers. A common one is called dB(A). The A refers to an average or weighted effect. The dB(A) is a single number which describes *very approximately* how loud and annoying a noise seems to a human being.

With this average measurement a jet plane landing 400 feet away has a loudness of 100 dB(A). A truck passing 50 feet away produces noise of 94 dB(A).

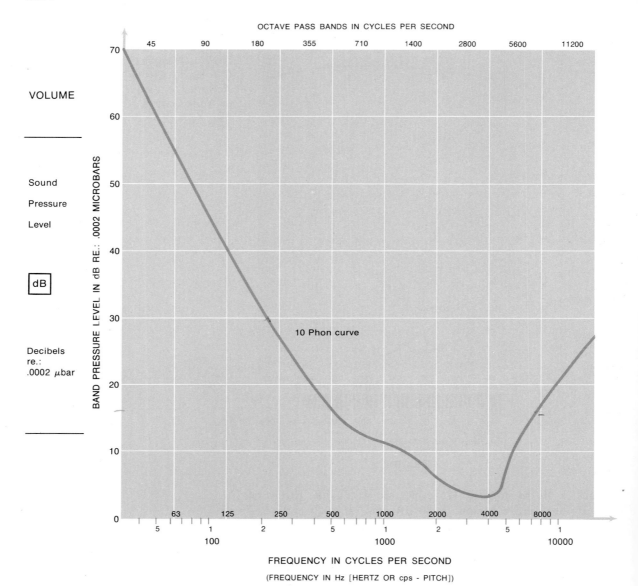

OCTAVE PASS BANDS IN CYCLES PER SECOND

10 Phon curve

FREQUENCY IN CYCLES PER SECOND

(FREQUENCY IN Hz [HERTZ OR cps - PITCH])

VOLUME

Sound
Pressure
Level

dB

Decibels
re.:
.0002 μbar

BAND PRESSURE LEVEL IN dB RE.: .0002 MICROBARS

Effects of Noise on People

How loud can noise be before it disturbs people?
Damage to hearing apparently occurs only when the noise is greater than 85 dB(A). The loudness can, of course, be much greater for short periods of time. Men who work near machinery which frequently causes noise levels above 85 dB should have their hearing checked regularly.

Fig. 5–41.
A curve showing the way the pressure changes as the frequency changes. To a listener, the sound seems to have the same loudness.
("Toward a Quieter City," Report of the Mayor's Task Force on Noise Control, New York, N. Y.)

How noise affects people is a difficult question. We can observe that deafness is common when men work in a very noisy environment. (Perhaps 10 million Americans are now seriously affected.) What we are really concerned about, however, is the long-time effect of noise on people. In the United States, hearing ability tends to decrease with age. Studies of African tribes who live in quiet environments have shown that people there still hear as well at age seventy as urban Americans in their twenties.

Other research indicates a need for caution. At the University of Tennessee, the poor hearing of a quarter of the freshmen suggested an experiment. A recording was made of a typical rock music session at a discothèque. For three months, this music was replayed at the same loudness for a guinea pig with one ear plugged—some days for as much as four hours, other days not at all. After 88 hours of music, many of the cells in the unplugged ear were destroyed. There was clear evidence that the hearing ability was seriously damaged by music of normal loudness. Of course, what happens to a guinea pig is not necessarily an indication of the effect on human beings. We need much more research.

Another direction of current research is interesting. Recent studies have shown that noise can move people from a deep sleep to a very light sleep even though they do not awaken. During a night, a sleeping man's brain waves are measured. Every time the man falls into a deep sleep, noise is produced. By morning, he believes he has slept normally, but he actually has not had deep sleep. We find that the man is unusually tired in the morning. Apparently, noise can prevent sleep from being really restful.

Furthermore, noise during sleep may prevent dreams. Major dreams ordinarily last about 50 minutes.* If the man does not have periods this long of reasonable quiet, his dreaming is incomplete. The danger here is not only in the inadequate rest, but also in the fact that people who do not dream might tend to be nervous and perhaps more likely to have emotional problems, possibly because they cannot work out their frustrations in dreams.† Can we foresee universal mental illness as the noise environment worsens?

These comments point out that we just do not understand the effects of the noise environment on people. The few studies that have been made suggest that the effects may be serious. There has been concern that excessive noise may even cause death. We really don't know what noise is excessive or what the effects are.

In such a state of ignorance, we should adopt all reasonable steps to reduce noise. To do this, we need to know the sources of noise (the system inputs). We need to know which sources are most significant in creating the total noise environment. How can people be shielded from this noise? In other words, we need to have a model for the noise-environment system. From this, we can decide where to focus our efforts to reduce noise.

*
The time happens to be exactly the length of the usual class period in college.

†
Most people are unaware they have dreamed during sleep. If you don't remember dreaming very often, there is no reason to be worried.

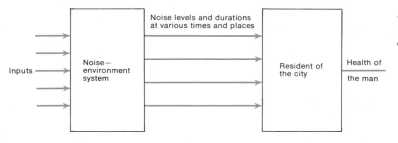

Fig. 5–42.
The noise
environment.

The Block Diagram of the System

The noise-environment system in a city can be represented by the block diagram of Fig. 5–42. The sources of noise (the inputs) cause various noise levels at different times of the day for our "average" man in the city. These combine to have some effect on his health. We really know almost nothing about the second (right-hand) block in Fig. 5–42. All we can really do is try to control the lefthand block and its inputs.

What are these inputs? That is, what are the major sources of noise for a city dweller? A few of the most important are:

1. Traffic. Here we include the sound of horns as well as of the vehicles.
2. Aircraft. Major sources are helicopters and, especially for those people near airports, jet planes landing and taking off. Sonic booms are occasional problems.

Fig. 5–43.
Some of the major sources
of noise. The chart shows the
levels associated with each source,
compared to the 85 dB considered
safe for hearing. (Photo by Vincent J.
Lopez for Department of Water
Resources. "Toward a Quieter City,"
Mayor's Task Force on Noise Control,
New York City)

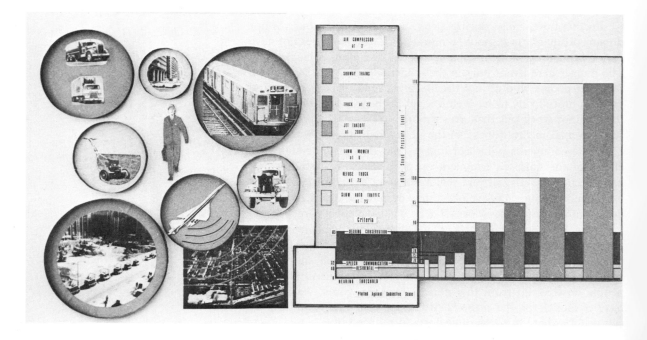

3. Construction. During the day, building and road construction equipment is a major source of noise. Compressors and pile-driving equipment are familiar to every city resident.

4. Refuse collection. Garbage trucks are especially noisy. Plastic bags (instead of metal cans) for storing the rubbish on the sidewalk have been a major help.

5. Heating, ventilating, and air-conditioning equipment.

6. Industrial factories and plants.

7. Sirens or alarm systems of emergency vehicles.

8. Subways. Here the tunnels and underground stations trap the noise. The new Montreal system has shown how the noise can be reduced by rubber wheels, better roadbeds, and sound-absorbent materials in the stations. In a very old system such as New York's, the subway is the major place where a large percentage of the population is in a noise environment which is likely to damage hearing directly.

9. Normal social activities. In some dance halls the noise level is 125 dB(A) which is just 15 dB(A) below the level we find standing 30 yards from a jet plane which is landing.

Once we identify these various inputs, we can consider steps to reduce the total noise level at the source (that is, reduce the inputs). Alternatively, if the input can't be controlled, we can try to change the system to reduce the noise levels at the man's ear. For

Fig. 5–44. Some of the noise reductions possible by city-government regulation. Noise from garbage trucks can be reduced 25% if new trucks purchased are designed to be as low-noise as possible. Rubber wheels on subway cars would cut the noise level by 80%. ("Toward a Quieter City," Mayor's Task Force on Noise Control, New York City)

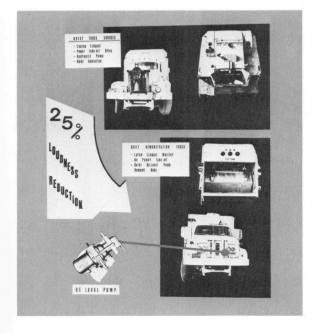

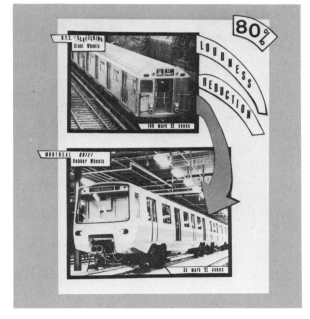

Fig. 5–45.
Another example of
possible city action (at
relatively low cost). Until
much quieter jack-hammers
are available, pavement
breakers should be surrounded
by portable acoustic panels.
(Photo by Vincent Lopez for
Department of Water
Resources. "Toward
a Quieter City,"
Mayor's Task
Force on Noise
Control, New
York City)

example, sound insulation in a city apartment can reduce the noise level inside, even if outside noises are uncontrolled. Anyone who regularly rides the New York subways should use ear plugs or ear covers similar to those worn by outside workers at jet airports.

The sound-environment system is a typical example of a social system. It is very complex (many inputs). There are many important parts of the system which no one understands. It is, however, a system which is of great importance to the quality of life in the broadest sense.

Figure 5–42 represents a general block diagram for the system. We could next proceed to describe the system in more detail. We might, for example, look at the noise inputs which affect our city

dweller during the night (when he is trying to sleep). How can these be controlled? What would be gained by a window air-conditioner to bring fresh air in (even during winter) without opening the window to bring in street noises?

Unfortunately, in a complex system like the noise environment, we know so little that it is difficult to find a detailed block diagram with numbers in each block. In such a case, we make measurements where possible, and then we have to guess at some of the cause-effect relations. The development of a noise-control program by the city government requires that we move from our general block diagram to a more detailed study, that we consider the costs of possible controls, and finally that we decide on a policy which reduces noise as much as possible within the allowable costs.

7 | MEASURING THE SYSTEM

A system model increases our understanding. From this, in many cases we can decide how to control the inputs or change the system in order to obtain better outputs. Man's control of his environment depends on this understanding. In decision problems, there is always a conflict. We can improve an output by making other outputs worse, by restricting personal freedom in order to change inputs, or by going to the trouble and expense of changing the system. We always want to find the least expensive and least annoying path to the desired output.

The noise-environment example illustrates this conflict in many ways. For example, we can decrease noise by banning sirens on ambulances. The "costs" are more accidents, more injuries to pedestrians, and more time in bringing emergency cases to hospitals. In order to make a logical decision in such a problem, we have to measure the system. If sirens are forbidden, how many more accidents will occur? How much time will be lost in delivering patients to hospitals? What will be the resulting increase in

Fig. 5–46. Another noise-reduction design. Though the aggressive cross-bar tread (D) is not particularly noisy when new, wear creates vacuum cups (E) which produce a high intensity whine persisting after other noise from the truck has subsided.

The tread shown as F has traction characteristics combining advantages of both cross-bar and rib treads. It is relatively quiet throughout its life. ("Toward a Quieter City," Mayor's Task Force on Noise Control, New York City)

deaths and disability? Finally, are these costs less than the value of lower noise level?

Even in this simple example, it is not easy to do experiments or make measurements. Often we have to make the best possible guesses. For example, to learn what a few minutes delay means in bringing accident victims to hospitals, we cannot purposely delay some people selected at random. About the best we can do is ask experienced doctors to estimate the value which should be given to time saved.

Experiments on Social Systems

In finding a model for a social system, it is often impossible to do an experiment to find specific data. In our noise-environment model, how can we determine the average noise from cars on a certain street? We cannot ban cars for a few days. Even if we could, this ban itself would change so many other activities in the city that we would not be measuring normal noise. When we try to make an unusual input to the system in order to measure the resulting outputs, we either change the other inputs or we alter the system. Then we cannot interpret easily the experimental results.

When we are modeling simple engineering systems, experiments are often useful. In the home-heating system of Section 3, for example, we can measure what happens when we don't use the front door at all for a day, when we stop using kitchen appliances, or when we turn off certain radiators. These are clearly defined experiments which allow us to describe the system quantitatively.

Part of the trouble with social systems is that the evaluation of the system depends partly on people's attitudes. Is the system performance annoying or pleasing? If we make a particular change, how much is public satisfaction increased? How much is this worth?

The answer to these questions lies in the subject of psychology. Asking people what they think about a controversial subject is often not adequate. People tend to answer according to what they believe they should say or what they would like to think, rather than the way they actually think. The phrasing of the question is also critical. For example,

> Do you (approve, disapprove) of the President's accomplishments so far?
> Do you (approve, disapprove) of the President's actions so far?

may elicit very different responses. The word "accomplishments" carries the suggestion that desirable things have actually been done.

To test public attitudes, experiments must be designed which try to measure these attitudes without the person realizing he is being measured, but with his responding in a natural way. For example, an attempt was made to measure the relative attitude of city dwellers toward their countrymen and toward foreigners.

This experiment was performed in three different cities (Paris, Athens, and Boston). In Paris three natives and three obvious foreigners were used to make the tests. The tester would pretend to pick up a piece of paper money, stop a passerby, and ask him if he had dropped it. Counter to the impression of some American tourists, Parisians treated foreigners more considerately than their own countrymen. Indeed, Paris was the only city where this was true.

Other, similar tests were also made in this study. Passersby were asked for directions or to mail a letter, store clerks were purposely overpaid to see if they would return the extra money, and taxi drivers were allowed to overcharge and take indirect routes. In mailing a letter, for example, Parisians again were more cooperative than residents of Boston or Athens and also treated foreigners better than their own countrymen.

Other ingenious methods have been devised to measure public attitudes. One is the lost-letter technique. To find the public attitude toward an organization like the Committee for Pure Air, for example, we would write 400 letters:

a. 100 letters to Mr. Williams Barnes, Chairman, Committee for Pure Air, P.O. Box 317, Boston, Mass., 02101.
b. 100 letters to Mr. William Barnes, Chairman, Committee for Better Education, P.O. Box 317, Boston, Mass., 02101.
c. 100 letters to Mr. William Barnes, Chairman, Committee for Overthrow, P.O. Box 317, Boston, Mass., 02101.
d. 100 letters to Mr. William Barnes, P.O. Box 317, Boston, Mass., 02101.

The first group (a) is the test, the (b) group is addressed to an organization which should be considered favorably by most people, (c) is for a group we would assume unpopular, and (d) is just to an individual.

In each envelope, we enclose a letter which seems important. (For example, we might state we will plan to be at the meeting two weeks hence and would enjoy having dinner afterward in order to discuss future actions.) The letters are then sealed and stamped, but not mailed. (Preferably, they are sealed in such a way that we can later tell if they have been opened and resealed.)

The 400 letters are now "lost," one by one, throughout the area where public attitudes are to be measured. Each letter is dropped face up in a spot where it obviously will be seen by people passing by.

A few days later, the mail is collected at Box 317 in Boston. Of the 100 letters in each category, the number returned seems to be a measure of public attitude. The more letters returned, the more the public likes the organization. The more letters that are opened, the more suspicious the public is of the organization.

These interpretations show up in the results of a survey in New Haven, Connecticut, to measure public attitude toward the Communist and Nazi Parties:

Letters to	% returned	Of those returned, % that had been opened
Friends of the Nazi Party	25	32
Friends of the Communist Party	25	40
Medical Research Associates	72	25
Mr. William Barnes (an imaginary name)	71	10

The results indicate what people think about the Nazi and Communist Parties. Apparently they are more suspicious of the Communists.

In this kind of test, we try to measure public attitudes. We hope people will respond according to the way they really think—not the way they feel they are supposed to answer.

8 | FINAL COMMENT

The system viewpoint often points out ways to improve the system. For example, many cities now have poison-control centers. When a person (usually an infant) swallows something poisonous, the mother can call the poison-control center. She describes the substance swallowed. The center operator types this information into a computer, which immediately types out the steps which should be taken (what to give the baby to drink, and so forth). The center operator relays this information to the mother.

Once this equipment is installed, it is easy to make a broader system study. The computer keeps track of all inquiries. The number of poisonings are sent regularly to Washington where national data are collected on each type of poison swallowed. From this study of the total-system inputs, it was found several years ago that many poisonings were occurring when the child found a bottle of baby aspirin. These pills were flavored to taste like candy, and the child would eat the complete contents of the bottle. As a result of this study, a law was passed which sets the maximum number of baby aspirin which can be sold in a bottle. Today you cannot legally buy a bottle containing enough baby aspirin to kill a child.

This system viewpoint is useful in many different areas. In economics, for example, a country's production may be described by an *input-output* model, also called a Leontieff model after the Harvard professor who developed the idea. The needed raw materials, labor, and other inputs are listed vertically at the left. The manufactured products are listed across the top (Fig. 5–47). Each entry shows the input needed for the output of that column.

Cold-war interindustry flows	Agriculture	Manufacturing	Household final consumptions	Gross totals
Agriculture		400	200	600
Manufacturing	200		600	800
Household labor and other factors	400	400	——	
Gross totals All numbers in billions	600	800		1,400

Fig. 5–47.
Leontief input-output
table. This figure just shows
the general form. If we are analyzing
a nation's economy, each industry
is listed twice, in a row and column.
Its row describes allocation of
its gross output as inputs for
other industries and for final
consumption. The column
shows inputs needed to
produce the output.
(P. A. Samuelson,
Economics,
McGraw-Hill,
8th edition)

An input-output array such as this is the basis for economic planning by the government (it has been used in the U.S.S.R., Israel, and India). Shortages of inputs can be anticipated and production can be predicted. Similar analyses are made by individual manufacturing companies as a basis for economic planning.

The system approach is also used in regional planning. Here the inputs are the future changes in population, jobs and housing, family income, college space needed, hospitals required, transportation needs, and so forth. Each of these systems must be coordinated by a central planning agency which makes sure that housing plans are coordinated with schools, hospitals, transportation, industrial development, etc. The development of one system without coordination with the others would result in chaos.

Thus, in the system approach to complex decision problems, we first find the inputs and outputs. We next find a model—a quantitative picture of the way the system relates these inputs to the outputs. The model of the system is the basis of the decision making.

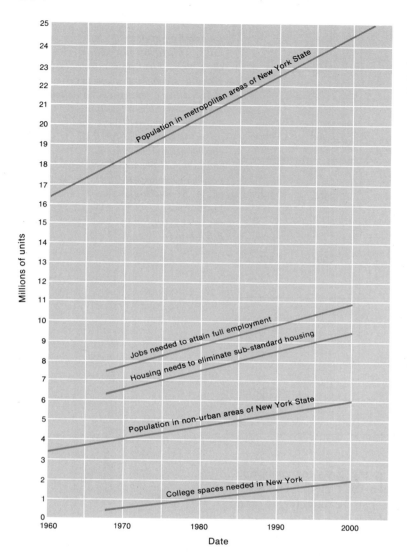

Fig. 5–48.
Examples of the inputs
for regional planning. ("Planning
for development in New York
State, "Office of Planning
Coordination)

Questions for Study and Discussion

1. The three news items cited at the beginning of the chapter are examples of dynamic systems. Explain what makes each example a system, and what makes it dynamic.

2. The examples of newspaper accounts of dynamic systems cited on the first page of this chapter were all front-page newspaper stories a few years ago. Look through today's paper for an example of a dynamic system which can be modeled. Explain what makes it a system and what makes it dynamic.

3. What are some steps which a government can take to ward off inflation? How would each of these steps have a negative effect on some segment of the population? List the steps and indicate the segment of the population which would be adversely affected.

4. The following block diagram and its accompanying graph is a model describing the effect of automation on the time needed to do a particular job:

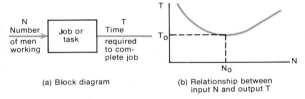

(a) Block diagram   (b) Relationship between input N and output T

Explain the reasons for the shape of the curve.

5. Suppose you are the owner of a small grocery store in the heart of a town. You are losing business to a large supermarket on the outskirts of a business district. You decide that the only chance for survival is to enlarge your store into a supermarket-type establishment.

What are the dynamic characteristics of your business with which you must be concerned in deciding whether to expand or go out of business?

6. Draw the block diagram which represents an input of students attending school and an output of graduates.

7. Draw a block diagram which shows the effect of report cards on the education of a student.

8. Draw a block diagram which shows the signal flow in a home-ventilation system which includes both heating and air conditioning.

9. Could the model of the population of the world be called a system (in the sense used in this chapter)? If so, what are the input and output signals?

10. What is usually meant by the phrase "the Mississippi River system"? Suggest a possible input-output relationship which would make it a system.

11.

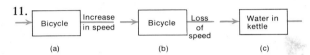

a) Give two different possible inputs for block diagram a.

b) Give a possible input for block diagram b.

c) Find a suitable input and output for case c.

d) Draw a block diagram to represent the electric lighting system in your bedroom.

12–14 The following diagrams show the input and output of a system, and a graph of output *vs.* input. From an analysis of the diagrams, tell as much as you can about the systems:

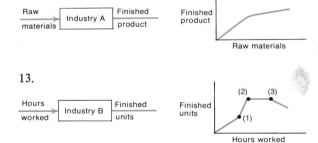

What action on the part of industry probably accounts for the change from (1) to (2)? What probably happened at (3)?

14.

Describe the system and make suggestions for leveling the curve.

Problems

1. A graph of an automobile trip, distance *vs.* time, is given below. Draw a graph of the velocity *vs.* time of the car during this same trip.

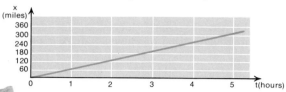

2. A distance-time curve for a 100-mile auto trip is shown below. Determine the velocity:

a) 120 minutes after the start.

b) 30 minutes before the end.

c) When the car is midway between the starting point and the destination.

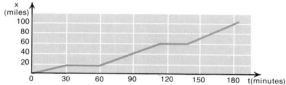

3. It is found during an acceleration of a racing car that it is 15 feet from the starting point at the end of the first second, 60 feet from it at 2 seconds, 135 feet in 3 seconds, and 240 feet in 4 seconds. Plot these data as a smooth curve and determine the velocity at 2 seconds and 4 seconds.

4. The current in a river has a velocity of 10 mi/hr. A motorboat on this river moves through the water at 30 mi/hr.

a) What will be the actual velocity of the boat with respect to the shore when going upstream?

b) What will be the actual velocity of the boat with respect to the shore when going downstream?

5. Dix Hills, New York, stores water for its residents in a large elevated storage tank. Water is poured into the tank from underground wells to replenish the supply as it is used. This added water flows in at a rate of q_1. The residents drain off water from the tank at a rate of q_2. If q_1 and q_2 (in gallons per hour) vary as shown in the figures, determine the volume of water in the tank at 10 P.M. if the volume at noon (the initial volume) is 16,000 gallons.

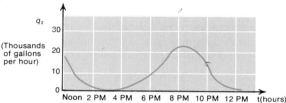

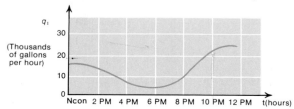

6. One of the instruments in a manufacturer's test-model of a new car is a very precise speedometer arranged to draw a graph (on a moving strip of graph paper) of speed against time. On a memorable day, the graph turned out to be a perfect semicircle. The curve left the 0-speed line, rose to a maximum height at 40 ft/sec, and returned to 0 exactly 20 seconds after it started. How far had the car gone? (Hint: Evidently it is necessary to find the area under the curve, and there is a handy rule for the area of a semicircle; but what do we use for a radius? Don't give up and fall back on the sum-of-rectangles method. What if we could use a single rectangle, 20 seconds long and R ft/sec high, where R is the *average* height of the semicircle? To get the average height, we need to change a semicircle with radius 40 (ft/sec) into a rectangle with base 80 (the diameter). How high must it be to have the same area?

7. Use the graph pictured in Fig. 5–39 and determine the *growth* rate for the island of Mauritius for the years 1945–1967. Plot this data on a new graph. Using the figure of 700,000 population in 1962, predict the population of Mauritius for 1980.

8. Use the graph pictured in Fig. 5–41, "Equal Loudness Contour," to answer the following questions:

a) A sound with a frequency of 500 cycles per second and a "volume" of 16 dB has the same loudness as a sound with a frequency of 2000 cycles per second and a volume of what?

b) Would a sound with a frequency of 8000 cycles per second and a volume of 15 dB sound louder or softer than a sound of 250 cycles per second and a volume of 30 dB?

Laboratory and Projects

I | MODELING A SYSTEM ON THE ANALOG COMPUTER

Part A

The purpose of this experiment is to model the input-output system pictured (Fig. 1) on the analog computer.

1. Wire the analog computer so that a constant signal of 0.5 volt is fed into one of the scaling units. The output of this unit will represent the amount the gas pedal is depressed.

2. Feed this output to one of the integrators. This will represent the system. Feed the output of the system to the meter.

3. Using the graph below as a guide, apply a signal of 0.5 volt (fully depressed gas pedal) and integrate this for five successive seconds. Read the speed on the meter (assume that 1 volt on the meter is 20 miles per hour).

Fig. 1.

Fig. 2.

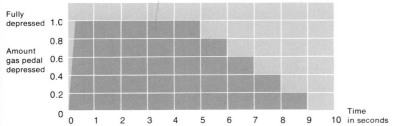

After the first 5 seconds, change the gas pedal to 0.8 of full depression (coefficient of scalor to 0.8) for one second, 0.6 for one second, 0.4 for one second, 0.2 for one second, and 0.0 for one second. Continue taking speed readings for five more seconds with the gas pedal at zero.

4. Record the speeds for each second.

5. Make a graph of speed *vs*. time.

6. How does this graph compare with that of Fig. 5–2 in the text?

7. If there is a difference, how do you account for this difference?

Part B

We will assume that the curve which you obtained in number 5 of Part A did not agree with the curve obtained in the text, and that we have not represented the system completely using just the integrator.

Let us assume for example that one characteristic of automobiles which we did not represent in our model is that of friction. Let us assume further that friction depends on speed, and that the greater the speed, the greater the friction. A block diagram to represent this situation is Fig. 3.

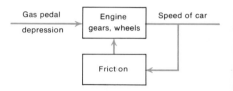

Fig. 3.

1. Let us now model the system on the analog computer. This time we feed the output of the integrator to another scaling unit which will represent friction. Since this friction has an effect opposite to depressing the gas pedal, we run our wire to an input to the scalor which has the opposite sign from that which we used for the gas pedal. The output of this scalor is then fed into the same integrator as that from the first scalor. Set the coefficient knob at about 0.5 and repeat number 3 of Part A.

2. Record the speeds for each second.

3. Make a graph of speed *vs*. time.

4. How does this graph compare with Fig. 5–2 of the text?

Part C

Optional

Figure 5–3 of the text indicates curves of cars with different rates of "pick-up." This can be simulated by changing the original signal to the first scaling unit.

1. Repeat number 1 of Part B using different values for the constant, one of 0.1 volt and one about 1.0 volt. Plot graphs for the data and compare them with the graphs of output in Fig. 5–3.

II | MODEL OF THE EFFECT OF MANPOWER ON JOB COMPLETION TIME

In connection with Fig. 5–6, there is a pictorial model of how the time required to complete a particular job changes with the number of men assigned to it. In this experiment we model the situation on the analog computer to gain some insight into the reasons for the speed-up and subsequent slow-down of the completion time for a job as more men are assigned to it. Use Fig. 1 as a block diagram to wire your computer. Simulate the conditions.

In this analog simulation we use the output of one of the integrators as the completion time (T) and each second of integration to represent the addition of one man (1 sec. = 1 man).

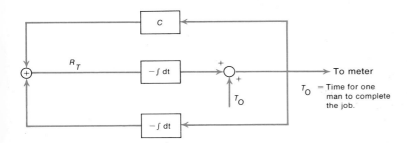

Fig. 4.
Block diagram
for manpower problem.

We now use the analog simulation to analyze the various aspects of our manpower problem. From Figure 4 we can see that we have a rate input (R_T) into an integrator where $R_T = \dfrac{\Delta T}{\Delta N}$ and depends on the difference between two factors. Notice that one of the factors is merely the product of the completion time (T) and a constant (C). Vary C on the analog computer and see how it affects your N vs. T curve. A chart recorder is an appropriate output device for this experiment.

1. What aspects of the manpower problem can be represented by the coefficient C? The other factor affecting R_T is the summation of T by the second integrator as more men are added to the job.

2. What aspects of the dynamics of the manpower problem does the summation of T represent?

3. Which factor increases the completion time of the job as more men are assigned?

4. How can a manager use this model to help him make decisions concerning the assignment of manpower?

5. What precautions should be kept in mind for this model?

III | MODELING GAS PEDAL VS. DISTANCE

In Experiment I we simulated the situation in which a change in the amount the gas pedal was depressed caused a change in the speed of the car.

In this experiment we shall follow that simulation one step further by feeding the velocity signal into another integrator which multiplies the velocity times time to give us the distance traveled.

Part A

Set up the analog computer as in Part B of Experiment I.

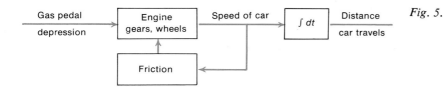

Fig. 5.

Feed the output into the second integrator.

■ 1. Use the same values of gas pedal depression as were used in Experiment I.

Time (sec.)	Gas Pedal Depression
1	full
2	full
3	full
4	full
5	full
6	0.8
7	0.6
8	0.4
9	0.2
10	0.0
11	0.0
12	0.0
13	0.0
14	0.0
15	0.0

■ 2. From the output of the second integrator read out the cumulative distance traveled from the beginning of the trip through each second until the end of the fifteenth second, and record.

■ 3. Graph these data of distance *vs.* time and compare the graph with Item 3 of Part B in Experiment I.

■ 4. How do you account for this difference?

Part B

Repeat the simulation by using the same constant values as in Part C of Experiment I.

■ 1. Plot graphs for the data and compare them with the graphs of Part C in Experiment I.

IV | SIMULATION OF A FALLING BODY

In Section 4 there is a discussion of velocity of falling bodies. Simulate this situation on the analog computer.

Part A

The initial model of this system will assume that there is no air resistance and the acceleration due to gravity is 9.8 meters per second each second. A block diagram is shown below:

Fig. 6.

A more mathematical model would be:

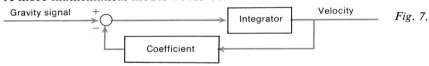

Fig. 7.

This assumes an initial velocity of zero and no other forces except gravity acting on the body. Wire this simulation on the analog computer and run it for 10 seconds. This data can be obtained by using the meter, CRO, or a chart recorder as the output device.

■ 1. Record the velocity at the end of each second for a period of 10 seconds.

■ 2. Graph this data as velocity *vs.* time. At what time was the velocity 20 meters per second?

■ 3. Does the velocity continue to increase throughout the 10 seconds?

■ 4. Repeat the simulation at five-second intervals for 30 seconds and graph the data. Does the velocity continue to increase throughout the 30 seconds?

■ 5. What is the velocity at the end of 30 seconds?

Part B

As a body falls through the air, there is considerable resistance to its fall. The system for return of spaceships to the earth must take this resistance into account.

The following block diagram indicates how this air resistance is added to the system:

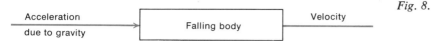

Fig. 8.

A model for the analog-computer simulation would be:

Fig. 9.

Wire this simulation on the analog computer and run it for 30 seconds with an air resistance coefficient of 0.2.

■ 1. Graph the velocity *vs.* time curve.

■ 2. Does the velocity continue to increase throughout the 30 seconds?

■ 3. What is the terminal velocity?

■ 4. Repeat for various coefficients (due to shape of body, density of air, etc.) and check against terminal velocity.

■ 5. What are the similarities between this experiment and Experiments I and III?

Part C

Using the second integrator to obtain distance traveled, wire the analog computer as in the block diagram below:

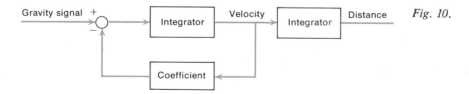

Fig. 10.

■ 1. If the acceleration due to gravity is 9.8 meters/sec each second and the air resistance coefficient is 0.1, how far will the body fall in 10 seconds?

■ 2. Assuming the same situations as in Question 1, how fast will the body be traveling at the end of 10 seconds?

■ 3. If the acceleration due to gravity is 9.8 meters/sec, but the air resistance is 0.4, how long will it take the body to fall as far as it did in 10 seconds under the conditions of Question 1?

V | ANALOG-COMPUTER SIMULATION OF A DYNAMIC POPULATION MODEL

The planning board of a midwestern city of 150,000 inhabitants is trying to attract manufacturing industries to locate their plants on the outskirts of the city. This plan will give the city additional income from corporations and property taxes. These industries, however, will effect a significant increase in the population, and the city must plan ahead for the increased demand for hospitals, schooling, housing, shopping facilities, and public utilities.

The future planning will be based on an estimate of the yearly population increase, and your problem as advisor to the planning board is to develop a population model which will predict future population changes. Use the analog computer to assist you.

You will base your model on the following assumptions:

a) The yearly rate of change of population, p', based on past experience, can be approximated by the equation:

$$p' = bp - dp + i - e$$

b = yearly birth rate
d = yearly death rate
i = yearly immigration into city
e = yearly emigration from city

b) The birth rate is, on the average, .05 per year.
c) The death rate is, on the average, .02 per year.
d) At the present time there is no net change in population due to people moving into and out of town. It is expected that the planning and construction of the first industries will take four years from the present time. At the end of the fourth year the net influx of population is expected to increase to 10,000 per year. The industrial development and growth is expected to last for 10 years. At the end of that time the rapid expansion will level off and the rate of influx and efflux will again be equal.

The analog-computer simulation for the population model is based on the block diagram in the text.

For this model let one volt be equivalent to 100,000 people and let one second of computing time be equivalent to one year of model time. Use a chart recorder as the output device.

1. Draw an analog-simulation diagram from the block diagram and make sure you understand the reason for all the connections.

2. Set the coefficients and wire up the simulation.

3. Obtain data and plot a graph of expected population at the end of each year for the next twenty years.

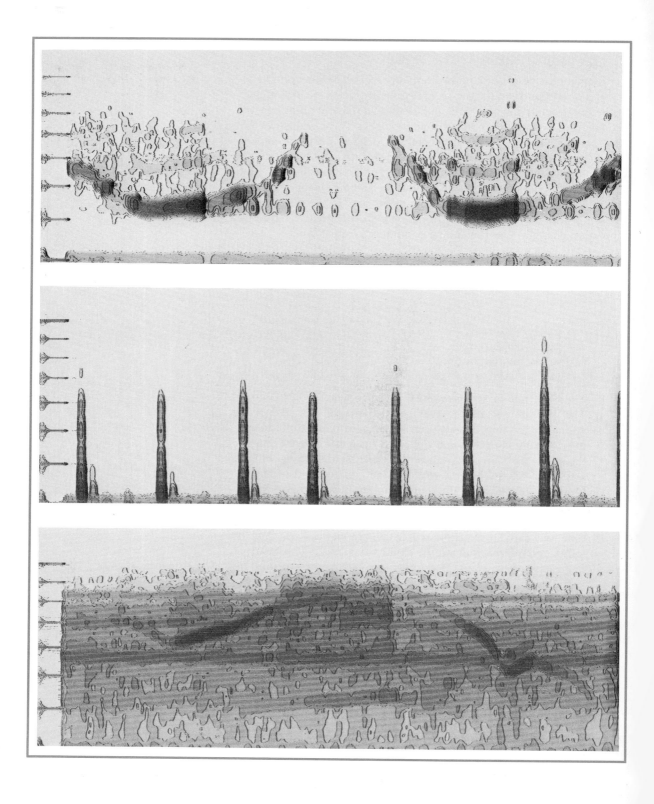

Patterns of Change

6

I | THE IMPORTANCE OF CHANGE

For a dull evening, nothing surpasses being paired with an individual who can talk about only one subject and then without any original or provocative ideas on that topic. The danger of falling asleep while driving is especially great on a modern thruway with little traffic, only gradual curves, and no billboards to arouse our interest or wrath. Every college student is painfully conscious of the effect created by a teacher who lectures (with great authority, perhaps) in a monotone—his voice never raised or lowered, with only an occasional pause for breath. Finally, almost no one is anxious to live in a community devoid of change, a town where no roads or parks are built, where there is no change in inhabitants, and where nothing "exciting" ever happens. In spite of the occasional, whimsical plea of the older generation for a return to "the good old days," very few people really want a life without change.

Change is essential in an interesting, exciting, and challenging life; the promise of change in the future stimulates and drives the individual to personal accomplishment. In the same way, situations of interest in applied science are normally characterized by change. We have to guess when and how changes will occur.

Indeed, we can even go further and state that the really interesting situations are those in which change occurs and in which *the change is unpredictable*. This feature of unpredictability adds enormously to the depth of interest. No one would watch TV if he could predict precisely what pictures would be shown for the

Sounds give clues to the behavior of both man-made and natural systems. Here three applications are shown for the sound spectrograms described in this chapter. Top to bottom, the whistle of a dolphin, clicking teeth, and a missile's whine. (Voiceprint Laboratories, Incorporated, Somerville, N. J.)

entire program duration. Interest is aroused because the future changes are uncertain, unpredictable.*

The idea of *unpredictability* is basic in communication. We only transmit information in the process of communicating if the message is unpredictable. As an extreme example, if the next page of this text included only a thoroughly familiar poem (for example, the words of the first stanza of the Star-Spangled Banner), the first few words would convey some information to the reader—the information that the poem was appearing. Once the reader recognized the poem, the remaining words would be totally predictable and would not convey any additional information.

In studying the man-made world, we are interested in this entire process of communication. Indeed, a major portion of our technological world has been created primarily to enhance the ability of people to communicate among one another. If we are to understand devices to assist in this process of communication, we must understand at least a little of the nature of the information which is conveyed.

In order to measure this information, we start from the idea that the amount of information contained in a message depends on the probability of that message. As an example of how information and probability are related, we consider a message or signal which consists of a sequence of 12 binary numbers, each either a zero or a one. Of all the possible messages we might send, one possibility is

$$0\ 1\ 1\ 0\ 0\ 1\ 0\ 1\ 0\ 0\ 0\ 1$$

Another possible message is

$$1\ 0\ 0\ 1\ 0\ 0\ 0\ 1\ 0\ 1\ 1\ 1$$

and so forth.

When we send this message, we have 12 different decisions to make: for each number in turn we can choose between 0 and 1. The man who receives our message thus obtains 12 different choices between two alternatives.

If for each of these choices, the 0 and 1 are equally probable alternatives (i.e., the receiver has no reason to suspect a 1 was more likely than a 0), we say that each number sent carries *one bit* of information. The bit is the basic unit of information; a simple choice between two equally likely alternatives is one bit. The total message, consisting of 12 successive choices, then carries 12 bits of information.

If we are using an alphabet with more than two symbols (e.g., the English alphabet of 26 letters plus a space, punctuation marks, and numbers), each letter in the message carries an amount of information which depends on the probability of that letter occurring: the smaller the probability, the greater the information. Thus,

* There is an interesting example of the importance of unpredictability. Television stations frequently broadcast sports events (football games, prize fights, etc.) a day or several hours after the actual occurrence. If one has learned the results from the newspapers or news broadcasts, interest is sharply diminished; the more detailed knowledge one has, the less the interest in the delayed telecast.

in a message using the English language, the letter E represents much less information than an X or Z, since E is a common letter (its probability is high).

In this chapter our interest focuses on *changing signals*, signals in which the nature of the change is the primary feature of interest. In particular, we wish to consider a few of the types of change which occur most often in the man-made world, how these changing signals can be used to understand the operation of a system, and finally some of the ways in which systems can be designed to control these changing signals. Our primary purpose in this introductory section is to emphasize that unpredictability and change are essential elements of information. When we communicate, we are interested in sending information to another person or to a machine. The signals we send (by voice, telephone, television, books, and so forth) must involve change.

In writing or speaking, we ordinarily do not want to maximize the information content. For example, if this book contained a maximum amount of information, every letter, word, sentence, figure and so forth should come as a surprise to the reader. Obviously such a book would be totally unreadable; any well-written book leads the reader gently from one thought into the next. This can be achieved only by a certain amount of redundancy— extra words and sentences to lead gradually into a new idea.

The redundancy of the English language can be illustrated very simply as follows. We write a logical paragraph. For each letter (one by one), we flip a coin. If the coin comes up heads, we leave the letter untouched; if tails, we erase the letter. Approximately half the letters are now erased. Now we give the paragraph with erased letters to a friend and ask him to fill in as many of the letters as possible.

The usual success in completing the paragraph suggests that about half the letters are really unnecessary. In a page, even more letters could be omitted, and indeed in a full novel or a textbook, a few pages missing would normally not be very serious. Similarly, when we speak, a large fraction of the sounds are redundant.

2 | COMMUNICATION WITH LANGUAGE

How is information communicated from man to man by the spoken or printed word? What are the properties of these man-made signals? How can we describe such signals? Can we hope to build machines which automatically translate written material into a different language? Can we build machines which recognize written material and give out sounds similar to those which a man would utter if he were reading aloud? Can we build machines which recognize spoken words? Will we ever in the future "dial" the phone merely by saying the number into the mouthpiece?

In this chapter, we cannot answer these questions. Many of them are still being argued by experts. We can only look at some of the characteristics of man-made speech signals.

Fig. 6–1.
"Red Man's Wireless,"
a painting by C. M. Russell
at the Heritage Center.
(*National Cowboy*
Hall of Fame)

Language

If we want to study some of the properties of human communication, we should first consider the language man uses to express his thoughts. There are between 3000 and 10,000 languages in the world today. The actual number depends on the degree to which different dialects are considered to be different languages.

The five major languages are Chinese, English, Hindi-Urdu (of India and Pakistan), Spanish, and Russian. More than two-thirds of the world's population speak one of these languages. In terms of printed material, English is by far the most important. For example, there are 2400 newspapers in English, compared with only 1000 in Spanish.

Actually, the number of different languages in the world is decreasing. If there are 4000 different languages today, there were more than 6000 different languages in 1800. We can anticipate an even more rapid disappearance of languages used only by small groups as travel and communication improve.

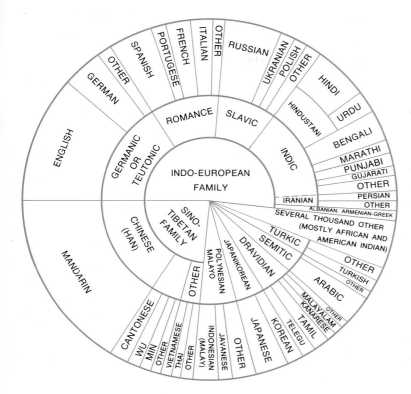

Fig. 6–2.
Principal
languages of the world.
(Quinto Lingo)

The ease with which young children learn to speak might suggest that the structure of a language is quite simple. By the age of five, most children speak fluently, although they may have an oversimplified picture of grammar. From this observation we can easily convince ourselves that our language has a logical structure. Perhaps there is a reasonable number of rules which describe the way sentences are ordinarily put together, the way words follow one another, and the way the language is built to match the capabilities of the human brain.

The hope that a reasonably simple structure for language does exist led to the optimistic feeling in 1955 that a few years of research could lead to computers programmed to translate automatically. For instance, we would be able to type an article written in Russian into the computer, the computer would then type out the English version. What information would be required for such automatic translation?

First, we need to store in the computer memory the translation of perhaps 30,000 Russian words.* But a word-for-word translation is certainly not very useful. Many Russian words have several possible English equivalents, and we must have rules for selecting one. Furthermore, word order must be rearranged if the English

*James Joyce in *Ulysses* uses about 30,000 different words. The average college student has a working vocabulary of 10,000 words, so our figure seems reasonable unless we want to handle many different scientific fields with their specialized vocabularies.

```
P* V NASTO45EE VREM4 MY NICEGO NE MOJEM SKAZAT6 O PRICINAX      IN PRESENT TIME WE NOTHING CAN SAY CONCERNING THE CAUSES
STOL6 SIL6NYX RASXOJDENII MEJDU 3KSPERIMENTAL6NOI I             OF SO STRONG DERIVATIONS BETWEEN EXPERIMENTAL AND THE
RASCETNYMI VELICINAMI *                                        CALCULATED QUANTITIES .
S TOCKI ZRENI4 SOZDANI4 ISTOCNIKA VYSOKOZAR4DNYX IONOV 3TO      FROM THE POINT OF VIEW OF CREATION OF SOURCE VYSOKOZAR4DNYX
RAZLICIE OKAZYVAETS4 PRI4TNYM , T. K. OBLEGCAET DOSTIJENIE      IONS THIS DIFFERENCE TURNS OUT TO BE PLEASANT , T. K.
NEOBXODIMYX ZAR4DNOSTEI *                                      FACILITATES ACHIEVEMENT NECESSARY ZAR4DNOSTEI .
TAK , NAPRIMER , REAL6NO SOZDAT6 ISTOCNIK S INTENSIVNOST6H      THUS , FOR EXAMPLE , REALLY CREATE SOURCE WITH INTENSITY
2.10,SUPER, 13 IONOV URANA ZAR4DNOSTI PLUS,24 V SEKUNDU ,      2.10,SUPER, 13 THE IONS OF URANIUM ZAR4DNOSTI PLUS,24 IN THE
KOTORYE MOGUT 3YT6 USKORENY DO NEOBXODIMYX 3NERGII NA          COURSE OF SECOND , WHICH CAN BE ACCELERATED UP TO NECESSARY
CIKLOTRONE S DIAMETROM POLHSOV 4 M *                           ENERGIES ON THE CYCLOTRON WITH THE DIAMETER OF POLES 4 M .
P* V ZAKLHCENIE AVTORY SCITAHT SVOIM PRI4TNYM DOLGOM            IN CONCLUSION THE AUTHORS COUNT ITS/THEIR PLEASANT LONG
POBLAGODARIT6 AKADEMIKA G.N.FLEROVA ZA INTERES K RABOTE I      TO THANK ACADEMICIAN G.N.FLEROVA FOR INTEREST TO WORK AND
E.D.VOROB6EVA ZA POSTO4NNUH PODDERJKU I POMO56 , VO MNOGOM      E.D.VOROB6EVA FOR THE CONSTANT SUPPORT AND HELP , IN MUCH
OBESPECIVWIE USPEWNOE VYPOLNENIE DANNOI RABOTY *               WHICH PROVIDED THE SUCCESSFUL COMPLETION OF THE GIVEN WORK .
```

is to be easily readable. (We won't even require that the translation be quality literature.)

Research over the decade after 1955 made one point clear: the structure of a language is very complex. From this work, it became apparent that the optimism of the early 1950's was premature. Truly useful, automatic translation has to wait until we have a much better understanding of language.

The complexity of languages is emphasized by the sharp differences from one language to another—differences which seem to point out major areas in which this man-made device was developed in very different ways by different civilizations.

Consider the way a noun is made plural. In English, the vast majority of plurals are formed by adding *s* after the singular form. There are, of course, exceptions for which we add *es*, change a final *y* to *ies*, or use a totally irregular form (women, fish). Regardless of these special cases, we can generally state that, when a noun refers to more than one, an *s* is added.

In contrast, the Chinese language has no required form change when a noun is converted from singular to plural. A number of languages have three different forms: one for singular, one for referring to two objects, and a third for referring to more than two.

Somewhat surprisingly, these are the only three classes into which languages fall if we compare them only on the basis of plural nouns. Each language has either one, two, or three forms for a noun.

How plurals are formed is just one example of the structural elements of a language and the way the rules vary from one language to another and, also, within a language from one word to another.

Structure of Language

We can see the complexity of language in another way. In order to obtain a picture of language, we need to understand that a sequence of letters making up a message is at least partially random

Fig. 6–3.
Example of the result of computer translation into English. The Russian typed into the computer is shown on the left. The "English" translation is on the right. Where "its/their" appears, this is merely the dictionary entry and has nothing to do with the algorithm on deciding which should be used. When a Russian word or term appears in the English translation, it means the term was not included in the dictionary and thus is reproduced in Russian rather than English. (Michael Zavechnak, School of Languages and Linguistics, Georgetown University)

Fig. 6–4.
Oriental calligraphy.
(TWA Ambassador)

(or unpredictable). If you can predict exactly the next sentence that follows this one, my writing the sentence and your reading it are obviously a waste of effort and time. Thus, the existence of randomness and unpredictability is necessary if the message carries any information.

We might first consider the kind of written message which would result with complete randomness, each letter equally probable, and each letter completely independent of what went before. The English alphabet of 26 letters is used, and we add a 27th symbol representing the space between words. If these 27 letters are equally probable and we choose a sequence at random, we obtain a message like*

XFOML RXKHRJFFJUJ ZLPW FWKCYJ
FFJEYVKCQSGHYD QPAAMKBZAACIBZLHJQD.

Similar sequences can be generated if we have a deck of 52 cards plus two jokers. We decide ahead of time that the jokers

* This and subsequent examples are taken from *Science, Art, and Communication* by John R. Pierce. © 1968, by John R. Pierce. Used by permission of Clarkson N. Potter, Inc.

correspond to a space, the red aces to A, the red twos to B, and so forth. The cards are then shuffled and one is selected to determine the first letter. The *complete* deck is then reshuffled and the second letter is determined. At each stage, the complete deck is used, so that the original probabilities ($\frac{1}{27}$ for each symbol) are preserved in each selection. Because of the difficulty of shuffling a deck perfectly, a long sequence will eventually tend to show uneven symbol probabilities, but the process is so tedious that few people want to continue for too long.

The message sample is obviously not English or any other language for that matter. In order to obtain something that begins to look like our language, we might attempt to perform the same experiment, but this time we use each letter according to its actual probability of occurrence in English. Table 6–1 shows these letter probabilities. The table simply means that an E occurs $\frac{107}{1000}$ of

Symbol	Probable number of occurrences in 1000 letters	Probable number of occurrences in 1000 letters if space is not considered
Space	187	——
E	107	132
T	85	104
A	66	82
O	65	80
N	58	71
R	55	68
I	51	63
S	50	61
H	43	53
D	31	38
L	27	34
F	24	29
C	22	27
M	20	25
U	20	24
G	16	20
Y	16	20
P	16	20
W	16	19
B	11	14
V	7	9
K	3	4
X	1	1
J	1	1
Q	1	1
Z	1	1

Table 6–1.
These numbers can be found by counting the letters in many different samples of English writing. We then calculate how many times each letter occurs in every thousand letters, on the average. To get a true picture, we would want to use typical samples of books, magazines, newspapers, letters, advertisements, and so on. (Reprinted by permission of Curtis Brown, Ltd. Copyright © 1942, 1970 by Inga Pratt Clark. The table we have used is an approximation derived from Pratt's table which gives the probabilities to more significant figures. For example, the probability of a Z is 63/100,000. For our purposes, it is easier to round off these numbers.)

the time (about $\frac{1}{10}$ of the time). Z, at the other extreme, occurs on the average only once in every 1000 letters. If we now write a message based on each of the letters appearing with its correct probability, we obtain a sample like:

OCRO HLI RGWR NMIELWIS EU LL NBNESEBYA
ALHENHTTPA OOBTTVA NAH BRL.

While there are no recognizable words, the letters seem less fantastic than in our earlier example.

We can generate a new sequence of this sort if we have a table of random numbers. (Table 6–2 shows a sample page from a book of random numbers.) Before opening the book to select a page, we decide how we will read the page. We can read to the right, downward, diagonally, any way we choose. Suppose here we decide to read in the normal left-to-right, line-by-line fashion. Furthermore, we will consider the digits in groups of three, the first three, the second three, and so forth. Finally, before selecting a page, we decide how various numbers from 000 to 999 will correspond to particular letters. Since the space occurs 187 times in every 1000, we might choose to let all numbers from 000 through 186

87 35	67 44	51 49	18 98	97 84	75 22	53 29	10 52	26 87	54 92
25 52	29 67	35 99	48 88	40 68	63 68	82 39	38 47	91 39	11 00
87 17	83 31	25 59	87 48	25 80	24 08	81 45	21 32	90 08	44 31
05 04	40 35	72 95	48 56	77 57	63 19	80 16	48 52	06 47	64 98
81 16	09 24	91 71	29 76	54 01	53 47	30 67	62 95	56 58	10 91
54 85	79 88	57 91	11 69	10 22	71 87	24 92	52 64	42 82	78 95
44 78	19 18	35 40	27 66	89 72	21 17	71 69	95 17	97 17	62 60
97 20	98 97	37 33	93 75	18 88	35 85	46 05	07 20	08 17	66 24
98 77	57 51	40 41	76 24	18 54	60 61	79 13	94 57	50 73	89 68
78 12	77 30	83 30	59 28	73 33	47 07	60 07	45 38	82 10	73 19
41 19	70 62	43 46	06 13	22 38	31 18	64 60	07 14	49 16	28 16
70 64	30 55	67 46	95 79	63 66	82 56	67 10	76 77	03 22	42 18
06 56	09 89	68 87	79 19	35 94	66 18	17 94	72 81	72 77	92 39
29 46	18 28	08 88	48 56	49 44	67 82	72 67	28 83	10 26	58 13
42 14	55 51	72 95	29 25	15 18	25 68	48 92	87 16	78 43	17 47
33 75	87 15	15 23	13 79	62 73	76 69	09 77	82 65	72 47	59 56
09 80	99 61	98 08	34 11	88 79	08 32	46 78	33 58	44 16	12 23
98 31	57 50	85 80	53 39	05 92	54 42	29 01	35 23	09 84	96 64
51 70	52 55	83 12	95 02	79 11	49 79	87 95	98 48	88 68	64 77
27 83	61 07	49 05	46 20	35 78	31 34	42 50	68 11	42 14	29 77
78 84	69 15	64 42	92 39	36 08	56 39	35 02	92 78	46 63	82 98
22 12	89 66	49 09	99 10	62 53	19 31	81 83	50 43	37 42	10 00
69 41	59 54	82 72	44 66	64 03	76 59	12 12	41 56	34 90	26 06
54 99	46 54	51 38	59 07	64 21	81 17	88 47	23 05	63 43	08 67
99 91	82 79	92 62	44 24	01 34	45 16	33 56	17 78	42 86	70 94

Table 6–2.
Tables of
random numbers
used to be generated
by looking at census
information and selecting
data which occur at random.
Since 1950, computers have been
used. A very large number (say 27
digits) is squared, another large
number is added, and the last nine
digits of the result are used. The
process is then repeated. ("Tracts
for Computers," edited from the
Dept. of Statistics, University
College London, No. XXIV,
"Tables of Random Sampling
Numbers," (2nd Series) M. G.
Kendall, B. Babington
Smith, Cambridge
University Press,
London, England,
1951, p. 20.)

correspond to a space; the next 107 numbers to an E; the next 85 to a T, and so forth. Then our rules for finding the letter corresponding to each three-digit number are summarized in Table 6–3.

We see that letters can be chosen according to their individual likelihoods, but the resulting message still would not look like English. Perhaps we can do better if we find a message which has both the correct letter probabilities and the correct probabilities of pairs of letters in sequence (called the digram frequencies). A sample message according to these rules is:

ON IE ANTSOUTINYS ARE T INCTORE ST BE S DEAMY ACHIN D ILONASIVE TUCOOWE AT TEASONARE FUSO TIZIN ANDY TOBE SEACE CTISBE.

In other words, this message has been chosen at random, with the use of a table of random numbers, but with correct letter and digram frequencies. We are beginning to get something that occasionally looks like English and is even in part pronounceable. There are several actual English words in this sample message.

The next step would be to form a message which also preserved the appropriate trigram (three-letter) probabilities.

IN NO IST LAT WHEY CRATICT FROURE BIRS CROCID PONDENOME OF DEMONSTURES OF THE REPTAGIN IS REGOACTIONA OF CRE.

In spite of the complexity, we are still far from English.

We can continue this study to understand the complexity of the English language by considering the formation of sets of words according to the appropriate word probabilities, then according to the probabilities of pairs of words, and so on. This sort of exercise leads ultimately to such word collections as,

> I forget whether he went on and on. Finally he stipulated that this must stop immediately after this. The last time I saw him when she lived.

where the words are selected with appropriate probabilities for every group of four in a row.

Even though such words are readable, they of course make no sense and convey very little in the way of information or emotion to the reader. They are convincing evidence that the meaning of a sentence depends not only on the grammatical form, but also on the total environment in which that sentence appears.

As a final example of this complexity, Pierce has cited the sentence: Time flies like an arrow. As he points out, these five words may have many different meanings. Three obvious interpretations are:

1. Time passes swiftly by, just as an arrow whizzes by.
2. Time flies (a special breed of flies) are attracted by an arrow.

Number	Letter				Table 6–3.

Number	Letter
000–186 inclusive	Space
187–293	E
294–378	T
379–444	A
445–509	O
510–567	N
568–622	R
623–673	I
674–723	S
724–766	H
767–797	D
798–824	L
825–848	F
849–870	C
871–890	M
891–910	U
911–926	G
927–942	Y
943–958	P
959–974	W
975–985	B
986–992	V
993–995	K
996	X
997	J
998	Q
999	Z

Notice that more than $\frac{1}{2}$ of the letters will be in this group of 5:

More than $\frac{3}{4}$ of the letters are in this group of 10.

More than 90% are in this group of 16.

3. An order: you are to time or clock flies as they go by, just as you would an arrow.

How can the computer, trying to translate the sentence into another language, ever decide which meaning is appropriate? Yet to the human reader, this sentence appearing in the middle of a story would almost certainly be perfectly clear.

3 | SPEECH

The preceding section leads to the conclusion that automatic or computer translation is not likely in the near future. Language is just too complex, but when we turn to speech signals we are a little more fortunate. While the detailed structure of speech is as complex as language, we can measure some very important properties of speech. From a knowledge of these properties, we can hope to make machines which "understand" very simple spoken commands and which recognize the speaker. Already we can speed up or slow down the speech signals so that the talker appears to be speaking faster or slower.

Speaker Recognition

In the summer of 1967, Israel and several Arab nations were involved in an intense, short war. In the midst of the conflict, Israeli officials called together reporters to hear a taped recording claimed to be a radio conversation between President Nasser of the United Arab Republic and King Hussein of Jordan. The speakers were discussing whether to blame the United Kingdom, or the United States, or both for the destruction effected by the Israeli air force. Translation of the alleged conversation reads:

NASSER: Will we say the United States and England—or just the United States?

HUSSEIN: The United States and England . . .

NASSER: By God, I say that I will make an announcement, and you will make an announcement, and we will see to it that the Syrians make an announcement that American and British airplanes are taking part against us from aircraft carriers . . .

HUSSEIN: Good. All right.

Throughout the world, the question immediately arose: Are these really the voices of Nasser and Hussein? In order to determine this, copies of the tape were brought to Lawrence G. Kersta, a New Jersey communications engineer who specialized in voice identification. Kersta compared the particular signals present when the alleged voice of Nasser spoke certain basic sounds and words with a similar analysis of recordings from Nasser's past speeches. Kersta was convinced that the voice alleged to be Nasser was indeed that of the U.A.R. President.

The accuracy of such speaker-identification schemes is a subject of great dispute among scientists, police officers, and computer designers. All agree, however, that if a highly reliable voice identification system were available, it could be of use in the protection of personal privacy such as the discouragement of threatening and indecent phone calls. In the next few paragraphs, we want to see how such a system may become possible.

The Speech Signal

As you listen to a friend talking to you, what signal reaches your ear?

At your ear, when sound is present, there is a rapidly changing air pressure. If we were to place a very sensitive pressure instrument (a microphone) at your ear, we would find the pressure

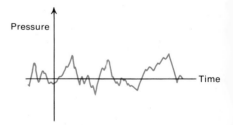

Fig. 6–5.
Pressure variation during a sound signal.

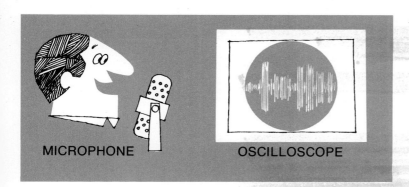

Fig. 6–6.
Viewing a sound signal.

changing as in Fig. 6–5. The exact way the pressure varies determines the particular sound you hear.

This pressure signal can be measured by the experimental arrangement shown in Fig. 6–6. The microphone is an instrument which changes the pressure signal into an electrical voltage. The voltage between the two wires from the microphone changes exactly as the air pressure changes at the microphone. As the oscilloscope spot moves from left to right on the screen, the voltage moves the spot vertically. Thus, the oscilloscope picture is a graph of the pressure changes with time.

The pressure signal is exceedingly complicated. The pressure rises and falls, but not at all in a regular way. The signal is so confusing, we cannot hope to identify the sounds by looking at a picture like Fig. 6–5. This sort of picture only convinces us that sound is indeed a series of complex, rapid pressure changes.

Fortunately, mathematics comes to our rescue. There is an important concept in mathematics which states that any real signal can be considered as the sum of simple signals. We may need to add up a great many of these simple signals—perhaps so many we never would want to try actually to do the addition, but the important fact is that the breakdown of a complex signal is possible. From this concept, we shall see that we can understand speech sufficiently to allow us to perform such operations as accelerating a speech signal. Furthermore, this concept is basic in our attempts to understand the sound-environment in which people live.

Before we discuss this idea further, we must understand the simple signals which can be added together to give speech, sound, and so forth. These simple signals are *sine signals* (also called *sinusoids*).*

Sine signals

Figure 6–7 is a plot of a sinusoid. The signal increases rapidly at first. The increase slows down until the signal is a maximum at time t_1. It falls until t_3, then starts to rise again. This cycle is

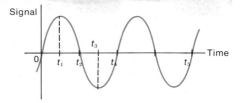

Fig. 6–7.
A sine signal.

*
The *Fourier theorem* states that any signal can be considered as the sum of sinusoids (sine signals) of different frequencies and amplitudes. Jean Baptiste Joseph Fourier (1768–1830), orphaned at eight, wrote sermons in his early teens, and became a math teacher at sixteen. In 1798 he traveled to Egypt as an adviser to Napoleon. His experiences there convinced him that the desert heat was healthy. In his later years, he dressed in heavy clothes and kept his room temperature at 90°. It is not surprising that his greatest work was done in the study of heat flow.

repeated. From t_4 to t_5 the variation is exactly the same as from 0 to t_4. Because the signal is continually repeating, we really need to show graphically only one *period* or *cycle* (Fig. 6–8).

There are two important numbers needed to describe our sine signal of Fig. 6–8. We need to state the *amplitude* or how big the signal is. Figure 6–12 shows various sinusoids of different amplitudes. (c) has the largest amplitude. If this were a sound signal, we would say (c) is the loudest of the three signals. In Fig. 6–13, we show a sinusoid of amplitude 8. The second important number which describes the sinusoid indicates the speed with which it passes through a complete cycle or period. To give a number to this idea, we talk about the *frequency* of the signal, or the number of complete cycles per second. In Fig. 6–14 signal (a) has a higher

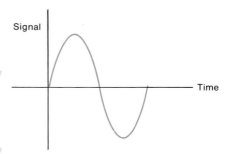

Fig. 6–8.
One period of our
sine signal.

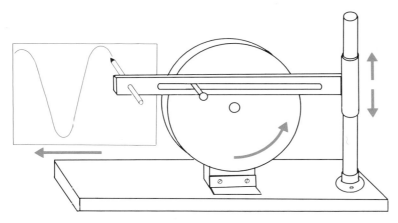

Fig. 6–9.
Production of
a sine signal. The
apparatus shown here can
be used to construct a sinusoid.
As the wheel is rotated, the pencil
moves up and down. If the paper is drawn
to the left at a constant speed, the sine
signal is traced.

Thus, the sinusoid is the vertical
position of P as the wheel rotates at
a constant velocity. If the wheel
turns at 50 revolutions/sec, the
sinusoid has a frequency of 50
cycles/sec.

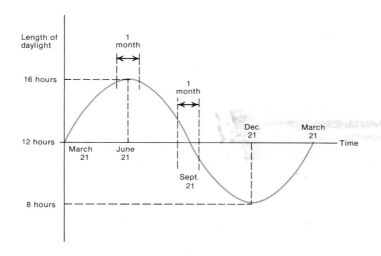

Fig. 6–10.
The sketch shows the
daylight hours during the year.
(The curve is approximately a
sinusoid.) The curve is shown for a
location in central United States, where
the longest daylight is 16 hours.

The sine signal changes very slowly
around the peaks, and rapidly near
the zeros. As a result, near
June 21 and December 21 the
time of sunset and sunrise
change very little each
day. Around March 21
and September 21,
the changes are
much greater.

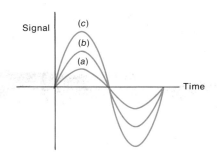

Fig. 6–12.
Sinusoids differing in amplitude or size.

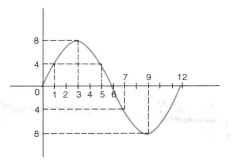

Fig. 6–13. *Proportions of a sinusoid. The peaks (maxima and minima) occur at regular intervals. The zero values occur midway between peaks. During the positive half cycle, for example, the sinusoid has ½ its peak value at moments one third of the time from the zero to the peak in either direction. This particular sinusoid is drawn for an amplitude of 8 and for a period (cycle duration) of 12.*

Fig. 6–11. *The tides vary in a sinusoidal fashion at any one location. They are partially caused by the gravitational pull of the moon and sun. They are influenced, however, by the earth's motion and by the motion of the water. There is no theory which allows prediction of how big the tides will be at a particular location or the time when high tide will occur.*

The picture shows Mont St. Michel in France, which is famous for the speed with which the tide changes. High tides rise as high as 50 feet and low tides recede to as far as 11 miles from shore. High tide travels into shore at a rate of 20 miles per hour. The top photo shows high tide. Bottom photo shows low tide. (French Government Tourist Office)

frequency than either (b) or (c). The frequency of the signal of Fig. 6–15 is 10 cycles/sec. This means that in each second, there are ten complete cycles.

For our purposes, then, a sinusoid can be described by its *amplitude* and *frequency*.

A sine signal in sound is what we call a *pure note* in music. The amplitude measures the loudness, the frequency indicates the pitch.

The terms *frequency* and *pitch* are closely related. The frequency is a term with a mathematical definition. A frequency of 440

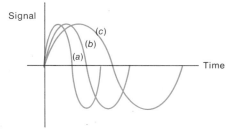

Fig. 6–14. *Sinusoids of differing frequency but the same amplitude.*

cycles/second means that the sinusoid repeats itself 440 times every second. For a sinusoidal sound signal, we can measure the number of cycles precisely and hence determine the frequency.

Pitch is the term used to describe the listener's judgment of the frequency. By international agreement, the musical note middle A is the sinusoid with a frequency of 440 cycles/second. It is not unusual, however, for a symphony orchestra conductor to have his musicians tune their instruments so that middle A is actually a few cycles/second higher. The important thing is that the instruments all be tuned to the same frequency.

Once the frequency for middle A is chosen, all other frequencies are fixed. If middle A is 440 cycles/sec, the A one octave higher is 880 cycles/sec. Each octave up corresponds to a doubling of the frequency (Fig. 6–16).* Four octaves above middle A is 7040 cycles/sec ($440 \times 2 \times 2 \times 2 \times 2$).

Likewise, once the frequency for middle A is chosen, all other notes are fixed (Fig. 6–17). The relationship among the frequencies for the various notes has developed in western civilization as shown in this figure. These are relationships to which our ears are now accustomed by listening to this man-made music from infancy on. When we hear Oriental or Indian music which is based on an entirely different man-made scale, we find it somewhat unusual.

A tuning fork generates a sound wave which is very nearly a pure sinusoid. Vibrating several tuning forks (of different frequencies) does not result, however, in particularly interesting music. The sound is not likely to remind you of any musical instrument. When we play middle A on several different instruments (piano, oboe, violin), why are the sounds quite different? Why don't they all sound like a tuning fork?

There are several reasons. First, and most important, each instrument generates not only the note being played, but also several *harmonics* (also called overtones). Harmonics are signals which are integral multiples of the primary note. For example, if middle A (440 cycles/sec) is being played, the second harmonic is 880 cycles/sec (twice the fundamental frequency), the third harmonic is 1320 (or three times 440), the fourth is 1760, and so forth. Each musical instrument gives a different set of harmonics. If the primary note were middle A with an amplitude of 2 units, two different instruments might give:

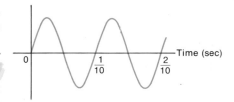

Fig. 6–15.
Sinusoid with frequency 10 cycles/sec.

* The word *octave* comes from the Latin *octo* for eight. In our musical scale, the eighth note is one octave higher than the first. The octave is divided into 7 parts by the notes.

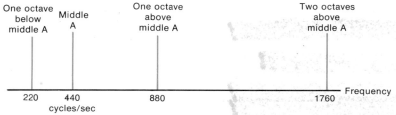

One octave below middle A — 220

Middle A — 440 cycles/sec

One octave above middle A — 880

Two octaves above middle A — 1760

Frequency

Fig. 6–16.
Octaves in sound.

Instrument A		Instrument B	
Harmonic	*Amplitude*	*Harmonic*	*Amplitude*
Third	0.5	Third	0.1
Fifth	0.2	Fifth	0.2
Sixth	0.3	Ninth	0.3
Seventh	0.2	Tenth	0.3

The two sounds would be very different to a human listener.

The same note played by different instruments also shows differences in the way the note starts and stops, and the manner in which the amplitude may change during the note. Such characteristics also depend on the person playing the instrument. One of the interesting research programs in recent years has attempted to determine what exactly are the differences between the sounds generated by an expensive violin and by one modestly priced. What features of the instrument cause these differences?

Such research is complicated by the fact that an expert violinist tends to play an expensive, high-quality instrument in a way that is different from the way in which he plays a cheaper instrument. A violin's sound output is particularly sensitive to the style of the player. As the bow is drawn across the string, the string sticks to the bow until it is pulled from its normal position, then slips back, sticks again, and so on. The string vibration is caused by this back-and-forth motion at the point of string-bow contact. The exact way the string vibrates depends critically on the angle between the bow and string, the pressure exerted by the player, the smoothness with which he draws the bow, the rosin used and the way it is put on the bow. Furthermore, to describe in detail any particular note played on a violin, we have to measure many

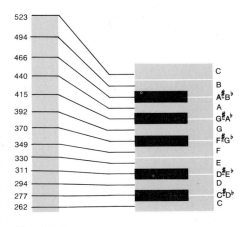

Fig. 6–17.

Frequencies of the notes which are used in a musical scale. The middle octave of a piano is shown. The standard is A at 440 cycles/sec. A sharp or B flat is the next note up. The frequency is about 1.06 times 440 or 466. B is 1.06 times greater still (or 494).

There are 12 equally spaced notes in each octave. The 13th is one octave above the first.

The white keys on the piano (C, D, E, F, G, A, and B) are the notes used in the C major scale. The black keys are for the five flats or sharps.

Fig. 6–18.

Louis Armstrong blows one of the notes that made him the world's foremost jazz trumpeter.

The curve is a drawing of oscilloscope pictures showing the pressure change when a pure note is played on the trumpet. The signal repeats itself regularly, but is not a sinusoid. The sound is the sinusoid plus a group of harmonics.

The exact signal depends on which harmonics are present and how large the separate harmonics are. These harmonics also depend on who is playing the instrument.
(The Bettmann Archive)

Box 6–1

The existence of harmonics can be demonstrated by a simple experiment. In a piano, each key controls a string which vibrates at the frequency of that note. When the key is idle, the string is clamped so it cannot vibrate. When the key is pressed down, the string is free to vibrate.

If you strike the key for middle C, that string vibrates during the time the key is depressed. The sound which is generated includes several sine signals: one at middle C (about 260 cycles/sec), one at the second harmonic (520), one at the third harmonic (780), and so on. The sound at the second harmonic will excite any string tuned to that frequency and free to vibrate.

To show this second harmonic, you can do the following. Depress the key for the C above middle C (one octave higher or 520) and hold it down. Wait until the sound dies out (eventually the string comes to rest). Now strike *momentarily* the key for middle C. After the middle C note disappears, you will hear the second harmonic note continuing for some time. The 520-cycles/sec string has been excited into vibration by the 260 note.

In a reasonably well-tuned piano and a quiet room, you can detect the third harmonic in the same way (the G one and one-half octaves above middle C).

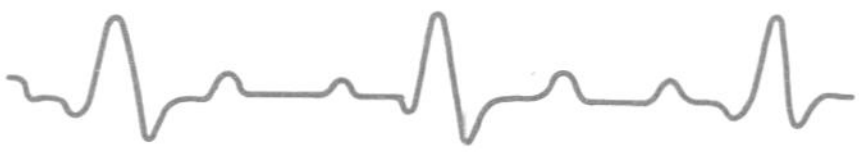

Fig. 6–19.
An electrocardiogram, measured as the electrical voltage between two points on the body. This is the electrical signal which is associated with each beat of the heart.

The signal shows that there are many harmonics present. (The signal is certainly very different from a sinusoid.)

In recent years, several different techniques have been developed to use a computer to analyze such a signal automatically. From measurements of the various peak values, the durations of the pulses, and the spacings, the computer is programmed to follow the same steps in logical diagnosis as used by a specialist.

different features: the amplitudes of the note itself and each of the harmonics, the way these amplitudes change during the note, and the start and end of the note. More than twenty different characteristics have been used to describe a note. It is not at all clear which of these are important in deciding how pleasant the note is to the human listener. Even if we eventually learn how to describe a single note, we then have the problem of understanding the rapid sequence of notes which occurs when music is played.

The preceding paragraphs begin to point out the complexity of sound. Even in the relatively trivial case of a single note played by a musical instrument, the signal is very difficult to describe. The attempts to program computers to generate sounds which resemble particular musical instruments have really not been too successful. Certainly no one is fooled by a computer which is supposed to sound like a violin.*

Research of this sort is important not only because of the understanding of music, but also because it provides a basis for a fuller understanding of human speech, how man hears, what signals are actually sent from the ear to the brain to allow man to recognize sounds, and how man generates sounds.

* Recently the French musician Risset generated a "trumpet note" which completely fooled listeners.

4 | SPECTROGRAMS

The most important characteristic of a note played by a musical instrument is the harmonic content: How many harmonics are present? This question is so important, we want a graphical way to display the answer. The display, called a *spectrogram*, is a picture that shows which frequencies are present as time progresses.

The simplest signal is the sound from a tuning fork. Here the sound is very nearly a sinusoid at a single frequency. Figure 6–20 shows the corresponding spectrogram. We plot frequency vertically and time horizontally. At time t_1 the tuning fork starts vibrating at 440 cycles/sec. This sound continues until time t_2, when it is abruptly ended.

In other words, the spectrogram shows the frequencies present at any time. The amplitude, or strength, of a particular sine signal is indicated roughly by how dark or heavy the line is. Thus, Fig. 6–21 shows the spectrogram when middle A is played on a flute; Fig. 6–22, the same note on other instruments.

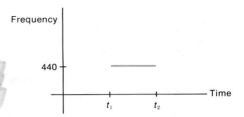

Fig. 6–20. Spectrogram of a tuning fork at middle A.

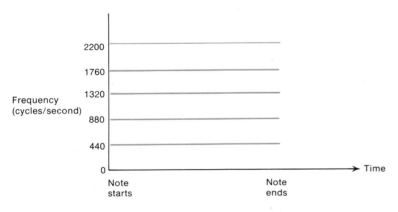

Fig. 6–21. Spectrogram for middle A on the flute.

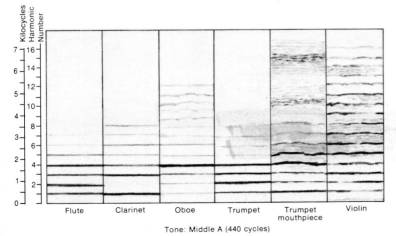

Fig. 6–22. A drawing of a spectrogram of the sounds of some common instruments. Depending on their shape, muscial instruments emphasize certain harmonics. (From the book Waves and the Ear *by W. A. van Bergeijk, J. R. Pierce, and E. E. David, Jr. Copyright (c) 1960 by Educational Services, Inc.; copyright (c) 1958 by J. R. Pierce and Edward E. David. Reproduced by permission of Doubleday & Company, Inc.)*

264

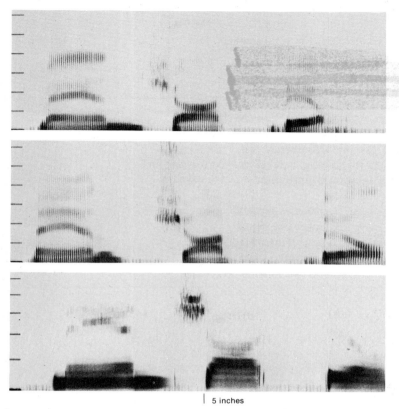

Fig. 6–23.
Typical speech
spectrograms. The
spectrograms represent
three different people
saying "man shot Bill." The
first two are males, the third a
female. The vertical scales are
marked in thousands of
cycles/sec. On the
horizontal scale, five
inches represent one
second. (Bell
Telephone
Laboratories)

| 5 inches

Spectrograms for Speech

As you speak, you are generating a sequence of air-pressure signals from your mouth to the ears of a listener or to a microphone if your words are being recorded or broadcast. Different sounds correspond to different spectrograms. Each sound you speak has certain frequency components, which change as the sound changes.

Figure 6–23 shows a variety of spectrograms for various speech sounds and different speakers. Obviously, speech represents many very different signals. Can we say anything about the characteristics of speech in general? As a first step, we might look at many different speech spectrograms. If we did this, we would soon notice one important fact. The important frequencies of speech never are greater than 3500 cycles/sec and the fundamental frequencies (not the harmonics) are far below 3500 cycles/sec. A bass singer has fundamental frequencies from 75 to 300 cycles/sec. A soprano voice ranges from 230 to 1400 cycles/sec. Just as with a musical instrument, however, the human voice generates not only the fundamental, but also many harmonics. It is these harmonics which cover the range up to 3500 cycles/sec.

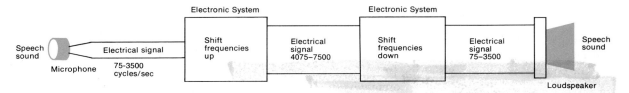

Sharing Communication Equipment

The fact that speech spectrograms show frequencies only to 3500 cycles/sec is one of the things which make modern communication possible. A telephone cable from Chicago to Denver, for example, can transmit signals with frequencies to about 2,000,000 cycles/sec.

Fortunately, with electronic equipment we can shift the frequencies of a signal. For example, we can take a conversation (75 to 3500 cycles/sec) and move this to the range 4075 to 7500 cycles/sec. Each component has its frequency increased by 4000 cycles/sec. The result is, of course, not intelligible, but then we can shift back down in frequency after it has been transmitted, so that we can recover the original signal. This process is shown in Fig. 6–24.

Once this idea of frequency shifting is accepted, we can see how 500 telephone conversations can be carried on one cable from Chicago to Denver at the same time. A system with three conver-

Fig. 6–24. Frequencies can be shifted up or down by any desired amount.

Fig. 6–25. Sharing of communication system. (a) Physical arrangement. (b) Frequencies of composite signal.

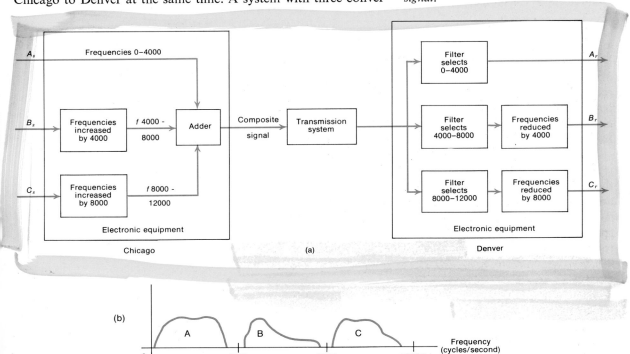

sations is shown in Fig. 6–25. We must have complicated equipment at each end to separate the 500 conversations. Even so, this equipment is much less expensive than a separate cable for each conversation.

The concept of frequency shifting (the technical term is modulation) is fundamental to modern communications. Exactly the same frequency shifting is used in radio and television broadcasting. As we tune our radio or change the channel selector in TV, we are changing the frequencies which are to be shifted downward to the original sound range.

Actually, AM radio stations usually broadcast a signal which contains original sound signals up to 5000 cycles/sec (music contains higher frequencies than speech). FM radio signals contain original sound signals up to 15,000 cycles/sec (nearly the maximum frequency the average person can hear). A television picture requires a signal which has frequencies out to 4,500,000 cycles/sec. Thus, each type of signal has a different spectrogram. All these systems, however, operate on the same general principles.

Changing Speech Speed

The spectrogram also suggests how the speed of speech can be changed. Suppose we want to tape record the reading of a novel so that blind people can enjoy the book. The average person can read aloud at about 100 words/minute. With practice he can reach 125 to 150 words/minute, although this is exhausting. A blind person would like to listen at 300 words/minute. Is there any way to speed up the reading?

The first thought would probably be to record the reading, then play back the record or tape at a higher speed. Playing a $33\frac{1}{3}$ rpm record at 45 rpm shows what happens with just a $\frac{1}{3}$ increase in playback speed. The "Donald-Duck effect" makes the speech irritating and unintelligible. In the speeding up, each frequency of the sound has been multiplied by $\frac{4}{3}$ (increased by $\frac{1}{3}$).

The spectrogram suggests a solution. A computer first finds the frequencies present in each short segment of the speech. Each of these frequencies is cut by a selected factor, say $\frac{1}{3}$ so the frequencies are now $\frac{2}{3}$ of the original values. The sum of the new frequencies is now recorded again. The recording is played back at $\frac{3}{2}$ the speed (so the frequencies are multiplied by $\frac{3}{2}$ and, hence, restored to their original values). The new speech signal sounds just like the original, but it is $\frac{3}{2}$ as fast. If the speaker was reading at 120 words/minute, the blind person listens at 180 words/minute. Yet he hears the *normal* tone of voice of the speaker. The result is called *accelerated speech*.

In recent years, accelerated speech has been used not only for reading to the blind, but also for recording lectures in college and in scientific work. Acceleration by a factor of 35% allows the listener to decrease the time required to hear the lecture without any real decrease in understanding. Indeed, some early tests indicate that understanding may actually be improved, perhaps because the listener has to pay closer attention.

Uses of Spectrograms

The speech spectrogram can be used in teaching children born deaf to speak normally. The deaf child is unable to copy the sounds spoken by others. As a result, he talks abnormally even though his lip movement may be nearly normal. In teaching such a child to speak, we find another person with a similar voice pitch and record his speech. As the deaf child speaks, his spectrogram is presented beside that of the normal speaker. By trial and error, the deaf child can try to make his spectrogram look like the one of the normal speaker.

This system has not been widely used so far. (At the Detroit Day School for the Deaf, it has been used successfully.) Simpler aids, displaying voice pitch and intensity, have also been used. All these devices help the deaf by giving them information about their own and others' speech. The devices replace, to some small extent, the missing hearing sense.

Spectrograms are basic research tools as we attempt to build equipment to recognize correctly spoken words or commands (e.g., feeding data into a computer by speaking, or telephone dialing by speaking a sequence of numbers), and to develop speaker-recognition systems.

Finally, spectrograms are being used to understand speech more fully. One of the goals is to build a computer which would "speak" normally. Different sound characteristics could be stored in the computer memory. The system could draw upon these to construct speech. The human quality of speech might be missing, but that would offer no major problem.

The new developments in speech spectrograms will help solve some of the difficult problems found in the synthesis of speech by man-made systems.

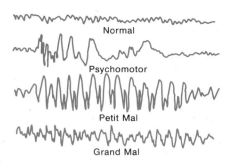

*Fig. 6–26.
The three
tracks of epilepsy
(complex sums of
sinusoids). A crucial
first step in the treatment
of epilepsy is identifying
which of the three main types
of seizure the patient has.
Doctors accomplish this with an
electroencephalograph machine. They
attach small, painless metal electrodes
to the patient's scalp. The electrical
waves given off by the brain are
magnified by the machine and written
on graph paper. The most frequent and
severe epilepsy is grand mal, "the
falling sickness," which involves
blackouts and convulsions.
Petit mal produces momentary
blackouts, with facial
twitching and a blank
expression. Psychomotor
epilepsy dims
consciousness and
often produces
brief amnesia.*

5 | SIGNALS RELATED TO SINUSOIDS

The sinusoid is a particularly simple signal to generate and to analyze. The sinusoid is important because many other signals are derived from it. In the following paragraphs, we consider radar and sonar as two important, yet relatively simple, signals which are related to sinusoids.

Radar Ranging

Radar* operates on the following principle. A radio signal of very short duration is transmitted. The sending equipment then

*Radar is a word derived from *radio detecting* and *ranging*. Initially developed during the 1930's to test radio reflections from the ionosphere (the layer of charged particles over the earth), radar became a vital weapon of World War II when it was refined for detection and location of ships and airplanes.

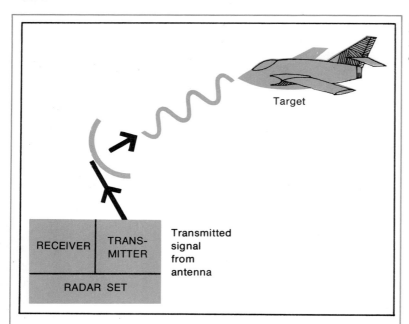

Target

RECEIVER | TRANS-MITTER

Transmitted
signal
from
antenna

RADAR SET

Fig. 6–27.
Radar system
operation.

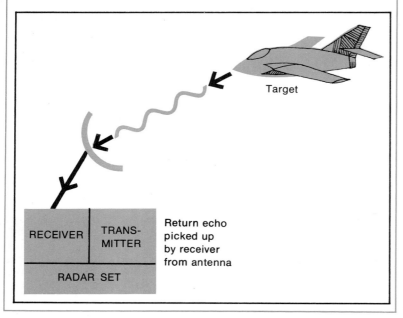

Target

RECEIVER | TRANS-MITTER

Return echo
picked up
by receiver
from antenna

RADAR SET

is turned off, and the receiver detects echoes returning from targets which reflect radio waves. Because radio signals travel at constant speed (the velocity of light, which is approximately 186,000 miles per second), the time between transmission of the signal and reception of the echo measures the distance from the radar set to the target (Fig. 6–27).

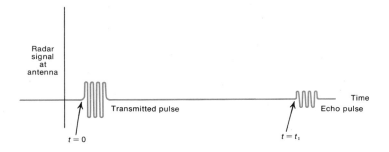

Fig. 6–28.
Antenna signals
in radar. The time
from the transmitted
pulse until the echo
pulse measures the
distance to
the target.

The operation is shown in a different way in Fig. 6–28. At a particular instant of time, which we call $t = 0$, a short-duration sinusoidal signal is transmitted from the antenna. At a later time t_1, an echo signal appears at the antenna. The echo is much smaller than the transmitted signal for two reasons:

1. Of the signal sent out by the transmitter, only a very small fraction hits the target and bounces back. Most of the signal misses the target altogether and either travels through space or bounces back from other targets.
2. Of the signal which bounces back, again only a very small fraction reaches the antenna; most misses the antenna and travels on through space.

As a result, the echo signal typically contains only 10^{-12} times the energy of the transmitted signal.

Even though the echo signal is very weak (just as it is when we listen to a sound echo bouncing back from a distant mountain), the receiver is able to detect this small signal. During the time from $t = 0$ to $t = t_1$, the radio signal travels from the antenna to the target and back again—a total distance equal to twice the range of the target.

The numerical relationship between the time duration t_1 and the range of the target can be derived from the known velocity of light. Because light or radio waves travel at 186,000 miles per second, an echo appearing one second after the transmitted signal ($t_1 = 1$ second) corresponds to a target at a range of 93,000 miles. An echo appearing one microsecond (10^{-6} second or one millionth of a second) after the transmission means the target is at a range of $\dfrac{93,000}{10^6}$ or 0.093 miles or $\dfrac{1}{10.75}$ miles. Thus, every 10.75 microseconds corresponds to one mile of range.

The relationship 10.75 microseconds/mile can be used to convert any measured time t_1 into the corresponding range. For example, an echo delay (t_1) of 90 microseconds means a range of

$$\frac{90 \text{ microsec}}{10.75 \text{ microsec/mile}} = 8.4 \text{ miles}$$

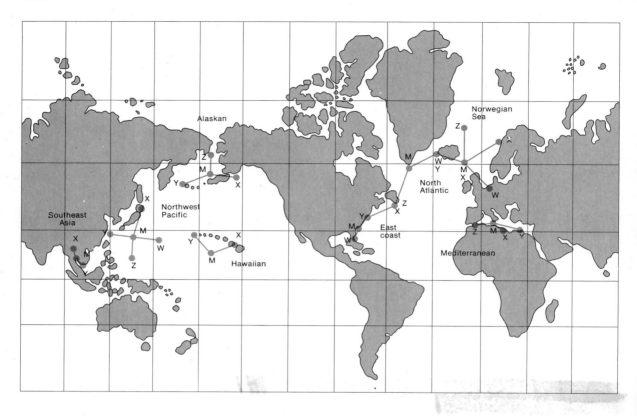

Radar Direction Finding

The preceding discussion indicates how radar is used to find the range of the target—by measurement of the time that passes before the echo is received. Location of the target requires determination of its *azimuth** and elevation angles. If radar is being used to locate a ship or object on the surface of the earth, it is only necessary to determine the azimuth angle of the target, because the elevation is known to be zero.

The basic technique for bearing measurement involves the use of a highly directional antenna which rotates continuously. A spherical or parabolic antenna "dish" is used with the signal transmitted mostly along the axis of the antenna (Fig. 6–27). As the antenna rotates, short pulse signals are transmitted frequently, so that in any direction several pulses are sent. When an echo is received, we know that the target is in the direction of the antenna at that particular moment.

Radar Measurement of Rate of Change of Range

If the radar system is being used to aim guns shooting at the target, we need to determine the velocity of the target as well as its position. The gun must be aimed to lead the target by the

*Fig. 6–29.
Broadcasting
stations for the
worldwide Loran
system. Loran (LOng
RAnge Navigation) was
developed during World
War II to permit ships and
airplanes to determine their
position. (Aerospace
Research, Inc. Updated
as of Aug. 17, 1970,
by Mr. L. Dennis
Shapiro of
Aerospace
Research, Inc.)*

*
The azimuth angle measures the bearing of the target on the surface of the earth clockwise from our reference direction, and the elevation angle measures how much the target is above the earth's surface.

amount the target will move during the flight of the bullet. The situation is identical with the problem of a quarterback passing to an end in a football game. The quarterback throws not to where the end is at the time he throws, but to the point he estimates the end can reach by the moment the ball arrives there.

The target velocity which we must estimate to predict target location when the bullet arrives has two components: the rates of change of the range and of the azimuth angle (again for simplicity we consider tracking a surface target, so elevation is not of interest). Either of these rates of change can be determined by measurement of the signal over a period of time and observation of the way it is changing. In the case of range rate of change, however, there is an alternative approach which is more accurate.

In 1842, Doppler* pointed out that if we move toward or away from a source of light, radio waves, or sound, the observed frequency depends on the relative motion of the observer and source. In particular, if we move toward the source or it moves toward us, the observed frequency of the signal increases. A whistle of an approaching train is higher in pitch than normal.

The reason for this apparent change in frequency is clear if we consider the transmission of sound from the train to our ear. If both the train and we are stationary, the train whistle sends out a sequence of air pressure peaks. Each of these pressure peaks travels toward us at the speed of sound (about 1100 feet/second). Our ear receives or senses the peaks as they pass by. Since we are stationary, the times from transmission to reception for each peak are the same.

If we move toward the train, each successive pressure peak is sensed by our ear a little sooner than it would be if we were further away and stationary. Hence, as we move toward the train, the successive peaks occur slightly closer together. Our motion toward the source tends to make the successive peaks in the sinusoid occur more rapidly; hence the frequency seems to be higher. Furthermore, the amount the frequency is increased measures directly our velocity toward the train. Hence, we can measure the frequency of the radar echo and from this determine the rate at which the target is moving toward or away from us.

Thus, radar can be used for sensing or the determination of the target's range, azimuth angle, elevation angle, and rate of change of range. In addition, the Doppler effect can be used to select (from all the targets spotted by the radar) only those targets which are stationary; more commonly, we are interested in just the moving targets, and we eliminate from the receiver output any stationary targets. For example, in flight control near a city airport, we want the radar screen to show only the moving targets (the aircraft), not the tops of tall, stationary buildings or nearby hillsides.

*Christian Johann Doppler (1803–1853), an Austrian physicist and mathematician, who was interested in the color (frequency) of light emitted by the stars. The Doppler principle permits astronomers to measure the rotation of stars and planets (one side is moving away from the observer, one side toward the earth, so the colors are different in the light received from the two sides).

Bat Navigation

One of the most impressive sights in the natural world is the exodus at sundown of thousands of bats from Carlsbad Caverns, New Mexico. After spending the daylight hours sleeping in the caves, the bats emerge from the mouth of the caverns, circle a few times, and then head off in search of insects for food. The awesome feature of this mass departure is the uncanny success of the bats in avoiding collisions. To the human observer, the situation seems comparable to that which would occur in the center of a major city if all traffic were able to move at 60 miles/hour with no accidents.

The success of the bat both in navigating among obstacles and in the capture of insects in flight depends upon an extremely refined sonar system.* During normal cruising in search of food, a bat emits short sound pulses about every 0.1 second. During the pulse of transmission, the sound signal is a sinusoid which varies in frequency from about 100,000 cycles/second down to 40,000 cycles/second (Fig. 6–30). The signal sent out is not strictly a sinusoid, because a sinusoid must be of constant frequency when transmitted. Rather it is a frequency-modulated signal (i.e., the frequency is modulated or changed during the transmission). In this case, the frequency is steadily decreased.

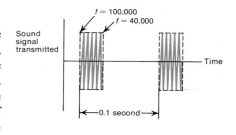

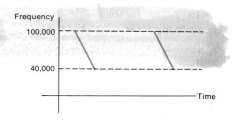

Fig. 6–30.
Bat navigation and food location.
a) Signal sent out by a bat.

b) Spectrogram for signal of (a).

Fig. 6–31.
Vikings located enemy shores and fortifications on the fog-shrouded European coast by shouts or hornblasts from their raiding long-ships, judging distance and size by echo return. (*Applied Technology, a division of Itek Corporation*)

* Sonar (an acronym derived from *sound navigation and ranging*) was highly developed during World War II for underwater detection (of submarines and of surface ships by submarines). In contrast to radar, sonar uses sound waves, commonly at frequencies slightly above the normal human hearing range (e.g., at 24,000 cycles/second).

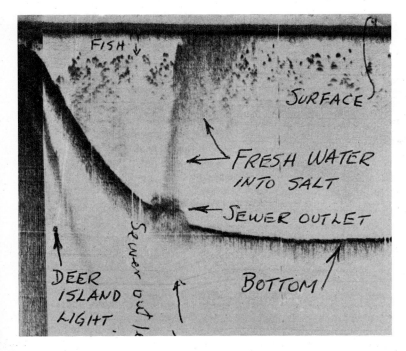

FISH

SURFACE

FRESH WATER
INTO SALT

SEWER OUTLET

DEER
ISLAND
LIGHT

Sewer out 1

BOTTOM

Fig. 6–32. Record made by a sonar system. The signal is sent downward from the surface. Echoes bounce back whenever there is a change in sound transmission. Thus, we can see fish as the sound signal reflects from the fish. Also, "fresh" water is different from salt, so we can see the fresh water dumping into the harbor from a sewer outlet. With such records, sonar is used to obtain accurate maps of a harbor and also to observe the flow of sewage lines (many of which were built so long ago that maps are unavailable). (Harold E. Edgarton and Martin Klein, M.I.T.)

Presumably, a changing frequency is used so that the bat can recognize the echo from his own signal (particularly when there are thousands of other bats in the vicinity). The frequency change also could permit the bat to recognize precisely which part of the transmitted pulse generated a particular part of the echo.

When the bat discovers a target (an insect), the pulses are transmitted more frequently (as often as 200 pulses per second just before the capture), the frequency of the transmitted sinusoid is lowered (varying from 30,000 to 20,000 cycles/second during a pulse), and the pulses are shortened. Once the target is captured and consumed, the normal cruising operation is resumed.

In the case of the bat, the signals are generated by the vocal cords, and the ears serve as receivers. The location of the target with respect to the straight-ahead axis (i.e., the angle to the right or left) is determined by the comparison of the echo signals received at the right and left ears, just as a human being determines the direction of a source of sound. If the object is in front and to the right of the bat, the echo is received at the right ear slightly earlier than at the left.

The radar signal is a burst or pulse of a sinusoid for a short period of time. The bat's sound signal is a pulse of a "sinusoid" of continuously varying frequency. In both cases, the sine signal is a basic starting point for the understanding of an important class of problems from the natural and man-made worlds.

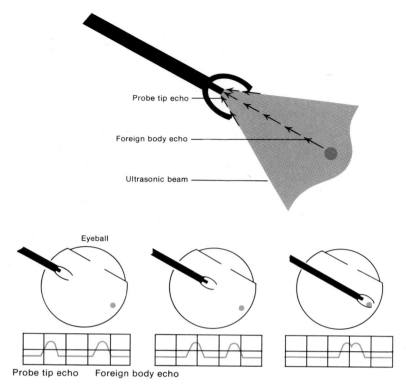

Fig. 6–33.
*Surgery by sonar was developed
in 1964 by Dr. Nathaniel Bronson. The
forceps emit a sonar beam which reflects
off the foreign object and the probe tip.
The difference in distance between the
tip of the probe and the object is
indicated on the oscilloscope.
When the two echos coincide,
the surgeon closes the
pincers and
removes the
object.
(Smith Kline
Instruments,
Inc.)*

6 | NOT ALL SIGNALS ARE SINUSOIDS

Many signals which occur in the real world are not sinusoidal in character and have very little relation to sinusoids.* While an objective of the chapter is to stress the significance of the sine signal, it is perhaps worthwhile to mention three examples of non-sinusoidal signals.

1. The price of a given stock on the New York Stock Exchange is not a simple sum of a few sine waves. If it were, we could look up the price over a long period of time in the past, evaluate the different sinusoidal components, and use this characterization to predict price variation in the future. While numerous "applied mathematicians" and "engineers" have in the past written articles arguing for such a possibility, it is perhaps sufficient to note that these individuals are still working for a living; none seems to be independently wealthy.

2. Figure 6–34 shows a set of four signals which are very different from sinusoids. These are typical signals measured in research on various safety devices in automobiles. The four signals shown are four different accelerations measured when a stationary car with two passengers is struck from the rear by another car moving at 23 miles/hour.

*
Fourier's theorem, mentioned in Sec. 3, states that any signal can be considered as the sum of sinusoids of different frequencies. If there are thousands of components, such a decomposition is not much help, however.

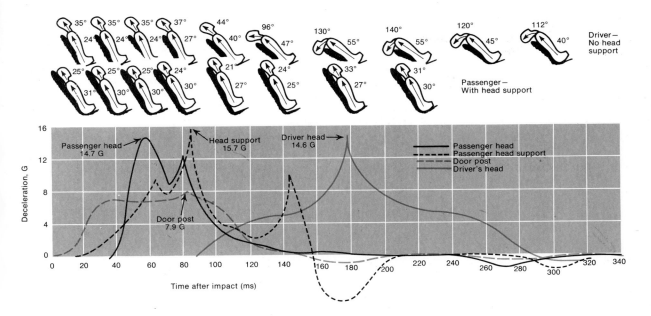

The four accelerations refer to four different measurements within the car which is struck. The dashed line is the acceleration of the door post (i.e., the frame of the struck car). At $t = 0$, the car is hit; shortly thereafter the door post (and car frame) accelerate. The maximum acceleration of about 8g (eight times the acceleration of a body falling freely) occurs about 80 milliseconds ($\frac{80}{1000}$ second) after impact.

The dotted curve is the acceleration of the head support (fastened to the passenger's seat). This part builds up in acceleration a little more slowly than the door post or car body, but reaches the maximum acceleration of almost 16g—about 85 milliseconds after impact. The passenger's head is even slower to start acceleration, but it pretty much follows the head support.

The most interesting curve shown is the curve which represents the acceleration of the driver's head (the $-\cdot-$ curve), since there is *no* head support behind the driver (the test was designed to investigate the value of head supports). We notice that the driver's head does not even start to accelerate until $t = 85$ milliseconds. By that time, the seat (or head support) and hence the driver's torso have already reached maximum acceleration. In other words, the driver's head does not start to move until the driver's torso and the seat have already moved forward a considerable distance (the car was struck from the rear).

The small pictures at the top of the figure show the result. The driver's head tends to remain fixed in space after his torso moves forward. Hence, the head falls backward with respect to the torso.

Fig. 6–34.
Kinematics of the
supported and
unsupported head in a
23-mph rear-end collision.
  a) The head of the passenger
    who has a head support.
  b) The head support.
  c) The door post (a part
    of the main frame of the car).
  d) The head of the driver
    who has no head support.
(Redrawn from "Traffic Safety,
A National Problem," based on research
by Derwyn Severy, U.C.L.A., The
Eno Foundation for Highway
Traffic Control, Inc.,
Saugatuck,
Connecticut,
1967)

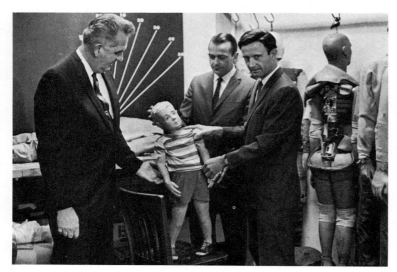

Fig. 6–35.
A dummy is used
in U.C.L.A.'s accident-
research project. Findings
have inspired collapsible
steering columns and head
supports—and the belief that 70 to 80
percent of auto injuries can
be avoided. (Institute of
Transportation and
Traffic Engineering,
U.C.L.A. Photo
by Bill Bridges)

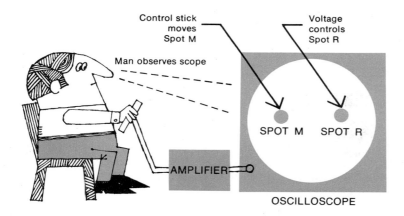

Fig. 6–36.
Human tracking
experiment (spots move
only horizontally).

When the torso is then slowing down (negative acceleration of the head support), the head is whipped forward. The strong shearing forces associated with the torso pulling out in front of the head cause the severe whiplash injuries.

Thus, the supported head accelerates in synchronism with the shoulders; the unsupported head is accelerated much later by forces transmitted through the neck. When a head rest is not used, severe binding and shear stresses are applied to the spine. The measurement of signals of this sort (on extensively instrumented dummies) indicates the extreme importance of head supports in avoiding serious injury in rear-end collisions.

3. Figure 6–36 shows an experimental arrangement for determination of the dynamic characteristics of a human being in a simple steering or piloting task.

In order to design airplane equipment to allow the pilot to control the aircraft, for example, we need to know how accurately and rapidly a man can steer or pilot. The test arrangement shown is used for this measurement.

The oscilloscope screen shows two spots at all times. One, called spot R (for the *r*eference), is moved back and forth horizontally in a random or irregular way. The other spot, called M (for the *m*an), is positioned or located as the man directs. The man has a control stick which he holds with his right hand. He can move this stick back and forth across in front of himself. If he moves the stick to the right, the spot M immediately moves to the right a corresponding distance.

The man is told to make his spot M follow as closely as possible the spot R which is moving at random, not under his control. The test arrangement is very similar to the driving-skill game often found at indoor amusement parks.

From experiments such as this (initiated by A. Tustin in England in the 1940's), it was found that the human being can be characterized by:

A time delay (0.2-0.4 second)—the man observes an error, but his response occurs a fraction of a second later.

Proportionality—the man moves his stick an amount proportional to the error between his spot M and the spot R. In other words, the bigger the error, the more he moves his stick.

Rate measurement—the man measures the rate of change of the error (the rate at which the error is increasing or decreasing) and uses this to predict the future location.

Integration—the man integrates the past error and corrects accordingly.

Thus, the human controller is a complex, dynamic system; his behavior in such a tracking task obeys complex laws or rules which depend on many factors—his degree of fatigue, his motivation to succeed, etc. Once we understand the way the man behaves in such a task, we can design automatic equipment which (for example) assists a pilot to land his airplane at high speed on the flight deck of an aircraft carrier which is pitching, rolling, and yawing in the sea.

In the experimental measurement of human beings in a control task such as just described, the target spot must be moved at random. If the spot is moved sinusoidally, or in a regular pattern of any sort, the man quickly learns the regular nature of the motion and he is able to anticipate accurately the future motion. The experiment then degenerates into a measure of the man's ability to understand the exact character of the motion and to adjust his joy stick accordingly.

7 | PERSON-TO-PERSON COMMUNICATION

In communication with another person, we must use one of our senses to acquire the information. Of the various senses, sight is the most effective for acquiring information. Indeed, perhaps 95% of the total information we receive comes to us through the eyes. Hearing, on the other hand, is the sense which we most often use for direct man-to-man communication through speech (although face-to-face conversation depends strongly on sight as well). Of the other possibilities, the sense of touch also provides the possibility of strong communication between people.

Sight

When we consider the use of sight for communication, the many capabilities and limitations of the eye are important. For example, the eye is relatively sluggish or slow in responding. Hence, movies which show only 24 separate frames, or pictures, each second are seen as a continuous picture, and television pictures which repeat at 60 times/second show no flicker. Actually, our ability to detect flicker depends on the brightness of the picture, but something like 50 pictures per second inevitably looks like a continuous picture. The most annoying flicker rates are those in the range from 10 to 15 cycles/second; possibly in this range there is a relation between the flicker rate and the electrical brain waves.

While the sense of sight is very limited in its ability to detect fast changes, the human being does have remarkable pattern recognition capabilities. A very brief glance at a picture is enough for understanding very complex inter-relationships among the objects. A human being can recognize as many as 2000 different shapes. The reading ability is best when the letters have sharp, pointed corners, and when the letter or symbol height/width ratio is between $\frac{3}{2}$ and $\frac{3}{4}$. This importance of distinct, easily read symbols is familiar to anyone who has looked at different kinds of type when buying a new typewriter.

Our knowledge of the characteristics of the sense of sight has been acquired over centuries of research. During the past two

When in the course of human events it becomes necess
When in the course of human events it becomes nece

When in the course of human events it becomes n
When in the course of human events it becomes n

When in the course of human events it becomes n
When in the course of human events it becomes n

*Fig. 6–37.
Various kinds of symbols available in commercial publishing. In a book or printed document, particular phrases can be emphasized by the use of distinctive lettering. Comparison of these lines indicates the differences in readability.*

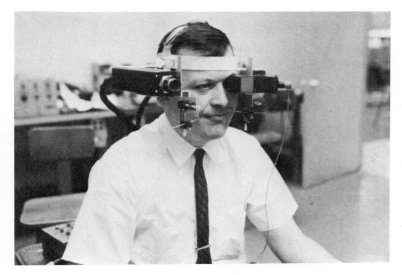

Fig. 6–38.
*Eye motion
analysis. The
individual is being
tested to determine how
his eye movement reacts to
a display. On the left is a small
television vidicon camera that shows
the area at which the subject is looking.
On the right is a device that shines a
light on the subject's eyeball and
picks up the reflection that is
superimposed on the field of
view seen by the TV camera.
Experiments of this type
are used in our attempt to
understand the way in
which human vision
collects information.
(Rome Air
Development
Center Display,
Rome, N. Y.)*

decades, research has been particularly intensive as we attempt to understand how different visual pictures stimulate particular nerves, and then how this information is coded for transmission to the brain.* Some indication of the complexity of the problem is given by the following situation:

Printed English contains about 10 bits of information per word.† We can read about 200 words/minute. In other words, in reading we acquire information at the rate of 2000 bits/minute or 33 bits/second. Yet commercial television signals can carry more than four million bits of information per second. Is our television system inefficient? Is most of the information capacity wasted? If we could design a television system well matched to the human eye (that is, a system carrying only information the human eye could use), we could have 4 million/33 or 120,000 times as many TV channels as we now have.

Hearing and Touch

In comparison with the eye, the ear is less accurate in detecting the direction from which a signal comes, but much better in noticing very small changes in sound. The ear can detect a sound break only $\frac{1}{3000}$ of a second long. For detecting brief signal changes, the sense of touch is somewhere between hearing and vision. The human being is able to detect an interruption of only $\frac{1}{100}$ second in a steady pressure applied to the skin.

Thus, we might communicate with a man by attaching to his skin buzzers or pressure devices at perhaps a dozen points around the body (points well separated and where the skin is sensitive to pressure). By applying different signals at the various locations, we can develop a complex code for sending information to him.

* Much of this present knowledge is described in highly readable terms in the book *Eye and Brain*, by R. L. Gregory, World University Library, McGraw-Hill Book Co., New York, 1966 (available in paperback).

† A *bit of information* was defined at the very beginning of this chapter. One bit is the information in a choice between two equally likely possibilities (for example, a 0 or a 1).

Such a form of communication is particularly important for people who are both blind and deaf. During the period of 1963 to 1965, there was a major epidemic of German measles in the United States. As a result, several thousand babies were born both deaf and blind. The sense of touch provides the only really useful method of communication with these young people. Communication by touch is also useful in special situations when we are working in darkness and in very noisy environments.

Actually, this idea of communicating through the sense of touch is hardly new. It was suggested by Jean-Jacques Rousseau in 1762. It is only in the last few years, however, that we have begun to understand the types of touch signals which can be understood by a man and the ways in which signals might be coded and presented to the man for easy and reliable understanding. Experimental systems have been built which give rates of communication

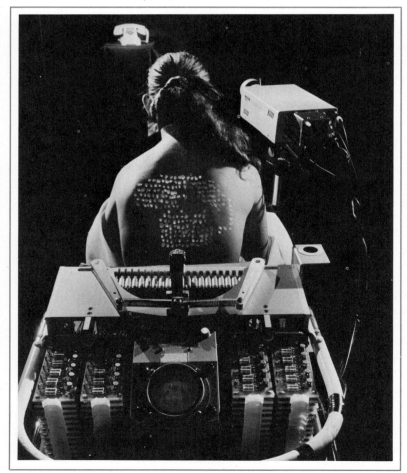

Fig. 6–39.
This young woman, blind since birth, is demonstrating how a new electronic system enables her to "see" a telephone with the skin of the back. The TV camera by her right shoulder picks up the image of the telephone on the distant table and converts it into the pattern of dots shown on the TV monitor in the foreground. Then, hundreds of tiny Teflon-tipped cones (not visible here) vibrate against her back allowing her to feel the dot pattern—illustrated in this photograph by fluorescent paint— and thus perceive the image of the phone. (Ralph Crane, Life Magazine © Time Inc.)

appreciably above those of telegraph transmission, but we still are far from matching the rate of communication which can be achieved by speech.

Other Forms of Communication

We have only touched the surface of this fascinating subject. We have, for example, not mentioned the possibility of communication with a sleeping person. We can now buy tape recorder and microphone equipment which can be set up in the bedroom. An alarm-clock system starts the recorder playing during the night after the man is asleep. Tapes are available which are advertised to overcome alcoholism, improve your willpower in dieting, or increase your self-confidence. A few years ago, there was considerable enthusiasm for the hope that a foreign language could be learned in this way.

Serious scientific research is being devoted to the study of how information can be placed directly into the nervous system where it would travel on to the brain. Each ear, for example, senses sounds through about 30,000 nerves (the visual system is much more complex). If we could determine the coding scheme, we might be able to build electronic equipment which would measure the sound signals and then excite the appropriate nerves directly for people who are totally deaf because of damage to the middle ear.

8 | FINAL COMMENT

In this chapter, we have seen that change is fundamental to communication. Changes are represented by *signals*, quantities which vary as time advances. Whether we are considering human communication or the performance of a system, the information is represented by a changing signal.

As an example of a changing signal, we considered speech. This led us to the important particular signal—the sinusoid. The sine signal is described by its amplitude and frequency.

The sine signal is important for three reasons:

1. Many signals are sinusoids, for at least a short time. The electrical voltage at the wall outlet is a sinusoid at 60 cycles/sec in most United States homes. Musical notes are the sums of a few sinusoids. Radar and sonar work with short bursts of sinusoids.
2. Almost any signal can be considered a sum of sinusoids.
3. We will use sine signals in our studies of portions of the man-made world.

Questions for Study and Discussion

1. Describe one situation which you have observed in the past two weeks in which lack of change resulted in boredom and another in which change resulted in interest.

2. Describe three major changes in your life brought about by technology. Be prepared to discuss these in detail.

3. On page 254 there is the statement "Time flies like an arrow" with three obvious interpretations. Try to devise a short statement which has three or more obvious interpretations.

4. Give an example of how predictability of a situation causes a loss of interest, and how unpredictability produces an increase in interest.

5. In Section 1 of this chapter we indicate that more than half of the letters could be eliminated from an entire paragraph (of logical text) and, because of redundancy, the paragraph could be reconstructed. Try this by eliminating a letter each time a head comes up, and every other time a tail comes up.

6. Give some examples of how technological improvements in packaging have caused problems in the disposal of solid waste.

7. List the sources of solid waste, other than private homes, within one mile of your house. Has waste disposal from these sources been improved or made worse in the past five years?

8. Why do we need to predict such things as production of solid waste or air pollution for from five to ten years in advance?

9. Does the informed citizen need to know what percent of solid waste is plastic, paper, rubber, etc., in order to be able to understand the basis for decisions on waste disposal made by government officials? Explain the basis for your answer.

10. What does the ability of a bat to find its way in the dark have in common with a newly developed method for removing foreign objects from the eye? Explain how this demonstrates that science and technology do work together to help man.

11. Describe three situations in which sinusoids are used in man-made systems.

12. Report on the latest developments in voice identification as described in popular and/or scientific periodicals.

Problems

1. a) What is the relationship among the period, frequency, and amplitude of a sine wave?

   b) If we doubled the amplitude of a sine wave which had an original frequency of 500 cycles/sec, what would its new frequency be?

2. Using Tables 6–2 and 6–3 and following the procedure discussed for finding letters at random, construct a statement of 50 characters (letters and spaces); compare it to the statements on pages 253 and 254. Which of these is most like your statement?

3. In a radar system, we cannot transmit a second pulse until all important echoes from the first pulse have been received. If the maximum target range of any significant echoes is 100 miles, what is the minimum allowable spacing between successive transmitted pulses? This antenna rotates through a full 360°; during this rotation, we wish to send at least 7200 pulses (20 every degree). What is the maximum allowable speed of rotation of the antenna?

4. Radar echoes have been observed from both the moon and the planet Venus. What length of time is required in each case for the radar echo to return?

5. A police radar is used to measure the speed of cars on a thruway. If the transmitted frequency is 3000

megacycles/second (3×10^9 cycles/second), the police radar antenna is stationary, and the car is moving at 80 miles/hour, what is the Doppler shift? The shift or change in frequency of the received signal is

$$2f \frac{v}{c}$$

where f is the transmitted frequency, c the velocity of light, and v the target velocity. If the equipment must measure car speed to an accuracy of $\pm 2\%$, what accuracy is required in the measurement of the difference of the transmitted and received frequencies? How would these answers be changed if the transmitted frequency were shifted to 3×10^{10} cycles/second? To 300 megacycles/second?

6. Plot and add two sine waves of the same amplitude which have frequencies of 15 cps and 30 cps.

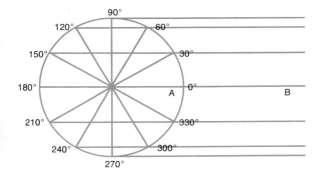

This group of problems shows you a rather easy way to construct the graph of a sine wave, and uses the method to emphasize some points, and to bring out some new information. Copy (or trace) the $1\frac{1}{2}$-inch circle with its six diameters, one every 30°; also the horizontal lines, making them run out a distance of 6 inches from the 0°, or 360°, point. Mark off on the line AB points $\frac{1}{2}$ inch apart, and at each point draw a perpendicular to AB, right across it. The corner of a pad or an envelope can be used to make the perpendiculars if you do not have a protractor. Label the points, in order, 0°, 30°, 60°, etc., to 360°. This is the basic framework.

7. Construct a sine wave 6 inches long. To do this put a dot where the two 30° lines cross (one perpendicular to AB, one parallel to it from the 30° point on the circle); repeat for the intersection of

the two 60° lines, etc. The dots mark the wave well enough, though there is no harm in sketching it in roughly if desired.

8. How would your wave differ if the circle had been only 1 inch in diameter instead of $1\frac{1}{2}$?

9. Draw another wave on the same basic framework, but with points $\frac{1}{4}$ inch apart on AB. How would you describe this wave, in terms of the first one?

Add the second wave and the first half of the first. The method is graphical. Every one of the lines perpendicular to AB crosses both curves. Measure the distances of these two crossing points from AB, add them, and mark a new point on the perpendicular at the sum of the distances from AB. This is done most easily by setting a compass to one distance, and using it to add this distance to the other. Pay attention to sign: all the crossing distances of the second half of the shorter wave are below AB and count as negative.

10. Suppose you had two waves, the one drawn in Problem 7 and a second identical to it but "180° out of phase." That is, the first point on AB would be numbered 180° for this wave, and the following ones 210°, 240°, etc., through 360° to 180° again. If these two waves were added, how would the result look?

11. Two tuning forks are sounded together; their frequencies are 30 cps and 35 cps. Plot the new wave formed assuming that the original amplitudes were the same.

12. The diagram on surgery by sonar on page 274 shows the oscilloscope pattern as the probe moves to remove a foreign body from the eye. Sound travels at the rate of approximately 5000 feet/sec in the fluid of the eye, and it takes 0.00001 second for the oscilloscope trace to cross the screen.

a) How far is the probe tip from the foreign body in the first picture?

b) In the third picture?

c) If the pictures were made at 5 second intervals, how fast did the probe move through the eye?

13. A string which has a natural frequency of 128 cycles/sec is set into vibration. What is the frequency of the second harmonic? Of the third? Of the fourth? What is the frequency of the note one octave higher than the natural frequency of the string? Two octaves higher?

Laboratory and Projects

I | PLOTTING TIDAL DATA

Part A

In the text, Fig. 6–11 shows the tidal situation at Mont St. Michel, France. If you live near a tidal water zone, join with some friends to measure the depth of water at the same spot near the shore at periodic intervals (every 15 or 30 minutes) for a span of about 14 hours. Plot the depth measured as a function of time.

■ 1. What is the shape of the curve which you obtain?

■ 2. What is the amplitude of the tide at the spot which you measured?

■ 3. What is the tidal frequency of the day you took your measurements?

Part B

Repeat the same measuring procedure starting 24 or 48 hours after the beginning of the set. Plot on the same sheet of paper.

■ 1. Has the shape of the curve changed?

■ 2. Has the position of the curve changed with respect to the first curve?

■ 3. Predict the time of high tide one week after your second set of measurements.

■ 4. How close was your prediction to the time of actual high tide one week later? How do you account for this?

■ 5. (optional) Describe a procedure which you would follow to make a model of the tides which would help you predict high tide more closely. How do the publishers of tide tables obtain the times which they publish years in advance?

II | USING THE CRO TO ANALYZE SOUND PATTERNS

In recent years music has been generated by completely mechanical means, computers and Moog synthesizers to name but two. Using an oscilloscope as pictured in Fig. 6–6 you will compare the wave forms obtained with as many different instruments as are available. You should obtain a tuning fork and some tape recordings of individual instruments or groups to round out your selections.

Part A

The CRO can only produce pictures of waveforms of electrical signals. To produce a picture of a non-electrical signal such as sound, we must use an instrument called a *transducer* to convert the non-electrical signal into the electrical form. Transducers that convert sound waveforms into electrical waveforms are called *microphones*.

Pictures of sound waves can be obtained on the CRO by following the procedure given below:

1. Plug a microphone into the Vertical (DC) terminal on the CRO.

2. Whistle a soft long note into the microphone and adjust the Vertical gain control to give a picture of suitable size. Adjust the Sweep frequency to obtain a good stationary picture of the periodic waveform (if possible). Change the loudness of your whistle and observe the change in the height of the waveform. Whistle a note of a different pitch, and observe that the frequency of the signal changes.

3. Speak various vowel sounds and adjust the Sweep control until you get a good picture.

4. Observe the sound waveforms produced by a tuning fork, a harmonica, a whistle, and other sound-producing instruments. In each case adjust the CRO Sweep controls so that you get a good clear picture of the periodic waveform.

5. In what ways do you observe the signals produced by these instruments to differ? (Careful observation will distinguish three ways.)

6. Of the instruments used, which produced the purest note?

7. Of the instruments used, which produced the most complex note?

Part B

Compare the notes produced by a signal generator (if available), a tuning fork, and the human voice.

■ 1. Which produced the purest shape?

■ 2. Note the differences in patterns when you talk in a "monotone" and when you speak in a more lively fashion.

■ 3. Note the change in pattern as a tape recorder or record player is played at the correct speed for the recording and when it is played faster and then slower than the correct speed.

III | SONIC DISPLACEMENT METER

In Section 5 of this chapter you studied two systems, radar and sonar, which used sinusoids. In this experiment you will make use of the sinusoids from a signal generator to develop some insights into the operation and design of a sonic displacement meter (a device which uses the movement of sound to measure distances between objects). Since the signal generator produces electrical signals, you will need a loud speaker (similar to those used in your TV and radio sets) to convert the electrical signals to sound signals.

You might have observed that sound travels at a much slower speed than light. If you sit in the bleachers at a baseball game you know that you hear the sound of the batter hitting the ball a fraction of a second after you see him hit the ball. You may have also noticed that the sound of thunder always reaches your ears after you see the lightning. The fact that sound travels at a relatively low speed (about 740 miles/hour or 345 meters/sec at room temperature) and the fact that we can convert between sound and electricity form the basis for the sonic displacement meter (Fig. 1).

Electrical signals from the signal generator are first converted to sound waves by the loudspeaker. The sound waves travel through the air and are picked up by the microphone which converts the sound waves back to electrical signals. These signals are then displayed on the CRO.

Part A

Operation and Design

The microphones which are supplied to you can be used both as microphones and as loudspeakers. (Note, however, that not all microphones have this characteristic.)

1. To observe this feature plug a microphone into the output terminal of the signal generator and turn the signal level

as high as it will go. Use a sinusoidal signal on the signal generator. Listen to the microphone as you vary the signal frequency from 200 cycles per second up to 20,000 cycles/second (cps).

You can now convert this sound back into an electrical signal by placing a second microphone near the first one.

2. Place the two microphones facing each other and about 2 inches apart and observe on the CRO the changes which occur in the electrical output signal from the second microphone as the signal frequency is slowly varied between 2,000 and 20,000 cps. Make the necessary adjustments in Sweep and Gain so that you clearly see the shape of the waveform on the CRO screen.

What happens to the amplitude of the signal on the CRO as you vary the frequency of the signal generator? Record your observations; they will be helpful when you try to answer Question 13.

We now use the two microphones as a simple device to demonstrate how sound waves can be used to measure the distance between two objects in air. Imagine that one object has the sending microphone (loudspeaker) rigidly attached to it and that the other object has the receiving microphone attached to it.

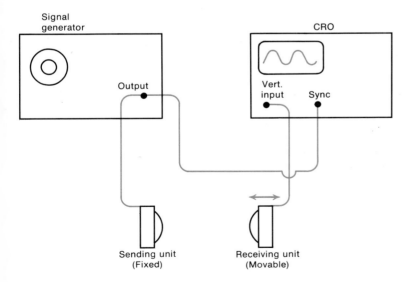

Fig. 1
Sonic displacement meter.

3. Connect the two microphones, the signal generator, and the CRO as shown in Fig. 1. Note that the signal generator output is connected to both the sending microphone and the synchronization terminal on the CRO.

4. Set the signal generator to produce a sine wave of 4000 cps.

5. Put the two microphones facing each other and 6 inches apart. Adjust the CRO controls so that several cycles of the signal from the receiving unit are displayed on the CRO screen.

6. Slowly move the two microphones together and, neglecting changes in the amplitude of the signal on the CRO, observe how the sine wave moves to the left. (In technical language, you are producing a phase shift.)

7. Use the peak of one of the cycles as a reference point and plot a curve of the movement of the sine wave on the CRO versus the distance between the microphones. Take readings for 1 inch changes in microphone displacement between 2 and 6 inches.

8. Repeat steps 5, 6, and 7 for a signal frequency of 2,500 cps.

9. The *sensitivity* of any measuring instrument is defined as the ratio of the change in the instrument reading to the change in the property being measured. From this definition you see that the sensitivity of our "displacement meter" can be obtained by measuring the *slope* of the graphs which were obtained in steps 7 and 8. Determine the sensitivity of the displacement meter for the 4000 and 2500 cps signal frequencies. Note that the sensitivity of the meter depends on the signal frequency. How can this dependency affect the design of a distance-sensing instrument?

10. From the data in step 9 and also from the fact that the sensitivity is zero when the signal frequency is zero, plot a graph of sensitivity versus signal frequency.

11. What signal frequency do you think should be used if we want a 10 *yard* change in microphone distance to be represented by a 1 inch shift on the CRO screen? (Assume that we have microphones which are sensitive enough to pick up signals 10 yards away.)

12. What is the maximum sensitivity that you can achieve with this instrument?

13. What input frequency would you pick for our distance-measuring device, taking into account the sensitivity factor and the resonance characteristic of our microphones?

14. Can you see any difficulty which you might have if you were trying to determine with the meter whether two objects are 6 inches or 12 inches apart?

Part B

Using the Sonic Displacement Meter to Measure the Speed of Sound

1. Set the Signal Generator to produce a good sine wave of 3000 cps frequency.

2. Place two microphones facing each other 6 inches apart. Connect the components as shown in Fig. 1.

3. Adjust the CRO controls to give a good picture of the waveform with several cycles. Since the sine wave which you see on the CRO is oscillating at 3000 cps, the distance between adjacent maximum points on the sine wave represents one cycle or $\frac{1}{3000}$ second.

4. Measure the distance between adjacent maximum points on the CRO screen and record it.

5. Move the microphones one inch closer together and measure the shift in the peaks of the sine wave on the CRO screen. Record this data.

    The time that it takes the sound wave to travel one inch can be calculated from the equation:

$$t = \frac{\text{shift in peak (step 5)}}{\text{peak distance (step 4)}} \times \frac{1}{3000}$$

6. Make the above calculation and determine the time for the sound wave to travel one inch.

7. The average speed of sound at room temperature is 345 m/sec. The speed of sound in air is affected by the temperature of the air through which it travels, increasing by about 0.1% for each 1°F increase in temperature. By evaluating how accurately you can measure the distances on the CRO and between the microphones, estimate the smallest possible temperature change that you think you could detect with the device.

IV | PERIODIC SIGNALS ON THE ANALOG COMPUTER

What happens to periodic signals such as sinusoids when they are sent through different parts of the analog computer?

In this experiment, different ways of sending periodic signals through the analog computer will be analyzed. Based on your newly acquired understanding of sinusoids and the operation of the analog computer, you will try to predict by means of diagrams the various outcomes of manipulating periodic signals. Then you can actually simulate (model) the described situations on the analog computer and check your predictions.

1. Describe the outcome of multiplying a sinusoid by a constant.

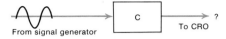

Fig. 2.

Simulate the situation in Fig. 2 on the analog computer and check your prediction.

2. Describe the outcome of adding a sinusoid to a constant signal.

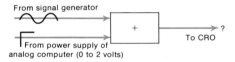

Fig. 3.

Simulate the situation in Fig. 3 on the analog computer and check your prediction.

3. Describe the outcome of adding two sinusoids of the same frequency but different amplitudes.

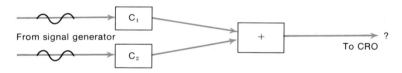

Fig. 4.

Simulate the situation in Fig. 4 on the analog computer and check your prediction.

4. By using one summing-scalor to add the signals from two different signal generators it is possible to demonstrate the types of signals which result from the addition of signals of different frequencies and waveforms.

V | ESTIMATING YOUR WRITING VOCABULARY

James Joyce had a working-writing vocabulary of 30,000 words; a college student uses 10,000 words. In order to estimate your writing vocabulary, proceed as follows:

1. Open a dictionary at random.

2. Count the number of words on the left page. Let this be A_1.

3. Looking at each word on this left page, decide whether you would ever use the word in writing. Count the number you would use. Call this B_1.

4. Repeat the above steps (1), (2), and (3) to find A_2 and B_2 for a second page.

5. Carry through steps 1, 2, and 3 five times in all to find A_1, $A_2, \ldots A_5$ and $B_1, B_2, \ldots, B_5$.

6. Calculate $(B_1 + B_2 + \ldots B_5)/(A_1 + A_2 + \ldots A_5)$. This is the fraction of words in the dictionary you use in writing.

7. From the total number of words in the dictionary, estimate your writing vocabulary. The same procedure can be used to find your speaking vocabulary and your reading vocabulary.

VI | SIMULATION OF A WATER-POLLUTION MODEL

One of the major problems facing our society today is the staggering rate at which we are destroying our natural resources, particularly the air and waters around our large cities. The air contains vast amounts of waste products from automobiles and factories. Many of our rivers and lakes have become virtually sewerage drains and their beauty and utility has been greatly diminished.

Rivers and lakes can actually digest or decompose significant amounts of untreated waste products without upsetting the living processes that occur within and around them. Pollution arises when there is unlimited and uncontrolled dumping of untreated waste into the water. The reason is that the decomposition process uses up oxygen which is dissolved in the water. If the amount of oxygen gets too low, the water will be incapable of further decomposition, fish will die, and the water will be unfit for drinking, swimming, and other recreational activities. Rivers, lakes, and the oceans continually replenish their oxygen level by drawing it from the air. The rougher or more turbulent the surface of the water, the faster it takes in a fresh oxygen supply. The water becomes polluted when the oxygen is used up faster than it can be replaced from the air.

Part of the solution to the water pollution, therefore, lies in our ability to predict, by means of a model, how the oxygen content of a particular body of water will change as waste products are dumped into it.

In order to formulate an oxygen model we must study the factors related to the following questions:

How fast do the waste products decompose in the water?
How fast is oxygen replaced in the water?
How is the amount of oxygen affected by the rates of decomposition of waste and replacement of oxygen?

You are going to study the relationships between variables of the oxygen model by doing the following three experiments:

Part A—Model of the decomposition of waste
Part B—Model of replacement of oxygen
Part C—Improved oxygen content model (combination of parts A and B)

In these experiments we will be using the unit (milligrams/liter) for our measurements of waste and oxygen concentrations. This is done for two reasons:

1. Most research literature on pollution studies uses this unit of measurement.
2. This unit of measurement is more convenient for modeling the water-pollution system on the analog computer.

It is, however, important to have a feeling of the size of the units that you are working with. So if you convert milligrams to ounces and liters to gallons, you will find that:

$$10 \text{ milligrams/liter} = .00133 \text{ oz./gallon}$$

In larger units, 10 milligrams/liter is 1.33 ozs./1000 gallons (approximately similar to putting a grain of sand into a bucket of water).

Part A

Model of the Decomposition of Waste

In order to develop a waste-decomposition model, we must identify the factors which affect the rate of change of waste in the water (R_w). One factor is the amount of waste which is in the water at a certain time (W). Other factors are the temperature of the water and the chemical composition of the waste, which can all be represented by a constant C_w. In general, it has a value between 0.25 and 0.75. Scientists using measurements from natural waterways and laboratory water tanks have found that:

$$R_w = -C_w W$$

From earlier experiments with the analog computer we know that if we are interested in learning the accumulation of something, we need to put into the integrator the rate of change of that item. In this experiment we are interested in knowing how long it will take for 50 milligrams of waste product (W_o) to decompose in a liter of water. Let one second of the analog computing time represent one day of real time and let one volt represent 10 milligrams/liter. Using the diagram (Fig. 5), simulate the decomposition of waste on the analog computer.

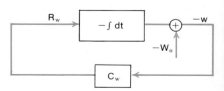

Fig. 5.

W waste at any time
W_O original amount of waste
C_w coefficient of decomposition
R_w rate of decomposition

1. How does changing the value of C_w affect the decomposition process?

2. For $C_w = .5$, how long will it take to decompose 50 milligrams/liter of waste?

In the previous model, we assumed a certain amount (50 milligrams/liter) of waste is being decomposed without new waste being added. In reality this is usually not the case. Let us improve our model by assuming that there is a constant rate of waste being dumped into the water (R_D). Our mathematical model would become:

$$R_w = R_D - C_w W$$

Use the pictorial model (flow chart) on the right to simulate the improved decomposition model:

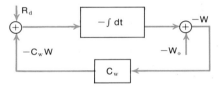

Fig. 6.

W waste at any time
W_O original amount of waste
C_w coefficient of decomposition
R_w rate of decomposition
R_D rate of dumping

3. Start by assuming that R_D is 10 milligrams/liter/day (1 volt).
   Use the simulation of this improved model and study the effect of varying C_w.

4. Now keep C_w at .5 and study the effects of varying R_D.

5. Will the wastes ever decompose completely?

6. What controls the amount of waste which remains in the water?

Part B

Model of the Replacement of Oxygen

Before studying how the decomposition of waste affects oxygen content (next experiment), we must first study how oxygen is replaced in the water. From studies of replacement of oxygen in the water it has been found that the rate at which oxygen is replaced (R_A) is proportional to the difference between the amount of oxygen contained in the water at any time (A) and the maximum amount (A_{max}) that the water can hold. Mathematically the rate of replenishment can be expressed as:

$$R_A = C_A (A_{max} - A)$$

where C_A is a coefficient which depends on the turbulence of the water. A_{max} depends on the temperature of the water. Tables 1 and 2 give typical values for these two constants.

Large ponds	0.4
Large lakes	1.0
Slow moving streams	1.5
Rapidly moving streams	3.0

Table 1. Typical values of C_A.

Temp (°F)	A_{max} (Milligrams/liter)
32	15
41	13
50	11
59	10
68	9
77	8

Table 2. Typical values of A_{max}.

Use the flow chart (Fig. 7), which is based on $R_A = C_A (A_{max} - A)$, and simulate the replacement of oxygen for different values of C_A and A_{max}.

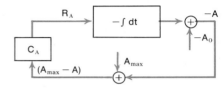

Fig. 7.

A oxygen at any time
A_O original amount of oxygen
A_{max} maximum amount of oxygen
C_A coefficient of turbulence
R_A rate of oxygen replacement

1. Assuming that we have 10 milligrams/liter (1 volt) of oxygen at the start (initial condition of integrator at 1 volt), and using $A_{max} = 15$ milligrams/liter, study the effects of varying C_A (values from Table 1).
2. Next keep $C_A = 1$, and study the effects of varying A_{max} (values from Table 2).

Part C

Improved Oxygen Content Model

Although oxygen is constantly being replenished in the water by the air, it is also being used up in the decomposition process. We now combine the models studied in the previous two experiments and generate a model which will monitor the oxygen content at any time. Remember that the rate of replacing oxygen (R_A) was mathematically expressed as:

$$R_A = C_A (A_{max} - A)$$

If we want to include the use of oxygen in the decomposition of waste, the mathematical model will become:

$$R_A = C_A (A_{max} - A) - A_w$$

where A_w represents the rate of use of oxygen by the wastes. This rate of usage directly depends on the rate of decomposition.

Therefore A_w is directly proportional to $C_w W$. Mathematically:

$$A_w = k \, C_w \, W$$

where k is a coefficient which represents the oxygen requirement (oxygen demand) for a particular waste product. Since our analog computer is limited to only three coefficient knobs we assume that $k = 1$. The final mathematical model which we use for our simulation of the oxygen content model is then:

$$R_A = C_A (A_{max} - A) - C_w \, W$$

Use the flow chart on the right to simulate the oxygen content model on the analog computer:

Part B
$R_A = C_A(A_{max} - A)$

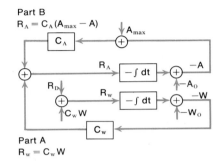

Part A
$R_w = C_w W$

Fig. 8.

W	waste at any time
W_O	original amount of waste
C_w	coef. of decomposition
R_w	rate of decomposition
R_D	rate of damping
A	oxygen at any time
A_O	original amount of oxygen
A_{max}	maximum amount of oxygen
C_A	coef. of turbulence
R_A	rate of oxygen replacement

We now use the model to answer certain questions which are related to the oxygen content in the water.

Based on the fact that fish begin to die when the oxygen content approaches 5 milligrams/liter (0.5 volt), answer the following questions:

1. At what rate can waste be dumped (R_D), if you want to keep the fish alive? Assume that $W_o = 50$ milligrams/liter (5 Volts), $C_w = .5$, $A_{max} = 15$ milligrams/liter (1.5 volt, value when $T = 32°F$), $C_A = 1.0$ (1 volt, value when body of water is a large lake), and $A_O = 10$ milligrams/liter (1 volt).

2. What is the maximum allowable value for W_O, if we want to keep the oxygen content above 5 milligrams/liter, if R_D is zero (no dumping of waste after initial amount), $C_w = .5$, $C_O = 1.0$, $A_{max} = 15$ milligrams/liter, and $A_O = 10$ milligrams/liter?

3. Study the effect of changing C_w, when W_O, A_{max}, A_O, and C_A are predetermined (choose values for A_{max} and C_A from Tables I and II from previous experiment). You might want to get data for W_O and A_O for a body of water near where you live, or else just choose some representative values. For some situations you may find that the oxygen content goes below zero (negative). This shows that our model has limitations and will only work within a range of values.

Feedback
7

I | A FEEDBACK SYSTEM

A man is standing in front of a table. A pencil is rolling across the table toward the edge. The man reaches out with his hand and grabs the pencil.

The *system* for picking up the pencil has operated successfully!

We probably will not read about the success of the system in tomorrow's newspaper. Launching of a spaceship to Venus is likely to attract more attention. Yet the system for grabbing the pencil is similar to the Venus shot. Both problems are very complex, and we are able to succeed in both because we use *feedback*.

What is feedback? To answer this question, let's construct a block diagram for our man-pencil system. Just as in Chapter 5, we will first decide what are the input and the output. Then we will look for the cause-effect relationships which move us from input to output.

The input is the pencil position. The man has no control over this; the pencil moves across the table. The man is in pursuit. The output is the position of his hand. Our system succeeds after this output has been made equal to the input: his hand is on the pencil. In Fig. 7–2, we have the start of a system model.

Now we have to describe how the system works. The man uses his eyes to measure both the input and the output. He watches the rolling pencil and the location of his hand. In particular, he sees

Immediate feedback to the student is an essential principle of education technology, ancient or modern. Knights in Charlemagne's time learned jousting skills with this teaching machine, called a quintain. Knight is charging at wooden figure on a pivot. If he strikes it squarely in the middle of the shield, it will fall over. If he strikes it incorrectly, off-center of the shield, it will swing around to apply feedback, with a club, doubtless highly educational. (15th Century Woodcut from Chronique de Charlemagne, *Collection of the New York Public Library, Astor, Lenox and Tilden Foundations)*

the difference between these two signals. The difference is the error between where his hand is and where the pencil is.

In the block diagram, we show this measurement of error by a *comparator* block. The man *compares* the pencil and hand positions. The result of the comparison is the error (Fig. 7–3). The eyes measure this error and the information travels electrically to the brain. There the man computes what he should do. How should he move his hand? Once a decision is reached, electric signals are sent out from the brain to the muscles of the arm and hand. In other words, the nervous system and brain accept information from the senses (the eyes here), and then decide what action to take to improve the output (Fig. 7–4).

These electrical orders from the brain cause the muscles to contract in just the right way to make the hand move toward the pencil. When the hand is at the pencil, the orders call for the muscles to contract, which will cause the fingers to close and grab the pencil. Thus, the complete system is shown in Fig. 7–5.

We said earlier that this system has *feedback*. From Fig. 7–5, we can now see what we meant by this term.

The output of our system is determined by the measured error. This error is sent to the brain where orders are chosen to change the output. In other words, the output results from the measured error (Fig. 7–6).

But in our system *the error depends on the output. This is feedback. We have really a loop of cause-effect relations:*

The error changes the output.
The changing output changes the error.
The error changes the output. And so on.

Fig. 7–1.
Feedback system
used for picking up
a rolling pencil.

Input	Output
Pencil position	Hand position

Fig. 7–2.
Input and
output signals for
our system.

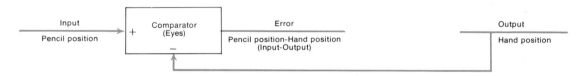

Fig. 7–3.
The system measures the error.

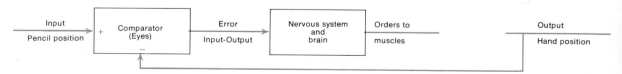

Fig. 7–4.
Block added to show decision making
on basis of error observed.

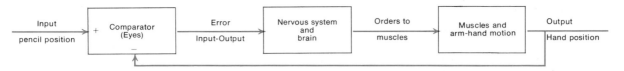

Fig. 7–5.
Model for man-pencil system.

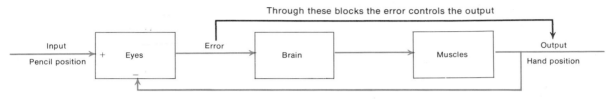

Through these blocks the error controls the output

Fig. 7–6.
An error causes the output to change.

Fig. 7–7.
A typical Eskimo man. The adaptation of man to his environment is an example of feedback. The adaptation occurs during growth of the individual. Adaptation also occurs as the gene pool of the population is changed over generations.

The average nose shape is related to the need to moisten the air that we breathe in. The Eskimo has a very narrow nose, compared to the length.

The Eskimo typically is small and squat, with unusually short limbs. In order to minimize the loss of body heat, the Eskimo has a relatively small surface area. In contrast, certain African inhabitants are often tall with very long limbs.

There is a relation in mathematics which states that in bodies of the same shape, the taller one has relatively less surface area. From this, we would expect people in colder climates to be taller than in warmer. This relation only holds when the shapes of people are the same. (The Bettmann Archive)

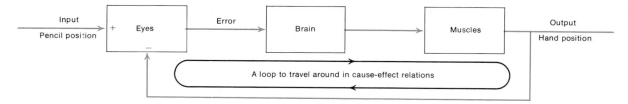

In other words, a system has feedback when the output depends not only on the input, but also on the output itself. In the block diagram of a feedback system, there is at least one loop we can travel around, always following the arrow direction (Fig. 7–8).

Fig. 7–8.
The feedback in the man-pencil system is shown by the loop.

Feedback Is Important

This chapter has two main purposes. First, we want to understand what feedback is and when it is present. Second, we want to study a few of the remarkable things we can achieve by using feedback.

Even before this study, however, we can see some of the advantages of feedback from our man-pencil system. Why can a man usually grab a rolling pencil? Because the output depends on the error, he is continually measuring the error and making corrections. If his hand starts off too much to the right, he observes this and changes the position of his hand. As his hand nears the pencil, the corrections become more and more refined.

The importance of feedback is clear if we try to do the same job without feedback. (We might blindfold the man.) As the pencil rolls across the table, he moves his hand toward the table, but not necessarily closer to the pencil. System success relies on the continuing measurement of error.

The feedback system automatically seeks the goal of reducing the error to zero: making the output equal to the input. In our man-pencil system, the *goal* is bringing the man's hand to the pencil. In the Venus space shot, the *goal* is to bring the space vehicle to the planet Venus. In steering a car down a highway, the *goal* is to hold the car in the middle of the lane.

In each of these cases the system has a clear goal. There is a measurement of the error: how far is the output from its goal? This error then changes the output toward the goal. This is feedback.

2 | GOAL-SEEKING

Have you ever thought about how many of your actions are directed toward achieving some goal? If you have, you quickly realize that almost everything you do is *goal-directed*, in some sense. These goals may be very simple and immediate. When you insert a dime into a soft-drink dispenser, your goal is to quench your thirst. Goals may be very complex and of long-term significance,

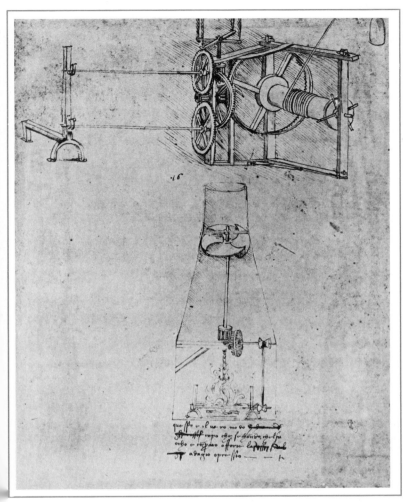

as when you decide to go to college to prepare yourself for a future occupation.

The purposeful activity of "seeking goals" is of course not unique to men alone, but may be found in all living things. Plants turn their leaves toward the sun and extend their roots toward moist and fertile soil. Salmon battle their way up rivers and through rapids to spawn their eggs in the same creek in which they themselves were hatched. Even the single-celled amoeba moves away from a disagreeable substance to seek a more pleasant environment.

In contrast with such purposeful activity as is exhibited by living things, inanimate objects appear to behave quite differently. Their behavior is based solely on *prior* causes. We do not think of inanimate objects as being influenced by any awareness of what may happen to them in the future.

Inanimate objects appear to differ from living things in another important aspect. With living things we recognize movement toward goals, toward maturity and death, toward satisfaction and reward, toward equilibrium, toward a conscious purpose. We do not think of inanimate objects as seeking "goals." For instance, a rock loosened by the morning's rain falls down the mountainside. We should think it a bit odd if someone suggested that the goal of the rock is to seek a lower elevation. If the rock should strike and kill a snake, only an irrational and superstitious person would assert that this was the rock's goal. Yet if you were to throw a rock at the snake and kill it, we would all agree that it was your goal.

Fig. 7–10.
Example of a feedback system. The hand that holds these scissors is a plastic-and-metal machine complete with motor, gears, strain gauge, and a sensor for measuring the velocity at which the finger closes on the thumb. Information from the gauge and the sensor is used in order to control the hand so that it automatically matches the wearer's intentions (which are detectable as changes in the electrical activity of muscle remnants). This degree of automatic control makes the hand the most sophisticated in general clinical use today.

In this case, the goal is the electrical signal in the arm muscles. The feedback system is the artificial hand, the output is the finger-thumb position. (Photograph Geoffrey Drury courtesy Ministry of Health, Roehampton, England, from Spare-Part Surgery by Donald Longmore, Aldus Books, London, 1968.)

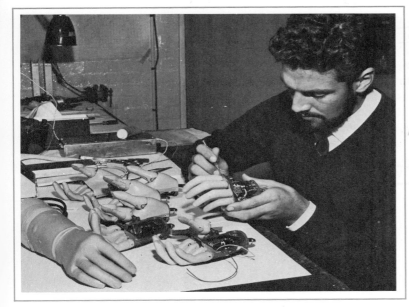

Fig. 7–11.
Stages in the
assembly of the
British hand; the
finished product is in
the left foreground. The
hand in the right foreground
shows the motor and the lead screw
assembly that moves the fingers. The
thumb has two fixed positions and
can be set by the wearer in
one of two fixed positions
before the start of any
given movement.
(United Kingdom
Atomic Energy
Authority)

Why does it make sense to speak of goals in describing the activities of living things, whereas it is nonsense to speak in the same way of inanimate physical objects such as the falling rock?

Perhaps you may conclude, as some people have done, that goal-seeking behavior is the unique characteristic of life. But is this always so? An elevator operator has the goal of bringing the elevator to rest at each floor so that the floor of the elevator cab is level with the floor of the building. In many older elevators, the speed of the elevator is controlled directly by the operator. At the end of the day when the elevator is heavily loaded with people on their way home, the operator must begin to bring the elevator to a stop before it reaches the desired level if he is not to overshoot his mark. If he does overshoot, he will reverse the motion until the floor of the elevator is at approximately the same level as the outside floor. With practice, a skilled operator can learn to adjust his actions in accord with the number of passengers. He stops the elevator at the desired level without a number of adjustments.

Modern elevators now have automatic mechanisms for accomplishing the same goal much more effectively. The mechanism built into the automatic elevator remembers at which floors it should stop to pick up passengers and at which floors to discharge passengers. It can also stop at the proper level more consistently than it could under the control of a human operator. The automatic elevator, thus, can be said to exhibit a goal-seeking behavior, despite the fact that it is completely inanimate. What then is the essential difference between the elevator and the falling rock that lets us describe one, but not the other, as a goal-seeking device?

The goal-seeking behavior of the elevator is made possible by the presence of a *feedback* arrangement in the elevator system.

The difference between *mechanistic behavior* (which is governed only by past causes) and *purposeful behavior* (which is guided by future or desired goals) has been a battleground of debate among philosophers, theologians, psychologists, and many others. Yet it is only within the past thirty years that man has discovered the concepts needed to understand some of the issues involved. The most important of these concepts lie at the heart of this course. Of them all, the central concept is that of *feedback*, which permits us to build machines and systems which display goal-seeking characteristics.

Box 7–1

GOAL-SEEKING DESCRIBED BY AN EQUATION
(*System of Fig. 7–12*)

If we call the input I and the output O, the error is

$$\text{Error} = I - O$$

The output is just K times this error, or

$$O = K(I - O)$$

This equation can now be solved for O by ordinary algebra:

$$O = \frac{K}{1 + K} I$$

The boxed equation describes the relation between the input I and the output O.

For example, if K is 4, the output is $\frac{4}{5}$ of the input. Regardless of how we change the input, the output is always $\frac{4}{5}$ as large.

When K is very large (e.g., 200,000), the output is practically equal to the input. In the elevator-control system, the output is the floor level of the car. We then use as the input the level of the floor at which we wish to stop. The feedback system brings the car floor to the desired level. The output automatically seeks the goal.

Fig. 7–12.
A simple feedback system.

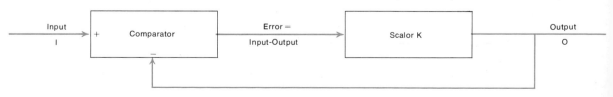

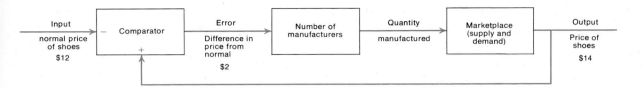

Fig. 7–13.
System
regulating the
price of shoes.

3 | FEEDBACK AS SELF-REGULATION

Feedback can be regarded as goal-seeking, and we can consider many feedback systems to be *automatic regulators*. These are systems which automatically try to hold the output at some desirable value.

Figure 7–13 shows the block diagram of a simple economic system. The output is the price of a certain product—for example, men's shoes of a particular quality. Suppose the normal or average price is $12. When the actual price in the stores is $14, the shoes are priced $2 above normal. The comparator output is then $2 (Fig. 7–13). This higher price causes more manufacturers to appear. People owning shoe factories start to make more shoes of this quality. The total quantity manufactured increases, causing an oversupply. The oversupply in the marketplace then causes the price to drop because the manufacturer must lower his price in order to sell his shoes. Now the price falls to $10—$2 below normal. The resulting lower profits cause some manufacturers to leave the business. The quantity manufactured falls, and the demand in the marketplace for the decreased supply results in higher prices. The price moves up toward $12. This is a remarkable system—it automatically adjusts the price to the normal $12. When the price is low, it automatically rises; when it is high, it automatically falls.

Figure 7–13 also illustrates one other factor which we often find in feedback systems: *instability*, or the tendency of the output to change out of control. The output may oscillate indefinitely—first high, then low, then high, and so on, or the output may simply grow and grow. This latter instability results because there is a loop within a feedback system. Signals can travel around and around the loop (Fig. 7–14). If we start with an error of $\frac{1}{100}$ and it is multiplied by 10 as it travels around the loop in one second,

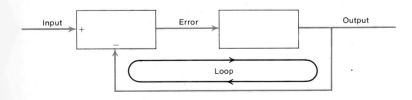

Fig. 7–14.
The loop
in a feedback
system.

the error is $\frac{1}{10}$ a second after the start. A second later it is one. After only eight seconds, the error is 1,000,000.

This tendency toward instability in feedback systems can be illustrated by familiar examples:

1. The relation between sleep and health is sometimes a feedback system. When you have a sore throat or nasal congestion, sleep is difficult. The lack of sleep or rest tends to make the cold worse, which in turn makes sleep even more difficult.

2. Economic systems frequently exhibit feedback. When a union as large as the auto workers obtains a sizeable wage increase, costs of automobiles tend to rise. Other costs rise as the added income of the union members results in more spending in other industries, hence greater demand for goods. The resulting increase in prices makes the union ask for further wage increases. A single union, of course, does not control this inflationary spiral, and the system tendency to oscillate can be controlled by changes in government spending and taxation.

3. The U.S.A.-U.S.S.R. armament race illustrates feedback. The United States learns of the development of the U.S.S.R. intercontinental ballistic missiles and launches a major program to develop more missiles than the Soviets possess. Learning of the United States missile arsenal, the U.S.S.R. then undertakes development of a major anti-missile system. In order to maintain the "balance of power," the United States must then launch an urgent program for an effective anti-missile system. Each of these steps is expensive for each country. In addition, each country is using valuable men in jobs which are not really productive.

4 | FEEDBACK FOR DISTURBANCE CONTROL

The examples of feedback in the last section were all cases in which feedback exists inherently in the system. In the armament race, feedback is inevitable since the United States' actions influence what the U.S.S.R. does. If feedback only occurred in such situations, and instability was the only consequence of feedback, there would be little point to our detailed discussion here.

Fortunately, we can intentionally use feedback to improve the way a system works. *Control over unwanted signals is a very important use of feedback.* Often unwanted signals affect our system output. These disturbances are unpredictable. They ruin system performance by causing undesirable output. When a system which includes feedback is operating properly, the output is continuously controlled. When the output changes because of signals beyond our control, corrections are made automatically.

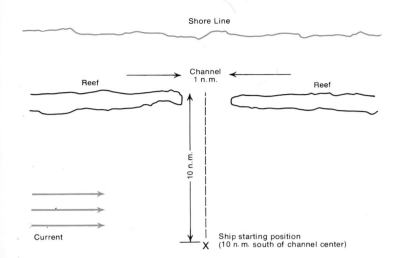

Fig. 7–15.
Navigating
through a channel.

For example, as a Navy pilot attempts to land his plane on an aircraft carrier flight deck, the primary input signal is his location relative to the deck. As he watches the flight deck from the cockpit, he continually adjusts his controls. There are also two secondary or *disturbance signals:* the wind gusts acting on his plane and the forces the sea exerts on the carrier. If these disturbances are too strong or too violent, successful landing becomes a most difficult task.

The problem of landing on a carrier is particularly complicated because it involves all three dimensions in space. Figure 7–15 shows a similar two-dimensional problem. A ship is located 10 nautical miles south of a channel opening through a reef. The navigator sights the buoys or light signals marking the channel and naturally sets a heading due north. A heavy fog suddenly sets in which hides the buoys from the men on the ship. The ship steams on at 10 knots (nautical miles/hour) with a north heading. The navigator, having correctly estimated the distance to the channel as 10 nautical miles, assumes that the ship will pass the reef in one hour—as indeed it will if there is no disturbance signal affecting the position of the ship. But what happens if a current of two knots pushes the ship eastward? The northerly motion is unaffected, but in one hour the ship is 10 nautical miles north and 2 nautical miles east of its original position, or 1.5 nautical miles east of the edge of the reef. Tragedy ensues.

In order to understand the system, a block diagram is useful (Fig. 7–16). The input signal is the heading the navigator orders for the ship. A disturbance signal (the current) adds to this heading to yield a signal which is the actual heading assumed by the ship. This actual heading determines the system output, the position of the ship relative to the center of the channel.

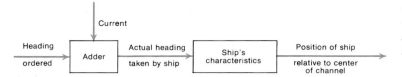

Fig. 7–16.
Block diagram
for navigation
problem.

The tragic ending can be averted if we can use feedback to counteract the effect of the current. For example, perhaps every 15 minutes the fog lifts long enough for the navigator to complete another sighting on the channel markers. After 15 minutes, the ship is at A which is 7.5 nautical miles south and 0.5 nautical mile east of the channel center as in Fig. 7–17. The navigator now orders a heading toward the channel center. Fifteen minutes later the ship is at B where another sighting is taken and a new heading adopted. (We still assume the navigator is stubborn. He refuses to recognize that his past errors might be the result of a constant current, for which he could compensate by aiming to the west of the channel center.) If this calculation is continued, we find that the ship follows the path shown in the figure, with sightings taken at locations A, B, C, and D.*

The block diagram of the system with feedback is shown in Fig. 7–18. Every 15 minutes, the navigator sees the channel. He adjusts the heading according to the difference between where he is and where he would like to be. In this example, feedback occurs every 15 minutes. We can do even better if we can use feedback continuously (i.e., no fog exists, so the navigator can sight the channel markings and adjust the ship heading continuously). Then the path followed by the ship is shown in Fig. 7–19. Initially the ship moves off the desired course because of the current, but the continual corrections result in final passage through the channel without difficulty. In this case, the feedback in the system is continuously compensating for the effect of the disturbing signal (the current). This navigation example illustrates in rather general terms a most important use of feedback. *Feedback can be used to reduce the effects of disturbance signals.*

Random Disturbances

The interesting feature of this property of feedback is that the effects of the disturbances can be controlled even when these signals are random and unpredictable. In our example, the current always flows at the same speed. In such a case, we do not require much feedback. After the navigator determines his position at the 15-minute mark, he can calculate the current. He then heads for an appropriate location west of the channel center so that, by the time he reaches the reef line, the current will have returned the ship to the desired midpoint in the channel. Feedback is needed

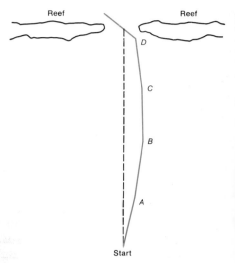

Fig. 7–17.
Path followed
by ship with feedback
every 15 minutes.

*
The path followed in Fig. 7–17 is most easily determined graphically if the diagram is drawn to scale. For example, once B is known, C can be found as follows. From B along a line toward channel center we mark off a distance of 2.5 nautical miles to determine a point which we can call C'. This is the point the ship would reach 15 minutes after leaving B if there were no current. C is 0.5 nautical mile due east of C'. Once C is known, D can be found similarly, etc.

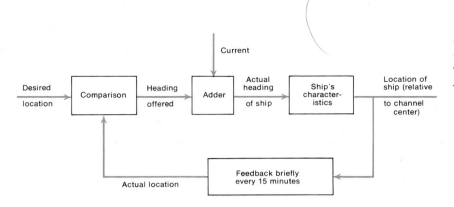

Fig. 7–18.
Block diagram
of system with intermittent
feedback.

only once so the navigator can determine the speed and direction of the current.

This role of feedback is also illustrated by an auto steering system. As we drive down a traffic lane, we continually observe our position and make corrections, since the disturbances are unpredictable. The need for feedback is clear if we consider what would happen if we were blindfolded.

As you have read this section, hopefully you have participated in a systems problem. The author and publisher have combined to generate a small input to the student's mind. The system output is your level of understanding of the section.

Because your mind may possibly have been subjected to disturbance inputs as you were reading (interrupting telephone calls, television, and so forth), we should attempt to control the effects of such disturbances with the use of feedback. A simple form of feedback in this learning system depends on the use of a set of questions to measure the system output (your degree of understanding). If the output is not at the desired value, the questions refer you back to an earlier portion of the section so that the input signals can be repeated.

Question 1. What is a primary use of feedback? If you are at all uncertain of the answer, a complete re-reading of the section is recommended.

Question 2. When feedback is used, what signal must be measured? Once this measurement is made, feedback usually involves a comparison. Why? If you are uncertain, the navigation example should be re-read.

Question 3 (Box 7–2). In the block diagram of Fig. 7–22, the gain K is to be chosen so that the part of the output due to x is the same for the two systems. Determine K. Which system results in less influence of u on the output? By what factor is the disturbance effect decreased? If you have any difficulty with this problem, it would be advisable to return to the discussion of Box 7–2.

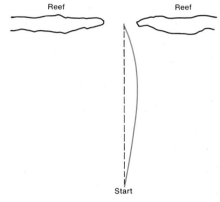

Fig. 7–19.
Path of ship
with continuous feedback
(but navigator not
estimating the current).

Box 7–2

EQUATIONS FOR DISTURBANCE CONTROL

We can describe disturbance control by algebra. Figure 7–20a shows a system with a disturbance input called u. The total output y is 20 times the main input plus 10 times the disturbance:

$$y = 20\,x + 10\,u$$

Figure 7–20b shows the same system (inside the dashed lines), but with feedback added with blocks A and B and the comparator. For this system, the output y is

$$y = 20[A(x - By)] + 10\,u$$

(20 times the input to the dashed box plus 10 times the disturbance.) Now if we use algebra to solve this equation for y, we obtain

$$y = \frac{20A}{1 + 20AB}\,x + \frac{10}{1 + 20AB}\,u$$

We can select A and B to give the desired behavior of the system. In our original system, we had $(20x + 10u)$. We might want the feedback to reduce the effect of disturbances by 40. Then we would like y to be $(20x + \frac{10}{40}\,u)$. Comparison of this with the last equation above shows that we want

$$1 + 20AB = 40 \qquad \text{(From the } u \text{ terms)}$$

$$\frac{A}{1 + 20AB} = 1 \qquad \text{(From the } x \text{ terms)}$$

Hence, A should be 40. Then we can find B from

$$1 + 20AB = 40 \qquad \text{when } A = 40$$

or

$$B = \tfrac{39}{800}$$

The final system is Fig. 7–21. By adding feedback, we can keep the x-to-y system unchanged, but reduce the effect of a disturbance by a factor of 40. Feedback permits control over unwanted signals.

We have now inserted the first type of feedback. Additional feedback is provided by your class discussions; with these various feedbacks, our theory tells us that it is certain that the student's level of uncertainty will be controlled regardless of any disturbance signals. When this system output is measured in the future by an exam, the change will surely be great.

This "example" itself emphasizes that the idea of feedback is important in people systems as well as machine systems.

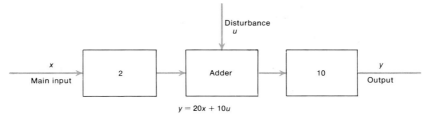

(a) Original system with no feedback

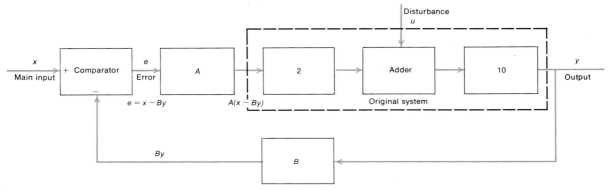

(b) System after feedback added

Fig. 7–20.
System to
describe disturbance control.

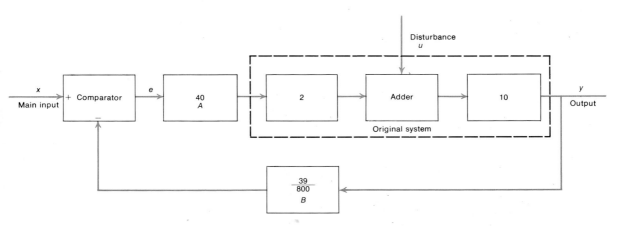

Fig. 7–21.
Final feedback system

$$y = 20x + \frac{10}{40} u$$

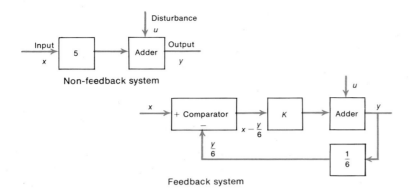

Fig. 7–22.
Two systems
with disturbance
input.

5 | AUTOMATIC COMPENSATION

Feedback has a second, very important advantage. *Feedback can correct automatically for changes in the system.* This characteristic of feedback is called the control of system sensitivity. How sensitive is the system to changes in its parts? Feedback can keep the senstivity small.

One of the most familiar examples of a feedback system is the regulation in the central heating system in a house. As we discussed in Chapter 5, a single furnace supplies heat to all parts of the house. The output of the system is the temperature. The input signal is the desired temperature for comfort. The thermostat measures the output and compares this measured value with the desired value (the input which is set manually on the thermostat). The system is shown in Fig. 7–23.

When the actual temperature drops below the desired temperature, the thermostat relay closes and the furnace is turned on. Heat then flows through the house. The actual temperature rise resulting from this heat flow depends on the thermal characteristics of the house (the windows and doors that are open, the insulation provided from the outside, and the degree to which air is circulating through the house). Regardless of these thermal characteristics, however, heat is supplied by the furnace until the actual temperature rises to about 3° above the desired temperature. At that point, the thermostat relay opens, the furnace is shut down, and the heat flow stops. The speed with which the changes occur depends on the particular form of the system. For example, if heat is transmitted through circulating hot water, heat continues to flow into the room from the radiators after the hot water stops circulating, until the water cools. If the desired temperature (the thermostat) is set at 72°F, the living room temperature near the thermostat fluctuates between about 70°F and 75°F and never exceeds these limits as long as the outside temperature is low and the furnace system is large enough to heat the house properly. The performance

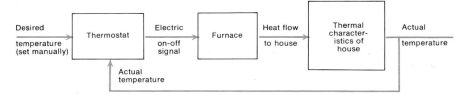

Fig. 7–23.
Elements of
household heating
system.

of the system is *completely independent* of the characteristics of the house and the outside temperature.

By using feedback, we have a system which is nearly perfect. We can even hold the front door open in the winter for a long time without losing control of the inside temperature.

A second example of feedback controlling sensitivity is shown in Fig. 7–24. It is a picture of a mechanical amplifier to increase the effective strength of a man. The machine is worn by the operator like an external skeleton, attached at the feet, forearms, and waist. As the man moves his arms, the exact motions are repeated by the machine powered by hydraulic motors (similar to the power steering and power brakes used in cars). The device can lift loads of 1500 pounds.

In order to permit the man to control the machine despite widely varying loads, a fraction of the load forces acting on the skeleton are applied to the man's arms and legs. For example, if the machine's arm hits an object, the operator feels a fraction of the force on his own arm. In this way, the machine becomes an extension of the man. The operator can use the machine in a normal way to move unusually heavy or cumbersome loads or objects that are particularly dangerous, such as bombs. The force feedback to the human being is the key to successful operation of machines of this type.

As in the last section, we attempt to add feedback to the learning system. In this learning system which involves the author, publisher, and reader, the characteristics of all three elements vary widely; we should attempt to ensure success by adding feedback in the form of a few simple questions.

Question 1. What are the two primary uses of feedback in system design (this section and Section 4)?

Question 2 (based on Box 7–3). For the system shown in Fig. 7–26, determine the percentage change in overall system gain when the amplifier gain falls by 20%.

Question 3 (based on Box 7–3). Figure 7–27 shows a more complex feedback system. Determine the change in overall system gain when the amplifier gain decreases by 90%. This is a rather unusual system, since it turns out that y is independent of the amplifier gain even though y is the output of the amplifier. Thus, even if the amplifier gain dropped to 0.001, the output y would be unchanged.

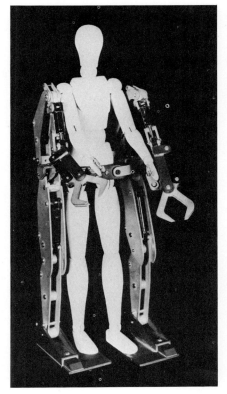

Fig. 7–24.
This is a
model of a set of
"mechanical muscles"
that will give a human being
the strength of a giant, and permit
him to lift a 1500-pound load while
exerting only a fraction of this force.
Attached to its operator at his feet,
forearms, and waist, the machine,
nicknamed HardiMan, will mimic
and amplify his movements.
(General Electric Company)

Box 7-3

SENSITIVITY CONTROL WITH FEEDBACK

The ability of feedback to correct automatically for system changes is explained by Fig. 7–25. For the system without feedback we have

$$y = 10x$$

The output is ten times the input.

For the system with feedback

$$y = 1000 \left(x - \tfrac{y}{10}\right)$$

If we use algebra to solve for y, we have approximately

$$y = 10x$$

Thus the two systems behave in essentially the same overall fashion. Each gives an output which is ten times the input.

Very often the gain of an amplifier varies during operation. If the total amplifier gain drops by 10%, what happens in each system?

The system without feedback is easy. If the amplifier gain falls by 10% (from 10 to 9), the output $y = 9x$: the output also falls by 10%. The feedback system behaves quite differently. The total amplifier gain is now 900 (10% less than 1000), and

$$y = 900 \left(x - \tfrac{1}{10} y\right)$$

If we solve for y, we find

$$y \cong 10x$$

The output is unaffected by the 10% change in amplifier gain. We have rounded off the numbers here. There is actually a very slight reduction in system gain. Thus, feedback makes possible satisfactory system operation, even when the characteristics of system components change rather radically in the course of time.

The extent of the value of feedback is even more apparent if we ask the reverse question: In the feedback system above, how much must the total gain of the amplifiers drop from 1000 before the output drops by 10%? If the amplifier gain is G (instead of 1000), the output is

$$y = G\left(x - \tfrac{1}{10} y\right)$$

If y is to be equal to $9x$ (rather than the normal $10x$)

$$9x = G\left(x - \tfrac{9}{10} x\right)$$
$$9x = G \tfrac{1}{10} x$$
$$G = 90$$

The amplifier gain must drop from 1000 to 90 (more than 90%) before the feedback system loses 10% in gain.

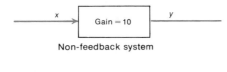

Non-feedback system

Fig. 7–25.
Two
comparable
systems.

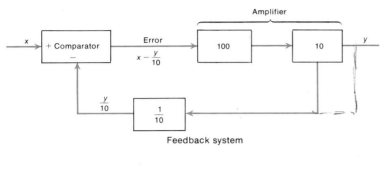

Feedback system

Fig. 7–26.
Simple feedback system
for Question 2. The overall gain
is the ratio Output/Input.

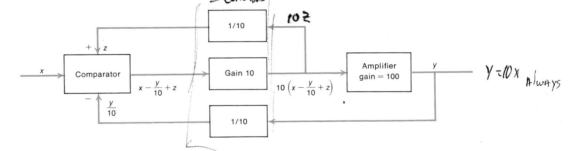

The system suggests some of the amazing things that can be done with feedback.

In this and the preceding section, the two basic purposes of feedback are described. We next look at a familiar feedback system to see these properties.

Fig. 7–27.
System
with two separate
feedback paths.

6 | A HUMAN FEEDBACK SYSTEM

Consider the system in a human being for the control of the internal body temperature. The system illustrates the use of feedback to give very accurate system performance even when there are large disturbance signals or large system changes.

The normal temperature of the core of the body (the internal organs and the central nervous system) is about 98.6°F. (The British consider the normal temperature to be 98°; in this country we use 98.6° since this corresponds to exactly 37° Centigrade.)

This temperature must be very accurately controlled. Cells of the central nervous system are damaged if the temperature rises as much as 7°F. An even smaller drop in temperature results in greatly reduced enzyme activity within the body. In the normally healthy individual, the temperature does not vary more than 2°F either way.

The core temperature is held almost constant even when the body is exposed to outside temperature changes of more than 100°F. We recognize the similarity between this system and the example of the last section, the control of temperature inside a house. In the human system, however, nature provides several different means by which temperature is measured as well as several different sources of heat, rather than the single thermostat and furnace common in houses.

In order to construct a model of our system, we must understand the basic elements of the system. From a thermal viewpoint, the body consists of three principal parts: (1) the core, (2) the skeletal muscles, (3) the skin.

In the control of core temperature, the human being uses several different methods to vary the heat in the core:

Chemical. The basal metabolic rate (BMR) is generated primarily in the core. The oxygen inhaled is brought by the blood to the cells where fat is stored. The oxidation or burning of this fat results in a release of carbon dioxide to the blood as well as a release of energy. When the BMR is measured during a physical examination, the patient is not permitted to eat for at least 12 hours in advance of the examination and must have complete rest during that time. The net oxygen consumption of the individual is then measured to determine the BMR, the rate at which the man is internally converting fat to energy. Metabolic temperature control is achieved through an endocrine gland which receives electrical signals from the brain.

Shivering. The muscles also provide a source of heat. If the sensors in the skin detect a sharp drop in outside temperature, electrical signals are transmitted from the brain to the muscles to order shivering. Here adjacent muscles (the same ones normally used for motion or useful work) operate in an uncoordinated fashion, with the result that there is very little useful work and most of the work is converted to heat.

Skin changes. The skin is used to effect changes in internal temperature in two ways. First, the blood flow to the surface of the skin can be controlled (this is called the vasomotor effect). When heat flow out of the body is to be decreased, less of the warm blood goes to the skin. Second, sweating leads to heat loss by evaporation, and is particularly important when the man is in a very hot place.

Thus, the human being has four ways to control the internal body temperature: variation of the metabolic rate, shivering, changing blood flow to the skin, and sweating. Each method is controlled by that part of the brain which regulates body temperature.

We now have described in general terms how the system works. We are ready to make a block diagram. The output of our system is the body temperature (the temperature of the core). Another output signal (Fig. 7–28) is the skin temperature. In addition to the four primary input signals to the body, there are disturbance inputs: changes in heat flow when the man exercises or when the outside temperature changes.

The primary input to our system is the desired core temperature (98.6°F). In case of illness, this input is probably increased to produce a fever. (Medical research has not yet indicated how this change is accomplished.) In the brain, the error is measured by comparing the actual core temperature with 98.6. This error is then used to change the core temperature. It is interesting to note that the human system of Fig. 7–28 also includes a measurement of skin temperature, in order to anticipate heat demands when the external temperature changes rapidly (as it does when one enters a hot or cold shower or moves from the inside to the outside of a house during cold winter weather). In our analogy to the home-heating system, the skin sensors correspond to outdoor thermometers connected to the thermostatic control system to anticipate sharp changes in the outside temperature.

The block diagram of Fig. 7–28 includes the main signals which make up the temperature-control system of the human being. What is the value of a diagram of this sort? How does such a description of the system aid the researcher in learning about the operation of the system? How might the model help in developing improved medical procedures?

Fig. 7–28. Thermal regulating system for temperature of body core.

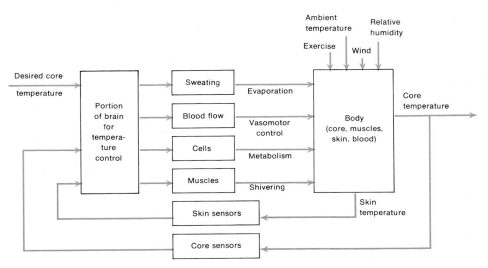

The answers to these questions are difficult unless we develop a much more detailed model. We should look at each block in Fig. 7–28. In each case, we would try to learn to describe the block mathematically. We might have to make measurements on men. We certainly would want to understand the physical laws which explain how things work. If we did this here, however, we would end up with a book on physiology, rather than a book on the man-made world. We can make certain comments just from the simple model.

First, the importance of feedback is apparent. The internal body temperature changes very little (typically less than a degree) when the man passes through radically different environments: feedback almost eliminates completely the effect of disturbance signals. Second, the core temperature is almost independent of changes in the body system. Major changes can occur (we might have a limb removed). Minor changes such as changing clothes are common. As a result of feedback, we have a system in which the performance is the same under all these conditions.

The model indicates basic questions, for example, how is the input changed (if it is) to cause a fever in a patient who is ill? Further study of this question may indicate that the input is not changed, but rather that the system loses effective control during illness. If the latter is true, perhaps major efforts should be made to hold down the temperature of an ill patient. On the other hand, maybe if we could learn more about the body, we would find that the fever is an important contributor to the body's power to resist disease. If so, we should not indiscriminately use drugs which reduce the temperature—at least not unless the body temperature approaches the danger range. When a child has a fever, the mother usually gives him an aspirin to bring the temperature down. Perhaps this is the wrong thing to do, at least if the temperature is not too high.

Furthermore, our model is only approximate. In the future, as we do more experiments, we will find new parts of the model. Then new experiments will be suggested. Hopefully, this process will lead to better and better understanding of the system. Thus, the feedback model is a key tool in the development of scientific understanding.

Finally, the model of this part of the human being may show us how to improve man-made systems. We saw that the human system measures skin temperature so it can anticipate large changes in the outside temperature. In the same way, we can build a better house-heating system if we measure outside temperature. When the outside temperature starts to fall, we could increase the heat flow inside to prepare for the cold wave.

Now let us consider a final feedback question. In hibernation what changes in the system would you expect? (The model might represent a woodchuck instead of a man.) What elements of the system would work during hibernation? Ordinarily, in hibernation the core temperature of the woodchuck drops to about 39°F. What advantages are there with this low temperature?

7 | INSTABILITY IN FEEDBACK SYSTEMS

Feedback is primarily of interest in goal-seeking systems. In the example of navigating through the reefs, the goal is to place the ship in the center of the channel. In the human temperature-control system, the goal is a constant core temperature of 98.6°F, regardless of surrounding temperatures or physical or emotional activity.

The success of feedback in controlling both disturbances and changes in the system involves certain disadvantages. We have already seen that feedback normally requires a more complicated system (the output signal must be measured and compared automatically to the desired value of the output in order to calculate an error signal which can be used to correct the output). In this section, we discuss a second, major disadvantage of feedback: *the possibility that the system may be unstable.*

In very general terms, a system is unstable if its output goes out of control. Perhaps the most dramatic example of instability is the hydrogen bomb; here a small detonation rapidly grows into an immense explosion. There are many other examples. Visitors to Bermuda observe the abundance of lizards on the island. Some years ago, a few lizards were brought to the island to control the mosquitoes; the lizards rapidly multiplied, the mosquitoes disappeared, and the islanders are now worrying about controlling the lizard population.

Perhaps a more familiar example is the instability occurring when a car goes from an understeer to an oversteer condition. Understeering means that as the car travels around a curve it tends to pull out of its curved path, or to increase its radius of curvature (Fig. 7–29). To follow the road, the driver must apply more and more force to turn the steering wheel. In the oversteer case, the automobile tends to decrease the radius, or to "tighten" the turn; to compensate, the driver must apply less force on the steering wheel during the turn.

The study of automobile behavior during turning is an exceedingly complicated problem since the motion depends on the road angle and surface and on the rapid actions of the driver. A dangerous situation occurs when an automobile moving at a certain speed passes from an understeer to an oversteer condition; the driver, if he is not aware of the change, tends to turn the steering wheel incorrectly to try to correct the car position. A problem in automobile design is to be sure that such a switch cannot occur suddenly and surprise the driver.*

*
There were hundreds of law suits against General Motors in connection with the Corvair car manufactured during a few years in the early 1960's. These were based in large part on the claim that poor engineering resulted in such a transition from understeer to oversteer. It was claimed that the sudden change made the car impossible to drive and accidents resulted. In the first case brought to trial with outstanding engineers testifying for both sides on the technical aspects, the judge decided for General Motors. The engineering problem in such a case is so complex that scientific, mathematical, or computer analyses and experimental tests often yield no clear conclusions.

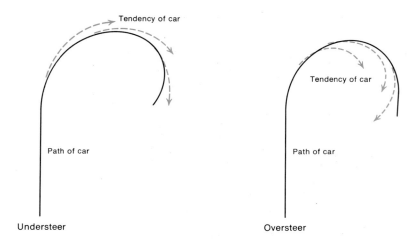

Fig. 7–29.
Path of
a car around
a curve.

Example of Instability

As another example of instability, we can return to our system of Section 1 of this chapter: a man picking up a pencil. As we saw there, the man observes the difference between the actual output and the desired value. Corresponding electrical signals from his brain to the muscles produce the required change in the output (Fig. 7–30). In addition to the main feedback through the eye, there is an additional feedback path directly from the muscular system to the brain: as the muscles are actuated, the individual feels the forces and resulting motion.* As soon as the hand touches the tabletop or pencil, there is an additional feedback path (not shown in the figure) which results from the sense of touch.

The above feedback system is normally excellent. It operates satisfactorily even when the disturbing signals are present (e.g., motion of the system when the task is performed inside an airplane or other moving vehicle) or when system characteristics change (the individual is tired, physically weak, or distracted mentally by other events).

Instability may exist, however, when individuals suffering from an illness called ataxia try this task. The hand starts to oscillate as it approaches the pencil. The man is unable to pick up the pencil because of the violent shaking of his hand. The loss of control comes from oscillation or instability within the feedback system.

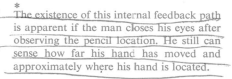

*The existence of this internal feedback path is apparent if the man closes his eyes after observing the pencil location. He still can sense how far his hand has moved and approximately where his hand is located.

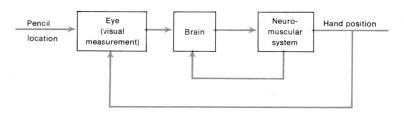

Fig. 7–30.
Feedback system
involved in picking up
a pencil.

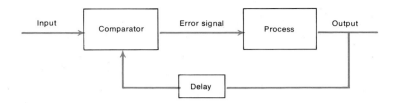

Fig. 7–31.
Feedback system
with feedback path
delayed.

Instability and Delay

There is no simple way to find when instability occurs in a feedback system. We can understand some aspects of the problem from the simple system of Fig. 7–31. Here the input is compared with the output, and the error causes the output to change to reduce the error. Suppose now that there is a delay as shown. The output is measured, but not sent to the comparator until a short time later. If the time delay is such that this output reinforces (or adds to) the input, the system may become oscillatory: the input causes a certain output, this is fed back and adds to the input, the even larger output is in turn fed back, and the output signal grows larger and larger.

The above description is certainly not exact or any real help in determining whether a feedback system will be stable or unstable; it does indicate that delay in the system is a key to instability. The delay (which we find in every real system) is important to instability in a feedback system. In general, the greater the delay, the more tendency there is for the system to become unstable.

In the examples considered previously, the cause of the delay is obvious. In the system for positioning the hand to touch a pencil, the delay comes from the time required for the brain to reach a decision and the time needed to move the muscles.

An Economic System with Delay

In inventory control, a company observes its sales of a particular product and then decides on a manufacturing schedule. If the company wishes to keep its customers happy, it must have enough completed products in its warehouses to satisfy rush orders. On the other hand, completed and unsold products in the warehouse represent dollars not working for the company. Thus, decisions are needed which will establish the best policy.

The feedback system is complicated by two delays: the delay in manufacturing (if the demand for the product rises sharply, time is required to increase the factory output because of the delays in obtaining raw materials and the time required for manufacture), and the delay in measuring changes in orders and demand for the product. Instability (or loss of control) results in very large changes in factory activity (and the resulting costs from large changes in the number of people working, the training of new employees, and so on). Instability may lead to warehouses bulging with unsold products or to an inability to meet customer orders (with the dissatisfied customers then turning to a competitor's product).

Instability in Biology

The population growth of hydra is a biological example of instability. Hydra are very small, freshwater animals which increase rapidly in numbers when the food supply is plentiful. One class of hydra can be fed exclusively on water fleas. As the hydra population in a closed container grows, the individual animals tend to become smaller (an automatic adjustment over generations to the population explosion). This regulating or control system may be unstable: the hydra become so small they are unable to eat the water fleas, and the entire population dies of starvation. In this case, the feedback mechanism (decreasing individual size with population growth) leads to the end of the system.

Such an extreme effect of feedback in a biological system probably results from the artificial environment created in the laboratory. In a natural environment, food sources would be available in at least small quantity. Some hydra would survive, even though the total population might decrease. This decrease in population would then result in larger animals, which would in turn thrive on the water fleas, and the total system would tend to oscillate around an average population and size.

Thus, in biology, feedback seems to cause instability when the natural environment is changed by man. With the rapid increase in technology in recent years, we now seem to be able to change the environment in very important ways.* Will instability result? If instability results, will this instability be permanently harmful to the human race? Unfortunately, we can only answer such questions if we understand the feedback system in detail.

Useful Instability

Finally, instability can be useful. We can use instability to obtain a sinusoidal signal. In the pendulum shown in Fig. 7–32, the ball (hung on a string attached to the ceiling at O) is moved to A and released. The motion is familiar: the ball swings down to B, then on to C where it reaches a maximum height, then back to B and on toward A. As the ball oscillates back and forth, a little energy is lost each trip because of air friction and friction at the pivot. Consequently, each return to the left is a little lower than the preceding, and the pendulum finally comes to rest at point B. Figure 7–33 shows the way the position of the ball changes as time passes.

If we wish to keep the oscillation going indefinitely, we can give the ball a very slight push each time it returns to the extreme left. We do the same thing when we pump a swing to keep it

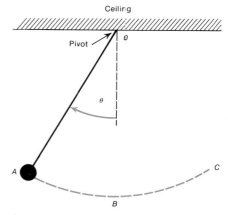

Fig. 7–32.
A swinging pendulum.

*The possibility of major success in weather control (particularly rainfall) is a particularly timely example. A more dramatic example is the current possibility of placing in orbit at an altitude of 22,500 miles (so that it sits over the same spot on earth) a large reflector to reflect sunlight back to the earth's surface during hours when the sun is below the horizon. The darkness of night would thereby be avoided—with the consequent, undetermined effects on nature as well as on man's living habits.

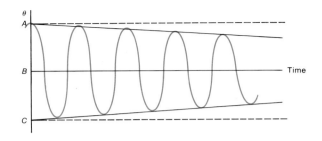

*Fig. 7–33.
Gradually
decaying oscillation
of a free pendulum.*

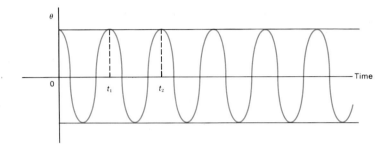

*Fig. 7–34.
Pendulum
oscillation with
regular excitation.*

moving. With this slight push in each cycle, the position signal has the form shown in Fig. 7–34.

If this were a ball swinging from the ceiling, control by a human being continually tapping the ball would be tiresome. Instead, we can automate the system by surrounding the area at A by an electromagnet, which is energized in such a way as to repel the ball at the top of each swing (the ball would have to be magnetized also, of course, because we are using the force between two magnets). Thus, at instants t_1, t_2, and so forth in Fig. 7–34, a pulse of current through the electromagnet gives the ball a slight push as it starts its downward swing: just enough of a push to compensate for the energy lost during the preceding cycle. If we wish, even the timing of the electromagnet current can be controlled by the motion of the ball so that the current starts slightly after the ball enters the electromagnet near the end of the travel toward A. The system, called an *oscillator*, is then entirely automatic and maintains constant-amplitude oscillation of the pendulum. The output is a sinusoid, as we discussed in Chapter 6.

An oscillator system is the basis for an accurate clock operated by a battery rather than by winding or from the electric power lines. The clock requires no winding and operates for the life of the battery. In this case feedback is used to yield an oscillation. What we want is an output which is constantly changing. We are really using feedback to obtain a controlled type of instability.

The main purpose of this section is to emphasize that feedback may be associated with instability. The amount of feedback which can be used in a system is often limited because we must have a stable system; otherwise, control is meaningless. When we use feedback in any system (economic, social, or man-made), we must be aware that the system may become unstable.

8 | FINAL COMMENT

Feedback is a way of looking at system problems. To understand and to design complex systems, it is often convenient to think of the system in terms of a block diagram. In a block diagram, feedback is represented by measurement of the output and comparison of this measured value with the desired value. The error is then used to change the output to reduce the error.

Feedback has been used by engineers for centuries. Perhaps the earliest engineered feedback system was the plumbing device developed by the Romans and used in much the same form today, the water-level control in the tank at the back of the common toilet. Here a ball floats on the water as the water level rises. When the water level reaches the desired height, the rising ball shuts a valve which stops the incoming water. When the water in the tank is released by flushing, the tank empties, a rubber stopper then covers the outlet, and the cycle repeats.

The first major work on feedback-system engineering was carried out at Bell Telephone Laboratories during the 1920's as the telephone engineers developed a system for long-distance telephony. Telephone conversations were possible across the United States only if amplifiers were used every few miles to amplify the voice signals. Once a large number of amplifiers were included, however, satisfactory system performance required that the gain of each amplifier should not change very much; otherwise, the volume for the listener might vary greatly. To build amplifiers constant over long periods of time, the engineers utilized feedback.

Feedback engineering made another major advance during World War II, primarily because of the importance of feedback systems in aiming large guns and radar antennas. In earlier wars, guns were smaller and targets moved very slowly. In World War II, antiaircraft gunfire required rapid aiming, which demanded more force than a man could provide. Machines had to be used. Feedback was essential if these automatic systems were to be accurate.

Since World War II, feedback engineering has continued to be a key part of modern technology. *Automation* involves automatic feedback control of decision-making processes, whether in factories, in traffic control, in the control of anesthesia during operations, or in the navigation of space vehicles. In addition, during the past twenty years the idea of feedback has been used in the study of biological, social, and economic systems. Automation is perhaps the most important technical development in recent years, and feedback is the central idea in automation.

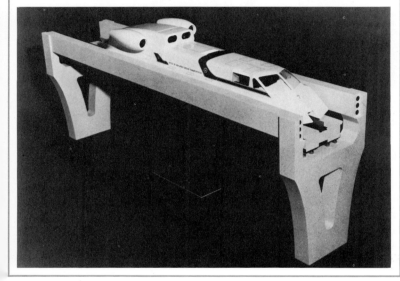

*Fig. 7–35.
Designs of
experimental
300-mph trains.
In both cases, the
train rides on an air
cushion. High-pressure
air is forced down from
under the vehicle, so that
the vehicle sits on this
air rather than on the solid
surface.*

*One of the major problems is
how much to bank such a train as it
goes around a curve. The banking angle
for maximum passenger comfort
depends on the speed and the
sharpness of the curve. In a
feedback system, the speed
and curvature are measured.
Air coming from the four
cushions underneath is
automatically adjusted
to give the proper
bank angle. Feedback
is essential if the
system is to
work.* (Aviation
Week and
Space
Technology)

Thus, it is appropriate to close this chapter with a description of an automated system. Figure 7–36 shows a diagram of the Alfred E. Perlman automated freight yard of the Penn-Central rail-road in Albany. When a freight train arrives at the yard from a distant point, it includes individual freight cars which must be sent to many different locations. In other words, the incoming trains must be broken up and the cars reassembled into outgoing trains bound for different destinations.

There are places for 7973 cars within the yard; on the average 3000 cars are switched each day. There is room for cleaning 100 cars at a time and repairing 113 others. There are 70 parallel tracks in the classification yard where cars are sorted and new trains assembled.

The entire system is controlled by a central digital computer. Each car is pushed by a locomotive over a hump 26 feet high. As the car then rolls down the incline, it is identified and its speed and

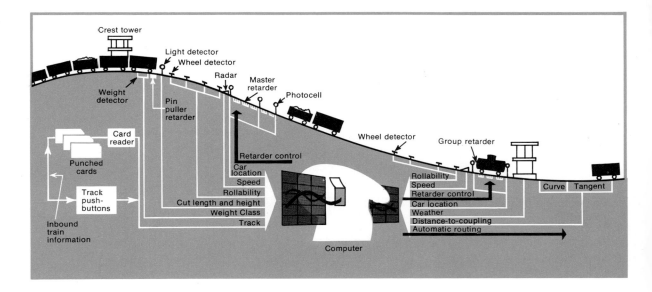

Fig. 7–36.
Humping operation
in automated railroad
freight yard.

weight measured. By the time it reaches the bottom of the incline, switches are automatically thrown to route the car onto a track where a train for its destination is being assembled.

As Fig. 7–36 shows, the speed of the car is controlled automatically to bring it smoothly into contact with the rest of its train, and the weather conditions are even fed into the digital computer since they influence the rate at which the car will slow down when it reaches the level. In addition to all the feedback instrumentation and computer control, the entire operation is monitored by six different television systems, 12 radio systems, and 15 teletype machines to communicate information to the men supervising this process. If the railroads are to provide low-cost, high-quality service, feedback is essential.

Questions for Study and Discussion

1. Your dog sees another dog and dashes off to battle. If you whistle to him and he obeys you, would you call this "goal-seeking behavior"? Give your argument in favor of your answer.

2. "Plants extend their roots toward moist and fertile soil." "Cells at the surface of the root differentiate to form . . . long extensions of the wall, the *root hairs*." Discuss the process which leads to the result stated in the first sentence above. Show how a feedback mechanism is involved.

3. An escalator or moving stairway must run at a speed which is as nearly constant as possible, whether nobody is on it or it is crowded with passengers.

   a) How can such control be accomplished?

   b) What is the input? The output?

   c) Illustrate with a block diagram.

4. In each of the following feedback systems tell what is the input signal and the output signal.

a) The automatic frequency control of a radio receiver.

b) A power steering gear for an automobile.

c) A machine in a sugar refinery to fill boxes each with just 2 lb of granulated sugar.

d) An automatic record player: the device to return the pickup arm after finishing a record.

e) An electric refrigerator.

5. Describe briefly at least one feedback mechanism in each of the examples given in Question 4.

6. One of the first engineering applications of feedback was the steam-engine governor invented by James Watt and diagrammed in simplified form as shown. The rotating shaft moves at the same speed as the main shaft of the engine, and therefore spins the iron balls. The balls, of course, tend to fly away from the shaft as they spin faster. The weight of the balls counteracts this tendency.

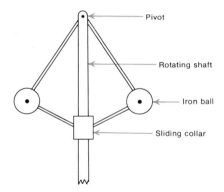

Describe a means to use this governor to keep the engine running at nearly constant speed, and illustrate the system with a block diagram.

7. Audio amplifiers (e.g., for a record player) sometimes "motor-boat," that is the sound from the speaker is disturbed by a series of popping noises. This is caused by oscillation of the amplifier at a low frequency. Suggest the cause and a suitable cure.

8. The temperature control of a shower is an interesting feedback system.

a) Construct a block diagram for the system assuming that the man standing in the water spray attempts to control the temperature of the water which reaches him by adjusting the hot-water control knob only. The knob is so placed that he can reach it while standing in the spray.

b) The simple block diagram which you have probably derived can be changed to include the fact that when the man changes the knob setting, he estimates the time delay before the temperature change will be detectable at the surface of his skin. Show this new block diagram if time delay was not included in your answer to (a).

c) The statement is made in the text that time delay is often associated with instability. Explain briefly how instability is likely to result if the man underestimates the time delay between the turning of the knob and an observable change in water temperature.

d) The stability problem of a feedback system is often complicated by parts of the system in which a change in input results in no change in output over significant ranges. This is one example of what the mathematician and the engineer call nonlinear behavior. For example, in many shower systems there is appreciable *backlash* in the knob: as the knob is turned clockwise, the amount of hot water decreases; but if we try to reverse direction (and move counterclockwise), the first 15° or so of motion result in no change in hot water flow. It is only after we have turned the knob through 15° that control is reestablished. This "slippage," or backlash, occurs every time the direction is reversed. Describe briefly how such a common malfunction may make effective temperature control more difficult (and may actually lead to instability).

9. In each of the following situations, decide whether feedback is or is not involved, and briefly explain your reasoning. (In each case, somebody had to take some action, and he, of course, is a bundle of feedback loops. Leave him out of the discussions.)

a) A golfer uses a 7-iron for an approach shot. His ball acquires backspin and hence does not roll off the green.

b) A weight is hung from a support by means of a spring. It is then pulled down an inch or two and set free, after which it bounces up and down for a long time but finally comes to rest.

c) A billiard ball rolls against the rail of the table and bounces back toward another ball.

d) A coin is dropped in the slot of a vending machine. The door opens and allows the buyer to pull out a bottle of soft drink.

10. In manufacturing thin sheets of metal, like aluminum foil, it is important to keep the thickness constant, but equally important not to stop the rolling machine to gauge the thickness. A mechanical device for continuous measurement would probably mar the sheet, even if it were able to make precise enough readings. The solution adopted is to place a radioactive source of alpha particles (a bit of the element plutonium is often used) at one side of the sheet, and a detector for alpha particles at the other. The rate at which the particles get through to the detector is exactly dependent on the thickness and material of the sheet. Show how such a gauge could be made part of a feedback loop to control the thickness of the sheet (use a block diagram if you prefer).

11. A famous medieval philosopher, Jean Buridan, told of a donkey placed between a bale of hay and a bucket of water. His hunger and his thirst were so exactly balanced that he starved to death because he could not decide which desire to satisfy first. Nowadays we might imagine that an oscillation resulted, with the consequence that he worked himself to death. Briefly explain how such an oscillation might be established.

12. Suppose in the pencil problem described in the first section, the man was blindfolded but the pencil was tapped loudly as it was placed on the table. Draw and explain the block diagram which represents this situation.

13. On more than one occasion an engineer has constructed an "automatic animal." The beast has motor-driven wheels and carries a battery. If it runs into an obstruction it stops, backs off, turns a little, and tries again. When its battery runs low, it rolls to its hutch, brightly lighted inside, and there plugs itself into a battery charger until it is powered up for more exploration of its environment. In general terms only, how can these accomplishments be built into the "animal"? Can you think of a way to equip it so that it could run to the edge of a precipice, stop, and go away from the precipice?

14. Describe the function in maintaining constant body temperature of:

a) the BMR (basal metabolic rate).

b) the muscles.

c) the skin.

15. Draw a block diagram which describes the feedback involved in an armament race between two countries.

16. Draw a block diagram which describes the use of feedback during a political campaign.

17. Describe the feedback system which you would propose to improve education in America, your state, and your school system, or your classroom.

Problems

1. Consider the navigation problem described in Section 4 (Fig. 7–15). Using a graphical construction, determine the path of the ship from its starting point 10 nautical miles from the reef if sightings are taken only every 25 minutes. Repeat the construction for the case where sightings are taken every 10 minutes.

2. In the amplifier system shown here, the output $y = 20$ volts. "Noise" is developed in the system which is represented by the addition of E_n, here 2 volts. What is the input voltage E_i under these conditions? What output voltage would be developed by this value of E_i if E_n were 0?

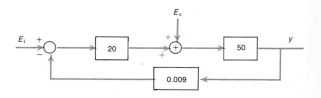

3. The connection at the comparator in the next system turns it into a "summer," that is the error signal is the *sum* of the input and the feedback. When the system is turned on, the amplifier, warming up, goes from a gain of 0 to a gain of 20. What

is the system gain when the amplifier gain has reached (a) 5? (b) 10? (c) 20? Do you recommend this connection?

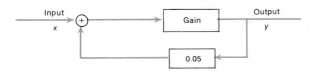

(4.) Find the gain of the system shown in the block diagram below.

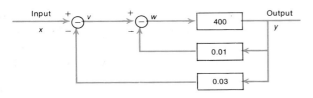

5. For the navigation problem discussed in Section 4, suppose that the current signal has the form:

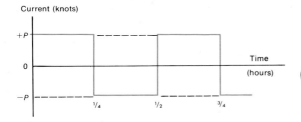

The current alternates between the values $+P$ and $-P$. Each quarter hour the ship moves north 2.5 n.m. At the beginning of each quarter hour, the helmsman measures his location and sets a course which would return the ship to the centerline in 15 minutes, if the current continued as during the previous quarter hour. Thus, in the figure below, if he is at A at the start of the 15-minute period, the course set would bring the ship to B_1 a quarter hour later if the current were $+P$. Actually the current changes to $-P$ and the ship ends up at B

rather than B_1. Determine the values of a, b, c, d, . . . when the ship starts from 0.

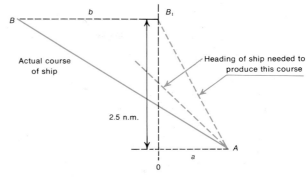

(6.) If you have an amplifier with a gain of 1000 which you wish to stabilize, what feedback constant B (Fig. 7–20) should you choose if you must have a system gain of (a) 100? (b) 10?

7. Refer to Fig. 7–28: "Thermal regulating system for temperature of body core." The ambient temperature and the relative humidity both rise suddenly. Describe the action of the elements of each of the feedback loops in maintaining the desired core temperature.

8. The amplifier for a certain hi-fi record player is represented by this block diagram.

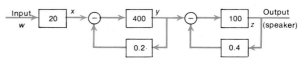

If the input voltage supplied by the pickup is 0.1 volt, what output voltage will appear at the speaker? It should be noted that in the last stage of amplification it is actually more important to give a large *power* output (in watts, or amperes × volts) than a large voltage output. However, this is a complication that need not be considered here.

Laboratory and Projects

I | THE TANTALIZER

Earlier in this course you used the Tantalizer to attempt to do various tasks while looking in a mirror. In this exercise you will look at the problems you had in terms of feedback.

Part A

1. Set up the tantalizer mirror as in the diagram below.

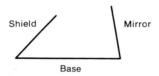

Fig. 1.

2. Place a piece of paper with the diagram below on the base and cover it with a piece of tracing paper.

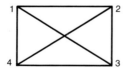

Fig. 2.

3. While looking in the mirror trace the figure around the perimeter 1, 2, 3, 4, and back to 1. Then trace the diagonal 1 to 3 and then 4 to 2.

   Most people who attempt this exercise do rather well on Lines 1-2 and 3-4, and have minor problems in starting the tracing from 2-3 and 4-1. The real problem seems to be in going from 1-3 and 4-2.

Part B

You can analyze the feedback involved in this tracing by answering the following questions.

■ 1. Describe how the direction of motion of the pencil compares with the direction of motion of its image in the mirror as you move the pencil from 1-2 or 3-4. Draw a block diagram for motion in this direction.

   Note that the error signal between the actual motion of the hand and the observed motion of the image is zero since they both move in the same direction at the same speed.

■ 2. Now describe how the direction of motion of the pencil compares with the direction of motion of its image as the pencil is moved from 2-3 or from 4-1. Draw a block diagram to describe this motion.

   What is the error signal between the actual motion and the observed motion of the image in the mirror?

■ 3. With the explanations of motion from 1 and 2 above in mind, explain the reason for confusion as the pencil is moved from 1-3 or from 4-2.

II | A TRIANGULATION PROBLEM

We simulate a navigation problem with the following procedure.

Part A

Draw a map of the classroom to scale and indicate on it the corners of the room. Assume that one of the edges of the room is a N-S Line. (This eliminates the need for a magnetic compass.)

1. One person should hold the theodolite flat and with the 0–180° line parallel to the edge of the room which was designated N-S. Another person should sight each of the corners of the room as shown in Fig. 3 and record the bearing.

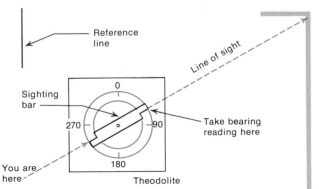

Fig. 3.
Taking a bearing with the theodolite.

2. Using a protractor, lay out on your classroom map a set of lines passing through the four reference points on the map with each line being drawn at the measured angle to the reference line.

3. Do all four intersect at the same point? If not, can you explain why they do not? Can you suggest improvements in the technique which would reduce the size of the error?

4. On the basis of a study of your error polygon make an estimate of your actual location and mark this on your map.

5. Use a tape measure to measure your actual distance from each of the corners of the room. Mark this position on your map. How does this compare with the position which you calculated by sighting?

   If the location of your position on the map is reasonably consistent by both methods, you are ready to start your navigation problem.

Part B

In this part of the experiment you will apply feedback to help you navigate through a simulated narrow channel.

One group of students will act as navigator, while two students (one supporting the theodolite and the other taking bearings) will act as the ship moving between two reefs.

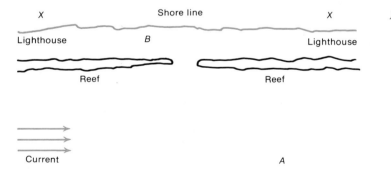

Fig. 4.

Figure 4 shows the waters to be navigated. The reefs will be represented by two tables, and the lighthouses by corners of the room. The navigators have a scaled map of the room with the reefs and lighthouses marked on it. The students representing the ship stand at *A* (predetermined by agreement between navigators and ship).

The navigators tell the ship what direction he should take to go through the reef. He takes three steps in the direction indicated and one step in the direction of the current. He now gives his new bearings to the navigators who calculate his new position and give him a new direction to take. The ship again takes three steps in the indicated direction and one in the direction of the current. The process is repeated until the ship reaches its destination B or runs into the reef.

■ 1. Compare the chart of this trip with Fig. 7–17 of the text. Consider both similarities and differences.

III | FEEDBACK FOR DISTURBANCE CONTROL

In Section 4, you studied the advantage of adding feedback to systems to control disturbances. You found out that, if properly used, feedback can be used to minimize the effect of disturbances. In this experiment we study this aspect of feedback with the aid of the analog computer. We start our study by looking at a very simple system.

In the simplest case, the output of a system is generally equal to its input multiplied (or amplified if you like that word better) by some number. We state this idea as

$$\text{output} = B \times \text{input} \quad \text{or} \quad y = B\,x$$

This mathematical model is represented by the block diagram below:

Fig. 5.

Since B is an amplification factor and each of the three summing scalors at the top of your analog computer can act as amplifiers, model the above diagram on your analog. (Hint: Set B to some known value and read the output y on your voltmeter. Any later "hints" won't tell you so much, so be sure you understand how to do this simple model.)

Let us assume that the above system represented the steering system of a car, where y is the motion of the wheels and x is the motion of the steering wheel. Another input which can affect the output is an external disturbance u, such as a wind. The pictorial description of this new situation, with wind as the new factor u, is shown in the block diagram in Fig. 6 below.

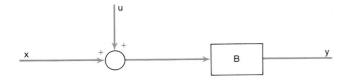

Fig. 6.

For those who are mathematically inclined, this changes the mathematical model to

$$y = Bx + Bu \quad \text{or} \quad y = B(x + u)$$

Keeping the same values for B and x, add a disturbance to your analog model and observe the effect on the output.

This disturbance, or "noise" as it is often called, is generally undesirable. We can reduce its effect on the output by using feedback. This involves the addition of two more amplifiers causing a modification of Fig. 6 as shown below. (A disadvantage in using feedback is that the models become more complicated but this cannot be helped.)

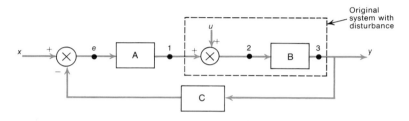

Original system with disturbance   *Fig. 7.*

An obvious question is, "Why two amplifiers?" The answer is that not only do we wish to:

1. reduce the effect of the disturbance, but also
2. we wish the output of the feedback system without the disturbance ($u = 0$) to be the same as the output of the original system (Fig. 5).

We are now ready to demonstrate this principle with a specific example.

Let us start by assuming that:

$$x = 1 \text{ volt}, \ u = .5 \text{ volt}, \text{ and } B = .5$$

After getting appropriate values of x, u, and B, use Fig. 6 to wire up the analog computer so that you can observe the effect of a disturbance signal. Notice with the above values of x, u, and B, the analog simulation has an output y of about .75 volt. Next use Fig. 7 to wire up the analog computer so that feedback can be added to the system. In this system, the feedback is added to minimize the effect of disturbance signals. Vary the constants A and C until the output y becomes .5 volt (output when there is no disturbance). Use the analog-computer simulation to answer the following questions:

1. What is the amplification factor of C which will eliminate the effect of the disturbance signal when $A = 5$, $B = .5$, $x = 1$ volt, and $u = .5$ volt?

2. What happens to the output y if we double the disturbance signal ($u = 1$ volt)? (Assume that the other values used in Question 1 have not changed.)

3. If the disturbance signal changed by a factor of ten (from $u = .5$ volt to $u = 5$ volts), what modifications of the analog simulation would result in maintaining the desired output ($y = .5$ volt)?

4. Does changing the input x affect the performance of the system? If yes, what would you have to do to the system to maintain its performance?

IV │ FEEDBACK CONTROL SYSTEMS

The applications of feedback and control are found in a multitude of situations in industrial operations and in our daily lives. These may be social, educational, electrical, psychological, mechanical, or many others. Mechanical systems in industry save money by being automated.

The water-level control in the water tank in your bathroom is an example of such a system (Fig. 8). Instead of building this system, we simulate its operation on the analog computer. We can then change the physical characteristics of the system by simply changing the values of the constants and coefficients of the analog computer.

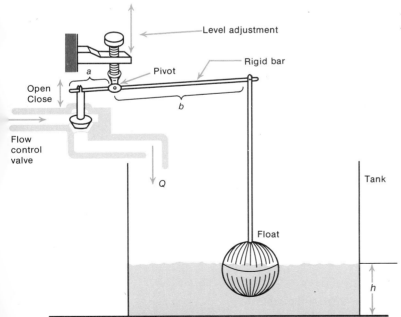

Fig. 8.
Liquid level control.

Part A

In Chapter 4, you used the analog computer to simulate the flow of water into a tank. Essentially you used the analog computer as a functional model of the following pictorial model (see Fig. 9).

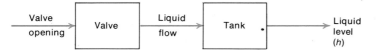

Fig. 9.

In this non-automatic system, a person has to watch the liquid level and must shut off the valve when the desired level is reached.

◼ 1. Modify the above block diagram (Fig. 9) to include the human feedback into the non-automatic system.

◼ 2. Wire the analog computer so that it simulates the operation of a non-automatic liquid flow system. Do not remove the wires, because you are going to build on this simulation later in this experiment.

◼ 3. Does your current simulation model the system which is shown in Fig. 8? If yes, explain; if no, what is missing from your simulation?

Part B

The previous system has the disadvantage that a man must be present constantly in order to turn the valve at the proper time. Since this is wasteful of both money and human talent, it would be helpful if a way could be found for this to be done automatically.

Now we modify the previous simulation to change it from a non-automatic to an automatic liquid flow system.

◼ 1. Use Figs. 8 and 9 to draw a block diagram depicting the operation of an automatic liquid flow system. Compare this block diagram with your answer to Question 1, Part A.

◼ 2. Use your latest block diagram to help you add the necessary wires to your analog simulation of the non-automatic liquid flow system to change it to an automatic system.

◼ 3. Match the parts of your analog simulation of an automatic liquid flow system to the parts of the system depicted in Fig. 8.

◼ 4. Which part of your analog simulation controls the final level of the liquid?

◼ 5. Which part of your analog simulation affects the time to fill a tank to a particular level?

V | FEEDBACK AND INSTABILITY (Classroom Demonstration)

In Section 7, a major disadvantage of feedback is discussed: the fact that sometimes feedback or the lack of it results in unstable systems. In this exercise, you will get a chance to participate in an activity which will help you to better understand the relationship between feedback and instability. These activities are all demonstrations of how feedback is involved in accomplishing even the simplest tasks. When you are finished with these activities, you may wish to describe others which also involve feedback.

Part A

Draw a reasonably straight line of about ten feet on the classroom blackboard. (The line should not be parallel with the edge of the blackboard.)

1. Give a blindfolded student *who has seen the line* a piece of chalk. Start him at one end of the line and ask him to retrace it.

2. Draw another straight line (not parallel with the first) and this time *do not let the student see the new line.* Start him at one end and ask him to retrace it.

3. If there is a difference in the student's ability to retrace the lines in 1 and 2, explain in terms of feedback why this difference occurs? (Where is the source of feedback?)

Part B

Along the length of the blackboard draw an irregular line which a blindfolded student has not seen.

1. Start the student at one end of the line and have a second student try to keep him tracing close to the line by instructing him to move his chalk up or down as he walks along the blackboard. (The observer must give a down instruction if the chalk is above the line and vice versa, but he should not anticipate the actions of the blindfolded student.)

2. Repeat Step 1 (do not erase the results), but have the second student signal by hand the up and down instructions to a third student (who should not be able to see the blackboard). The third student then calls out the instructions to the blindfolded student.

3. Repeat Step 2 with the instructions being relayed through three, four, and five students (do as many as time permits).

4. Compare the resulting tracings and explain the reasons for the differences in the student's ability to retrace the line.

5. Draw a labeled block diagram of the system for each of the above exercises.

VI | UNIQUE SPRING SCALE (Classroom Demonstration)

The basic elements of a feedback system involve an input, an output, a comparator, an amplifier, and a feedback loop (see Fig. 7–12, page 304). In this demonstration experiment we are going to illustrate the operation of feedback systems by developing a unique type of spring scale. It measures forces or weights without having the weight move as the spring scale deflects. This demonstration is also an illustration of how the addition of feedback to a system can correct automatically for changes within it (in this case, we are dealing with changes in amplifier gain).

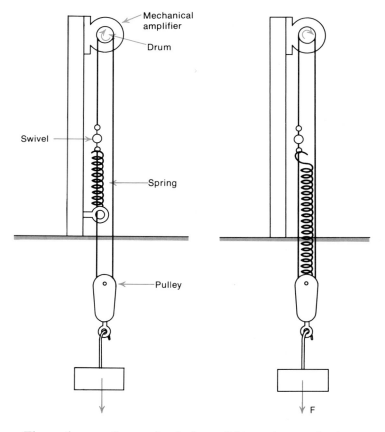

Fig. 10.
Spring scale without feedback.

Fig. 11.
Spring scale with feedback.

The scale uses the mechanical amplifier and a special loop of string with a spring, a pulley, and a swivel.

1. Clamp the mechanical amplifier on one end of a table and lay the loop around the top of the pulley as shown in Fig. 10. One end of the spring should be hooked to the bottom of the swivel and the other end to the large eye bolt. Note the location of the swivel with respect to the direction of rotation of the drum in Fig. 10.

2. Hang a mass of about 0.05 kg (.11 lb) from the pulley. Make one wrap of the string around the drum. Grasp the side of the loop opposite the swivel to form the wrapping. Do not allow turns to overlap.

3. Turn on the motor. Observe the deflection of the top of the spring and note the position of the pulley. Add more weight or pull down on the pulley and notice that the pulley remains at the same height, while the spring deflection increases. What is the difference between the operation of this device and the operation of a simple spring balance? In what situation do you think this device would be better than an ordinary spring scale?

There is one major disadvantage to the spring scale. The reading of the scale will change if the amplification factor changes. This change of scale reading may be caused by changes in the number of turns around the drum or changes in friction (e.g., coating the drum with grease or wax). We will change the number of turns to demonstrate this disadvantage.

4. Remove all extra wraps from the drum. Just let the string hang over the drum. This is equivalent to half a turn. Hang a mass of about 0.05 kg from the pulley and read the position of the top of the spring. Turn on the amplifier and measure the deflection of the top of the spring.

5. Place one additional wrap on the drum (there will now be $1\frac{1}{2}$ turns) and measure the deflection.

6. Measure the deflection of the spring with $2\frac{1}{2}$ wraps on the drum.

7. Plot a curve of spring deflection versus number of wraps.

8. From your curve, what do you predict the deflection would be if you could put $3\frac{1}{2}$ wraps on the drum? You can see how seriously a change in friction affects the reading on the scale.

We can represent this scale by the block diagram below

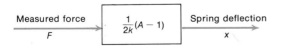

Fig. 12.

where A is the gain of the mechanical amplifier and k is the spring constant.

We now demonstrate how feedback can be used to reduce the sensitivity of this scale to changes in the gain of the amplifier. Since the gain of the mechanical amplifier is directly related to the number of wraps around the drum, we are in effect demonstrating how feedback can be used to compensate for changes in the amplifier gain.

To produce the feedback effect, we simply unhook the end of the spring from the large eye bolt and hook it to the top of the pulley as shown in Fig. 11.

The output signal from the scale is the deflection of the top end of the spring. The movement x of the top of the spring, however, causes a force kx to pull up on the pulley. This force is the *feedback* signal.

Note that this feedback force, which is pulling on the pulley, opposes the input force F which is pulling down on the pulley. The pulley therefore acts as the comparator.

A block diagram of the scale with feedback is shown below.

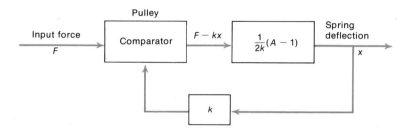

Fig. 13.

Using the block diagram above, show that the scale reading can be represented by the equations:

$$x = \frac{1}{2k}(A-1)(F-kx) \quad \text{or} \quad x = \frac{(A-1)F}{k(A+1)}$$

Notice that when the amplifier gain A is considerably larger than unity, the terms $(A-1)$ and $(A+1)$ are approximately equal to A; the output of the scale then is $x = F/k$. Hence, the reading is independent of the value of A.

We have now derived two equations for the scale:

a) $x = \dfrac{1}{2k}(A-1)F$ without feedback

b) $x = \dfrac{(A-1)}{k(A+1)}F$ with feedback

9. If you had two scales, one with feedback and one without feedback, how would the forces applied to the scales compare if both indicated the same spring displacement? (Assume both scales have the same values for A and k, and that A is greater than 1.)

10. Hook up the scale for feedback as shown in Fig. 11 making $1\frac{1}{2}$ wraps of the string around the drum.

11. Hang a mass of about 1 kg from the pulley. Turn on the amplifier. Add more weight or pull down on the pulley and notice that the pulley remains at the same height. Observe the movement of the spring.

12. Hang a 0.05 kg weight from the scale. Can you detect any movement of the spring? What would you have to change in the scale design to enable it to measure small weights?

   We now will see how the scale with feedback reacts when the friction is changed.

13. Put only $\frac{1}{2}$ wrap on the drum.

14. Hang a 1 kg mass from the pulley and measure the unstretched length of the spring.

15. Turn on the amplifier and measure the deflection of the spring.

16. Measure the deflection of the spring for $1\frac{1}{2}$ and $2\frac{1}{2}$ wraps.

17. Plot a graph of spring deflection versus number of wraps.

18. Compare the shape of this curve with the shape of the curve plotted in step 7.

VII | MODEL OF A HOME-HEATING SYSTEM

In Fig. 5–10, a block diagram of a heating system is depicted. In this exercise, we shall use this diagram to develop a functional model of the system on the analog computer.

Part A

The first step is to develop a block diagram relating the "real" system to the analog-computer system.

1. Using the different systems of the AMF analog computer as models of units of the heating system, draw a block diagram which models the operation of a home-heating system.

Part B

Now use your block diagram and wire your analog computer so that it becomes a functional model for a home-heating system.

1. Check your analog-computer model to see if it has the same general characteristics as an actual heating system (change thermostat setting, etc.).

2. Is your analog-computer model completely representative of the actual system being modeled? Explain.

Stability
8

1 | INTRODUCTION

When we design systems to cope with the world in which we live, we usually have in mind some desired operating state or condition. This desired state may be very simple: we may wish to place a book on a table so that it remains stationary. Or the desired state may be quite complicated: we may wish to place an astronaut into a precisely calculated orbit.

Sometimes, describing the operating state may be easy: we want the book to remain in a fixed and unchanging position. At other times the description of the operating state may be difficult: we must compute the astronaut's precise orbit with the aid of high-speed computers. But in either case, the design must involve additional information. We must know if the operating state is *stable* or *unstable*. An unstable design may produce disastrous effects. A man in a satellite sent into an *unstable* orbit would be catastrophic. His distance from the earth would either increase or decrease until tragedy occurred.

In this chapter we will probe into the concept of stability. We will find that unstable and stable systems differ in behavior if a disturbance is applied to them when they are in equilibrium. In many systems, we can deal with the phenomenon of stability in a quantitative manner by applying our techniques of modeling and analysis. By applying these techniques we can determine when a system changes from a stable to an unstable state.

2 | SKYSCRAPERS BEGET SKYSCRAPERS

There is a common saying, "skyscrapers beget skyscrapers." What does this mean? Until the beginning of this century, office buildings were generally a few stories in height (often only three or four stories). During the early 1900's, skyscrapers began to appear in several of our major cities. The trend toward multi-story buildings rapidly accelerated until today a large percentage of urban office buildings are at least 30 stories high.

The New York skyline,
1937 (Wide World Photos) and 1970 (J. Paul Kirovac).

The advantage of a tall building is that we can centralize many operations. Workers can live near their place of business, thus cutting down transportation costs. With taller buildings, land can be used more efficiently. Buildings such as the Empire State Building, more than 100 stories high, occupy only about $\frac{1}{100}$ of the land that would be needed for all the tenants if they were to be arranged on a single floor.

Some disadvantages, however, are associated with the height. As a building is made taller, the foundations must be made sturdier to support the added weight. With the all-masonry buildings that existed at the end of the last century, gigantic foundations were required for tall buildings. Elevators were non-existent, and people simply would not or could not walk up more than a few flights of stairs.

A natural competitive state of equilibrium existed. Every office building owner realized approximately the same profit, and all office buildings in a large city were generally about the same in height. The profit equilibrium was stable. Any attempt to disturb this equilibrium with the construction of a taller building proved wasteful since no one would walk up many flights of stairs. Hence the income necessary to pay for the additional cost of the foundations could not be found. A building of fewer stories did not realize the income of the buildings of the normal height.

But toward the end of the century two changes occurred. First, the cheap fabrication of steel beams made possible tall, lightweight, steel-frame buildings. Second, Otis invented the elevator, and electrical motors became available for their operation. The equilibrium state for three- or four-story buildings could have continued to exist because the new inventions could have been ignored. But such equilibrium was now obviously unstable.

The first skyscraper brought its owner large profits from a small plot of land. This produced an increase in the value of the land on which the skyscraper was built. Owners of vacant land were no longer willing to sell at the price that prevailed when buildings were only a few stories in height. They could wait for a skyscraper builder who would be willing to pay more for the same plot because his building would realize a larger return for his investment. With increasing land prices, only a skyscraper was profitable. Each skyscraper encouraged new, and for a time, higher skyscrapers. In New York, for example, the instability was especially severe. (In addition, the bedrock of Manhattan made skyscraper building easy.) Recently, in cities such as London where the foundation conditions for tall buildings are much less favorable than in New York, the "skyscraper begetting skyscraper" effect has become quite noticeable.

The preceding example really describes two different systems. The first system is one that existed during the last century. This was a stable system: if anyone tried to build a skyscraper, he found it unprofitable, and buildings were held to three or four stories.

With the invention of the elevator and the availability of low-cost steel, the system was changed; it was now unstable. After the first skyscrapers were built, all construction had to be sky-scrapers. Land became too valuable to permit low buildings. Today in city after city we find low buildings being torn down and replaced by skyscrapers. Even apartment houses are frequently 20 stories high. Urban universities are characterized by a set of structures towering above the campus.

Thus, discoveries in technology changed our urban construction system from stability to instability. In the new system of the twentieth century, a new operating point had to be found in order to achieve stability once again.

Two features associated with the stability or the instability of a system are illustrated by our example. These are:

1. In a stable system, a disturbance which moves the system will lead to forces that tend to return the system to the original state.
2. In an unstable system, a disturbance will cause forces that tend to drive the system even further from the original state.

Again and again, we meet these characteristics: the tendency of a stable system to restore itself and the tendency of an unstable ystem to "snowball" away from the equilibrium. A knowledge of these characteristics is essential for the creation and control of many systems in the man-made world. Although a qualitative appreciation of these characteristics is enough in many instances, in others we need a quantitative mastery: a knowledge of the numerical values of the quantities that may change the system from a stable to an unstable state. Later in this chapter, we apply modeling and analysis to some systems in order to illustrate the method of approach to such problems.

3 | STABILITY IN TRAFFIC FLOW

The snowballing effect in unstable situations is illustrated by a variety of examples taken from our normal life. Some of the most frustrating examples are provided by automobile traffic in a city. Anyone who has driven from the center of a city to the suburbs during the evening rush hour is familiar with the sharp transition from instability to stability in this system. Typically, as we move away from the city, traffic is bumper-to-bumper and moving at a very low average speed (perhaps 10 miles/hour in badly congested situations, even on a freeway or expressway). This slow rate of movement, with frequent stops, continues for some distance, even though a few cars leave the freeway at each exit as the individual

Fig. 8–1.
A common type of
model used in the planning
of traffic facilities in a city. This
is called a "desire-line diagram."
To construct such a diagram, we
mark the start and finish of each
trip. These two points are then
connected by a line. Thus, the
thickness of a solid mass of
lines represents the number
of trips along this direction.
(Traffic in Towns, *Her*
Majesty's Stationery
Office, London,
United
Kingdom)

drivers turn off for their homes. There comes a time, however, when a few cars leave an exit and, as a result, the average traffic speed rises rapidly, perhaps to 50 miles/hour. The departure of just a small percentage of the total number of cars changes us from an unstable system to a stable system.*

In this example, the location at which the system becomes stable depends not only on the traffic density, but also on such factors as the road condition, the weather, and the location of accidents which block certain lanes of the expressway. An accident or stalled car may cause extreme instability, but the stable situation after we have passed the accident may actually result in a lower total time to reach home. Ice on the road during winter months causes radical changes in the driving habits of motorists and, hence, major changes in the conditions for stability.

The mathematical determination of what traffic density will cause instability is difficult. The model includes not only the physical characteristics of the highway, but also the manner in which drivers behave under different conditions such as how closely

---

*
On the author's road home from work, this transition usually occurs at about the same point. A few miles further on, the expressway narrows from six lanes to four (two each way); the system is then again unstable, and it is another ten miles before stability returns.

they follow the car in front, and the extent to which they weave in and out of lanes. The problem is so complicated that it is usually easier to determine the characteristics of a given highway by *measurements* of automobile speed as a function of traffic density, time of day, weather, road conditions, and so forth.

Once such measurements are made, we have a model which permits the traffic engineer to determine the maximum number of cars which can be on a given stretch of highway if the system is to be stable (and, for example, average speed is to be at least 40 miles/hour). This model is the basis for modern systems for control of traffic flow on major city expressways. The number of cars on the controlled stretch of road is measured automatically by one of the following means:

1. Pressure transducers in the pavement and actuated by a car going over them.
2. Current-carrying loops of wire over the road. When a car passes under the loop, the magnetic field is changed because of the steel in the car. This change in magnetic field is detected electrically.
3. Photoelectric transducers (electric eyes). A light beam shines across the road on a device which generates electricity when illuminated. The passage of a car breaks the light beam and hence can be measured electrically.
4. Radar.

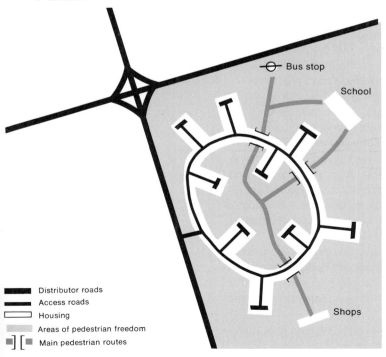

Distributor roads
Access roads
Housing
Areas of pedestrian freedom
Main pedestrian routes

Bus stop
School
Shops

Fig. 8–2.
One city-planning approach to the reduction of traffic congestion. This is known as the Radburn plan, since it was first used in Radburn, N. J., in 1928. In this approach, a central section exists free of vehicular traffic. In addition, there are pedestrian paths entirely separate from vehicular roads, so that people walking are protected from traffic and vice versa.
(Traffic in Towns, Her Majesty's Stationery Office, London, United Kingdom)

This information on the number of cars on our controlled stretch of highway is sent to a computer in which are stored the measured model and also the current data on weather, time of day, and location of accidents or stalled cars. The computer then transmits appropriate control signals to the traffic lights at each entrance to the highway. When the automobile density is approaching the point of instability, cars trying to enter the highway are held back by a red light. The duration of the stoplight is controlled by the computer signals. The closer the system is to instability, the longer the delay for motorists waiting to enter the highway.

Limitations of the System

The primary limitation of such a traffic-control system comes from the need to win public acceptance. When a traffic-control system was installed on a portion of the Eisenhower Expressway in Chicago, the waiting period for each motorist was controlled up to a maximum of 14 seconds. If there are longer waiting periods (and of course there may be few cars within sight on the expressway ahead), the motorist tends to go through the red light or complain about the ridiculous traffic engineers. In general in a democratic form of government, it is exceedingly difficult to enforce laws which the public finds unreasonable.

This unwillingness of the motorist to accept control exists even though his own personal welfare is best served by obeying the control. Successful traffic control is even more difficult to achieve when, as commonly happens, an entire group of motorists are best served if everyone obeys the regulations, but any one motorist can disobey and benefit if the others obey. The temptation is then strong for the individual to think only of his own welfare. Once enough people are selfish, the traffic-control system collapses.

Thus, a basic limitation of an expressway traffic-control system arises from a lack of public understanding of the technology.

A second major limitation of such an expressway control system is the interaction of this system with other parts of the total automobile transportation system of the city. Any given expressway can clearly be kept stable if we prohibit any cars entering and enforce the regulation with enough policemen. Indeed, stability would be ideal if we kept all cars off the highway. An immediate result, however, is to clog the parallel highways and the streets feeding into the expressway.

Thus, it is not possible to treat only a small part of the overall highway system. We cannot break the complex system down into a large number of simple systems, since rapid and smooth traffic flow in one portion often means extraordinary congestion in the neighboring portions.

This interlocking of all parts of the total system is actually much worse in most cities of Europe than in the United States.

Most American cities have grids of north-south and east-west streets. Consequently, one-way streets can be used, and motorists can easily seek alternate parallel routes.

In the typical European city, and to some extent also in Boston, the streets tend to radiate outward from the center and to meander in various directions which have evolved over centuries from the old footpaths and horse trails. Furthermore, streets in Europe are usually much narrower than streets in the United States. Finally the construction of expressways within European cities is often impossible because of the scattered historical landmarks which must not be destroyed. When an expressway interchange may require 60 acres or more of cleared land, there are few parts of downtown London where such highways could be built.

As a result of these factors, urban traffic congestion in many European cities is already extremely serious, even though the percentage of the population owning cars is still far below that in the United States. The nations of western Europe are now approaching the period when most of the population can afford automobiles. As a result, it appears that some of these cities will within a few years display instabilities more severe than any encountered so far in the United States, particularly since road construction is already so far behind the current needs.

In this section, we have described in general terms a system in which stability is a most important goal. The model in the case of our expressway control must be measured by observation of actual performance of the system. The conditions under which instability occurs are determined by this observation. Fortunately, there are also problems in which we can work with a mathematical model in order to detect instability.

4 | THE BLACK DEATH

One third of the people in the world dead, 36% of the universities closed, 50% of the clergy dead, Europe devastated in only three years. Towns deserted, farms lying idle, and gloom over the entire continent. This is the picture in 1351 at the end of the Black Death, the bubonic plague that swept from Asia throughout the civilized world in the great epidemic of the fourteenth century.

Some of the most interesting stability studies are those which describe the spread of epidemics. In this section, we take a brief look at the bubonic plague epidemic, the Black Death. This major catastrophe illustrates the kind of instability modern epidemiologists attempt to anticipate and prevent by the use of models.

Bubonic Plague

Bubonic plague is a disease spread primarily by fleas which usually live on rodents. The disease is spread among the rodents (often rats) by the fleas, and humans contract the disease from fleas, especially when the rat population decreases because of disease. The fleas then seek human hosts.

Actually, the bubonic plague of the fourteenth century had three different forms. The first, involving bubons, or boils, at the lymph nodes in the groin or armpits, is spread primarily by fleas. In the second form, in which pneumonia is involved with disease of the lungs, transmission can occur directly between human beings. The third form, involving the nervous system, is highly contagious and causes death within hours.

The first recorded instance of the plague occurred in the Near East in the eleventh century B.C. Apparently, there are regions of northern India and central Asia where the plague is continually latent. Worldwide epidemics have started there at least three times during the last 2000 years.

The first great pandemic (an epidemic covering the civilized world) occurred in the sixth century A.D. Lasting fifty years, "Justinian's plague" involved the Roman world and certainly was one factor in the disintegration of the Roman Empire.

The latest pandemic started in the middle of the last century in China, reached Hong Kong by the start of this century, and then spread throughout the world from the Chinese ports. With an estimated 10 million deaths in India, the epidemic was most serious in Asia. With the concept of quarantine established in Europe by about 1720, the emphasis on sanitation and hospitalization during the last century, and improved urban sewage-disposal and rat-control systems, the epidemic did not seriously affect the western world. In more recent years, particularly with the use of antibiotics, bubonic plague has been well controlled with only 200 deaths per year over the last decade.

The great pandemic of the fourteenth century started in northern India. It is not known why the plague began its spread. Apparently the rodent population which normally carries the disease was forced to migrate, because of food shortage, excessive rain, earthquakes, or other reasons.

The plague reached western Europe in 1347 and primary effects lasted for the next three years during which more than 30 million Europeans died, or a third of the population. Actually the epidemic followed the normal pattern and recurred over the next two centuries before finally subsiding, but the term "Black Death" usually refers to the first catastrophic three years.

While the plague undoubtedly was brought to Europe in several ways, historians often spotlight the primary source as the Tartar siege of the Crimean port of Kaffa in 1347. As so often happens in war,* thousands of men were brought together in unsanitary conditions, in this case in a region infested by rats. The Tartars began dying by the thousands† and finally dispersed.

---

* Until this century, war casualties were primarily from disease rather than battle. In the Napoleonic wars, for example, the British lost 26,000 in action, but 194,000 from disease.

† They actually catapulted corpses into the city in the hope of infecting the defenders, although such an action was totally unnecessary.

Fig. 8–3.
*Approximate chronology
of the Black Death's rapid sweep
through Europe in the middle of the
14th century is indicated on this map,
which shows the political divisions as
they existed at the time. The plague,
which was apparently brought from
Asia by ships, gained a European
foothold in the Mediterranean
in 1347; during the succeeding
three years only a few small
areas escaped. (From
"The Black Death,"
William L. Langer.
Copyright © 1964
by Scientific
American, Inc.
All rights
reserved.)*

Once the siege was lifted, ships immediately sailed for Genoa, Italy. When the ships reached Genoa, they were sent away and then carried the plague to Sicily, Spain, and northern Africa.

The chronology of the epidemic gives a picture of the plague fanning out from the Italy-Sicily start northward, westward, and eastward:

1348	N. Africa, Spain, France
1349	Austria, Hungary, Switzerland, Netherlands, northern Germany, England
1350	Scandinavia, Scotland, Ireland

In any one location, the response curve followed a common pattern of growth to a peak, then a gradual fall-off as a significant fraction of the population had developed immunity. (For example, Fig. 8–4 for the city of London.)

The size of the catastrophe is emphasized by the death rates, which varied from $\frac{1}{8}$ to $\frac{2}{3}$. Investigations of the death toll in Europe at that time are reported by W. Langer in the February, 1964, issue of *Scientific American*:

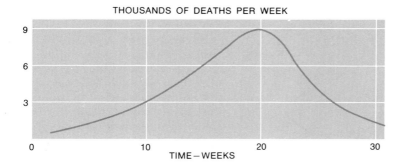

THOUSANDS OF DEATHS PER WEEK

TIME — WEEKS

Fig. 8–4.
Death rate figures for London during the Black Plague.

As reported by chroniclers of the time, the mortality figures were so incredibly high that modern scholars long regarded them with skepticism. Recent detailed and rigorously conducted analyses indicate, however, that many of the reports were substantially correct. It is now generally accepted that at least a quarter of the European population was wiped out in the first epidemic of 1348 through 1350, and that in the next fifty years the total mortality rose to more than a third of the population. The incidence of the disease and the mortality rate varied, of course, from place to place. Florence was reduced in population from 90,000 to 15,000; Hamburg apparently lost almost two-thirds of its inhabitants.

Langer further reports that

It is now estimated that the total population of England fell from about 3.8 million to 2.1 million in the period from 1348 to 1373. In France, where the loss of life was increased by the Hundred Years War, the fall in population was even more precipitate. In western and central Europe as a whole the mortality was so great that it took nearly two centuries for the population level of 1348 to be regained.

In many areas, the population did not return to its pre-plague level until 500 years later. The tragic extent of the epidemic was largely a result of the totally inadequate state of medical knowledge. The only treatments were bleeding, or the smelling or eating of aromatic herbs and unusual foods.

Popular Response

The response of most of the people to the plague was strongly fatalistic, in accord with the historical era. People believed that the disease was spread by elements in the air and, hence, moved according to the winds. There was widespread acceptance of the idea that the plague was God's sentence on a sinful population and, hence, that there was no reason to attempt to fight it.

As a consequence of this attitude, some groups of people (particularly in central Europe) became flagellants, publicly beating themselves. The hope was that God would be satisfied with this self-inflicted punishment.

A more tragic consequence of the ignorance of the way in which the disease spread was the widespread belief, particularly in France and Germany, that the Jews had poisoned the drinking water. The resulting mass burnings and massacres (16,000 in Strasbourg, 12,000 in Mainz, and so on) were not rivaled in barbarity in history again until the 1930's. Sixty large and 150 smaller Jewish communities were exterminated, with major shifts in the Jewish population (for example, to Poland and Lithuania).

Effects of the Plague

Because of the scarcity of carefully prepared histories of the fourteenth century, it is impossible to determine in detail the effects of the disaster. Clearly, the most profound aftermath was the deep impression the epidemic left on the people. The disease often dragged its victims through days of intense suffering, accompanied by extensive vomiting and a pervading stench surrounding the victim. When the bubons broke, black blood was discharged and the patient often died lying in a pool of the liquid. With an average of two people per household affected, few individuals in Europe escaped without direct contact with the plague.

In economic and social terms, the resulting shortage of farm labor tended to break down the power of the landowners and certainly was a factor in such social upheavals as the Peasants' Revolt in England in 1381. The tendency toward overpopulation in 1300 (at least in terms of farmland under cultivation) was abruptly reversed, wages of farm workers rapidly doubled, and long-term economic changes were initiated.

There was a particularly heavy toll among priests and monks, with mortalities as high as 50 percent. In part, this resulted from the fact that clergymen typically lived near graveyards; also, the monasteries were particularly tight communities in which the disease tended to hit everyone. As a consequence of the high mortality rate, new priests were ordained with essentially no training and selectivity, and there was a marked decline in the intellectual activity among the clergy which had provided such important leadership during the Dark Ages. Historians have argued whether the plague can be cited as a cause of the Reformation. Regardless of this question, however, the inability of the Church to protect the people from the plague certainly led to a diminished influence over the lives of the populace.

From the viewpoint of *The Man-Made World*, the primary interest in the Black Death is the way in which the spread of the plague conforms to the simple epidemic model. The study of past epidemics is the basis for refinement of such models so that the quantitative picture can be used to guide public health officials in future crises. For example, in the fall and early winter of 1968, an epidemic of Asian flu swept across the United States. Because of the problems associated with isolating the virus and developing a suitable serum, vaccine supplies were very limited and had to be restricted to the aged and people particularly susceptible. Vaccine was allocated to regions according to the model estimates of when the flu would strike particular locations, in order to optimize the effectiveness of the limited supply. In such a case, an accurate model for prediction is essential if treatment is to be effective.

Unfortunately, in the 1968 example the predictions based on epidemic models were not very good. In New York City, for example, the prediction was that the epidemic would peak about January. Vaccine was scheduled for late December. Actually, the peak occurred in mid-December and by January the epidemic had almost passed.

5 | EPIDEMIC MODEL

Some years ago a person suffering from smallpox was accidently permitted to enter the United States. To prevent the outbreak of an epidemic, an immediate search was launched to locate all persons who had been in contact with this individual since his arrival. The possibility of a rapid spread of the disease from an extremely limited beginning was well understood. The need to quarantine or isolate infected individuals to prevent widespread infection is really a process to prevent an unstable situation from developing. A single infection acts as a disturbance of the entire community which is originally in equilibrium. The city of Aberdeen in Scotland, for example, was affected by a typhoid fever epidemic in 1964. Strict measures were taken to quarantine the entire city to prevent a further spread of the disease.

This suggests that an epidemic has the mark of an instability phenomenon, with its sudden and abrupt change from one state to another caused by a disturbance. In the smallpox case, the disturbance was a single infected person. In the typhoid fever case, the initial disturbance was traced to an imported shipment of tainted fish.

For a number of years public health officials have developed mathematical models of epidemics in order to obtain deeper insight into the epidemic mechanism. For example, if many people in a community are immune to smallpox by vaccination, what would happen if a single infected person were to appear? He may infect some of the others. But if these in turn come in contact with only immune persons, no epidemic will develop. On the other hand, if each infected person infects three or four others, the epidemic will grow exponentially.

Another factor is that people who recover become immune. If they recover quickly enough, they may not infect enough additional people to expand the epidemic. How important are these factors relative to one another in the growth of an epidemic? Should the public health officials concentrate their resources on quarantine or on vaccination?

Such are the questions that one wishes to answer with the help of a model. In many situations, answers are difficult to secure because the phenomena are complex and influenced by unpredictable events. Nevertheless, the insights that the models have provided have been useful and have spurred further research.

We consider only the simplest models of an epidemic. Let us first analyze the following, somewhat artificial, situation. We assume that in a total population of N persons, everyone is either infected with a certain disease or is susceptible to catching the disease from those who have been infected. This model thus omits the possibility of individual recovery and subsequent immunity, death, or isolation. This is an oversimplification for most diseases, although for certain mild upper respiratory infections with a long period of infection, the model does have validity.

We denote the number of infected persons by i and the number of susceptible persons by s. We are interested in the rate of increase in i, that is how many new people will become infected each day. Let us denote the rate at which new people become infected by R_i. We expect this rate to depend on the number of persons who are already infected. The greater this number, the greater the probability that they will transfer the infection to others. But we can also expect that the number of newly infected persons on any day will depend on the number who are susceptible. Again, the greater the number of susceptible persons, the greater the number of people who will be infected.

Hence a reasonable model should have the number of newly infected increasing both with the number of presently infected, i, and the number who are susceptible, s. But what is the precise form of this dependence? We could, for example, assume that a certain fraction (1%) of the infected and 2% of the susceptible will produce newly infected persons each day.

Would such an assumption make sense? As a first, and simple, test of whether this model is useful or not, we might see what happens when i is zero. If we have no infected, s is just N (the total population). The model then says that 2% of the total population would predictably develop the disease despite the complete absence of carriers. Clearly the proposed model of the epidemic is not realistic.

Since the above model is apparently inadequate, we consider other ways in which R_i might increase with both i and s. Hence, we assume that the number of newly infected per day is some fraction of the *product* of the number of infected and the number of susceptible. If we assume that the fraction is 1%, the number of newly infected each day is given by the equation or model

$$R_i = 0.01si$$

This equation or model is free of the objections of the initial model. If there are no infected persons in the population ($i = 0$), there will be no newly infected. Likewise, if there are no susceptible ($s = 0$), there will be no newly infected. We may assume that the population consists *only* of infected or susceptible people so that $N = i + s$. Our new model of an epidemic is in fact generally used, although it is a very crude mathematical model. With this model we can now study the spread of the contagious disease and predict the nature of the epidemic.

This model can be represented in the form of a block diagram (or analog computer program) as shown in Fig. 8–5. We require the scalor, subtractor, adder, and integrator we used in earlier chapters, and also, one additional element, the multiplier for which the output is the product of the two inputs (in our case the output is si).

This block diagram appears rather formidable at first glance, but we can grasp its meaning with a step-by-step examination of its structure. We should remember that we are searching for the day-by-day value of the number of infected persons. We want to predict the growth of the epidemic and discover whether the peak will occur before conditions in the community become dangerous.

If the total population is N and the number of persons who have already been infected is i, then the number of people who remain to be infected is $N - i$. This represents the number of people who are susceptible, according to our model, and is implemented by the subtractor of the block diagram.

Because the value of i (the number of infected people on any day) is exactly what the analog computer produces as its output, we bring the output back to the subtractor. The output of the subtractor is then a numerical value for the number of people s who are susceptible to the disease.

Our mathematical model for the epidemic informs us that the rate of increase involves the product of s with i and the factor .01. We therefore tap the line which supplies the value of i and feed this value into a multiplier along with the value s. This output from the multiplier represents the product si. Multiplication in

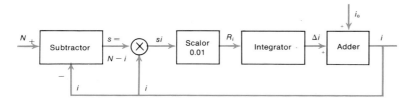

Fig. 8–5.
Block
diagram of
epidemic model.

Fig. 8–6.
Signaling a virus to stop.
The mask worn by this New York policeman directing traffic at the height of the 1918 influenza epidemic offered little protection from the disease, for the tiny viruses that cause it could easily filter through. The 1918 scourge ranks with the Black Death as one of the most destructive holocausts ever to sweep the earth; in a few months it took some 20 million lives. (U. S. War Dept. General Staff photo no. 165-WW-296G-7 in the National Archives.)

the scalor unit by .01 produces the output .01si or the desired value of R_i.

The factor R_i is the daily rate of increase in the number of infected cases. If we send this information into an integrator, we can sum up the total number of newly infected cases after any period of time. This is the change in i (that is, Δi) which, when added to the initial value i_o, gives the quantity i or:

$$i = i_o + \Delta i$$

A meter attached to the output of the system and read at intervals of time can be used to graph a curve which simulates the course of the epidemic.

A Numerical Solution

The block diagram of Fig. 8–5 can be used in the laboratory for experimental determination of the growth of the epidemic. Alternatively, we can use the diagram as a guide for numerical calculations, if we want to study a specific example. If we carry through in succession each step indicated by the block diagram, we can find the way in which i changes each day.

Box 8–1

NUMERICAL STUDY OF FIRST EPIDEMIC MODEL

The block diagram (Fig. 8–5) shows the different relations involved in the model. As a specific example,

$$N = 100 \qquad i_o = 1$$

During the First Day

$s = N - i = 100 - 1 = 99$ — We first calculate the number susceptible (the subtractor).

$si = 99 \times 1 = 99$ — The output of the multiplier is si.

$R_i = \frac{1}{100}si = 0.99$ — The rate of increase of infected is found.

$\Delta i = R_i \times 1 = 1$ — One additional person is infected during the first day.

$i_1 = i_o + \Delta i = 2$ — At the end of the first day, two people are infected.

During the Second Day

$s = N - i = 100 - 2 = 98$ — And we go through the entire procedure again to find i_2, the number infected at the end of the second day.

The calculations of the box can be summarized in tabular form.

During day number	i at start of day	$s = 100 - i$	si	$R_i = .01si$	$\Delta i = R_i(\Delta t) = R_i(1)$	i at end of the day
1	1	99	99	0.99	≈ 1	2

Then, on the second day, the number of infected (new i) will be the number from the first day (old i) plus the newly infected (Δi) or $1 + 1 = 2$. Hence, on the second day, we can repeat the calculation. A continuation of this calculation gives Table 8–1.

During day number	i at start of day	$s = 100 - i$	si	$R_i = .01\, si$	$\Delta i = (R_i)(1)$	i at end of day
1	1	99	99	0.99	1	2
2	2	98	196	1.96	2	4
3	4	96	384	3.84	4	8
4	8	92	736	7.36	7	15
5	15	85	1275	12.75	13	28
6	28	72	2016	20.16	20	48
7	48	52	2496	24.96	25	73
8	73	27	1971	19.71	20	93
9	93	7	651	6.51	7	100
10	100	0	—	—	—	—

Table 8–1. Epidemic calculation.

Figure 8–7 shows the epidemic of Table 8–1 in graphical form.

The desirable state of a population is, of course, one in which no disease exists. We see from Table 8–1 or Fig. 8–7 that, according to our model, this desired state is unstable. A disturbance in the form of one infected person sets off a chain reaction which increases the movement away from the stable condition. In our example, this movement from the initial stable conditions continues until the 7th and 8th days, after which the rate at which people become infected begins to decrease as we approach the situation in which the entire population is infected.

Thus, the equations or the equivalent block diagram make up our simple model for the growth of an epidemic. From the model, we can construct numerically (in a step-by-step manner) a solution for any given initial conditions. The block diagram is particularly useful if we want to build an analog computer to find the response experimentally. (This would be especially desirable if we wished to determine i for many different values of N, of the initial i, and of the coefficient of si in the equation for R_i.) In addition, the block diagram describes pictorially how to organize the calculations if we wish to determine i for a single example, as in Fig. 8–7.

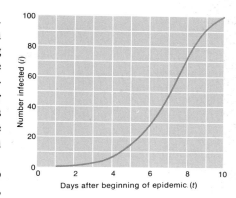

Fig. 8–7. Plot of simple epidemic model given in Table 8–1.

6 | AN IMPROVED EPIDEMIC MODEL

The preceding model is just too simple to be realistic. We know Fig. 8–7 really makes no sense. By the end of the epidemic, there certainly will be some people who have not caught the disease.

One obvious problem with our model is that no one ever recovers, or even dies. Everyone who catches the disease stays infected. Can we model an actual epidemic if we allow for recoveries? How do we change our model to permit recoveries? A recovered person is certainly not infected or susceptible. Hence, we need a third category within our population. The population N then is the sum of three parts:

s susceptible (people who can become infected)
i infected (who have the disease)
r recovered (or immune)

The next question is how many people have recovered? How does the number r change each day? Here we have several choices. We might say the number recovering each day equals the number infected ten days earlier, if the disease lasts ten days. The duration of the disease varies from person to person, and we might just assume some fraction (say 10%) of those infected recover each day.

With these assumptions about the recovered group r added to our assumptions of the last section, we can now find an "improved" model. Let us try to build up the block diagram step-by-step.

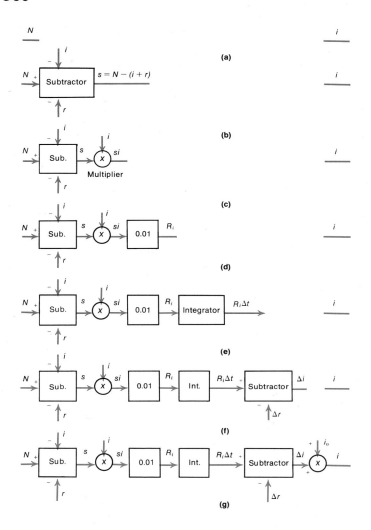

Fig. 8–8.
*Block
diagram for improved
epidemic model.*

Building the Block Diagram

Figure 8–8 shows the 11 steps in drawing the block diagram for our "improved" epidemic model. The following paragraphs are lettered to correspond to the parts of the figure.

a) Just as in the last section, the population N is the input. The output we hope to find is the number of infected i.

b) We first find the susceptible s (the total population minus the infected and recovered). Here we assume the signals i and r are available. We will have to find them later.

c) We need the product si.

d) The rate at which i is changing is $0.01si$, just as in the last section.

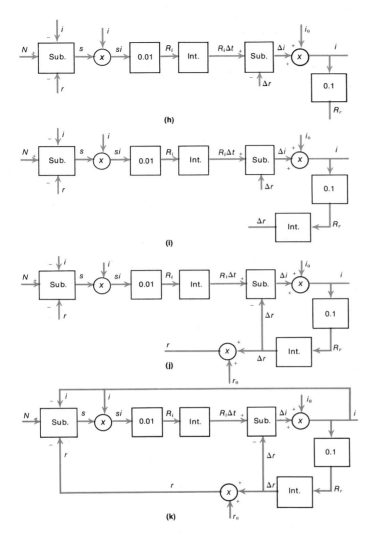

Fig. 8–8. (cont.)

e) The increase in infected is the integration of R_i if we consider only those catching the disease.

f) When we include people recovering, the actual increase in i is the number newly infected ($R_i\Delta t$) minus the number recovered Δr.

g) We now have found the total i on any day. The block diagram is complete except we have used r and Δr. We must now find what these quantities are.

h) We assume the recovery rate R_r is 10% of the infected.

i) The change in r is just the integral of R_r.

j) The actual r is the starting value (r_o) of those immune at the beginning plus the change.

k) The block diagram is now complete. This figure represents all details of our model. Notice that the block diagram is moderately complex. Actually, this is quite a simple system. In the modeling of an airplane wing vibration, housing growth in a city, or water pollution problems, it is not unusual to have several hundred integrators (instead of our two). Then the block diagram would be perhaps a hundred times as complicated.

In spite of the complexity of Fig. 8–8k, the construction of the block diagram is a step-by-step procedure. We work from the input to the output. Then we go back and find the new signals (r and Δr) which we introduced.

Numerical Calculation

Once we have the block diagram, we can calculate the behavior of an epidemic. The block diagram shows we need three different input signals:

N the total population
i_o the number of infected at the start
r_o the starting number of recovered, isolated, and immune people

Table 8–2 shows the results of the calculation when we start with 51,000 people, 1000 of whom are infected and none recovered.

Day	i	s	r	$R_i = 0.01is$	$R_r = 0.1i$	Δi Infections	Δr Recovered
1	1	50	0	.500	.100	.400	.100
2	1.400	49.500	.100	.693	.140	.553	.140
3	1.953	48.807	.240	.953	.195	.758	.195
4	2.711	47.854	.435	1.297	.271	1.026	.271
5	3.737	46.557	.706	1.740	.374	1.366	.374
6	5.103	44.817	1.080	2.287	.510	1.777	.510
8	9.118	39.604	2.278	3.611	.912	2.699	.912
10	14.888	31.740	4.372	4.725	1.489	3.236	1.489
12	21.208	22.119	7.673	4.691	2.121	2.570	2.121
14	25.544	13.284	12.172	3.393	2.554	.839	2.554
16	26.355	7.281	17.364	1.919	2.636	−.717	2.636
18	24.449	3.987	22.564	.975	2.445	−1.470	2.445
20	21.373	2.320	27.307	.496	2.137	−1.641	2.137
22	18.119	1.464	31.417	.265	1.812	−1.547	1.812
24	15.114	1.000	34.886	.151	1.511	−1.360	1.511
26	12.496	.732	37.772	.091	1.25	−1.159	1.250
28	10.276	.568	40.156	.058	1.028	−.970	1.028
30	8.422	.463	42.115	.039	.842	−.803	.842
32	6.889	.392	43.719	.027	.689	−.662	.689
34	5.627	.342	45.031	.019	.563	−.544	.563

Table 8–2. Improved epidemic calculation.

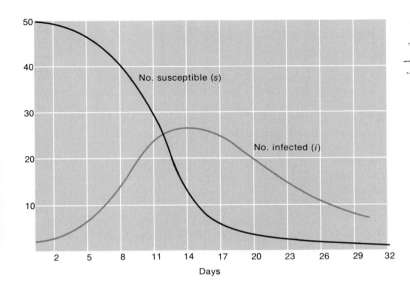

Fig. 8–9.
Plot of
epidemic given by
Table 8–2.

The numbers given are all in thousands (so we round off at the third number after the decimal point). We find the table by calculating the first row from left to right, then going to the second row, etc. Notice that at the dashed line we start giving results every other day. This is merely to shorten the table; we had to calculate values for each day before we could move on to the next day.

Plots of *i* and *s* are shown in Fig. 8–9. If the desired state is a disease-free population, then, in accordance with this refined model, such a state is unstable. One infected person triggers a "snowballing" away from the desired state. But our model predicts that not all of the susceptible people will develop this disease. The epidemic dies out before everyone is infected. This "improved" epidemic model, allowing recovered people, shows behavior which is similar to an actual epidemic.

Final Comments

The equation derived in Box 8–2 for the critical number of susceptibles quantitatively relates three factors that we know play important roles in an epidemic. Reduction in the number of susceptibles *s* to a value less than the critical value prevents the growth of an epidemic. This often suggests the need for vaccination and immunization. In some South American countries, for example, the malaria-preventative drug atabrine is added to the table salt in small quantities. Although not everyone uses table salt and different people use it in different amounts, it is hoped that enough immunity will be given to reduce *s* below the critical value. The rate at which the susceptible become infected, represented by the factor f_i, can be kept down by keeping susceptible

Box 8–2

MATHEMATICAL CONDITION FOR STABILITY

We know that the introduction of a single infected person into a population need not necessarily trigger an epidemic. Does our model reveal this possibility? What does our model predict about the conditions under which an epidemic will occur?

To explore these questions, we define f_i as the fraction of the product (si) that become infected in a single day and f_r as the fraction of infected persons who recover in a single day. (In our example, $f_i = 0.01$ and $f_r = 0.1$.) If these numbers f_i and f_r are changed, when is the system stable?

We can write the equations of the model in the form:

Change in recovered, $\Delta r = f_r i \Delta t$
Change in infections,
$$\Delta i = (f_i si - f_r i)\Delta t = (f_i s - f_r)i\Delta t$$

We note that the sign of the righthand side of the first equation is always positive. This indicates that the number of persons who have recovered will always increase. On the other hand, the sign of the second equation varies with the sign of the term in parentheses $(f_i s - f_r)$.

If the number of susceptible persons in the population is large enough, $(f_i s - f_r)$ is positive, and an increase in the number of infected persons results. However, if the number of susceptibles is small, the quantity $(f_i s - f_r)$ is negative: an epidemic will not be initiated by the addition of another infected person.

According to our model, the borderline between instability and stability, that is, the condition which separates the epidemic from the non-epidemic response to an infection, is that in which:

$$f_i s_{cr} - f_r = 0$$

where we denote the critical number of susceptibles by s_{cr}. For a large number of susceptible people, the rate at which people become infected will be greater than the rate at which people recover and become immune or die. If the number of susceptible people is sufficiently small, the rate at which people recover or die will exceed the rate at which they become infected, so that no epidemic will occur.

Solution for s_{cr} gives

$$s_{cr} = \frac{f_r}{f_i}$$

This is a relationship for the *critical* number of susceptibles in terms of the factors f_i and f_r. In the above example, $f_i = 0.01$, $f_r = 0.1$ and therefore $s_{cr} = 10$, and an epidemic develops.

The importance of the values f_i and f_r is apparent if we change f_r to 0.6 and leave f_i at its original 0.01. Then

$$s_{cr} = \frac{0.6}{0.01} = 60$$

With an initial value of s of 50, the system is stable. We can compute the outcome of this condition in terms of the following table:

i	s	r	$f_i si$	$f_r i$	Δi	Δr
1	50	0	.500	.600	-.100	.600
.900	49.5	.600	.446	.540	-.094	.540
.806	49.054	1.140	.395	.484	-.089	.484

No epidemic results.

We note from the equation

$$\Delta i = (f_i s - f_r)\, i\Delta t$$

that the model has a self-limiting property. Even if s initially exceeds s_{cr}, as i increases we find that s must decrease. Eventually the factor $(f_i s - f_r)$ becomes negative, after which the number of infected begins to decrease. If this occurs with sufficient speed only a small proportion of those susceptible to the disease will become infected.

persons away from contact with the infected ones through quarantine and isolation. Finally the rate at which people recover, f_r, is also important. We can expect to find that diseases which are contagious for long periods are more difficult to control than those with short periods of contagion.

We have considered only simple models of epidemics in this section. In more refined models, the laws of chance are included, such as various probabilities that an infected person will come in contact with another who is susceptible.

7 | LAW OF SUPPLY AND DEMAND

It is a commonly accepted truism that the "supply always adjusts to meet the demand." For example, when medical evidence recently indicated that it was healthy to be slim, demand for weight reduction drugs increased. In response to this demand, the drug manufacturers increased the supply of such drugs.

The "law of supply and demand" clearly involves the concept of a static operating state. In the example of the weight reduction drug, we may imagine the existence of a number of "demanders" who consume the drug at the same rate at which it is supplied by the manufacturers. If for some reason the demand for the drug were to decrease, we believe that the supply would also decrease until the attainment of a new equilibrium state.

But as we have already seen, it is also necessary to know about the *stability* of the operating state. In economic systems, as in all other situations involving people, this is a very complex question and it is influenced by many factors. Complicated mathematical models have been devised to deal with problems such as the supply-and-demand drug example. For simplicity of treatment, we focus attention on a small economic system consisting of a single supplier and a single buyer.

We consider the following situation. Mr. I. M. Crafty is a toy manufacturer. Mr. Crafty makes one toy and his principal outlet is Discount Stores, Inc., a chain that depends on high-volume sales. To cut costs, Discount Stores stocks as little as possible in warehouses, where rental and deterioration add to expense. Discount's policy is to place orders every Tuesday for an amount that can be sold in a single week. Their estimate of the weekly sales volume is based on the current price quoted by the manufacturers.

Mr. Crafty on the other hand sets his price on the basis of the most recent order from Discount Stores. Every Wednesday he sends letters to his customers announcing the new price.

What will happen? There is obviously a possible operating state where Discount orders the same amount Tuesday after Tuesday and Mr. Crafty sets the same price Wednesday after Wednesday; but will this operating state remain stable?

In order to consider the stability question, we consider a possible change in the normal pattern of operation. For example, Discount might decrease its toy order. Mr. Crafty then may increase his price in an attempt to maintain his total profit. Discount may feel that it cannot sell as many toys at the higher price and may further decrease its purchase order. If this action by Discount causes a further increase in price by Crafty, it is apparent that the system is unstable.

We might start out assuming a different disturbance in the normal operation of the system. Discount may place an even larger order than normal to take advantage of the high-volume price discount. Mr. Crafty could then increase his price because his toy is now in greater demand, which may result in a cut in the size of the order. In this case, the system seems to be stable. Clearly the situation is not simple; a multitude of possibilities exists.

A Quantitative Model

How can we apply modeling and analysis to gain an insight into this complicated affair? We can first try to model Discount's strategy of buying. Probably Discount will buy a greater amount the lower the price set by Crafty. If we plot price versus the amount that Discount will buy, the simplest curve of this sort would be a straight line as shown in Fig. 8–10. Here, for a given price P, Discount will buy an amount A.

Let us now examine and model one of Mr. Crafty's possible selling strategies. Recognizing that Discount Stores has increased its purchases because his price to them is low enough to gain a good profit on the resale of the toys, Crafty may attempt to induce a larger sale to Discount by a further reduction in price. The larger his sales to Discount, the larger the profit he hopes to earn. A simple straight-line model of this strategy is displayed in Fig. 8–11.

Figures 8–10 and 8–11 represent the quantitative model for the system. In this particular case, the two straight lines indicate the rules of behavior assumed for the two parts of the system: the manufacturer and the buyer. Obviously, in an actual problem the

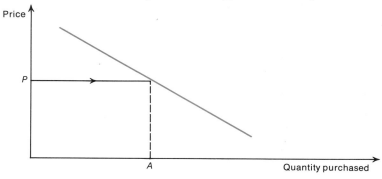

Fig. 8–10. Discount Stores' buying strategy.

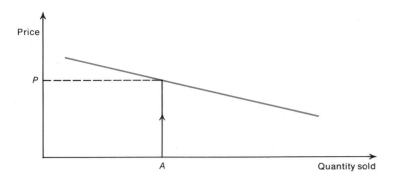

Fig. 8–11.
Crafty's
selling
strategy.

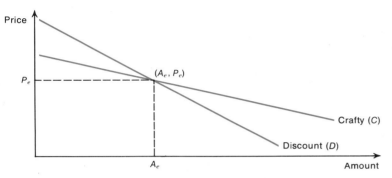

system analyst would have to determine each of these strategies by study of the operation of the two organizations; we merely assume these curves in order to illustrate the determination of stability.

Stability Analysis

On the basis of our straight-line models of Crafty's "supply policy" and Discount's "demand policy," what will happen? When will equilibrium of supply and demand occur? We can easily find this and also analyze for stability graphically. In Fig. 8–12 we plot both Crafty's selling strategy, labeled C, and Discount's buying strategy, labeled D. Equilibrium occurs at the intersection point A_e, P_e. At this price P_e, Crafty is willing to supply amount A_e week after week, and Discount is willing to buy this amount at the stated price. For a smaller purchase, Crafty insists on a higher price. Discount also expects to pay a higher price for a lower amount, but not necessarily the price Crafty expects. Likewise for an amount greater than A_e, Crafty's and Discount's expectations differ. Only for an amount A_e are they both in agreement on the price.

The point (A_e, P_e) represents what we call an *equilibrium point* or *equilibrium condition* (this is the reason for the subscript e). The system behaves properly as long as its operation rests at this

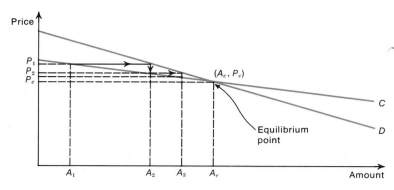

Fig. 8–13.
Stability of
supply-demand point for
Discount and Crafty.

point—as long as Crafty sets the price P_e and Discount continues to order A_e.

Stability of the system depends on what happens if a change or disturbance is introduced which momentarily moves operation away from equilibrium. For example, a fire in the toy department of one of Discount's stores compels them to reduce their order to A_1 in Fig. 8–13 (or a new buyer for Discount might underestimate the number of toys his firm could sell and reduces his order).

Then Crafty sets his price at P_1 in the same figure. At price P_1 Discount is willing to buy an amount A_2, as shown on the demand model in the graph. For this larger purchase A_2, Crafty offers a lower price P_2, which encourages Discount to increase its purchase further. This produces a cycle of adjustments until the point A_e is reached.

In other words, the system is stable. (We assume that both parties maintain constant strategies—the straight lines do not change.) After a disturbance initiated by Discount's low order of A_1, the system gradually returns to the equilibrium point. Indeed, we can show that for the strategies of Fig. 8–12, regardless of the nature of the momentary disturbance away from equilibrium, the system always returns to (A_e, P_e).

An Unstable System

On the other hand, if the strategies are those shown in Fig. 8–14, where the Crafty curve C is steeper than Discount's curve D, the system is unstable. If Discount now places an order for an amount A_1, which is less than the equilibrium purchase A_e, the amount purchased decreases; then Crafty increases his price. At the new price, Discount reduces his order. This cycle is repeated until Crafty has priced himself out of business. If in any one week Discount were to increase its order above A_e, the prices will continue to drop until Crafty loses all profits. The equilibrium is unstable in this case, as indicated by the arrows shown in the figure.

Examination of Figs. 8–13 and 8–14 shows that the stability criterion is a very simple one in terms of our model. For stability, *Discount's buying curve must be steeper than Crafty's selling curve.*

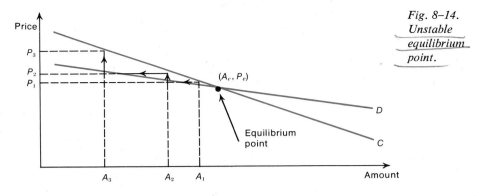

*Fig. 8–14.
Unstable
equilibrium
point.*

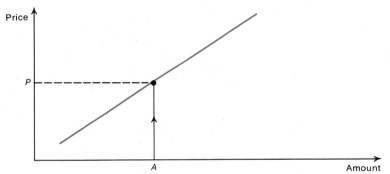

*Fig. 8–15.
Crafty's
improbable selling
strategy.*

Stability with Other Strategies

The preceding stability criterion or requirement is particularly simple; we need only determine the relative values of the slopes of the two strategy lines. The determination of stability is slightly different if we consider other strategies for either part of the system (either Crafty or Discount). For example, Crafty may believe that when Discount increases his purchase, the toy sells early during the week and is apparently in great demand. He may, in fact, ask a higher price. This rather unusual selling strategy is shown in Fig. 8–15.

Likewise, Discount may observe that their customers have peculiar buying habits: the customers believe that when an item is high priced, the item must be of high quality and therefore they buy at an increased rate. This buying curve is shown in Fig. 8–16.

Let us consider the stability of equilibrium when either Crafty or Discount adopts an unlikely strategy. For example, if Crafty adopts his improbable strategy, we have the case shown in Fig. 8–17. In this case a reduction to A_1 of Discount's order causes a cyclic pattern of buying and selling. The astute customer in Discount's store notices that for some reason the price of Crafty's toy fluctuates between P_1 and P_2 every two weeks. This cycle is in fact another equilibrium state, in addition to the point A_e, P_e.

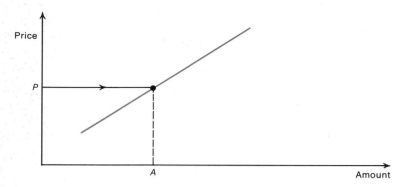

Fig. 8–16.
*Discount
Stores' unlikely
buying strategy.*

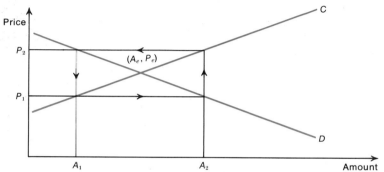

Fig. 8–17.
*Supply-demand
when Crafty adopts
his unlikely strategy,
and Discount adopts his
likely strategy. (The
slopes have the same
magnitude.)*

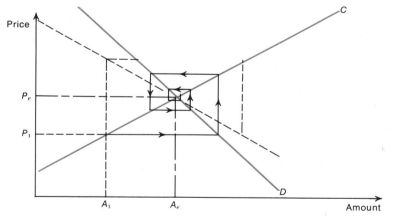

Fig. 8–18.
*Stable supply-demand
when Crafty adopts his
unlikely strategy, and Discount
adopts his likely strategy.
(Slope magnitude for* D
is greater.)

In Fig. 8–17, the slopes of the C and D lines are *exactly* the *same* in magnitude, although the slopes have opposite signs (C positive and D negative). The result is the cyclic behavior shown. If the initial disturbance or change is smaller (nearer A_e), the cycle is smaller as well.

If the magnitude of the slope of the D curve is slightly greater than that of the C curve, then the situation shown in Fig. 8–18 occurs. (The equal slope case is shown in dotted lines for comparison.) Figure 8–18 indicates that point A_e, P_e is stable.

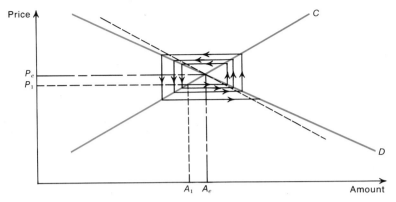

Fig. 8–19.
Unstable supply-demand
when Crafty adopts his unlikely
strategy. (Slope magnitude of D *is*
less than the slope magnitude of
C.)

Figure 8–19 demonstrates that if the magnitude of the slope of the *D* curve is less than that of the *C* curve, the equilibrium point A_e, P_e is *unstable*.

We have thus considered two special cases. Two other possibilities remain: one in which Discount uses his unlikely strategy (Fig. 8–16) while Crafty uses his likely strategy, or both Discount and Crafty go to their unlikely strategies.

With a graphical analysis of the two remaining possibilities similar to those outlined above, it can be shown that for all cases the stability criterion depends upon a very simple rule: *The equilibrium point, A_e, P_e, is stable if the magnitude of the slope of Discount's buying strategy exceeds that of Crafty's selling strategy.* With the use of a simple model we have thus discovered a fact that might not otherwise be apparent: namely, the *slope or rate at which Crafty and Discount change price with respect to the quantity purchased* determines whether or not the equilibrium point is stable.

When the equilibrium point is stable, a slight disturbance brings about "forces," in this case psychological or economic forces stemming from Discount's and Crafty's business sense, which tends to restore equilibrium. On the other hand, for an unstable situation, a slight disturbance evokes forces that drive the price and amount even further away from equilibrium.

Final Comment

The example of this section is included not because of its importance in any actual analysis of economic systems, but rather because of three important points:

1. The analysis illustrates the phenomena of stability and instability. A system is operating normally (at an equilibrium condition). A slight disturbance or change in operation occurs. The system thereafter may
   a) Return to the equilibrium condition.
   b) Oscillate indefinitely.

c) Diverge further and further from equilibrium.

The first system is stable, the last is unstable, and the second is a situation on the borderline between stability and instability.*

2. The example demonstrates the insight which we often can obtain by relatively simple analysis. Here we learned that stability depends only on the magnitudes of the slopes of the two strategies. A seemingly minor change in one strategy may convert a stable system to an unstable situation which results in disaster.

3. Finally, consideration of the example indicates what measurements we need to make on the system to determine stability—what features of the model are relevant. In our case, stability is determined by the two strategies. To simplify analysis, we have assumed these are represented by straight lines; actually the analysis is not really any more difficult if we assume curves.

Thus, the example is included because of the insight it yields into the stability phenomenon, rather than for a standard procedure that will predict the marketplace happenings to within a few percent. Our model omits all probabilistic considerations. It does not consider that Crafty and Discount will probably quickly guess each other's strategy and revise their original strategies accordingly.

Nevertheless, situations such as those pictured in our graphs are observed to occur in a free market economy. The interested reader is referred to *Economics*, 8th edition, by P. A. Samuelson, for a more extended discussion of the so-called "dynamic cobweb" of Figs. 8–17 and 8–18 and its relation, for example, to the statistics of the production of corn and hogs and the observed "corn-hog cycle."

8 | INSTABILITY IN PHYSICAL SYSTEMS

The detailed, quantitative study of instability has its roots in the exploitation of physical systems for man's benefit, for example, systems such as bridges, buildings, and electrical controls. The wheel with an off-center weight is a simple example. On level ground, two equilibrium conditions are obviously possible, as shown in Fig. 8–20.

From experience, we know that the equilibrium condition shown in Fig. 8–20a is stable and that shown in part (b) of the figure is unstable. A slight disturbance of the wheel when it is in the condition shown in (a) merely causes the wheel to rock back and forth until it settles into its original position. On the other hand, a slight disturbance of the wheel when the wheel is in condition (b) causes the wheel to turn further from its original posi-

*Whether we call this special case (Fig. 8–17) stable or unstable is arbitrary. If the oscillating system is useful, we often call it stable; if useless, unstable.

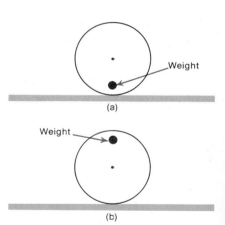

Fig. 8–20.
*Stable
and unstable weighted
wheels.*

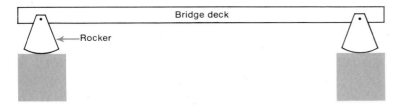

Fig. 8–21.
Bridge deck
on rockers to allow for
thermal expansion.

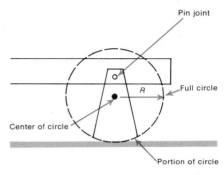

Fig. 8–22.
Details of
rocker and pin
joint.

tion so that the weight moves to the bottom. The wheel then rocks
back and forth until it comes to rest as in condition (a). Thus, state
(a) represents stable equilibrium, while (b) depicts an unstable
situation.

A similar condition exists in a less obvious form in the following situation. In bridge construction, the road deck must be permitted to expand and contract as the temperature changes. To
permit this to occur safely, the bridge deck design allows the bridge
to vary its length at the points of support. In one case, the bridge
deck was supported by rockers placed under the two ends, as shown
in Fig. 8–21. While this bridge was in the process of construction,
collapse occurred, which killed several workmen and incurred
costly damages.

Collapse could have been avoided if the designer had been
familiar with the principle illustrated in Fig. 8–20. The bridge deck
is supported by rockers which can tilt as the length of the deck
changes. This is shown in Fig. 8–21 and in detail in Fig. 8–22. The
bottom of the wedge-shaped rocker is shown as a portion of a
circular arc. The circle, from which the arc is cut, is centered at
a point below the pin joint which connects the girders with the
rocker. With no connection to the bridge deck, the only forces
acting on the rocker are its own weight Mg and the reaction force
from the earth (Fig. 8–23). The weight Mg acts from a point below the center of the circle, called the *center of gravity* of the rocker.
The rocker itself is stable.

When the deck is pinned to the rocker, the much larger deck
weight, W, acts *above* the center of the circle, to produce a situation similar to that in Fig. 8–20b, where the weight of the wheel
is negligible in comparison to the applied weight W. It was the
result of a condition of this type that a small disturbance sent
thousands of dollars of material and effort crashing to the ground.

The effect of a disturbance on the equilibrium of the structure
which has just been described can be examined in detail. For simplicity, we assume that the weight of the deck, W, can be considered as concentrated at a single point (the center of gravity).
In Fig. 8–24, (a) shows a slight disturbance of the wheel when the
center of gravity is *below* the center of wheel, (b) when it is *at* the
center of the wheel, and (c) when it is *above* the center of the wheel.
In all cases there will be a reaction force W' from the earth on the
wheel. In both (a) and (c) the forces W and W' create a torque

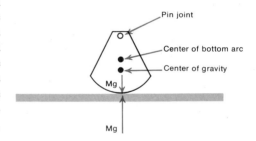

Fig. 8–23.
Forces on
the unloaded
rockers.

which tends to turn the wheel. In (a) the unbalance tends to rotate the wheel counterclockwise to *restore* it to a position where the center of gravity is at its lowest position. On the other hand, in (c) the unbalance tends to rotate the weight clockwise to turn the wheel further from its original position. In the borderline or *critical* case (b), a disturbance creates no torque.

This simple example, which can be readily demonstrated with rockers made of wood and a wooden deck loaded with weights, displays the same characteristics that we have met in our epidemiology and economic examples. In the stable configuration (a), a disturbance away from the equilibrium state evokes torques that tend to restore the system to equilibrium. In the unstable state (c), the evoked torques tend to drive the system further from the equilibrium state.

As mass is added to the bridge deck, both the magnitude of the effective weight and the height of the center of gravity (Fig. 8–24) are increased. With a sufficient added mass, the vertical position of the center of gravity is *above* the center of the wheel. At this point, the system passes suddenly and abruptly from a stable to an unstable condition.

An example of instability in a physical system which is of great engineering importance is aerodynamic or wind-induced instability of structures. This sort of instability has come to the fore only relatively recently as the result of sophisticated structural design of bridges and airplanes.

This effect can be illustrated with a small electric fan and a spring supported half-round section (Fig. 8–25). When the airstream from the fan is directed at the flat side of the section, the section starts to oscillate due to the periodic shedding of eddies. If the frequency of oscillation is close to the natural frequency of the spring support system, the oscillation of the section grows with ever increasing amplitude until either the section strikes the frame or the springs collapse.

Aerodynamic instability has been the cause of bridge and airplane failures in which millions of dollars and many human lives

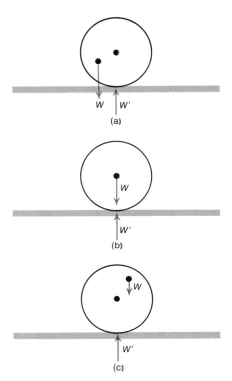

Fig. 8–24.
Slightly disturbed wheel when the weight is below, at, and above the wheel's center.

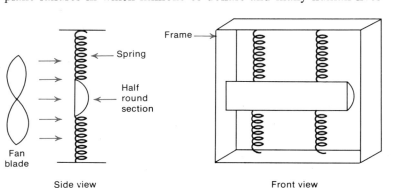

Fig. 8–25.
Sketch of model showing aerodynamically induced instability.

have been lost. For example, the collapse of the Tacoma Narrows Bridge was caused by resonance due to the periodic shedding of eddies. The bridge developed torsional oscillations with amplitudes up to 40° before it broke. The effects of such instability are particularly noticeable in the midwestern part of the United States. Here the broad plains allow steady winds to build up. The common result is highway stop signs snapped off by unstable oscillations, or the destruction of overhead telephone lines oscillating in the wind (called "galloping conductors" by engineers who attempt to introduce damping devices to control the amplitude of oscillations).

9 | USES OF INSTABILITY

Thus far we have discussed the phenomena of stability and instability, with examples in which instability represented an undesirable feature. We cannot, however, draw general conclusions about the overall desirability of any phenomenon. In some societies the social structure is often stratified. Individuals are not permitted to attain social positions which are considered "above their natural station in life." A society of this type may be considered highly stable.

In the colonies which preceded the establishment of the United States, a handful of people were confronted with the task of developing enormous natural resources. A growing population was beneficial. In terms of a population model, this growth was a phenomenon typical of instability. Under these conditions instability is a desirable feature.

As we have seen, a system which moves further from its initial state when it is disturbed is an unstable system. If this change is rapid, and produces a sharp demarcation between the initial and final condition, it may have useful applications. Perhaps the simplest application of this behavior is displayed by a mechanical switch, such as a light switch, shown in Fig. 8–26. In part (a), the switch bar is in an unstable equilibrium. It is impossible to keep the switch bar in this position. Instead, the switch bar prefers one

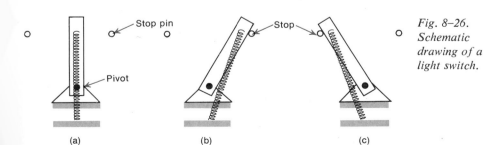

(a)   (b)   (c)

Fig. 8–26. Schematic drawing of a light switch.

of the two equilibrium positions on either side, shown in (b) and (c). By using the property of an unstable equilibrium, we produce a device that is definitely either on or off.

Thus, in a particular system we may seek either stability or instability, although in the latter case it is usually desirable to realize instability in a controllable fashion. For example, corporate management customarily attempts to realize continually increasing profits rather than to operate at a stable equilibrium point.

An even more obvious system in which stability is undesirable is the economic status of the world's population. Here it is obviously essential to move away from our present equilibrium point if significant amelioration of current poverty conditions is to be effected. Experience in many countries has shown, however, that the system is relatively stable; we cannot change the equilibrium point markedly by building new, improved housing for the underprivileged. Indeed, when this has been attempted in some sections of the world where illiteracy is common, the new housing rapidly degenerates into a new slum area.* Job opportunities and education must force a change in the structure of the system so that the equilibrium point becomes unstable.

* For example, in one such large development in Latin America, 50% of the bathtubs had been ripped out and sold within six months of occupancy.

In spite of these examples, we most often attempt to design and control systems in such a way that they are stable. Today in the United States a high percentage of the federal tax and economic policies are designed to maintain stability in the value of the dollar and to realize a controlled growth in national economic activity. Perhaps the most dramatic recent example of financial instability occurred in Germany in 1923. Following the fall of the German empire in 1918, food shortages, foreign occupation, costs of rebuilding, and inability to manufacture resulted in severe drains in gold reserves. As more and more paper money was printed, the value of the mark declined until, in 1923, four trillion marks were equivalent to one dollar. This highly unstable inflation destroyed the savings and foundations of the middle class, bankrupted many companies, and created the atmosphere in which Hitler was able to achieve power following the economic difficulties of 1932 (when 6,500,000 workers were unemployed in parallel with the severe depression in the United States).

Many other examples can be cited of systems in which stability is of vital importance—both in man-made systems, such as aircraft, and in natural systems interacting with man-made elements, such as ecological systems.

10 | FINAL COMMENT

In creating systems to serve man's ends or in coping with the world as we find it, it is not enough to seek a desired operating state. We must also concern ourselves with the stability of such a state.

If the state is unstable, a small disturbance initiates a rapid shift from the desirable state. Forces—physical, psychological, economic, or biological—are evoked which tend to move the system away from the desired state. On the other hand, if the desired state is stable, the evoked force tends to restore the system to the desired state.

The phenomenon of stability surrounds us. Traditionally man has studied stability in connection with physical systems where the stability phenomenon could be examined in its simplest form. But as we have observed in such diverse examples as epidemiology and economics, stability plays an important role in much broader areas of human activity. It is only recently that man has learned to apply the quantitative techniques of modeling to study the phenomenon of stability in other than physical systems.

Questions for Study and Discussion

1. What does the phrase "superhighways beget superhighways" mean? Explain in detail.

2. In the urban traffic example given in the text there are examples of stability and instability. Describe these situations and indicate for each one what the original disturbance was, and also the forces which tended to drive the system toward stability or away from stability.

3. Give an example of an unstable system which is not given in the text. Point out in your explanation the disturbance which started the instability, and the forces which tend to drive the system further from the operating state.

4. Describe briefly how the computer is used to maintain stability on a controlled stretch of highway.

5. What is one fundamental limitation of an expressway traffic-control system?

6. List the forces which tend to drive an epidemic toward instability.

7. List the forces which tend to drive an epidemic toward stability.

8. Construct a block diagram for an epidemic in which the following situation prevails:

   N = total population
   i = number infected
   s = number susceptible
   r = number recovered and immune
   M_o = number who leave the population during the epidemic
   M_i = number of new people who move into the population during the epidemic
   R_i = rate of infection = $0.02si$

9. The textbook uses the rate of infection as $0.01si$. For a particular disease, how would this number (0.01) be determined? How could the use of computers provide a more reliable number?

10. Would a casual examination of the curve in Fig. 8–7 for the first 3 days give an indication that an epidemic is imminent? Why? How about the first 5 days?

11. Experience with graphic models indicates that for the first 5 days the graph in Fig. 8–7 is exponential. If it were actually exponential, it would grow

steeper on the 6, 7, 8, 9, and 10th days rather than flattening out as it does.

What happens during an epidemic to cause that flattening?

12. If all people who were infected recovered and were immune 10 days after the infection, would the graph pictured in Fig. 8–7 look any different? What would the graph look like if the time base were extended to 20 days?

13. Explain (as you would to a person your age who is not taking this course) how to construct a block diagram to find the population of a region if the following situation prevails:

N_o = original population
P_i = number of people moving into the region each year per 100 population
P_o = number of people moving out of the region each year per 100 population
R_b = number of children born each year per 100 population
R_d = number of people who die each year per 100 population
N_t = total population at the end of any given year

14. Suppose you had determined f_i and f_r in the epidemic model by examining statistics for a smallpox epidemic in Asia. Could these values be used to predict the course of a smallpox epidemic in New York City? Give reasons for your answer.

15. Which is more effective, as far as the critical number of susceptibles s_{cr} is concerned, increasing the recovery-isolation factor f_r by 50%, or decreasing the infectious factor f_i by 50%?

16. When Henry Ford brought out the original Model T "Tin Lizzie," he set a price that was fantastically low, and at the same time established a wage scale in his factory that was much higher ($5 a day) than elsewhere in industry. This brought about instabilities in the automobile business in general. Describe one or two of these. Why did they not snowball indefinitely?

17. Water impounded behind a dam is in unstable equilibrium, in the sense that if the dam should break, the water would rush downhill and perhaps do great damage to structures or people in its path. How do engineers employ this instability of impounded water for useful purposes?

18. Compare the desired state, the disturbance which moves the system, and the forces which tend to return the system to the desired state in each of the following situations:

a) a smallpox epidemic

b) a gasoline price war

c) a bridge rocker support

d) an urban riot

e) a college student strike

f) soil erosion

g) an international "situation" between two countries which border each other

h) a sudden gust of air causes an airplane to "bank" suddenly

i) depletion of minerals in soil by "one crop" plantings

Problems

1. Fill in the information in a table similar to Table 8–1, epidemic calculation, for the first 4 days if all factors were similar to those described in the text except that the population N is 1000 instead of 100. What differences are noticeable?

2. From the data in Table 8–1, epidemic calculation, plot a graph of the number of people susceptible versus time for 10 days.

3. Question 13 in the "Questions for Study and Discussion" at the end of this chapter outlines a population situation for a region:

N_o = original population 40,000
P_i = people moving in each year 0.6/100
P_o = people moving out each year 0.4/100
R_b = birth rate 1.7/100 population
R_d = death rate 0.9/100 population

Find total population at the end of years 1, 2, 3, 4. Graph this and predict the population at the end of 10 years.

If you have a computer or desk calculator available, find the population at the end of each of 10 years.

4. Two nations, A and B, are glowering at each other, each threatening war. The commanding general of nation A reads his agents' intelligence reports each week and for the next week calls up an additional number of troops equal to 10% of B's strength. The commanding general of nation B does likewise. Formulate this situation as a system of equations, letting N_A and N_B be the number of A and B's troops.

5. Assume in Problem 4 that A and B each start with 1000 troops at the border. How many troops will each have after 10 weeks?

6. In an epidemic situation what is the critical number of susceptible if $f_i = 0.001$ and $f_r = 0.6$?

7. Assume that you are the president of Discount Stores. You believe that it is important to present an image of stable prices. You find that Crafty's toy prices have been fluctuating wildly. The accounting department has provided you with the following data:

Date	Crafty's price	No. bought
7/1	$2.00	250
7/8	2.50	188
7/15	1.88	265
7/22	2.65	169
7/29	1.69	289

How would you instruct your purchasing department in order to stabilize Crafty's price?

8. Discuss the stability of the following Crafty-Discount supply-demand curve.

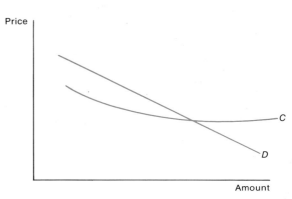

9. Find the equilibrium points and discuss the stability of the following Crafty-Discount supply-demand curves.

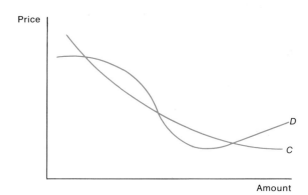

Laboratory and Projects

I | DYNAMIC STABILITY DEMONSTRATION

In this experiment you are to complete the design of a simple three-wheeled cart. The cart has two fixed wheels and one swivel wheel as shown in Fig. 1.

You are to determine the best point at which to attach a string which will be used to pull the cart. The best location will be the one for which the cart will move in a straight line when the string is pulled in a straight line and for which the wheels do not slide or drag along the floor.

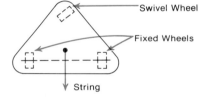

Fig. 1.
The three-wheeled cart (*top view*).

1. Use a piece of string about 3 feet long and attach it to various points on the top of the cart. Observe the motion of the cart when the string is pulled slowly along a straight line. Observe the motion when the swivel wheel is at the front and when it is at the rear. Include points directly above the swivel wheel, along the axis of rotation of the fixed wheels, and at the center of the triangle. The string can be attached with a tape, a thumbtack, or a small screw.

2. Record your observations and your final decision as to the optimum location of the point of attachment.

The three-wheeled cart serves as a good model for such real life systems as a child's tricycle and the landing gear of airplanes. The observations which you made about stability character can be applied to these systems.

3. Identify as many systems as you can which could be modeled by means of the three-wheeled cart.

II | ANALOG SIMULATION OF AERODYNAMIC STABILITY (Classroom Demonstration)

Your teacher will demonstrate a unique type of stability phenomenon which is caused by aerodynamic forces. From the explanation of the phenomenon you will see that aerodynamic forces are produced when the half cylinder is put into motion and that the direction of these forces tends to assist or alter the oscillation depending on the shape of the windward surface of the cylinder.

1. Use the analog computer to simulate the motion of the unstable system when it is given some initial displacement.

2. Use the analog computer to simulate the motion of the stable system when it is given some initial displacement. For what other systems might the simulation serve as a valid model?

3. What modification in wiring, if any, are necessary to change the analog simulation from an unstable to a stable system?

4. Two scalors are used in this simulation. What physical characteristics of the actual system (Fig. 8–25) do they represent?

III | DEMONSTRATION OF STATIC STABILITY

Perhaps you have noticed while riding in a car or a train that many bridges seem to be supported by pins and rockers placed under the ends of each section. In bridge construction, whether the bridge is made of concrete or steel, the bridge deck must be permitted to expand or contract due to temperature changes. (Fig. 8–21.)

The purpose of the rocker is to tilt as the length of the deck changes. In Fig. 8–22, notice that the bottom of the rocker is a portion of a circular arc. The circle, from which the arc is cut, is centered at a point below the pin joint which connects or supports the bridge deck with the rocker. The object of this experiment is to determine whether the center of gravity should be (a) above the center of the wheel, (b) at the center of the wheel, or (c) below the center of the wheel.

1. Place a piece of wood about 16 to 18 inches long on two unstable rockers. Place a 1-kg mass on the "bridge deck." Give the bridge a slight push to the right or left. Observe and note the result.

2. Repeat using the stable rockers.

Machines and Systems for Men

9

I | INTRODUCTION

Any subject which has a great effect on the everyday lives of people soon leads to popular jokes, cartoons, and stories. Automation is no exception. One of the classic stories goes as follows:

Passengers for the Los Angeles-to-New York flight were all seated, the airplane doors were closed, and the motors started. The soft music was interrupted by a recorded announcement.

"Ladies and gentlemen, welcome to the *first totally automated* passenger flight. We will be landing in New York in four hours. The entire flight will be automatic. There is no pilot aboard, but the plane is under the complete control of an on-board computer. There is absolutely no cause for alarm. This computer is backed up by a communication link to a ground computer, which will watch your flight at all times.

"We assure you that there is absolutely no reason to worry. Every part of the aircraft-control system has been thoroughly tested. The aircraft engines, autopilot, and navigation system are absolutely reliable. So, sit back, relax, and enjoy your flight since nothing can go wrong, can go wrong, can go wrong, can go wrong, can go wrong. . . ."

While the story is extreme, it does illustrate a primary danger in modern technology. All too often, we become so enamored of gadgets, we lose all perspective on what the devices are really for. Technology should be used to give man a more pleasant, fuller, and richer life. Technology should relieve man of tasks which are tedious or difficult. It should perform tasks man cannot accomplish. It should be used to substitute for man when such a substitution is logical.

The problem in our story is obvious. Man is uniquely able to adapt to unexpected events. We cannot foresee *every* possible happening and build the machine to take care of all these eventualities. We must have men in the system—intelligent men who can respond logically to the circumstances.

While our story is extreme, we can think of many instances where technology is over-used. We call this situation "technological overkill." It is analogous to using a 40-foot crane to pick up a pencil from the sidewalk. As we look at the wonders of our age

Delprad's Aire Velociped.
(*The Bettmann Archive*)

of technology, can we seriously feel that the average housewife's life is better because of the electric can opener? Or is the labor saved more than offset by the annoyance of trying to have it repaired when trouble develops?

Not too many years ago, auto water temperature was indicated by a thermometer on the dash panel. On hot summer days, in heavy traffic, the driver could see the temperature rising. When the temperature was approaching a dangerous point, the driver could shift to neutral and race the motor to cool it. Many of today's cars have only the "idiot light." There is no indication of water temperature until the boiling point is reached. Then the light goes on—often too late for the driver to do anything. This "modern" system does not match the machine and the man. Once the man could observe the way the temperature was changing and take corrective action. In the modern system, these human capabilities are not used.

In this chapter, we will look in more detail at this important problem of matching machines and systems to men. What should we ask technology to do? In other words, what are the things man cannot do? Where does he need to look to machines for help? How can we fit machines and men together in such a way that we maximize the benefits of technology?

To begin to understand some of these problems, we look at different shortcomings of a human being. In each of the following sections, we focus on a different human limitation and then consider how technology can be used to solve the problem.

2 | MAN AS A CONTROLLER

Control systems are those in which we are trying to keep an output signal in a desired position. For example, a common control problem is steering a car. Here the driver observes the location of the car in the road and then makes corrections by turning the

Fig. 9–1. A simple resonant system.

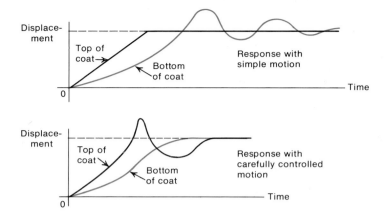

Fig. 9–2.
Damping the resonance.

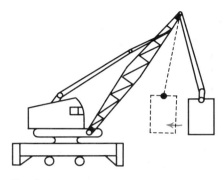

Fig. 9–3.
System analogous to Fig. 9–1.

steering wheel. As he makes turns, he tries to keep the output signal (the car position) in a desired location.

Figure 9–1 shows a man involved in a simple task. He is holding his jacket at the top. He wants to move the jacket sideways to a new position. If he merely moves the top to the new, desired location, the bottom oscillates around its final position. Eventually the oscillation dies down and the system comes to rest in the new position. We can plot the position of the bottom of the coat (Fig. 9–2).

If the man does this over and over again, he soon learns how to avoid the oscillations. He finds that he can give the top an extra push to get the bottom moving. Then he jerks the top back hard to brake the bottom. With a little practice, he can learn exactly how to move the top so that there are no oscillations at all (Fig. 9–2).

We find exactly the same problem in construction work (Fig. 9–3). When a huge crane moves a large weight, it may be very important that the load *not* swing back and forth. Such oscillations may damage nearby buildings. Unfortunately, the required motion of the top depends on the weight of the load. (This could have been shown by filling the jacket pockets with heavy objects.) In our crane example, the human operator can't practice a few times with each different load. Consequently, we have to use automatic technology in place of the man. *The control task is too difficult for the human being.*

Control of an Unstable System

Man also has trouble controlling an unstable system. In Fig. 9–4, the broom stands, upside down, on the man's hand. As the broom starts to fall, the man tries to move his hand horizontally in such a way that the broom stays up. After a few seconds, he loses control, and the broom falls. Here the system we are trying to control is unstable. In physics, we would call this an "inverted

Fig. 9–4.
A man trying to keep a broom upright. The same experiment can be done with any long stick (e.g., a blackboard pointer).

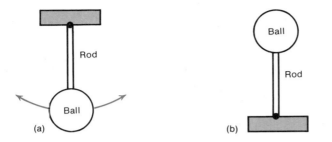

Rod

Ball

(a)

Ball

Rod

(b)

Fig. 9–5.
Pendulums
a) Normal pendulum.
*After a slight movement
of the ball left or right, it
oscillates back and forth around
the vertical position. The ball finally
comes to rest at the starting position
(before the original disturbance).*
*b) Inverted pendulum. As long as the
rod is exactly vertical, the ball
will not move. The slightest
disturbance, however, results
in the ball rapidly falling.
This system is unstable.
It cannot recover from a
disturbance, no matter
how small.*

pendulum" (Fig. 9–5). Without control, any small disturbance causes the ball to fall and it never returns to its original position.

If technology is substituted for man, we can control such a system without difficulty. Indeed, we can control several brooms simultaneously. Figure 9–6 shows a system which was actually built at Stanford University to demonstrate the control of such unstable systems. In this case, there are two inverted pendulums (rods A and B). The position and velocity of each rod are measured at all times. These signals are used to move the cart back and forth in such a way that the two rods are kept in an upright position.

The idea of controlling an unstable system is of practical importance. Airplanes have an unstable oscillation at a very low frequency and the altitude of the airplane tends to change at a very low rate (a cycle every ten seconds or so), with the amplitude of the oscillation increasing with time (Fig. 9–7). The oscillation in this case is so slow that the human pilot or the autopilot can easily correct for it and keep the altitude constant.

When the unstable frequency is higher, the human being is unable to respond fast enough to keep control. Figure 9–8 shows another example of an inherently unstable system. Here a satellite carrying men is tied by a rope to another space vehicle. (The latter might be the final stage which is separated from the main vehicle after the desired orbit is reached.) One of the problems associated with men in orbit for long periods of time is the effect of "weightlessness" (e.g., the calcium normally in the bones tends to dissolve). In order to create an artificial gravity, the two vehicles are spun as indicated in Fig. 9–8. This spinning motion causes an effect on the man which is like gravity. The speed of rotation is con-

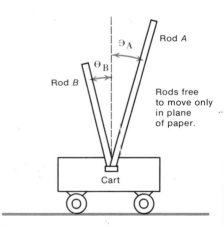

θ_A Rod A

θ_B

Rod B

Rods free
to move only
in plane
of paper.

Cart

Fig. 9–6.
Two rods on a movable cart.

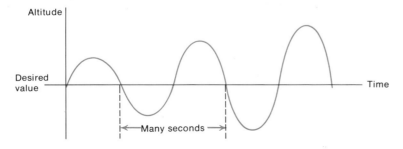

Altitude

Desired
value

Time

Many seconds

Fig. 9–7.
*Unstable motion
of an airplane. Because
of the very low frequency
(the long period), the
oscillation is easily
counteracted by
the pilot.*

trolled by reeling in or playing out the rope between the two vehicles. The control problem is to change the rope length in such a way that the rotational speed is constant.

Such a system is unstable. Furthermore, the frequency of oscillation is so high that a man cannot control the system. The rope has to be changed in length by a machine operated by a computer, which rapidly calculates the required rope length and then drives a motor which adjusts the rope.

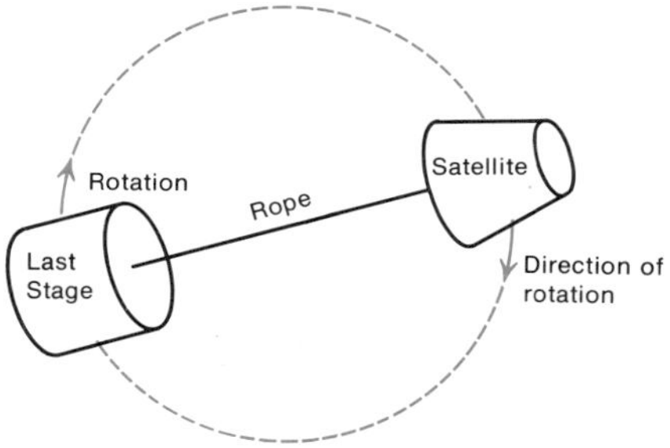

Fig. 9–8.
Two tethered satellites.

Control of a Hovering Aircraft

A VTOL (vertical take off and landing) aircraft can be positioned by changing the direction of the jet exhaust. When the aircraft is stationary over one spot (Fig. 9–9), the jet exhausts point straight down. The reaction force lifting the aircraft is just enough to offset gravity.

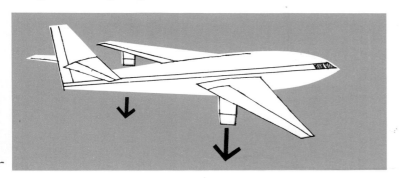

Fig. 9–9.
One vertical take-off and landing aircraft has a jet exhaust which can be rotated. When the jets exhaust downward and the motors are going full, the aircraft rises straight up. Once the desired altitude is reached, the jets are rotated so they give a forward thrust. The decrease in lift is offset by the usual flow of air around the wings.

Suppose the aircraft is hovering at one spot and the pilot wants to move horizontally to a new location (Fig. 9–10). Through his controls, he can apply a force tending to rotate the jet exhaust. This force eventually causes the exhaust tubes to start turning.

This system can be modeled as shown in Fig. 9–11. The pilot's signal, asking for lateral motion, causes this movement through

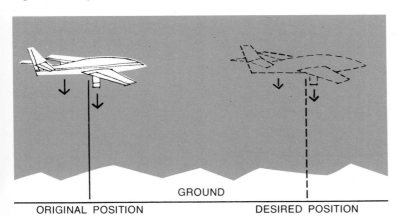

GROUND

ORIGINAL POSITION   DESIRED POSITION

Fig. 9–10.
Changing the location of our hovering VTOL aircraft.

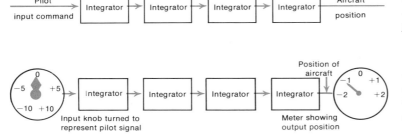

Fig. 9–11.
Model for the horizontal position of the VTOL aircraft.

Fig. 9–12.
Laboratory test set-up for simulating control of VTOL aircraft.

four integrations. This model is derived directly from Newton's laws* in physics.

In Fig. 9–12, a man turns a dial to adjust the input signal, while he watches the meter measuring the output. The output is initially constant at -1. The man's job is to apply an input which will bring the output to $+1$ and hold it there. When we try this in the laboratory, we find that the man just can't learn to control the output. A typical set of trials is shown in Fig. 9–13. Apparently no matter what signal the man applies, the output runs away. It is not surprising that the VTOL aircraft is so difficult to fly.

When an automated system is used to control the output, the necessary signal to give the desired output is calculated from the equations describing the system. In this particular example, it turns out that the required input signal is as shown in Fig. 9–14. There just is no way the human being could learn to decide on this input signal. (Fortunately, there are some problems which we can solve with mathematics that we cannot solve by guesswork or by practicing on the system.)

The examples just mentioned are three cases in which man is not able to perform control tasks. In each situation, the tasks can be so difficult that man can never succeed, no matter how long he practices or what special training he receives. Part of the limitations of man come from the behavior of man as a system. As we saw in Chapter 1 and again in 6, the human controller has a dynamic response. For example, there is always a time delay of 0.2 second or more. Man's ability to predict a future signal is limited.

These characteristics can be traced to the time for signals to go from the eye to the brain and then to the muscles, and also to the type of computations made in the brain. As a result, man is unable to control a system which is moving too rapidly. If the oscillation, for instance, is at a frequency greater than one cycle/second, man tires rapidly. Above 1.5 cycles/second, he loses control completely. Even when the signal is changing slowly, man may be unable to control the system if the necessary control system is very complicated. The VTOL aircraft example illustrates this limitation. In the 1950's, when power steering for cars first ap-

*
Isaac Newton (1642–1727) was a remarkable man. After being a good but undistinguished student in college, he graduated at the age of 23. He wanted to stay on, but Cambridge University was closed for two years because of the plague. So he spent the two years discovering new theorems in math (and numerical analysis), developing the calculus (the most important math development in many centuries), discovering the law of gravitation, explaining the motions in the solar system, developing his laws of motion, inventing the reflecting telescope, discovering that white light can be broken down into all colors, and so forth. After the University reopened, Newton was made a professor at the age of 26. He lived almost 60 more years, but never again approached his early productivity.

We might draw two conclusions: (1) Schools should close more often. (2) Don't waste creative men by making them professors.

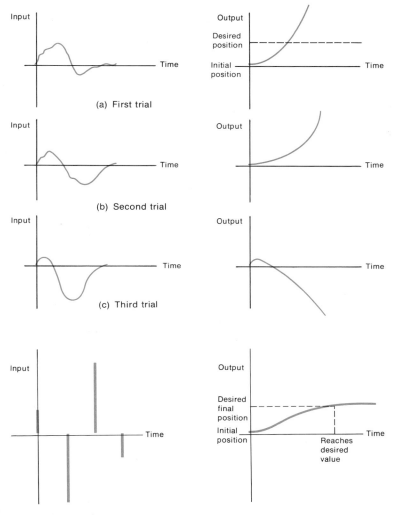

(a) First trial

(b) Second trial

(c) Third trial

Fig. 9–13.
*Typical results
when a man tries
to control the system
of Fig. 9–12. In the first
trial, he applies a positive
signal, then a negative one to try
to brake. The output runs away in a
positive direction.*

*In the second trial, he uses a smaller
positive signal and more braking. The
system still runs away positively.*

*In the third trial, he uses a very little
positive signal and a lot of braking—so
much braking that the output now runs
away in a negative direction.*

*No amount of practice seems to
lead to the right combination of
positive signal to get the output
started and negative
signal to brake.*

Fig. 9–14.
*An input signal
which does give the
output we want. A positive
pulse is used to start motion. A
big negative pulse brakes. A positive
pulse is used to offset the braking in part.
The input signal ends up with a small
negative pulse.*

*This kind of input signal
is not at all obvious. No pilot
is likely to learn to fly this
way. An autopilot is
absolutely essential.*

peared, some of the steering systems for trucks became unstable when the truck made a sharp turn at moderate speed. The system was so complex, the driver was unable to bring the vehicle under control. Several severe accidents resulted before the power steering system was changed.

Thus, man may be unable to control a system because of the speed at which signals are changing or because of the complexity of the control system.

3 | MAN IN COMMUNICATION

The difficulty of accurate communication is illustrated if we have a small group of people, each speaking two or more languages. We give the first person a sentence in English. He relays it to the second person in a different language (say French). After a small number of translations, we come back to English. Very often there is no obvious relation between the sentences at the beginning and end.

One problem is that each person interprets the sentence in his own way. As the sentence travels from person to person, even the main idea may change radically. The chance for change is increased even further by the fact that a listener may not hear a word or phrase correctly. We are particularly likely to misunderstand a word if we expect a certain message. Finally, errors arise because other noises or sounds interfere with our hearing correctly. The unwanted sounds "mask" or drown out the message.

In ordinary conversation between two people, we use enormous amounts of redundancy to correct such errors. For example, we repeat the same thought several different ways if it is important. We use inflections in our voice, facial expressions, and even hand motions to help communicate. These actions enrich the conversation, and they also help to correct misunderstood words or phrases.

In spite of these redundancies, voice communication between people is filled with errors. How often have we been told a telephone number, address, or other information and found later we had the wrong data. In some cases, we were told wrongly; in others, we failed to remember; but often we just misunderstood.

In today's complex systems, we often cannot afford any chance of error. Stock prices given to a broker must be correct. Otherwise, he will be buying or selling foolishly. A few such mistakes and our broker is bankrupt (or rich). As another example, when we are controlling high-speed trains in a railroad network, we cannot allow errors as we send information on the position and speed of each train to a central computer.

Can we communicate data with no chance of error?

Automatic Error Correction

When we want reliable communication, we ordinarily send information in binary form. That is, we use a sequence of 0's or 1's. For example, a typical signal might be:

$$1\ 0\ 1\ 1\ 0\ 1\ 0\ 1\ 1\ 1\ 0\ 1\ 0\ 1\ 0\ 0$$

This sequence of 1's and 0's, reading from left to right, carries the information of the signal. Electrically, this signal can be sent as shown in Fig. 9–15. Regularly timed pulses are sent. During

Fig. 9–15.
Pulse train
representing the
binary signal.

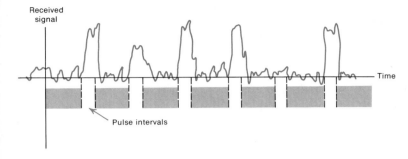

Received signal

Time

Pulse intervals

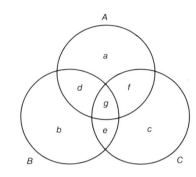

Fig. 9–16.
Received signal when
noise is added. (*Only the first
six pulse intervals of Fig.
9–15 are shown.*)

Fig. 9–17.
*Three overlapping circles
with the regions* a *through* g *defined.*

any space for a pulse, a pulse being present represents a 1, no pulse signifies a 0.

When noise is added during transmission, the received signal has the appearance of Fig. 9–16. During the second pulse interval, the signal should be zero, but the noise happens to be positive during this period. There is the possibility that this looks like a pulse to the receiver. If these first six binary digits

1 0 1 1 0 1

represent the number 45 in the message, the error in the reception of the second digit means the receiver interprets the number as

1 1 1 1 0 1

or number 61 in the decimal system. (We are assuming that the above are binary numbers, converted in the usual way to their decimal equivalents.)

In order to avoid this sort of occasional error, we can introduce redundancy as follows:

1. The signal to be transmitted is first grouped in sets of four digits. We initially consider the first four

1 0 1 1

2. Next we draw a set of three overlapping circles called A, B, and C (Fig. 9–17). In this figure, there are seven distinct, closed regions. Two such regions are g common to all three circles and f common to A and C.
3. In place of d e f g we insert *in order* the four numbers of our message, 1 0 1 1 (Fig. 9–18).
4. We now fill in the numbers for a, b, and c by choosing each as either 0 or 1 in such a way that the total number of 1's in each circle is even. In other words, in Fig. 9–18, circle A already includes three 1's; hence, we must choose $a = 1$ to make the total number of 1's even (here four). Similarly, $b = 0$ and $c = 0$ (Fig. 9–19).

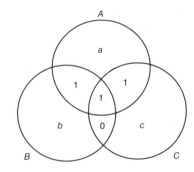

Fig. 9–18.
The digits of the signal are inserted.

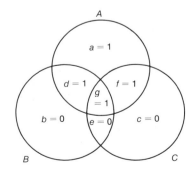

Fig. 9–19.
Determination of transmitted signal.

5. We now transmit the seven-digit sequence,

a b c d e f g

or in our case

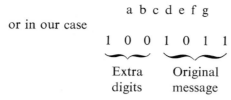

Each block of four digits in the original message is handled in this way. In other words, for every four message digits, we actually transmit seven digits, three more than necessary.

These three redundant or extra digits permit *automatic correction* at the receiver of any single error in the seven-digit block. The receiver must, of course, be familiar with the *coding* used at the transmitter. Then at the receiver we reconstruct the three-circle diagram of Fig. 9–19. If any one digit is received incorrectly, one or more of the circles contains an odd number of 1's.

As a specific example, we can return to our earlier problem and assume the second digit of the message is wrong. That is, the receiver obtains

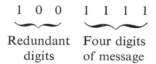

The receiver checks the validity of this message by constructing the diagram of Fig. 9–20.* Inspection reveals that there is an odd number of 1's in circle B and also in circle C. Therefore, if there is only one error, it must be the element common to circles B and C, or element e. Hence, the received signal can be automatically corrected to

1 0 0 1 0 1 1

The communication code described above (known as the Hamming code) uses three extra digits for each four message digits. The same procedure can be used if we draw four regions overlapping in all possible combinations (this is a complicated geometric construction). There are then four extra digits for each 11 message digits. Similarly, five extra digits compensate for 26 message digits. As the message block is made longer and longer, however, there is a growing possibility of two simultaneous errors: a situation which the system cannot correct.

This example is interesting because it is used in practical communication systems. It is primarily of value, however, to indicate the simple way in which technology uses redundancy to improve communication among men and machines. With technology, we can have better performance than is possible in direct man-to-man communication. With the code described above, we can *auto-*

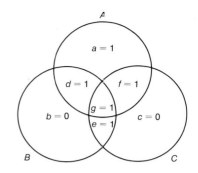

Fig. 9–20. Received signal including error in e.

---

*
Of course, electronic equipment does not construct a diagram and look at it. Instead, the receiver is designed to measure $a + d + f + g$, $b + d + e + g$, and $c + e + f + g$. Each of these three sums should be even. In the case of our received signal, the three binary sums are 1 0 0 , 0 1 1, and 0 1 1; the last two are odd; the error is in the element common to B and C, but not to A, or element e.

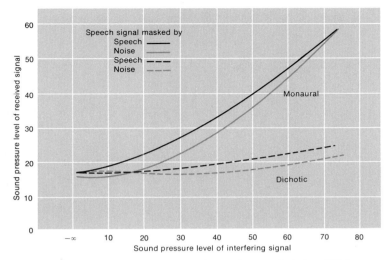

Fig. 9–21.
Understanding
speech in the presence
of noise. The vertical axis
shows the intensity of speech
which can just be understood when
noise is present. The horizontal
axis shows the intensity of the noise or
a disturbance signal.

The monaural curves refer to the test
with one ear only used. Two cases are
shown: one when the disturbing signal is
speech, the other when it is a random
background noise. (From
Man-Machine Engineering, by
Alphonse Chapanis. © 1965 by
Wadsworth Publishing Company,
Inc., Belmont, California 94002.
Reprinted by permission
of the publisher.)

matically correct any single error in each set of digits. With more complex codes, we can correct any one or two errors. Thus, we can build communication systems which are essentially error-free.

Communicating in Noise

In parallel with his use of redundancy, man has a remarkable ability to "hear" only particular portions of the sound reaching his ear. For example, two men can carry on a conversation even in an extremely noisy room. Figure 9–21 shows the human capability. We can understand even when the noise intensity is very much larger than the message sound. This characteristic is called the "cocktail party effect."

In spite of these remarkable capabilities of the human being to communicate in the presence of interfering signals, if the noise becomes too large and the signal too weak, errors occur. If communication is to be accomplished under such circumstances, it is necessary to slow down the rate and limit the vocabulary being used. One procedure which has proven very effective is called *word spelling.* Each word in the message is spelled, but with each letter represented by an agreed upon code word. The International Civil Aviation Organization has developed the following code for such purposes:

Alpha	Hotel	Oscar	Victor
Bravo	India	Papa	Whiskey
Charlie	Juliet	Quebec	Xray
Delta	Kilo	Romeo	Yankee
Echo	Lima	Sierra	Zulu
Foxtrot	Mike	Tango	
Golf	November	Uniform	

To send the word STOP, we say "Sierra Tango Oscar Papa."*

* In the choice of the words in such a code, we desire to meet several different criteria. Obviously each word must be readily identified with its letter (we would not use *Cease* for C both because the initial C sounds like an S and because of the ease of confusion with Sierra). Furthermore, the words must be easily recognized when spoken by people of many nationalities. Once the code is agreed upon, however, communication is made easier because the listener has to decide what the received word is from among only 26 possibilities.

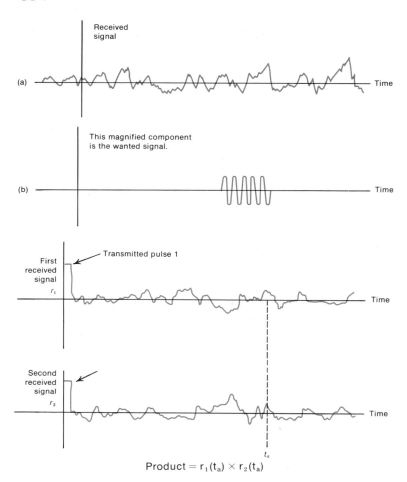

Fig. 9–22.
*Signal detection
problem.*

(a)

Received
signal

Time

This magnified component
is the wanted signal.

(b)

Time

Fig. 9–23.
*Processing
of received
signals.*

First
received
signal
r_1

Transmitted pulse 1

Time

Second
received
signal
r_2

Time

t_a

Product = $r_1(t_a) \times r_2(t_a)$

Even with these techniques, there are cases in which man's ability to detect signals in noise must be helped by technology. For example, when radar measurements are made of the planet Venus, a short burst of energy is radiated from an antenna pointing at the distant planet. This radio energy travels to Venus, a small fraction of the energy hits that planet, a small fraction of that is reflected back toward earth, and yet another small fraction of that reaches the antenna. Because of the noise signals which are also received by the antenna, the noise may be a thousand times larger than the signal.

We are then faced with the problem shown in Fig. 9–22. On an oscilloscope, the received signal simply looks like noise. Somewhere buried within this signal, however, there is a very small component which, if it were magnified, would look like part *b* of the figure. This is the echo.

If a regular series of transmitted pulses are used, each received signal (r_1 and r_2) has the form shown in Fig. 9–23. Comparison of two different received signals reveals that the noise in one is unrelated to the noise in the other; however, the echoes are the same in the two cases. Thus, if we multiply the two received signals together (r_1r_2) at similar instants of time (Fig. 9–23), the noise tends to average out (if r_1 is positive, r_2 is as likely to be positive as negative). The echo parts of the two signals are exactly alike, however: when the echo signal in r_1 is positive, so is that part of r_2. When this multiplication process is carried out with signals received after many different transmitted pulses, the echo part of the product becomes larger and larger compared to the noise part. With enough manipulation of the data in this fashion, we can identify echo pulses which are very much smaller than the average noise level.*

Thus, technology can be used to detect a signal of known shape which is swamped by a very large noise signal. This use of electronic equipment has been developed only during the last twenty years, primarily as a result of attempts to build radar equipment to obtain echoes from distant planets and very small objects in the sky.

In communication man works under very severe limitations Errors do occur. Noise tends to mask the signal. In addition, as we mentioned in Chapter 6, the rate at which man can communicate is limited. If we wanted to send the latest prices of 1000 stocks to a friend across the country, we certainly would not call him on the phone and read them one-by-one. Even at four a minute (with time allowed to repeat in order to avoid errors), the job would take over four hours. By that time, the prices would have changed. We would use automatic equipment which could easily send the information in a few minutes.

* This problem of finding a signal in noise also occurs in other situations. When an attempt is made to determine whether a person is dead (for example, in order to remove a kidney or the heart for a transplant operation), one of the criteria is that the electroencephalogram should show no brain waves, even within the normal noise of the instrument. In a different direction, current attempts to detect signals from possible civilizations in other parts of the universe require the detection of signals of unknown frequency amidst the noise of space and the receiver. Economists and businessmen frequently search for periodic or sinusoidal variations in profits or economic indices in order to detect oscillations which should be taken into account in business decisions.

4 | MAN LIMITED BY ENVIRONMENTAL NEEDS

Man is seriously limited by the environment he needs. This limitation has been extensively studied in the manned space program. The decision by the United States to focus efforts on manned exploration of the moon meant that the space vehicle had to provide an environment in which man could survive and work for a period up to two weeks. As a result of the need for the presence of man, the interior of the space vehicle was supplied with oxygen and nitrogen so that men could breathe without helmets. Temperature within the vehicle was controlled to avoid extreme cold or heat. In a longer space mission, artificial gravity will be used to avoid weightlessness. In other words, the presence of man leads immediately to a much more complex space vehicle. In addition, of course, we then place great emphasis on returning the vehicle safely to the recovery area.

These same environmental needs of man are severe limitations in other places. Robots have been developed to work in radio-

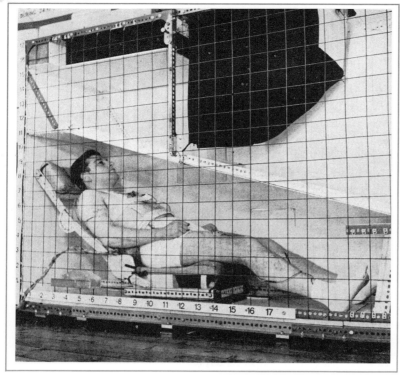

*Fig. 9–24.
How stressful
is long confinement
in small places? Ask an
astronaut or any commuter.
But what is the very least amount
of room needed? To find out,
Dr. Reginald Whitney of England's
Medical Research Council has been
putting volunteers in cages like this for
24 hours at a time. The man's position
is varied between lying and sitting while
he is given tasks to do and his mental
alertness is checked. Meanwhile, heart
rate and skin temperature are monitored,
urine is analyzed, and psychological
tests are administered. A subject can
kick his way out of the cage.
None did, which augurs well
for crowded space capsules
and led experimenters to
suggest that minicars
could be made more
mini.* (Science and
Technology,
Jan. 1968)

active areas—for example, to position objects within an operating nuclear reactor. Robots are currently being made to carry fire hoses and chemicals into burning buildings where there is a danger of the structure collapsing. When such fire fighting equipment is in general use, we should avoid some of the tragedies besetting firemen today.

Part of the "environment" for man is his acceleration. (Gravity is one example.) Here the need for human comfort limits the rate at which trains can be accelerated to a high speed or elevators can descend. If the acceleration is too great, man is uncomfortable. Repeated acceleration and deceleration (as in a subway train reaching high speed between stops) can cause severe discomfort and eventually nausea. (Usually the limit in a subway or bus arises first from the desire not to injure passengers who are standing, or trying to stand.)

This need for a pleasant environment also limits high-speed trains riding above ground. Most people like windows in a train. If there are windows and the train moves at 300 miles per hour, there must be an appreciable amount of clear space on either side of the train. If trees, houses, and so forth are too close to the windows, they will move by so fast the passenger will become carsick. (One designer proposed to use artificial windows, through which movies show a landscape moving by at a reasonable speed, perhaps 60 mph.)

Thus, the environment for man includes all of the signals which he senses, some of which are noise, temperature, pressure, acceleration, speed. We also must include the air he breathes and the stress under which he works. Man is remarkably adaptive to changes in this environment. He can work, for example, in environments with a very wide range in temperatures. Man can even perform useful work during high accelerations in rocket travel. When the limits are reached, however, machines must be used to replace men.

5 | MAN AND SENSING

Man's personal tie to reality is through his senses. Human observers can sense, or perceive, many quantities. Light and sound are particularly significant since they provide us our best means of personal communication, by sight and hearing. The human body has a large number of very refined sensing devices. They are built into some two square yards of body surface which make up the interface between our internal and external worlds. Specialized sensors that respond to different signals are scattered over this surface. It is through these sensors, including the eyes and ears, that all information about the external world comes to consciousness.

In addition to the eyes and ears, there are other sensors distributed over the body, such as those for the detection of pressure, heat, and cold. In addition, there are internal sensors which inform us of the positions of various parts of the body (for example, we know where our arms are even with our eyes closed). There are sensors which tell us how fast a limb is moving and the difference between up and down. Many people have a very good sense of compass direction which helps them, for example, to estimate how much they have turned while blindfolded. Some people have a remarkable sense of elapsed time. The list goes on and on, totaling many more than the five senses we are customarily said to have.

Many of these senses are extremely delicate or sensitive. Our normal sense of hearing can detect a sound so feeble that it moves the most sensitive part of the inner ear only some 10^{-11} centimeters. This distance is roughly one-thousandth the diameter of a hydrogen atom! If this sheet of paper is four-thousandths of an inch thick, then it is a billion times thicker than the distance the inner ear mechanism moves in responding to a faint sound.

Signals Man Cannot Sense

There are many quantities, however, which man cannot sense directly. Radio waves and X-rays pass through our bodies without any sensible effect. Try as we may, we cannot sense directly the television signal sent out from the broadcast station or the high-

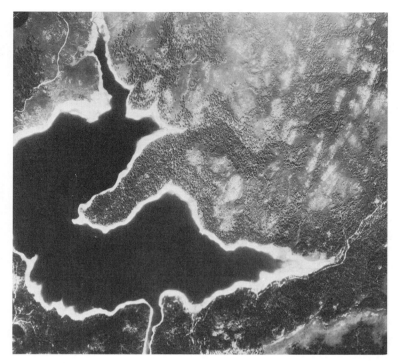

*Fig. 9–25.
View of the Bucks Lake
area of the Sierra Nevada
was obtained with panchromatic
film, meaning film that is sensitive
to the entire visible spectrum.
Such a photograph is particularly
useful for estimating the
density of vegetation
and for identifying
certain species of
vegetation. (Dr.
R. N. Colwell,
School of Forestry,
University of
California at
Berkeley)*

pitched screams which certain bats use for locating food. We can and do, of course, create sensing devices that transform an insensible signal into one that can be sensed. Television sets, for example, change broadcast signals into light and sound.

Some animals can sense signals which man cannot. Porpoises and some fish can hear sounds too high in pitch for man to hear. Some migratory birds and animals, as well as homing pigeons, are believed to sense the earth's magnetic field directly—a useful ability when navigating to places beyond the line-of-sight.

Among the most impressive modern sensors are those used in space vehicles or high-flying aircraft to observe the earth. Photographs, taken with special cameras, show a surprising amount of detail, as Fig. 9–25 indicates. The ability to inspect equipment, airfields, missile installations, and the like from space is, according to government experts, one of the major hopes for effective disarmament. Countries will seldom, if ever, allow on-site inspections by observer teams even to confirm that treaty terms are being kept. At the same time, nations are reluctant to sign agreements without the ability to confirm that all parties are conforming. Overhead photography from satellites and, in some cases, high-flying aircraft provides a technique for inspection which is often acceptable to all parties. The ability to know that no other country is deploying its missiles, ships, and other forces to attack can indeed be an influence for peace in the world. Satellite photography is useful too in weather forecasting, surveying of natural resources, and in locating sources of pollution.

Selective Sensors

Sensors in space and in high-flying aircraft are able to reveal much more than the usual camera might show. For example, objects of a specific color can be picked out and highlighted in a picture by using a filter of that color in front of the camera lens. The filter allows light of the desired color to pass through, but absorbs light of other colors. In a photographic print, the object can be made to appear light against a dark background.

A series of nine pictures of the same scene taken at the same time with a nine-barrel camera is shown in Fig. 9–26. Each barrel had a different filter so nine different views result. (The photos are

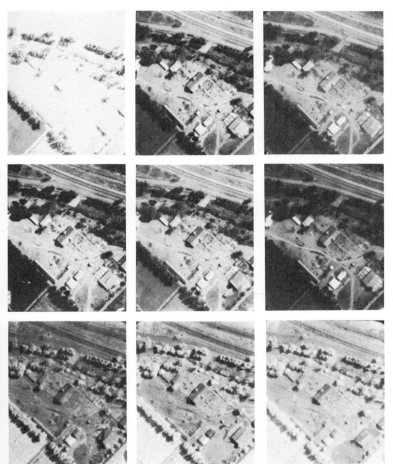

Fig. 9–26.
Multiband exposure
was made with a multiband
camera. The wavelengths of energy
(*that is, colors in the visible part*)
represented range from .38 micron
(*top left*) *to .9 micron* (*bottom*
right), *which covers not only*
the visible spectrum, but
also part of the ultraviolet
and near-infrared regions.
(*Dr. R. N. Colwell,*
School of Forestry,
University of
California at
Berkeley)

Fig. 9–27.
The same picture taken with two different filters. (Dr. R. N. Colwell, School of Forestry, University of California at Berkeley)

printed in black and white even though they were taken with colored filters.) Note that the various trees, buildings, and ground features show up quite differently in the different photographs. This happens because each object reflects different colors of light. In fact, objects tend each to have a *signature:* a pattern of light that they reflect. This can be determined by shining white light (which contains all colors equally) on the object and then measuring the color or colors of the light reflected from the object.

The importance of the "tone signatures" of objects is very well illustrated in Fig. 9–27, which shows two photographs taken with different filters of the same scene. The left photograph shows underwater detail to about a depth of 20 feet and highlights a road through the upper part of the picture. The right photograph emphasizes waterways rather than the road and shows certain plants and vegetation as bright white areas. The combination of pictures gives much more information than either one alone.

It turns out that objects reflect not only light, but heat and radio waves as well. In fact, light, heat, and radio waves are all closely related; they are all examples of *electromagnetic* waves. The only difference among them is their frequency.

Thus, objects reflect infrared (heat) and radio waves just as they reflect light, that is, some frequencies are reflected and some absorbed. Objects have an infrared and radio signature just as they have a visible light signature. With proper sensors, infrared and radio pictures can be obtained. These can reveal more than visible light alone can.

Infrared photography makes use of special film sensitive to the infrared portion of the spectrum. One principal advantage of infrared photography is that these waves penetrate smoke and haze and so can be used under conditions where ordinary photography would give poor results. Another use is in studying land features and vegetation. Infrared pictures taken at night will show objects which are hotter than their surroundings. In Fig. 9–28, three infrared pictures of the Yosemite Valley are shown. In one

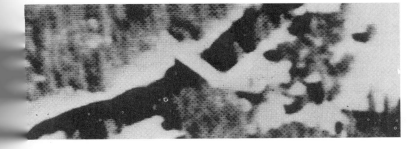

Fig. 9–28. Infrared pictures of part of the Yosemite Valley. (Dr. R. N. Colwell, School of Forestry, University of California at Berkeley)

the camera was equipped to show fires in the valley. Most are campfires, but even single charcoal briquettes were visible. The second picture shows also the heat from certain vegetation, recorded at lower frequencies. A daytime infrared picture of the same site is also shown in Fig. 9–28. Timber shows clearly here by reflection of the sun's infrared radiation.

All of the above demonstrate that appropriate sensors can reveal a great deal about the real world; they can extend man's own senses, enabling him to perceive objects and events beyond his native capacities. This possibility opens many avenues. One example is food-resource management to make adequate food available. Today only about 10% of the world's land area is cultivated; an additional 21% can be farmed. To make food needs and food production balance, more information is required in order to determine what crops to grow where and what the likely yields will be. The techniques discussed above, adapted for surveying, can help provide this information. They can provide, for example, mapping of soil and water temperatures, mapping of surface water, mapping of disease and insect invasions, assessment of crop vigor, determination of soil type and soil moisture, and mapping of major soil boundaries. This information could be invaluable in locating and planning usage of land for food production. More broadly, the information could provide data quickly on where existing crops are threatened, what crop resources are actually available in famine areas, and on what crops could be grown where. Sufficient manpower is not available to collect such information by personal inspection.

Final Comment

Aerial and space photographs are just one example of the use of technology to extend man's sensing ability. We can give other illustrations. For example, we often need to measure temperature with an error much less than one degree. Man can certainly tell when an object is hot or cold; he cannot tell whether a temperature is 121.2° or 121.3°. Yet this accuracy is necessary if the Apollo navigation system is to work properly.

Automated factories require that we sense the condition of products being used in the manufacturing process in order to be sure of the quality of the output product. Inside a boiler of an electric generating station, we must measure temperatures continually to avoid explosions. In a high-speed train system, tragic collisions are prevented by sensing the distance between trains. In the automated highway of the future for intercity travel, cars will be moved at high speed to their destination. Light beams (from lasers) will be sent out frontward from a car to measure the distance to the next car ahead. We must detect at once a decrease in this distance in order to brake soon enough to avoid accidents.

Finally, air-pollution sensors throughout the city warn of danger early enough to shut down incinerators, ban auto traffic, or take other safety action.

Thus, whenever we build a technological system to replace man in a particular task, that system must include sensing. In sensing, we measure the signals which represent the important behavior of the world with which the system interacts. Sensing is an important area in which devices are often much better than man.

6 | PROSTHETICS

In the preceding sections of this chapter, we have seen a few ways that technology is used to extend man's abilities: to do things which man cannot do. In a parallel direction, we try to use technology to give a normal life to handicapped people. One measure of a civilization is the compassion that society has for its less fortunate members. Among this group are those who have lost arms or legs, are deaf or blind, or are in some way physically unable to pursue normal activities.

In recent years, widespread publicity has been given to the attempts to develop an artificial heart and an artificial kidney. The transplant of a human heart emphasized the pressing need for an artificial heart, because the possibility of finding the more than 100,000 heart donors required annually in the United States alone is obviously remote.*

The high cost and scarcity of artificial kidney machines have accelerated engineering attempts to develop a portable, low-cost device that could be used in the patient's home. More than 20,000 people die needlessly each year in the United States because kidney machines are not available. These could be provided at a national cost of about 100 million dollars. The problem here is the fascinating one of national priorities. Should we spend 100 million dollars more each year? If we decide to spend 100 million dollars more on health each year, should we put this money into kidney machines? Who is to decide the answers to these questions? Shall we leave the decision to Congress? How do we avoid simply giving in to the group which makes the most noise? Such questions are of such fundamental importance for the future of our society that the proposal has been made that everyone's home phone should be tied to a computer network so that the people could vote on questions of national priorities.

From such broad social questions, let us return to our consideration of _prosthetic_ devices: devices to replace parts of the human body. Perhaps the most familiar are the technological devices to help people who have impaired senses, particularly hearing and sight.

*

Even if enough donors could be found so that all weak hearts could be replaced by transplanted human hearts from individuals who died of other causes, there would be an argument for the use of an artificial heart instead. The human body tends to reject organs transplanted from other human beings more than artificial mechanical devices. The principal argument for using transplanted hearts is that engineers are not at all sure they can build an artificial heart which will work reliably for perhaps 15 years.

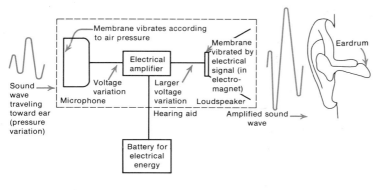

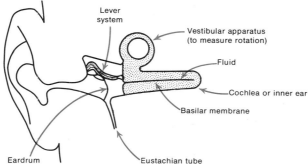

Fig. 9–30.
Diagram
of ear parts
(cochlea is shown
straightened out; it is
actually a three-turn
spiral about 35
millimeters
long).

Hearing aids are exceedingly common among older people. The hearing aid consists simply of a microphone to detect the sound waves and an electronic amplifier and small loudspeaker to generate a sound signal which is an enlarged replica of the incoming signal (Fig. 9–29). The system is powered by batteries which are often concealed inside the wearer's clothes or behind the ear. A volume control permits the man to turn down the sound when he is anxious for quiet (as during a boring lecture or conversation) or when he is in a particularly noisy environment.

The successful use of a hearing aid requires that the man have at least a little hearing capability left: he cannot be totally deaf. All the hearing aid does is amplify the sound signals so that they can be heard by the listener. The man still detects the sound signals in the usual way.

The human ear is a complex device which operates in the following general manner. Sounds are really pressure changes in the air. When the sound signal reaches the eardrum (a thin membrane), the pressure changes cause the eardrum to vibrate or move back and forth: the louder the sound, the greater the back and forth motion; the higher the pitch, the faster the motion (Fig. 9–30).

Just behind the eardrum are located various bones formed to act as a lever. This lever multiplies the pressure changes and reduces the amplitude of motion by a factor of about 16 (just as a lever permits a man to lift large weights, or two people of radically different weights to balance on a seesaw). The stronger pressure changes are then sent on to the liquid filled inner ear.

In the middle of this inner ear is another membrane (called the basilar membrane) which vibrates. Different parts of the membrane are particularly sensitive to notes of different pitch. Thus, the way in which this basilar membrane vibrates depends on the pitch or frequency of the original sound signal (from about 20 variations per second to almost 20,000 variations per second, the range of sound frequencies the normal young human being can hear).

As the basilar membrane vibrates, small hairs going from this membrane to another nearby membrane which is stationary are forced to bend. This bending of the hairs somehow causes an electrical voltage to appear in the nerve cells which are close to the hair cells. There are about 30,000 such nerve cells in a human ear. Which nerves are excited electrically and the rate at which they are excited depend on the hair bendings, which in turn are determined by the vibration of the basilar membrane.

These electrical signals then travel to the brain through the nervous system. In the brain the original sound is recognized by the nerve signals arriving from the ear.

Research is directed toward an attempt to design devices which can receive the sound signals and generate the appropriate electrical signals to be inserted directly into the nerves leading from the ear to the brain, in other words, a device which replaces man's sensing

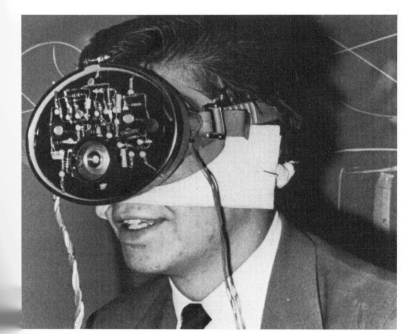

Fig. 9–31.
Newly-developed
amauroscope ("amauro-"
means blind) to "see."
It "catches" light. The light
triggers electrical impulses
that are sent to the brain
through wires attached
to the skull. (Wide
World Photos)

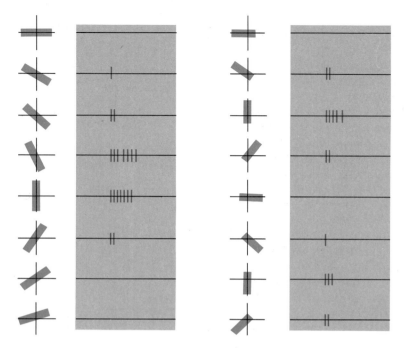

Fig. 9–32.
Response of visual area
of cat's brain to simple shapes.
In this experiment by physiologists
Hubel and Wiesel, a cat was shown
the various lines indicated at the left.
The electrical activity of single cells in
the brain of the cat was measured. The
spokes indicate the firing (or electrical
discharge) of the cell.

Such experiments demonstrate that a
particular nerve cell responds differently
to different shapes. These studies are an
early step toward understanding how
the brain and eye combine to recognize
visual forms, shapes, motions, colors,
and so forth. Similar experiments
show, for example, that certain
brain cells are fired electrically
when the cat sees motion
in particular directions.
(Figure from R. L.
Gregory, Eye and
Brain, *McGraw-Hill*
Book Company,
N.Y., 1966.)

of the sound signals. If such a device can be developed, hearing capability can be restored for people who are totally deaf due to non-functioning of the ear's sound-sensing system.

In order to construct a device of this sort, the engineers and biologists must understand the detailed operation of the sound-sensing system of the ear. How do the electrical signals depend upon the sound waves arriving at the ear? For each possible sound signal, what electrical signal should be delivered to the nervous system? How can this be done by a man-made device which is small enough to be portable (or wearable by the man)? In other words, how can we design a device which matches the compactness, yet remarkably flexible performance of the human ear?

Problems of this nature are even more complex when we turn to a consideration of the eye and the way in which a man sees. Eye glasses are similar to hearing aids in that they assist the man with impaired vision, but they are not useful for the totally blind. Information is sensed in vision by the light passing through the pupil (the opening at the front of the eye) and falling on the retina, which consists of a very large number of light-sensitive cells having varying properties. The electrical signals generated by the retina are transmitted by the optic nerve to the brain. Thus, the design of an artificial eye requires an understanding of what electrical signals result from specific light inputs and in what patterns the human eye operates. For example, how does the eye perceive different shapes, colors, light intensities, and so forth?

Box 9–1

PERSONAL IDENTIFICATION

(*A human capability machines can't match*)

Have you ever wondered how you identify a particular person? By sight, of course, since humans are good face recognizers, and no two faces are exactly alike. By voice, too, but often we can be fooled since many voices tend to sound alike. Fingerprints uniquely define a person, and toe prints, too. Some muscular-coordination styles are partially helpful, as in gait or in handwriting. Skull structure and tooth geometry, as well as dental work, are highly individualistic; X-rays have been used advantageously to type, classify, and even identify individuals. Cell chromosomes are, of course, unique for each person, but so far we lack the means to read that code adequately. Some bodily characteristics are, for adults, reasonably stable and useful as tags; examples are bone structure, height, skin pigmentation, basic fingernail shape, eye color, and odor. Some occupations and habits stamp us with telltale signs, like skin stains and blemishes from photographic processing, hard manual labor, and smoking. Teeth are deformed by betel-nut chewing. Certain kinds of chemical specificities in addition to blood type and immunological responses can be useful, but our knowledge of these is still relatively primitive. Last, but not least, are all of the mental characteristics and knowledge, quite different for every human being.

Generally, to identify a person positively requires a combination of these characteristics, unless one uses the only proven test, namely, fingerprints.

The difficulty of achieving such scientific understanding is a consequence of the remarkable capabilities of the human senses, whether we consider hearing or vision. The human being, for example, has an astonishing ability to recognize patterns. At a large party in a small room, a man can converse with a friend even though the background noise is very great. The man has the capability of picking out a desired message from a noisy background.

The pattern-recognition talent of man is more simply illustrated by the two drawings of Fig. 9–33. If we are told that the first includes a capital letter and the second a numeral, most of us can find these symbols with no difficulty. (They are an E and a 3.) To build equipment which accomplishes this same feat is extraordinarily difficult; we are not even sure what part of this pattern recognition occurs in the eye and what part in the brain. Yet this sort of pattern recognition is exactly the kind of task which must

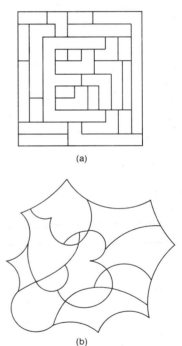

Fig. 9–33.
Two well-known signs, submerged in "noise." (*Colin Cherry*, On Human Communication, *M.I.T. Press, Cambridge, Mass., Second Edition, 1966.*)

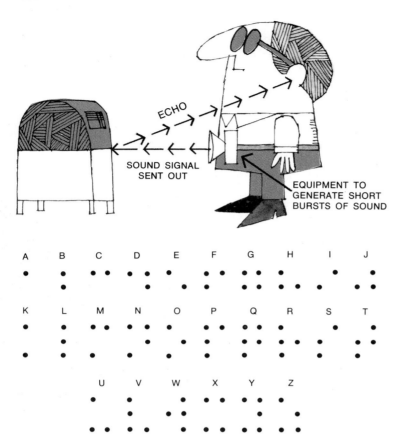

Fig. 9–34. An obstacle detector for the blind.

ECHO

SOUND SIGNAL SENT OUT

EQUIPMENT TO GENERATE SHORT BURSTS OF SOUND

Fig. 9–35. *The Braille alphabet. In this system, combinations of dots (up to six per letter) are used to represent letters and short words. The trained blind person runs his fingers over the raised dots. With practice and experience, most individuals can achieve reading rates of 30 words/minute (a sighted person reading 60 pages per hour is moving at a rate of about 300 words/minute).*

be accomplished if we are to build machines which can automatically convert from spoken words to written text, or machines which can automatically scan cloud pictures taken from satellites in order to predict future weather trends.

A normal man receives most of his information about the world around him through his eyes (for this reason, television is a vastly more effective means of communication than radio). Because of this complexity of the sight mechanism, modern technology is making a strong effort to develop devices for the blind which perform only very specialized tasks. To assist the blind in walking about, efforts are being made to develop light-weight, simple obstacle detectors (for example, small sonar sets which send ahead sound signals and detect any echoes which bounce back from objects— the system sketched in Fig. 9–34 and similar to that which bats use to navigate and to capture insects for food).

Another important scientific effort is in the development of devices to help the blind person to read ordinary printed material (magazines, newspapers, and books).* The expense of publishing in Braille (Fig. 9–35) severely limits the variety of reading material

*
In the United States the scientific research effort for the blind has really not been a large national undertaking. Where in the mid-1960's we spent about $200 each year per prospective patient on research in heart disease and cancer, we spent only $1.25 per prospective patient on research and development of devices for the blind.

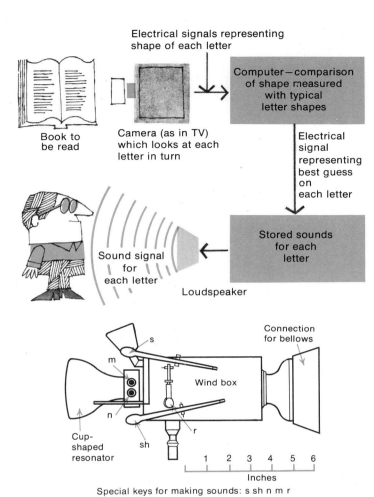

Electrical signals representing shape of each letter

Book to be read

Camera (as in TV) which looks at each letter in turn

Computer—comparison of shape measured with typical letter shapes

Electrical signal representing best guess on each letter

Sound signal for each letter

Stored sounds for each letter

Loudspeaker

Fig. 9–36. A reading device for the blind.

Connection for bellows

Wind box

Cup-shaped resonator

Special keys for making sounds: s sh n m r

Inches

Fig. 9–37. Von Kemplen's talking machine was powered by bellows (not shown) pumping a wind box, which in turn blew a reed in the throat of the cylindrical cup. Vowels were made by the operator's hand, which partially covered the mouth of the cup, while the nasals and fricatives were made by special attachments. (McGraw-Hill Yearbook of Science and Technology)

available in this medium.* Recent research has included electronic devices to scan letter by letter, then to compare the shape of each letter in a computer with typical letter shapes to find the most probable letter, and then to generate from a loudspeaker or earphone an oral statement of the letter (Fig. 9–36). By speeding up the rate at which the successive letters are "spoken" and by training the blind person to recognize shortened versions of each letter, reading speeds comparable to those achieved with Braille can be obtained with ordinary magazines and books.

We have considered only devices to help individuals who are deaf or blind. Equally dramatic advances are being made in other devices. As examples, we might mention:

1. The artificial larynx, or voice-box, to permit a man who has lost this organ to speak again.

*
Braille was originally developed for military purposes—in order to communicate written orders at night.

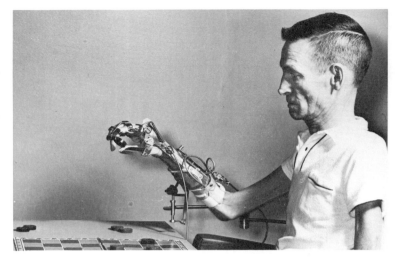

Fig. 9–38. Myoelectric control system for partially paralyzed patient ("myo-" means muscle). The attachment shown allows the patient, with a paralyzed lower arm, to move his "artificial" hand in a normal way. Electrical signals from the nervous system of the patient are used to actuate the electromechanical device. (Dan Antonelli, Rancho Los Amigos Hospital, Downey, California)

2. Artificial arms and legs. Current research is directed to designing limbs which can be moved in a normal way by the usual electrical signals from the brain (Fig. 9–38).

3. The heart pacemaker, a device inserted under the skin of the stomach with wires going to the heart. The device generates electrical signals which regularly shock the heart and thereby cause the heart to beat, for patients in which beating is impaired or too slow or irregular. By 1970, pacemakers had been placed in more than 100,000 people in the United States. New devices were being used at the rate of 15,000 a year. Once the pacemaker is in place, life expectancy is about 90% of what it would be if the man had never had the heart trouble at all.

4. Arterial replacements. The replacement of a diseased or weakened section of the aorta with an artificial artery is now a relatively common operation.

The successful development of each of these examples depends upon a scientific understanding of the detailed way in which the human part operates. The technological possibility of building replacement elements for the human body thus stimulates the medical researcher to learn in greater detail about the biological phenomena. Thus, our knowledge about science grows in parallel with our ability to build more and more sophisticated man-made devices.

7 | MATCHING TECHNOLOGY TO MAN

Throughout this chapter, our emphasis has been on using technology for the benefit of man. When we speak of the rest of the twentieth century as the *age of technology*, we mean that this technology to serve man is appearing in more and more of our lives. Man's work and recreation, his communication and travel, his education and enjoyment of the arts—in all directions, technology properly used can enrich life.

The key to success here is in the idea of technology *properly used*. What is implied by this phrase?

Meeting Man's Needs

First, we must know man's needs.

The subject of nutrition is an obviously important example where we need to know the needs. Discussions of the world population problem usually dramatize the problems of famine, particularly in Asia. Any planning for new sources of food or new methods of food distribution must be based on knowledge of man's daily requirements to assure health and the ability to work and live normally. Yet our scientific knowledge today does not even include the number of calories an adult must eat each day to avoid losing weight. There is no general understanding in detail of why some people are overweight.

The problem is difficult because nutritional needs are apparently highly individualistic. In other words, what one person needs may be very inappropriate for another. In England, the disease of rickets was controlled by giving babies vitamin D in their food. Soon after the program started, it was found that certain babies

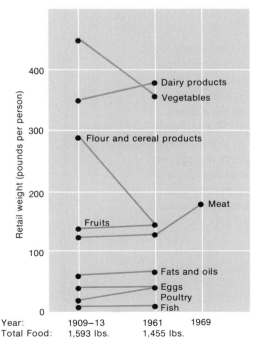

Fig. 9–39. Changes during this century in the foods Americans eat. During the 50 years shown, the total quantity of food per year dropped by 10%.

Source: U.S. Department of Agriculture

were dying from kidney trouble, traced to too much vitamin D. When the vitamin D was decreased, rickets reappeared in other children. The solution lies in an individual nutritional program. Just as in the overweight problem, each person must be considered as an individual.

Furthermore, nutritional needs must be determined very early. We now know that the maximum size of various organs of the body is determined before birth. The brain reaches its full number of cells by the age of two. Experiments with rats suggest strongly that a shortage of milk during the first few months of human life could lead to a small number of brain cells and, possibly, changes in mental behavior. Scientific experiments on people have not been performed, but there does seem to be some relation between nutrition in infancy and mental development. There certainly is a relation between nutrition in early childhood and the way the adult later produces fat. The overnourished child develops a lifelong desire for extra food.

In other words, the problems of nutrition are an example of an important area where the possible benefits of technology are severely limited by our inadequate knowledge about the needs of man. We continually seek new methods of food production and distribution. These developments can never provide enough food to prevent starvation and malnutrition unless we learn how to measure accurately minimum human needs and then base our planning on this system knowledge.

Fitting Man's Characteristics

To use technology properly, we must also design the system for the user. The machine must take into account man's capabilities and limitations, his likes and dislikes. The problem displayed in Fig. 9–40 is a simple example of this entire area. Here are pictured the tops of four different cooking ranges, each with four burners labeled A, B, C, and D. Each burner is controlled by the knob with a similar label. For each of the four models, tests have been made to determine how easily people can learn to associate the proper control with a particular burner. The results are startling in their differences. System I (which happens to be the arrangement seldom used in commercial products) proves to be very much simpler to learn than the others, and the complexity increases sharply from I to IV in the diagram.

Figure 9–41 shows another example of technology which does not seem to be matched to the human user. The top part shows the way the manufacturer designed the heat controls on an electric range, while the bottom portion shows the way they should be designed to match the normal expectations of a housewife (as determined by surveys of women). The housewife expects the heat to increase as one moves from left to right, she anticipates 3 to

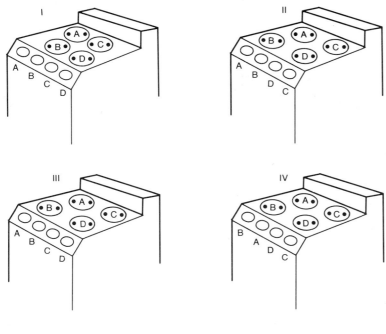

Fig. 9–40.
Models
of four stoves.
(From Man-Machine
Engineering by Alphonse
Chapanis. © 1965 by
Wadsworth Publishing
Company, Inc.,
Belmont, California
94002. Reprinted
by permission of
the publisher.)

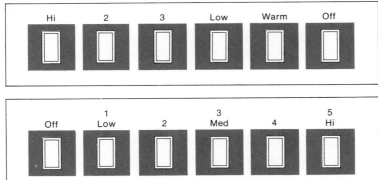

Fig. 9–41.
Pushbuttons for
an electric stove as
the manufacturer
originally designed them
(above), and the stereotypes
which housewives had about these
pushbuttons (below). (From
Man-Machine Engineering by Alphonse
Chapanis. © 1965 by Wadsworth
Publishing Company, Inc., Belmont,
California 94002. Reprinted
by permission of the
publisher.)

represent more heat than 2, and she assumes "Warm" means more heat than "Low." In other words, the original design contains a variety of sources of significant or even dangerous errors in use of the technological device.

Figure 9–42 is a final example of the contrast between poorly designed devices and devices matched to human characteristics. A micrometer is used for precise measurements of thicknesses of a solid substance. Each micrometer is set at a reading of 0.398 inch. The top unit, which was standard in laboratories until very recently, is read with the two dials in combination, with the dial around the rotating handle used for the fine interpolation between divisions on the principal scale (shown as 0, 1, 2, 3, 4).

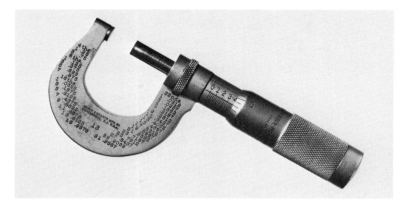

Fig. 9–42.
Micrometers
(a) Conventional
design. (b) Designed
for easy readability. (From
Man-Machine Engineering *by*
Alphonse Chapanis. © *1965 by*
Wadsworth Publishing
Company, Inc., Belmont,
California 94002.
Reprinted by
permission of the
publisher.)

These examples emphasize that the usefulness and value of technology depend not only on the ability of the device to do something which is meaningful for man, but also on design so that the device is suitable for man's use. Fortunately, in recent years manufacturers have been developing a growing appreciation of the importance of this part of technology, and we now frequently see extensive consumer testing of new devices in order to ensure satisfactory performance. There has been a major effort to improve highway road signs, but it is still relatively easy to find signs indicating a turn-off after it is too late to move safely into the turning lane. New railroad cars and airplanes are being designed from a knowledge of typical physical size of the individual American traveler. Laser obstacle detectors to assist the blind to move about are being designed on the basis of research on the needs and wishes of the blind.

To conclude this section, one example illustrates the potential seriousness of poor design. Figure 9–43 shows a possible factor in the disastrous airplane collision over the Grand Canyon in 1956—a crash which took 128 lives. As indicated by the figure, visibility from the cockpit of many modern airplanes is severely limited,

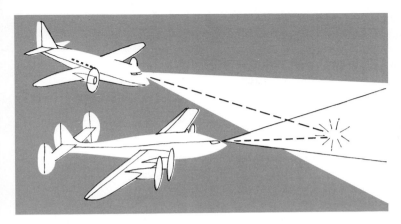

Fig. 9–43.
*Poor visibility to the rear
and below probably caused the
Grand Canyon crash. Although the
exact positions of the two aircraft
are not known, this drawing shows
how the limited fields of vision from
the converging DC-7 (behind) and
Super-Constellation could have
prevented each pilot from seeing
the other plane until too late.
The triangles show the approximate
field of vision from each airliner.
(From* Man-Machine
Engineering *by Alphonse
Chapanis. © 1965 by
Wadsworth Publishing
Company, Inc., Belmont,
California 94002.
Reprinted by
permission of
the publisher.)*

and it is entirely possible that the pilots could not see the other plane until just before the impact.

As technology becomes more elegant and more of everyone's life is directly concerned with technological systems, the best design of equipment becomes more significant. The automated highways and trains of the future must be both entirely reliable and excellently matched to the characteristics of the human users.

Controlling Technology's Side Effects

Finally, if technology is *properly used*, the side effects must be anticipated. We are all familiar with the new, expensive, eight-lane highway which suddenly ends at the city line. Cars move into the city at a delightful speed. Indeed, the trip is so easy that many more people now drive. But there is nowhere for the extra cars to go (or even to park). This is an example of planning we might call ill advised. Unfortunately, the problem is often much more serious and difficult. With the rapid introduction of new technology, we can expect dangerous side effects unless a major effort is made to carefully plan ahead. We have already seen such effects in the thermal pollution from nuclear power plants. It was only recently that nuclear energy was widely heralded as bringing to an end the air pollution caused by coal and oil plants. Now, only a few years later, we find serious ecological changes in the waters near nuclear stations—even massive fish kills.

The problem of predicting side effects of new developments is called *technology assessment*. In an assessment, we attempt to anticipate the effects not only on the system in which the technology is used, but also on related systems. For example, the proposal has often been made to require all taxis in a city to be electric cars. The direct hope is to reduce air pollution.

What are some obvious side effects? First, the cost of taxi rides would rise. We should estimate the increase, then predict the number of people who would (as a result) bring their own cars into the city. The electric energy used to recharge the taxi batteries would have to be generated and distributed. What changes would this require in the electric utility? Would electricity rates rise? If so, how many businesses would move out of the city, with resulting unemployment? How serious would be the traffic problem because of stalled taxis? Where would battery-recharging stations be located? Who would man these stations? How would traffic flow be affected? Would people who use taxis regularly move to the suburbs?

Clearly, we can go on and on. The new technology which at first seems so attractive actually has far-reaching effects. Unfortunately, as the ecologists have emphasized, such effects are often obvious only after they are irreversible.

8 | FINAL COMMENT

In this chapter, we have considered briefly a few of the problems associated with the human use of technology. From these discussions, it is apparent that we really cannot separate science and technology from many other fields:

Biology (for its study of man's abilities and limitations).
Psychology and behavioral science (for the study of man's attitudes, likes, learning ability, etc).
Social sciences (for the evaluation of the effects of technology).

In the first nine chapters we have met a variety of the concepts which are the foundation of modern technology. In the last six chapters, we will sharpen our focus. We will concentrate on one product of modern technology: the computer. From this study, we will gain an appreciation of how we can assemble an astonishingly complex man-made system on the basis of very simple rules, how we can utilize technology to expand man's scope, and how modern technology can be designed to serve man's needs.

It is appropriate that we select the computer for this focus. The computer is a monumental achievement of twentieth-century man. It illustrates dramatically the crucial importance of a logical, quantitative approach to system understanding. Finally, the computer, still in its infancy, will profoundly influence individual lives as well as the complete social structure during the remainder of this century. In other words, the computer is central to the interrelationship among man, technology, and society.

Questions for Study and Discussion

1. What are some of the criteria we use when we evaluate the way in which a device is designed for a specific human purpose? Give at least two examples of such devices.

2. In the area of locomotion, what is the major advantage which the machine has over man?

3. In the area of communication, what advantages does man have over the machine?

4. In the area of control, what is the major advantage a machine has over man?

5. What are the major sources of error in electronic transmission of a binary signal?

6. Describe the "cocktail party" effect in communications.

7. The word spelling code used by the United States Armed Forces during World War II had the following word alphabet:

Able	Jig	Sugar
Baker	King	Tare
Charlie	Love	Uncle
Dog	Mike	Victor
Easy	Nan	William
Fox	Oboe	X-ray
George	Peter	Yoke
How	Queen	Zebra
Item	Roger	

What advantages does the ICAO code discussed in the chapter have over this one?

8. What is the purpose of multiplying two radar signals received from a distant object, such as another planet? What medical application does the same procedure have?

9. Figure 9–8 shows a system of two tethered satellites. Explain the purpose of this arrangement — how it works and how it is controlled.

10. Below is a block diagram of a human control system. Using the diagram, explain how a person maintains balance when riding a bicycle, a skateboard, a surfboard, or backing his automobile out of the driveway.

11. Give two examples in which a technological system is superior to a human. In each example point out the specific parts of the system in which the machine is superior.

12. Give two examples in which the human is a vital part of a technological system. In each example point out the specific task which can be done better by the human.

13. What is the major purpose of the automatic error correction code described in Section 3?

14. Section 5 lists many quantities which man cannot sense directly. Prepare a list of quantities which man can sense and for which he seems to need no help from technology, as well as a list of quantities which man can sense, but for which he sometimes uses technology to aid his sensing.

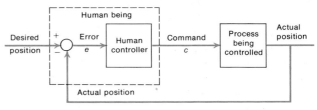

Laboratory and Projects

I | MAN AS A CONTROLLER

In the tantalizer laboratory experience in the previous chapter you demonstrated how delayed or confused feedback interfered with man's ability to control a system as simple as tracing a straight line. When the signal that man is trying to control is dynamic, the problem becomes even more complex.

Part A

A low frequency square wave is generated by the analog computer. This signal is subtracted from a normally controlled signal produced by a second analog computer, and the output of this system is fed to the cathode ray oscilloscope or to the chart recorder. The operator's objective is to cancel the output of the first analog computer by operating the second analog computer. Start with a low frequency and increase until the operator loses control completely.

1. At what frequency did each of six operators lose control?

2. Are all people equally capable of controlling signals?

Part B

Docking a Boat

It is often necessary for man to control systems which are not regularly repetitive such as the square wave of Part A. One such situation is that of docking a boat. The boat which we will simulate on the analog computer has the following characteristics:

The acceleration of the boat is controlled by adjusting the angle of the propeller blades.

The drag resistance between the hull and the water is proportional to the speed of the boat within the range of speeds in which we operate the boat.

The rudder is fixed so that the boat moves in a straight line. Figure 1 shows the wiring of the analog computer for this simulation using the characteristics listed.

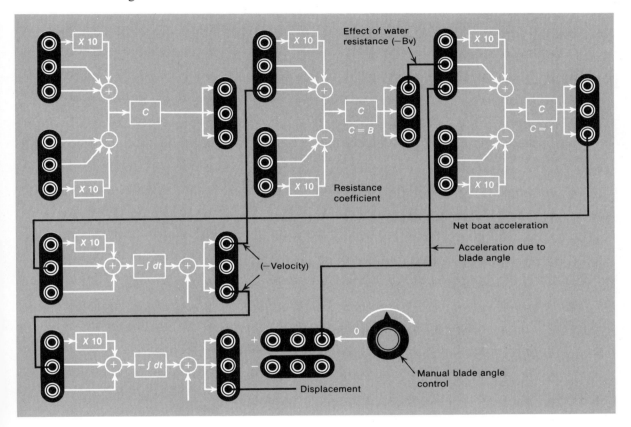

Fig. 1.
Boat simulation.

1. Wire the analog computer for the boat simulation (Fig. 1). Wire the displacement output to the meter, or the CRO, or the chart recorder.

2. Assume that the numerical coefficient of resistance between the hull of the boat and the water is 0.7 times the velocity of the boat.

3. Set the initial conditions of both integrators to zero. Put the integrator switch in the continuous (manual) position. The constant knob is now the manual blade angle control which controls the forward acceleration of the boat. Pick a number on the dial of the meter, or a spot on the face of the CRO, or a specific line on the chart recorder as the dock.

4. Use the constant knob (blade angle control) and try to pilot your boat to the dock across the lake.

5. Repeat the preceding exercise, assuming a change in the coefficient of resistance due to a different hull.

6. You can simulate a current from behind the boat or from the front of the boat by setting a very small initial condition on the first integrator. Try this. What do you notice about the effect of an independent current on the ease of controlling the position of a vehicle?

II | LEM LANDING SIMULATION

In Section 2 we saw several vivid examples of man's limitation as a controller. In this experiment, you are going to control the acceleration due to the ascent engine of the LEM to try to land it softly on the moon. Essentially the displacement (altitude) of the LEM depends on the difference between the accelerations of lunar gravity and the ascent engine. Near the surface of the moon the lunar acceleration is approximately 1.6 ft/sec². The ascent engine's acceleration is variable and is controlled by you.

Part A

Figure 2 is a block diagram from which a simulation of the landing of the LEM on the surface of the moon can be made.

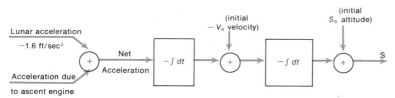

Fig. 2.

Notice that the lunar acceleration has been assigned a negative sign and the ascent-engine acceleration is positive. You can now use the analog computer to simulate the landing of the LEM on the surface of the moon. Start by assuming that your initial altitude is 1000 ft. (let 1 volt = 100 ft.), set S_o = 10 volts. To simplify matters, represent the surface of the moon by letting V_o = 0 volts and S = 0 volts. The main purpose of this experiment is to provide you with an experience which illustrates man's limited ability to control dynamic systems. So if you have difficulty in figuring out the wiring for the analog computer, use Fig. 3 to set up the LEM landing simulation. Now you are ready to see if you can make a *soft* landing on the surface of the moon. After several trials on the simulation, answer the following questions:

1. Did your landing technique improve with successive trials? Was there a limit to the improvement?

2. What happens to your ability to control the movement of the LEM when you add an initial velocity?

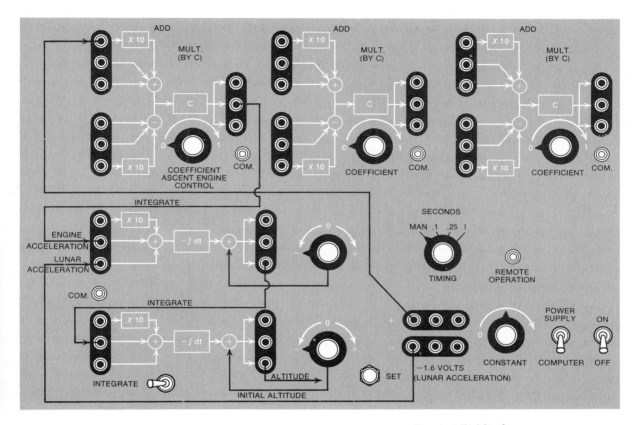

Fig. 3. LEM landing simulation.

Part B

Simulation of a More Complex System

In the previous simulation we assumed that you had direct control of the acceleration of the ascent engine. This is not the case in the real situation; you can only control the rate of change of the acceleration. We invite you to try a new simulation based on Fig. 4.

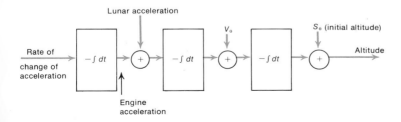

Fig. 4.

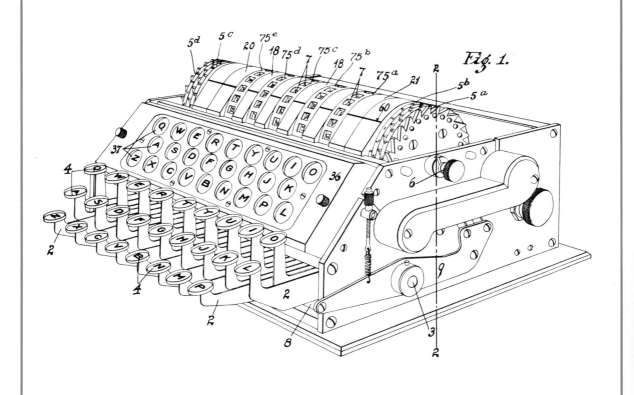

Fig. 1.

The Thinking Man's Machine

10

I | THE KEY TO MAN'S SURVIVAL

Why has man succeeded so well in battling nature for survival? Surely the secret does not lie in the strength of his body or the acuteness of his senses. On every side he is outclassed by one or another creature in strength, endurance, speed, or sharpness of senses. All in all, when man first appeared on our planet, he may well have been a good candidate for extinction. As it turned out, exactly the reverse was true.

Man could think creatively. He saw how to make tools. Tools made him a more capable creature, better able to stave off his enemies and to live in his environment.

Early man devised physical tools such as sharp stones to scrape skins, fur clothing to keep him warm, shelter to keep out the elements, and weapons with which to kill and to ward off attack. Primitive tools led to modern tools—today's jet aircraft for long-distance travel, television for seeing at a distance, and pacemakers to aid the human heart, among many others. Tools have served man well in the struggle for survival.

However, in controlling and changing nature, man has created another threat to his existence. It is the threat of overpopulation and pollution of the environment. Man has not always been able to foresee the long term results of his actions, nor has he tried

Edward Hebern's "Electric Code Machine," U.S. Patent 1,683,072. This code machine invented by Edward Hebern in about 1921 is a symbol manipulator. By typing a message on the keys, a person can secure an enciphered version to keep the text secret. As each key is pressed, the enciphered letter lights on the front panel. The code can be varied by resetting the wheels at the top. Machines to handle letters and numbers are closely related to digital computers. (The Codebreakers, published by Macmillan Company, 1967)

often enough. In the past 25 years, man has developed a tool which may be the key to the foresight and planning necessary to preserve and improve the complex societies in which he now lives. Far from regimenting his life, this new tool can provide individuals with an increased range of choices. This new tool is the electronic digital computer. Of course, computers create their own hazards. For example, they can be used to invade personal privacy or to "prove" an untruth. However computers are used, they *will* be used. Informed people are our major resource in seeing that the computers benefit mankind.

Digital computers deal with symbols, language, and information; more simply, ideas. Computers are subtle tools that can aid man in thinking and remembering.

Computers represent a whole new family of tools for dealing with complex ideas and masses of information. They can carry man another step upward on the ladder of human achievement if he uses them wisely. In this chapter, we shall see how a few simple ideas combine to yield this immensely powerful effect. Computers are having a major impact on our way of life; this impact will be much greater in the years ahead. This section of *The Man-Made World* will not make the reader a computer expert, but upon completing it, he will be among the few who will know enough to help in making computers the servant of man.

We should be aware, however, that the ideas behind computers are strange and are not part of most people's everyday experience. Most of us find, however, that only a little time and reading is needed to begin to think in the necessary way to be able to use and understand computers. In fact, it turns out that young people as a group seem more capable of thinking in these new ways than parents, teachers, and adults. The computer seems not only to be the "thinking man's machine," but indeed an agent of constructive change for society.

2 | MAN AS A SYMBOL MAKER

Symbols are the basis of almost all thought. Examples of symbols are plentiful: letters and numbers, musical notes on a page, and spoken words. A symbol is something that stands for something else. Letters stand for speech sounds, numbers for "how many" objects, and notes on a scale for musical sounds. There is another use of the word *symbol*. The cross and the flag are symbols of this different kind. The cross stands for Christianity and all the ideas behind it. The flag stands for patriotism and loyalty. Such symbols stand for whole sets of human values and can be powerful influences on people. The symbols we will talk about here are much more limited in what they represent. Yet they too have changed the world for they are the basis of communication and knowledge. Let us look at some other examples of this kind of symbol.

About five thousand years ago the Egyptians devised symbols called *pictograms*—sketchy pictures—then developed them into their famous hieroglyphs, meaningful markings to record their historic events, business dealings, and other community affairs. The pictograms were pictures showing the object symbolized. For example, a man was represented by a humanlike figure. Later the pictures were used to represent syllables pronounced the same as the name of the object pictured. This principle was the basis of hieroglyphics and was the beginning of alphabets. The most powerful feature of any alphabet is that literally millions of words and syllables can be composed from a few symbols (there are already over 500,000 English words composed from the 26 letters in our alphabet and new ones are added almost every day). By 1200 B.C. alphabets had been developed by Phoenicians and Greeks.

Symbols are the basis of language. A language is a system of arranging symbols so that they convey meaning. (See Fig. 10–1a, b, c.) Usually, the system involves arranging the symbols in some order according to the so-called "syntax" of the language. The proper order of words in the English language (and most others, too) is vital to understanding. A misordered sentence can be quite a puzzle. For example, "country is all men of aid Now for the time their good come to the to" is a misordered, but familiar sentence. Though we don't usually think of numbers as a language, they have the properties of a language. The numerals $0,1,2,3 \ldots 9$ are symbols and their meaning depends upon their position in a sequence, that is, whether they appear in the units, tens, hundreds, or thousands position. Algebraic equations, too, are formed according to a language—the language of algebra. Over the centuries man has developed hundreds of spoken and written languages involving a great variety of symbols. They are the basis of his communication across the years, across thousands of miles, and across the gap between minds.

The utility of symbols depends upon physical ways to handle them—the pen to write, the book to record, movable type for printing, the printing press, the typewriter, the abacus, the cash register, and the copying machine. During the past 25 years man has developed that superb ultimate tool for handling symbols—the electronic digital computer. A computer can handle all manner of symbols, not just numbers as is often thought by the uninitiated. This ability depends upon using electric currents, switches, and magnetic fields to represent symbols.

We are so accustomed to particular symbols and the fixed meanings we attach to them, that we are inclined to forget that symbols are in themselves meaningless. A symbol means only what we agree it means, only what its users decide it shall mean.

Δεδομένων δύο σημείων A καί B, σταθερῶν ἤ κινουμένων πρὸς ἄλληλα, τὸ πρόβλημα τῆς ἀποκαταστάσεως μιᾶς ἠλεκτρικῆς ἐπικοινωνίας μεταξὺ των, δηλαδὴ τῆς μεταδόσεως ἑνὸς καθωρισμένου σήματος (πληροφορίας) ἀκουστικοῦ, ἠλεκτρομηχανικοῦ ἤ ὀπτικοῦ χαρακτῆρος δι᾽ ἐκμεταλλ εύσεως τῶν ἰδιοτήτων τοῦ ἠλεκτρομαγνητικοῦ πεδίου, ἀφορᾷ εἰς τὴν ἐπιστημονικὴν καὶ τεχνικοοικονονομικὴν ἐπιλογήν (ἐντὸς ὡρισμένων ὁρίων ἅτινα θέτει, ἐν πολλοῖς, ἡ φύσις τῆς πληροφορίας) τῶν κάτωθι παραμέτρων καὶ συντελεστῶν.

(a)

Fig. 10–1.
a) Greek text,
b) choreographic scores,
c) mathematical derivation.

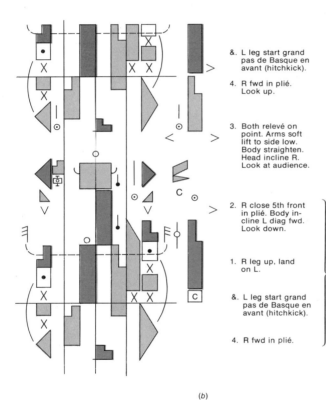

&. L leg start grand pas de Basque en avant (hitchkick).

4. R fwd in plié. Look up.

3. Both relevé on point. Arms soft lift to side low. Body straighten. Head incline R. Look at audience.

2. R close 5th front in plié. Body incline L diag fwd. Look down.

1. R leg up, land on L.

&. L leg start grand pas de Basque en avant (hitchkick).

4. R fwd in plié.

Arms circle in, from side high, up, to 5th low en avant, L arm ahead of R, ending with R wrist above L.

(b)

First Two Moments of T_1

$$E(T_1) = \sum_{i=1}^{N} \pi_i \int_0^\infty t\, dH_i(t) = \sum_{i=1}^{N} \pi_i i\, \theta_1 = \theta_1\, \alpha,$$

$$E(T_1^2) = \sum_{i=1}^{N} \pi_i \int_0^\infty t^2\, dH_i(t) = \sum_{i=1}^{N} \pi_i [i(\theta_2^2 - \theta_1^2) + i^2 \theta_1^2],$$

(c)

For example, the symbol "8" has no intrinsic meaning, no fixed connection with the "concept of eightness." Indeed the Romans represented this concept by an entirely different symbol: VIII. Likewise, the symbol X may mean an alphabetic letter, or a Roman ten, or an approaching road intersection, or the operation of multiplication. That we can make a symbol mean what we please gives us great power. It leaves us free to pick whatever symbols may best serve our purposes, and it means that a computing machine can at one instant act as an arithmetic calculator, and an instant later be a traffic simulator or an automated librarian. This transformation depends upon assigning different meanings to the internal symbols of the machine (the electric currents, magnetic fields, and switches) in the various cases.

Symbols and language as a means for communication depend on an agreement between the communicators on the meaning of symbols. Nowhere is this more evident than in the case of codes and ciphers where the sender of a communication arranges a special set of symbols so that only an intended person will be able to read the message. Let us examine this field of secrecy based upon symbols.

3 | CIPHERS AND CODES

The word *cipher* means *secret writing*, and more particularly, it means secret writing in which each letter of a "plain text" or original message is disguised by substituting a different letter (or numeral), or in which the letters of a message are rearranged to disguise its meaning. *Codes*, on the other hand, substitute whole words or phrases for a message. We will discuss only substitution ciphers here, because these are the most widely used today, and because they best illustrate that symbols are what we make them.

It is hard to overstate the importance of ciphers because they have determined the course of history. Many battles and even wars have been won or lost because of ciphers. The Battle of Midway in 1942 was won by the U.S. Navy's outnumbered forces because they knew where and when to attack the Japanese, having read the enemy's secret naval messages. Later in World War II, the key Japanese commander, Admiral Yamamoto, was killed when his plane was shot from the sky, again because a crucial message had been deciphered by the U.S. Navy. Indeed, the whole course of the war hinged upon the fact that, unknown to the Japanese, their cipher had been broken by U.S. Intelligence. Today every government uses ciphers and codes to conceal its own communications, and employs professional cryptographers to try to decrypt other governments' messages that fall into its hands. Significantly, computers and other machines play a vital role in these activities.

Ciphers are very old. An early Greek, Polybius, originated a now widely known cipher in which pairs of numbers are substituted for letters. According to his method, the letters are arranged in a square, and the rows and columns numbered as shown in Table 10–1. Note that there are only 25 letters in the square; *U* is used both for itself and *V*. To encrypt a letter, substitute the numbers of the row and column in which it appears. The first digit selects the column and the second digit represents the row. For example, for *E* substitute 15; for *M*, 33; for *W*, 52. The number along the top of the square must be put first.

	1	2	3	4	5
1	A	F	K	P	U
2	B	G	L	Q	W
3	C	H	M	R	X
4	D	I	N	S	Y
5	E	J	O	T	Z

Table 10–1.
A
Polybius Code
Table.

This simple substitution cipher can be effective in concealing a very short message, but it is easily broken by a person skilled in cryptography if the cipher is used for long messages, or used often. The cryptographer would notice how often each number appeared in a long sequence, and how often certain numbers appeared next to each other. These statistics, together with his knowledge of the language in which the message was written, would enable him after a few trials to fill in the square and so be able to read any message encrypted using that table. In English, for example, the most frequent letter is *e*, followed by *t*, *a*, *o*, *n*, *i*, and *r* in that order. Furthermore, *e* falls adjacent to more of the other letters, even rare ones such as *x*, *q*, and *z*, than any other letter. These facts alone make it easy in most cases to spot which cipher number stands for *e*. Many other similar relations lead quickly to the total solution.

The cryptographer's job can be made much more difficult by using a cipher system invented by a French nobleman, Blaise de Vigenère, in the sixteenth century. He, like the Greeks, arranged the letters of the alphabet in a table, today called a Vigenère Table. A sample is shown in Table 10–2. To encipher a message, we first pick a "key word," that is, any word that may come to mind; for example, *BROTHER*. Then we take the message and write the key word over it:

Table 10–2. The Vigenère Table.

Clear letters

	A	B	C	D	E	F	G	H	I	J	K	L	M	N	O	P	Q	R	S	T	U	V	W	X	Y	Z
A	a	b	c	d	e	f	g	h	i	j	k	l	m	n	o	p	q	r	s	t	u	v	w	x	y	z
B	b	c	d	e	f	g	h	i	j	k	l	m	n	o	p	q	r	s	t	u	v	w	x	y	z	a
C	c	d	e	f	g	h	i	j	k	l	m	n	o	p	q	r	s	t	u	v	w	x	y	z	a	b
D	d	e	f	g	h	i	j	k	l	m	n	o	p	q	r	s	t	u	v	w	x	y	z	a	b	c
E	e	f	g	h	i	j	k	l	m	n	o	p	q	r	s	t	u	v	w	x	y	z	a	b	c	d
F	f	g	h	i	j	k	l	m	n	o	p	q	r	s	t	u	v	w	x	y	z	a	b	c	d	e
G	g	h	i	j	k	l	m	n	o	p	q	r	s	t	u	v	w	x	y	z	a	b	c	d	e	f
H	h	i	j	k	l	m	n	o	p	q	r	s	t	u	v	w	x	y	z	a	b	c	d	e	f	g
I	i	j	k	l	m	n	o	p	q	r	s	t	u	v	w	x	y	z	a	b	c	d	e	f	g	h
J	j	k	l	m	n	o	p	q	r	s	t	u	v	w	x	y	z	a	b	c	d	e	f	g	h	i
K	k	l	m	n	o	p	q	r	s	t	u	v	w	x	y	z	a	b	c	d	e	f	g	h	i	j
L	l	m	n	o	p	q	r	s	t	u	v	w	x	y	z	a	b	c	d	e	f	g	h	i	j	k
M	m	n	o	p	q	r	s	t	u	v	w	x	y	z	a	b	c	d	e	f	g	h	i	j	k	l
N	n	o	p	q	r	s	t	u	v	w	x	y	z	a	b	c	d	e	f	g	h	i	j	k	l	m
O	o	p	q	r	s	t	u	v	w	x	y	z	a	b	c	d	e	f	g	h	i	j	k	l	m	n
P	p	q	r	s	t	u	v	w	x	y	z	a	b	c	d	e	f	g	h	i	j	k	l	m	n	o
Q	q	r	s	t	u	v	w	x	y	z	a	b	c	d	e	f	g	h	i	j	k	l	m	n	o	p
R	r	s	t	u	v	w	x	y	z	a	b	c	d	e	f	g	h	i	j	k	l	m	n	o	p	q
S	s	t	u	v	w	x	y	z	a	b	c	d	e	f	g	h	i	j	k	l	m	n	o	p	q	r
T	t	u	v	w	x	y	z	a	b	c	d	e	f	g	h	i	j	k	l	m	n	o	p	q	r	s
U	u	v	w	x	y	z	a	b	c	d	e	f	g	h	i	j	k	l	m	n	o	p	q	r	s	t
V	v	w	x	y	z	a	b	c	d	e	f	g	h	i	j	k	l	m	n	o	p	q	r	s	t	u
W	w	x	y	z	a	b	c	d	e	f	g	h	i	j	k	l	m	n	o	p	q	r	s	t	u	v
X	x	y	z	a	b	c	d	e	f	g	h	i	j	k	l	m	n	o	p	q	r	s	t	u	v	w
Y	y	z	a	b	c	d	e	f	g	h	i	j	k	l	m	n	o	p	q	r	s	t	u	v	w	x
Z	z	a	b	c	d	e	f	g	h	i	j	k	l	m	n	o	p	q	r	s	t	u	v	w	x	y

Keyword: *BROTH ERBROTHE* (key)

Message: *BEGIN TOMORROW* (plain text)

To encrypt the message, enter Table 10–2 at the column corresponding to the first key letter and at the row corresponding to the first message letter. The cipher letter is found at the intersection of the row and column. This process is then repeated for the entire message. The one above yields:

cvubu xfnffkva

Note that in this cipher the same message letter is not represented by the same cipher letter unless two instances of that letter fall under identical key letters. (For instance, the two *R*'s in *TOMORROW* are represented by *f* and *k*. However, the first two *O*'s in *TOMORROW* are both represented by *f* since they both happen to fall under *R*'s in the key.) Thus the relative frequency of letters and letter pairs in the language appears to be of little help to a person trying to decipher the message. The intended recipient, however, has both the table and the key word, and so can recover the message by reversing the encryption process.

For a long time, it was thought that a Vigenère cipher was undecipherable without knowledge of the key word, until someone realized that cipher letters spaced by the length of the key word follow the statistics of the clear language (since letters so spaced are encrypted using the same substitutions, as specified in one column of the Vigenère Table). This realization has led to the use of long key "words"—actually chaotic sequences of letters chosen at random or generated by some set of rules usually embodied in a machine. The longer the key word, the more secure the cipher. If the key is as long as the message and is never reused, the cipher cannot be broken without knowledge of the key itself. Systems of this kind are at the base of most countries' secret communications.

We complete our brief discussion of ciphers with the reminder that the letter and number symbols so familiar to us are only a convention agreed upon by people who desire to communicate with each other.

4 | SYMBOLS AND MACHINES

The importance of symbols in our civilization lies not in the symbols themselves but in what can be done with them. Numerals are important because they can be used to calculate according to the laws of addition, subtraction, multiplication, and division. Thereby numerals are vital to commerce, industry, government, and even the home budget. The ability to work with numerals (arithmetic) is as essential for making one's way in the world as are writing and reading. The ability to work with symbols is sometimes called symbol manipulation, and arithmetic is not the only example. For instance, the process of encrypting and decrypting a message according to the rules of the Vigenère Table involves manipulation of the letter symbols of the key, message, and table. Indeed, as we have said, symbols are tools of thought and knowledge. We now add that symbol manipulation is the act of making the tools effective.

The influence of symbols and their manipulation has grown to massive importance in our society as mechanized symbol manipulators, among them digital computers, have become widespread. Today the results of computer-assisted studies are at the base of important governmental and business decisions. There is much social controversy about the merits of such decision making. All agree, however, that computers, and indeed all mechanized symbol manipulators, should be aids to human thought, not replacements for it.

We often hear computers called *data processors* or *information processors*. What is being processed? Symbols, of course, but what form do these symbols take? Since a machine is a physical device, the symbols it handles must be in some physical form. In the

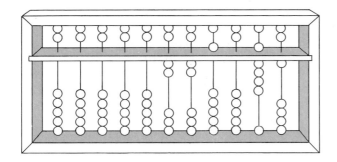

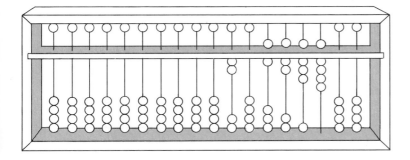

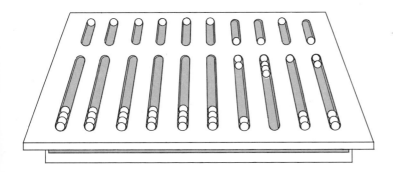

Fig. 10–2.
The abacus.
(V. X. Darnowski,
Computer Theory
and Uses, N.S.T.A.)

a) *Chinese abacus. In the simple Chinese abacus, beads move on rods. They are held in two compartments, one holding two beads and the other, five. Each bead in the smaller compartment is assigned a value of five and those in the larger are counted as one each. The beads have a counting value when pushed toward the counting board dividing the compartments. In this illustration, the beads are set for a total of 225,091.*

b) *Soroban, a Japanese abacus. The bead in the smaller compartment has a value of five; those in the larger each count one. The value of the number shown is 20,678,900.*

c) *Roman counting board. The Romans used a form of the abacus which was a counting table with grooves on its surface. Small stones, or calculi, were placed in the grooves to indicate values. The stones had the same values as the beads in the soroban. What is the number shown by the setting of stones in the illustration?*

abacus, numbers are represented by the position of numbered beads (Fig. 10–2). In mechanical adding machines numbers are often represented by the position of numbered cog wheels, as illustrated in Fig. 10–3.

In each case we assign a number to a bead or to a cog, then we represent the particular number in which we are interested by positioning the beads or the cogs. There is more than one way to apply this approach. For example, we might do the same thing with a row of electric lamps, assigning a number to each lamp. Then we could represent a particular number by switching particular lamps on or off.

In principle, this technique is used to represent numbers in a digital computer. It turns out, however, that it is less expensive,

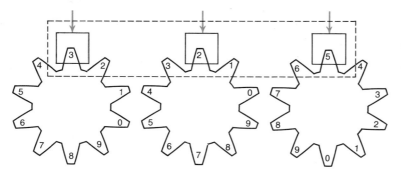

Fig. 10–3.
Cogged wheels
storing numeral 325.
(*T. H. Crowley,*
Understanding Computers,
McGraw-Hill Book
Company)

more reliable, and altogether better to build computers from elements which have only two states, *off* or *on*, for example. Some examples of two-state computer elements are shown in Fig. 10–4. A switch can be either off or on, electric current through a transistor can be arranged to be either off or on, and the magnetic field in an iron core (doughnut-shaped) can be in either one direction or the other. Such elements make up all electronic digital computers today.

How can numbers and letters be represented, or as we say coded, by such two-valued elements? (The word "code" here is used in a different sense from that in Section 3, Ciphers and Codes.) There are many different possibilities; let us examine two. First, let us label the two states of switches, transistors, or magnetic cores as 0 and 1. Then any representation in terms of 0 and 1 can in turn be represented by two-valued elements. The common English numbers and letters coded into 0 and 1 are shown in Table 10–3. Note that it takes six 1's and 0's to specify each letter or number so that each will have its own code. The 0's and 1's in such a code are often called "bits"—an abbreviation for binary digits. Thus the code of Table 10–3 is shown as a six-bit code. It is described as alphanumeric because it includes both alphabet letters and numbers.

Another way of coding numbers and letters into binary form is incorporated into the punched card, commonly called an *IBM*

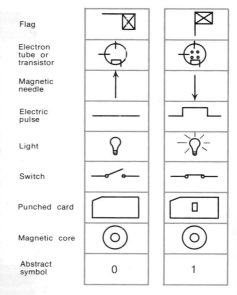

Flag		
Electron tube or transistor		
Magnetic needle		
Electric pulse		
Light		
Switch		
Punched card		
Magnetic core		
Abstract symbol	0	1

Fig. 10–4.
Binary
indicators.

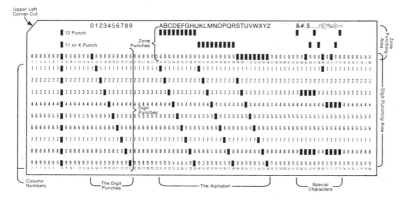

Fig. 10–5.
Typical punched
card.

card. Here the binary element is the existence of a hole in the card or not, usually called *punched* or *unpunched.* A typical punched card is shown in Fig. 10–5. The code is as follows. Each of the

0	000000	↑	100000	*Table 10–3.*	
1	000001	J	100001	*Six-bit*	
2	000010	K	100010	*alphanumeric*	
3	000011	L	100011	*code.*	
4	000100	M	100100		
5	000101	N	100101		
6	000110	O	100110		
7	000111	P	100111		
8	001000	Q	101000		
9	001001	R	101001		
[	001010	-	101010		
#	001011	$	101011		
@	001100	*	101100		
:	001101	)	101101		
>	001110	;	101110		
?	001111	'	101111		
(space)	010000	+	110000		
A	010001	/	110001		
B	010010	S	110010		
C	010011	T	110011		
D	010100	U	110100		
E	010101	V	110101		
F	010110	W	110110		
G	010111	X	110111		
H	011000	Y	111000		
I	011001	Z	111001		
&	011010	←	111010		
.	011011	,	111011		
]	011100	%	111100		
(	011101	=	111101		
<	011110	"	111110		
\	011111	!	111111		

columns represents one number or letter. Each column can have at most three punches, and there are twelve possible places for punches in each column. In effect, each character is represented by twelve bits vertically (if we equate a hole in the card with 1 and no hole with 0) in which only three bits at most can be 1. This particular code requires twelve bits rather than the six of Table 10–3 because of this limitation. Note that the character corresponding to the pattern of punches appears on the top row of the card. This printing is produced by a special device at the same time that the card is punched or later by a device which "reads" the pattern of holes and prints the appropriate character.

These examples of coding into bits are as vital to mechanized symbol manipulation as the Egyptian invention of hieroglyphics was to that civilization. The principle permits a computer to manipulate any symbol or sequence of symbols that can be coded into bits, ranging from English to Sanskrit, through shoe sales data to choreographic patterns and complex equations.

5 | DIGITAL COMPUTERS: WHAT GOES IN AND WHAT COMES OUT

We have seen that symbols can themselves be symbolized in physical forms. But what else is required so that computers can manipulate these forms to a useful purpose? The answer brings us to the heart of computing. The computer user must supply the machine not only with symbols in a form that can be "read" (punched cards, for example), he must also supply a set of instructions to the computer saying exactly what is to be done with the symbols. The set of instructions for any particular task is known as a *program*. To make a computer perform, the user must put symbols (sometimes called *data*) and a program into the machine, which then supplies the desired results automatically by processing the symbols according to the program. This relationship is illustrated in Fig. 10–6c. A typical example is the use of a computer to encipher automatically a message in the Vigenère cipher. Recall that there were three essentials in this process.

1. Vigenère Code Table
2. Key
3. Message

With these, we saw how to arrive at the encrypted message. A machine, or computer, to carry out this operation would have the same three inputs and one output as shown in Fig. 10–6a. In the earlier explanation of the Vigenère enciphering, instructions for the student were written out in English sentences. These told students how to manipulate the three inputs to obtain the output. In the computer case, instructions make up the *program*. So the complete diagram of a digital computer for encryption is as shown in Fig. 10–6b.

The words and sentences of the student explanation of encryption can be represented in the binary symbols and fed to the computer, but computers cannot follow the explanation written in English. Computers are similar to parrots in that they can repeat words and sentences, but they do not understand their meanings. Computers require instructions in special forms if they are to act on them. These forms themselves we will take up later. Even more vital is that computer instructions must call *only* for operations which the computer can perform.

Computers can perform a wide variety of operations. They can add, subtract, multiply, and divide numbers. They can compare

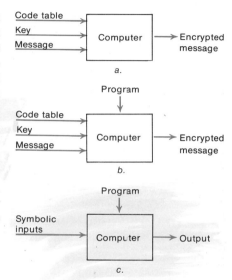

Fig. 10–6.
(*a*) *To encrypt a message by computer (or manually) requires three inputs.* (*b*) *Computer encryption also requires a program of instructions.* (*c*) *A digital computer to perform any task requires inputs and program to obtain output results.*

two symbols to determine if they are the same. They can do many other operations, depending upon the particular computer. Large, modern computers typically can perform 20 to 100 or more operations. Instructions for a particular computer must specify tasks which can be done with its operations.

As an example, let us assume that a certain computer can perform the following operations:

1. Compare two symbols to determine if they are the same.
2. Read symbols from an input.
3. Look up symbols in a table.
4. Print out symbols on a typewriter.
5. Determine when there are no more input symbols.

A program for this computer to encrypt a message is indicated in *flow diagram* form in Fig. 10–7. A flow diagram is a kind of block diagram which shows the sequential steps in a program. Flow diagrams are a convenient way of sketching out programs,* but they are not the form in which programs are ordinarily supplied to computers. They are useful in constructing programs. To read them, merely begin at the input, following the arrows through the boxes which indicate the successive operations to be performed. In Fig. 10–7, the first step is to read a message symbol. Next, the computer determines if the symbol is a letter or number. (It could be a space or a punctuation mark, but these are usually ignored in encryption.) If the symbol is a letter or number ("Yes"

*
A flow diagram is a convenient way of writing down an algorithm to perform some useful task. As explained in Chapter 2, an algorithm is a series of steps for carrying out a task or solving a problem. The Königsberg-bridge solution presented in Chapter 2 is an algorithm and could be represented by a flow chart. So could a cookbook recipe for making fudge.

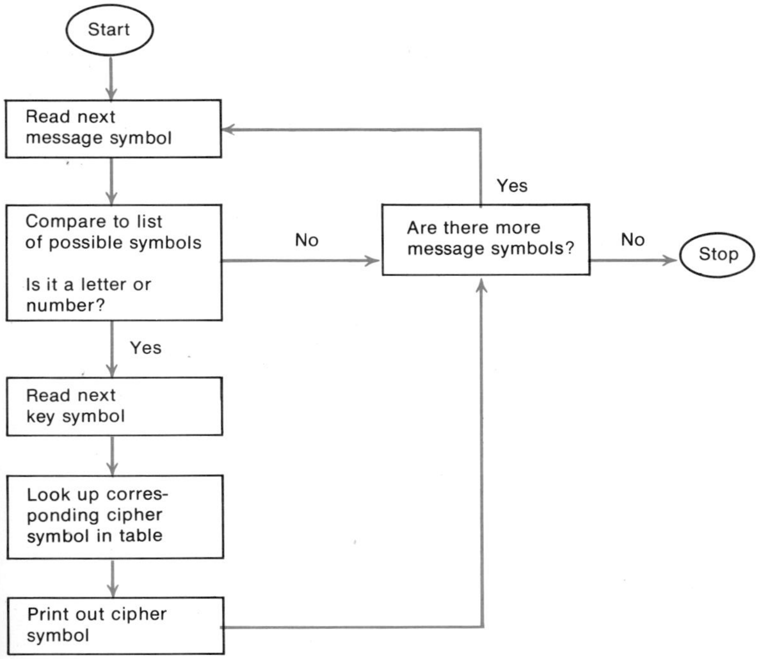

Fig. 10–7.
Flow diagram
of instructions for
enciphering
message.

branch of diagram), then the computer reads a key symbol. After obtaining both key and message symbols, the computer looks up the corresponding cipher letter in the Vigenère Table and prints it. The computer then reads the next message symbol and repeats the process. The enciphering steps are skipped if the symbol is not a letter or a number ("No" branch from second box). Before repeating the process, the computer checks to see if there are more message symbols; if not, it stops ("No" branch from right-hand box). Thus, the computer runs through the loop as many times as there are message symbols. When it finally stops, the ciphered message will have been printed on the output typewriter.

So with this program, written in an appropriate code, the computer will take in a sequence of message symbols and produce a sequence of ciphered symbols. In addition to the program, recall that both a Vigenère Table and a key word or a list of key symbols had to be supplied. As we said, they are usually supplied in binary form, for example using the code of Table 10–3. Also, the computer must have a copy of this code since it must separate letters and numbers from punctuation and other marks not to be enciphered. *All* of this information is essential for computer encryption.

Notice that in the flow diagram of Fig. 10–7, there are two kinds of boxes. One specifies an operation to be performed; these boxes have one input and one output. The other box asks a question and the next instruction to be performed depends upon the answer. These boxes have one input (or more) and two outputs, one for the "yes" answer, and one for "no." The use of these "decision" instructions enables the computer to process a long string of message symbols without having a new sequence of instructions

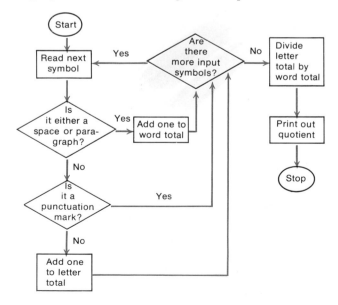

Fig. 10–8. Flow diagram for program to compute average word length in a text.

for each one. The computer "loops" through the same instructions as many times as necessary. To emphasize such decision points, the corresponding boxes are usually drawn in diamond form rather than square as in Fig. 10–7.

Let us examine the program flow diagram for another example: one concerning the analysis of English text. Suppose we have the text of a book in punched card form including all the punctuation, space, and paragraph marks. One step in analyzing the author's writing style might be to find the average word length in his book. To do this, a computer would have to count the number of words and the number of letters, then divide to obtain the average. A corresponding flow diagram is shown in Fig. 10–8. Here the decision boxes are drawn as diamonds. The program has the form of a loop so that the same instructions are used over and over again to process the string of input symbols. The output from this program is a single number, the average word length. (To obtain other results, such as the length of the longest word, additional program steps would be required.) The same basic computer operations listed earlier for computer enciphering are sufficient for this task.

Again, we should emphasize that a flow diagram is not a program, but can be translated into one if the operations in the boxes and diamonds can be carried out by a computer. It is not possible, for example, today for a computer to "recognize the identity of a person from his voice." We just don't know what properties of a voice are unique to an individual. So a flow diagram containing a box with this instruction could not be translated into a program.

On the other hand, neither do boxes have to contain instructions that can be completed in one computer step. For example, many small computers are not able to multiply in one step; however, they can usually add. To multiply on such a computer requires several steps of addition. A flow diagram for multiplication of two numbers by successive additions (known as The Idiot Multiply) is shown in Fig. 10–9. This program merely adds the larger of the two numbers together a number of times equal to the smaller number. It does this by counting the number of times it adds until that number equals the smaller number. This is not the most efficient way to multiply by adding. (Neither is it the most inefficient, but it almost is!) It does show, however, that a "multiply" box in a flow diagram can be translated into a program by incorporating multiple steps as illustrated in Fig. 10–9, even though the computer may not be able to multiply directly.

A program for a computer is a sequence of steps each of which the computer can carry out, or *execute* as we commonly say. The sequence of steps, if correctly formulated, leads to the desired result, provided that the necessary data, tables, and definitions also have been fed to the computer. Planning of the program steps and providing the additional information are vital to using a computer.

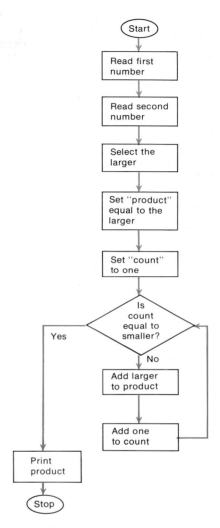

Fig. 10–9.
The Idiot
Multiply.

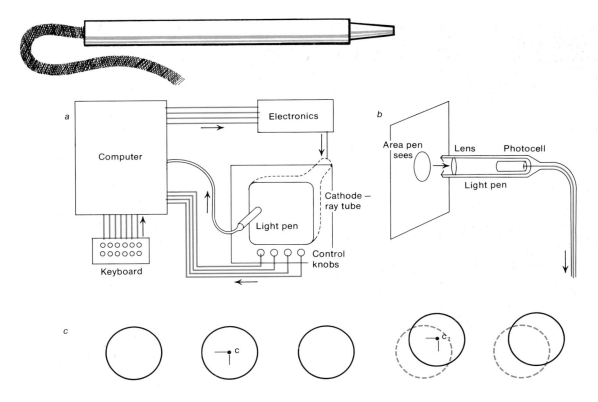

6 | DIGITAL COMPUTERS: WHAT'S INSIDE?

Input and Output

From the examples of the last section, we can see what elements are necessary inside a digital computer. Symbols coded into binary form are suitable for computer processing, as was said earlier. We saw how such symbols were recorded on punched cards. Punched-card reading equipment is commonly used to transfer symbolic information from cards into a computer. Readers of this kind sense the positions of holes punched in the cards. This same principle is used for reading punched paper tape, and magnetic sensing is used to read symbols from magnetic tape. Electric typewriter keyboards can also be used to provide symbolic input to computers. Each press of a key generates an electrical signal which is sensed by the computer. (Computers can operate the typewriter to produce printed output.) There are many other input techniques, but they all involve transfer of symbols from outside to inside computers. Similarly, there are many output techniques: printers, typewriters, and even pictures displayed on cathode-ray tubes. Some modern input and output terminals are illustrated in Fig. 10–10.

When symbols are fed into a computer, they go into a *memory* where they are retained for future reference. When symbols are

Fig. 10–10 a.
The light pen, top, is another input device for drawings. The pen holds a photocell which responds to a point of light displayed by a computer. The light pen allows the user to point at a piece of a drawing being displayed from memory as in b. The user can erase, change, or move a figure and he can also draw on the screen with the aid of a program which follows the end of the pen as shown in c. The program then computes the position of the pen's center and records that point in memory. By recording a series of such points, the computer can take in a line or curve drawn by the user. (Sol Mednick)

*Fig. 10–10 b.
The majority
of computer input
today is by punched
cards through a card reader.
Holes are sensed by light and
photo cells or electrically. The
reader pictured at the top can
handle 1000 cards per minute. A
newer input technique is the Rand
table, at the bottom, which accepts
direct input of hand drawings. Its
surface is a mylar plastic sheet coated
with 1024 copper lines in each of the
X and Y directions. Each line
carries a unique repeated pattern
of electrical pulses. These are
picked up by the electrical
stylus as it is being used to
draw. The computer can
tell where the stylus point
is by the dual pattern of
pulses it receives. (Top
photo Sol Mednick; bottom
photo Control
Data Corporation)*

Fig. 10–10 c.
*Simultaneous
input and output
provides a young student
with immediate feedback. He
is presented with a problem above
and responds below, selecting one of two
possible answers by pointing with a
light pen according to instructions
provided over his earphones. The
computer then informs him
"correct" or "incorrect." The
next problem is selected on
the basis of his reply.
(Black Star Publishing
Company, Inc.)*

Fig. 10–10 d. Most computer output is by means of a printer or typewriter. However, computers can print for the blind through a Braille printer. (Sol Mednick)

taken out of a computer, they are transferred from memory to the output device. These flows of data, reference tables, and other information are indicated in the block diagram of Fig. 10–11a which is the beginning of a general diagram for the major elements of any digital computer.

In modern computers, input and output are often used together. The user may, for example, want to look at a picture of the data in the computer's memory at the same time he is entering additional data. This kind of arrangement is very convenient when a computer is used to edit documents, such as letters and manuscripts. The document is typed into the computer memory in a preliminary or first-draft form. The user (perhaps a secretary) then might edit this version, deleting words, adding phrases or even paragraphs, correcting spelling, to bring the text to a final form. In doing this, it is especially convenient for the user to view the text on a TV-like cathode-ray tube so that he can indicate the changes directly and actually see the proposed final version immediately. The final copy can be produced automatically by an output typewriter (see Fig. 10–11b). The TV-like display of the text is an example of computer output where the symbols are kept in memory during the process of output. Even after the final version has been typed, a copy may be retained in the memory for future reference or duplication.

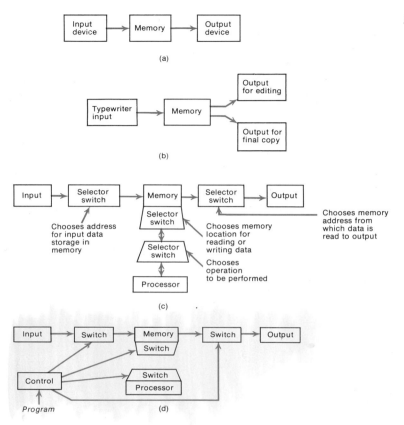

Fig. 10–11.

Memory

Computer memory itself can be of several different varieties, a basic difference being how rapidly data can be put in and taken out of these memories. The fastest, but also the most expensive today is transistor ("semi-conductor") memory. Bits of information can be written or read in about 50×10^{-9} second (50 nanoseconds). Somewhat slower (500-1000 nanoseconds, equivalently stated 0.5 to 1.0 microsecond) is magnetic core memory composed of small iron-powder rings (the iron powder is pressed into solid form) that can be magnetized in one direction or the other to store bits. Magnetic discs and tapes and photographic memories are still less expensive and slower, but come in much larger sizes for storing massive amounts of information. Table 10–4 gives an idea of these memory speeds and sizes. Often a computer has several kinds of memory, fast memory for often-used items, slow memory for reference files used only infrequently.

Perhaps the most important aspect of memory, however, is how it is organized. Each bit is stored separately in a core, a transistor circuit, or as a magnetized spot on a tape, for example.

Memory	Access time to any stored bit (seconds)	Typical size		Cost ($ per bit)	Table 10–4.
		Bits	Typed pages		
Transistor (solid state)	50×10^{-9}	128,000	8	1	
Typical magnetic cores	$0.5\text{–}1.0 \times 10^{-6}$	5×10^6	250	0.1	
Magnetic drums	2×10^{-6} to 30×10^{-3}	130×10^6	800	0.01	
Magnetic disc	10×10^{-6} to 0.5	10^9	5000	0.001	
Photographic	0.25 to 10	1000×10^9	A library of books	0.00001	

However, several bits are usually recorded or read as a unit. One such unit is called a *byte*. A byte consists of 8 bits. One letter, number, or punctuation mark can be stored in a byte using the code of Table 10–3. (Since this is a 6-bit code, there will be 2 bits left unused in a byte size unit. These are sometimes used to store check bits.) Several bytes (from 2 to 12) are usually stored together forming what is called a *word*. Memories are arranged so that many words can be stored without interfering with each other but at the same time so that each can be found easily.

This arrangement is created by organizing the memory some-what like a hotel letter box (see Fig. 10–12a). An array of boxes is constructed, each with an *address*. In the letter-box case, each "pigeon hole" has an address which gives first the floor number, then the room number. Some of these addresses are shown point-ing to their corresponding pigeon holes. Note that the addresses only say *where* letters or notes are to be found; the address says nothing about the contents of the pigeon hole.

A magnetic-core memory is similarly arranged. Cores are strung on a grid of wires as illustrated in Fig. 10–12b. They are addressed just as the hotel mailbox is. In writing bits into the memory or reading bits from the memory, a particular core is selected by putting electrical signals on the X and Y wires which cross at that core. For example, to select core 32, a signal is sent along X-wire number 3 and Y-wire number 2. The state of core 32 is thereby either sensed to find out if a "one" or "zero" is stored, or mag-netized to the correct direction to store a "one" or "zero." These processes are illustrated in Fig. 10–13. Note that wires in addition to those shown in Fig. 10–12b are required. Typical core memory planes hold many more than 64 bits; a 16,384-bit memory plane is shown in Fig. 10–14.

Notice, however, that a memory plane can hold only one bit at each location. To store a byte or a word in one location, we need additional cores. They are provided by stacking identical core planes as indicated in Fig. 10–15. The four memory planes are

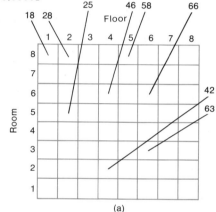

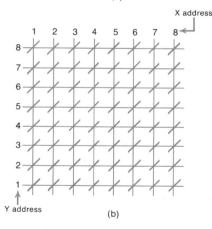

Fig. 10–12.
(a) Mail boxes for a hotel with eight floors and eight rooms on each floor. (b) A 64-bit core memory plane, addressed as the hotel mail boxes.

444

1 Store

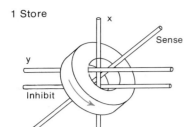

x
Sense
y
Inhibit
0 Stored

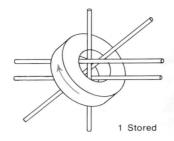

1 Stored

2 Read

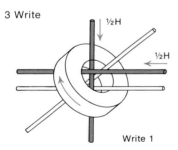

No pulse
½H
½H
Read 0

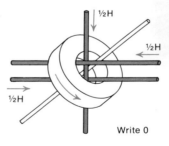

Pulse
½H
½H
Read 1

3 Write

½H
½H
Write 1

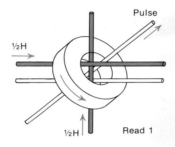

½H
½H
½H
Write 0

4 Store

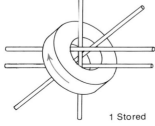

1 Stored

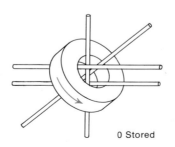

0 Stored

*Fig. 10–13.
Direction of
magnetization
shown after current
switches core.*

*Operation of a
magnetic core memory
requires changing the
direction of magnetization
in a core. One direction
represents 0, the other 1.
Currents to drive the core are
carried on wires (X and Y) specified
by its address. Each wire carries only
half the current needed to change the
magnetization direction. In reading, the
currents are directed so that only a
core storing 1 is reversed. This
reversal produces a pulse in the
sense wire. Cores storing 0
do not produce a pulse.
In writing, the current
flows so as to reverse
magnetization from its 0
position. To write a
zero, the same
currents plus an
"inhibit" current
are used. The
core then will
remain in its
0 state.*

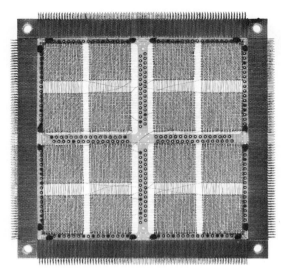 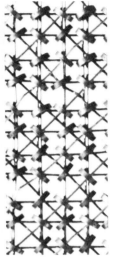

Fig. 10–14.
The magnetic core
memory plane shown at
left contains 16,384 individual
ferrite rings, shown in detail at right.
They are each one-fiftieth of an inch in
diameter and can be switched from one
direction of magnetization to the other
in about one microsecond. The plane
as shown is about ¾ size. A full
computer memory might
consist of 32 planes to form
a stack organized to hold
16,384 words of 32 bits
each. (William Vandivert)

stacked so that a four-bit word is stored in every addressed location. Each bit of the word is stored on a different memory plane.
The cores on one plane are not connected electrically with those on
other planes. When writing or reading a word, all four bits are
selected simultaneously by putting signals on the X and Y leads
with the same address on every plane. This action can write a
four-bit word into one location. In reading, the four sense wires
(see Fig. 10–15), one from each plane, register simultaneously the
four bits (ones or zeroes) when the X & Y leads are activated.
This action reads out a four-bit word which appears on the four
sense wires. Later, in Chapter 14, we will see in detail how this
memory action takes place. For now, it is sufficient to understand
that memories are organized as mailboxes; each pigeon hole has
an address and can store, or remember, several bits. This collection
of bits is known as a word or sometimes a byte if each location
holds eight bits. Other forms of memory (see Table 10–4) are also
organized in this way, but their read and write operations are
different and depend upon the particular mechanisms employed in
those memories. It is not important that we understand these in
detail here.

Processor

As Fig. 10–11 indicates, we have now discussed three of the
five principal units inside digital computers. The next one to be
examined is the processor. It is in the processor that the various
computer operations are performed. Symbols supplied from the
computer memory, for example, can be added, subtracted, multiplied, or compared there. The processor in a modern computer
can perform many other operations as well. (The part of the

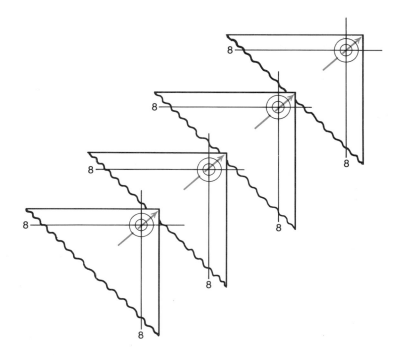

Fig. 10–15.
Four core planes stacked so that a four-bit word can be stored. Each core has the same address, 88, on its own plane. So when the 8-8 leads for all planes are pulsed at the same time, the four bits appear simultaneously on the four sense wires, one from each plane. A four-bit word can be written into the memory similarly.

processor that adds, subtracts, multiplies, and divides is known as the *arithmetic* unit.) These are carried out by special electronic circuits, each designed for a particular task. The processor operates on symbols supplied from the memory. For example, suppose the processor is to add a number to a second number. The sequence of events is illustrated in Fig. 10–16. After the two numbers, in symbolic form, have been put into memory, then the first number is brought into the processor and held in an *accumulator register*, which is merely a group of electronic circuits which can store the number temporarily. The second number is brought into the same accumulator through a circuit which produces the sum of the two in that register. Finally, the sum is put back into memory and printed out. A third number could have been added to the sum and actually a long string of numbers can be added in this way Notice that particular words are selected from the memory and they go to specific processor circuits which carry out the desired operations. So, in the computer block diagram of Fig. 10–11c, we have added the processor and shown it communicating with memory through selector switches. The one on the memory must be set to the proper memory address to obtain the desired word. The one on the processor must be set so that the word from memory goes to the circuit for carrying out the desired operation. In Fig. 10–11c, we have shown the input-output connection with switches also, since input data must be stored at specific memory addresses, and similarly output data must come from certain addresses.

Control: Instructions Run Computers Automatically

How are these switches set and how is the appropriate processor operation determined? From the program, of course. The instructions in the program must specify the addresses and the operations. The actual switching is performed through the computer *control* which is the final box in our block diagram. Control makes the operation of the computer automatic. It must take instructions, interpret them, and see that they are carried out by setting switches and causing actions to proceed. This is indicated in Fig. 10–11d, and with the control unit our computer block diagram is complete.

Let us look in detail a bit more at the automatic cycle of a computer. The heart of the control unit is the *instruction* decoder. Its job is to translate program instructions into actions. Each instruction must specify an operation to be performed, and it must also specify a memory address where the appropriate data are to be found or are to be stored. So, instructions have two parts:

Specifies *which* operation is to be performed	Specifies *where* in memory to find or to store data

OPERATION CODE	ADDRESS

Just as in the case of symbols, these pieces of information are coded in binary form (the actual code for a particular computer is designed in by the manufacturer). In the following discussion, however, we will use decimal numbers for convenience, since binary numbers are difficult for people to read and remember. Operation codes needed for adding two numbers are:

Operation Code	Operation
0	Take number from input and store it in memory.
1	Clear the accumulator of any previous numbers and transfer a number to it from memory.
2	Transfer number from memory to accumulator and add it to number already there.
6	Transfer accumulator contents to memory.
5	Transfer number from memory to output.
9	Halt the computer.

These codes plus appropriate memory addresses can form program instructions for a simple computer which, in spite of its simplified form, can illustrate all the major features of real computers. A cardboard model of this computer has been constructed, so it is

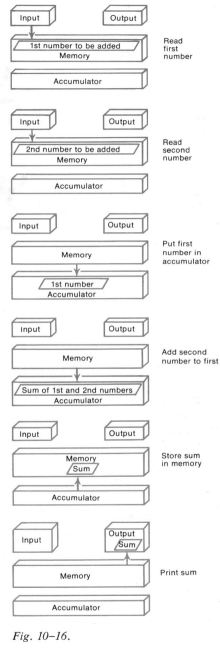

Fig. 10–16.

known as CARDIAC©. It is sometimes used to demonstrate the
basic actions of a digital computer.* This computer model has 100
memory slots numbered 0 to 99. Each slot can store three decimal
digits. So, instructions for CARDIAC take the form of three-digit
numbers. The first digit is the op code, the second two digits are
address. Assume that two numbers are to be added according to
the plan of Fig. 10–16, and that the two numbers are to be stored
in memory cells 11 and 12. The sum is to be put in cell 13. Then the
program for adding is:

The cardboard version of CARDIAC may
be available to you using this book. If so,
you can use CARDIAC to carry through
programs step-by-step, thereby seeing directly what they do. All programs in this
book use the CARDIAC operations.

011	Take first number from input and store in cell 11.
012	Take second number from input and store in cell 12.
111	Clear accumulator and transfer contents of cell 11 to it.
212	Add contents of cell 12 to contents of accumulator.
613	Store contents of accumulator in cell 13.
513	Print contents of cell 13.
900	Halt.

Through the instruction decoder, one after another of these instructions, beginning with 011, is translated into action.

The instruction decoder breaks each instruction into its two
parts. The address part controls the memory switches to connect
the designated memory cell. The op code part controls the other
switches so that the input, output, or processor register (whichever
is specified by the op code) is connected to the memory. With the
correct memory cell connected to the appropriate register, the
operation called for by the instruction can be executed. So, as
one instruction after another flows through the instruction register,
the computer performs the task specified by the program (reading
two numbers, adding them, and printing out the sum in this case).

Instructions for CARDIAC, like instructions for all computers,
must be in symbolic form and supplied in correct sequence to the
instruction decoder. To do this, we might punch the three-digit
decimal instructions (in binary form) on cards and have the computer read them one after another using a reader attached to the
instruction decoder. However, this technique is slow and not very
flexible. (For example, using the same instructions many times in
a program loop would be difficult.) It is more convenient to store
the instructions in a memory before beginning to use them. That
way instructions can be read rapidly and as many times each as
needed. As we shall see later, it is important, too, for the computer
to be able to alter its own instructions.

In CARDIAC, we will put instructions into the same memory
as the data, but of course into different cells to keep them separate.
Instructions will be brought from the memory to the instruction
register, a register which holds the three-digit number while it is

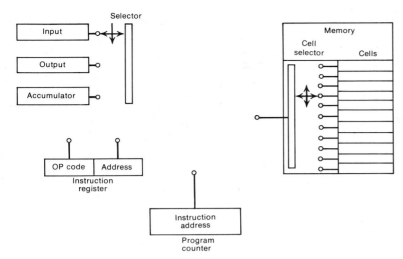

Fig. 10–17. *Both data and instructions are stored in computer memory. In its automatic operation, the address of an instruction is put in the program counter. The instruction stored at that address is brought into the instruction register for decoding and execution.*

being translated into action by the instruction decoder (see Fig. 10–17). To bring an instruction into the instruction register, control must know the memory address where the instruction is to be found. The address of the first instruction in a program is supplied to the control unit. This address is put into a *program counter* (sometimes called an *instruction counter*). The address in the counter is changed by control so that it always contains the address of the *next* instruction to be brought into the instruction decoder for execution.

Note that the *addresses of instructions* appear in the *program counter. Addresses of data words* appear in the right-hand part of the *instruction register.* In this way, instructions and data are each used for their own purposes. Since instructions and data are both three-digit numbers in CARDIAC, they cannot be separated merely by inspecting the contents of the memory. The only distinction between them is how they are used in the computer's automatic cycle. Sometimes a program error or a computer error can cause a data word to be treated as an instruction, or the other way around. Then strange results can flow from the computer. Fortunately, such results are typically so strange and meaningless that the user knows something has gone wrong. He can then correct the program or repair the computer, whichever is appropriate.

The principle of putting the program into memory and using control circuitry to find and read the instructions is the essential feature of *stored-program* computers. Stored-program computers are tremendously flexible. They can be adapted to tasks in business, government, literature and art, science, and an almost unlimited number of other fields. This flexibility makes stored-program computers the basis of the so-called "computer revolution."

We have now discussed briefly the essential features of a stored-program computer. Later we shall see further how the insides of such computers actually work. For now, it is important to recognize that the automatic operation of stored-program computers is based on a repeated cycle. This cycle proceeds over and over until the program has been completed. One cycle is as follows:

1. An instruction is brought ("fetched") into the instruction register from the memory address specified by the program counter.
2. The address in the counter is changed to that of the next instruction (usually it is in the next larger address so the counter address is increased by one).
3. The instruction in the instruction register is carried out through the instruction decoder.

The cycle then repeats starting at the first step.

The steps in this cycle are illustrated in Fig. 10–18. We have now completed our review of the essential units and operations in digital computers. It is rather surprising that the seemingly miraculous computer feats described so often in newspapers and magazines are based on such a repeated cycle of simple events. Yet, this fact is probably the most important one concerning computers. If a task is to be done by a computer, that task must be reduced to a sequence of simple steps, steps which can be done by a computer using electronic circuitry.* Therefore, a task that has been done by computer represents a landmark, one that provides a firm foundation on which to build further technique and knowledge. Thus, the computer provides a tool for men and women to test their thinking and to find out just where their ideas can lead. The impact of this ability on human thought and societies is the most important influence of computers.

Even before computers, wise men understood the importance of making ideas precise. The famous Elizabethan Francis Bacon once said: "Writing maketh an exact man." By this he meant that to write accurately we must think accurately. Bacon's words apply even more to computer programming. If a writer makes a mistake in his manuscript, the human reader usually automatically exercises judgment, overrides the error, and makes his own interpretation of what the writer intended to say. Sometimes a reader correctly interprets what the writer intended and the error causes no harm. Often, more often than we think, the reader comes up with a different meaning. Much misunderstanding can result. In contrast, the computer can make no interpretation. It can only act upon what the programmer has written. Sloppy thinking by the programmer soon shows up in the computed results. The consequences of sloppy thinking are brought home quickly. "Computer programming maketh an exact man."

*
The famous English mathematician A. M. Turing, in the late 1930's proved an important fact. In effect he showed that the range of what computers can do is not limited by the machine itself but only by man's ability to reduce important tasks to sets of simple instructions. Turing actually proved that any task which can be done at all by any computer can be done by a simple idealized computer. These elementary computers are known today as Turing machines.

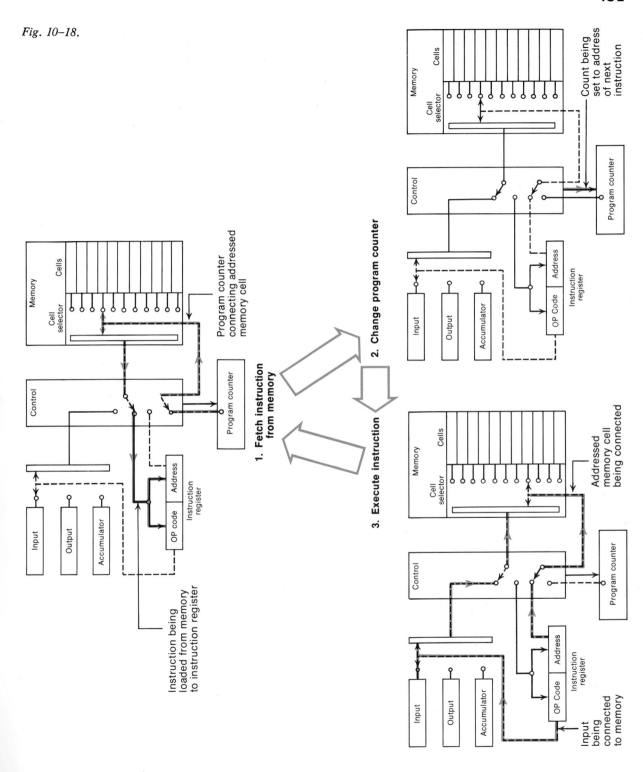

Fig. 10–18.

Memory

Cells

Cell selector

Count being set to address of next instruction

Control

Program counter

Input

Output

Accumulator

OP Code | Address

Instruction register

2. Change program counter

Memory

Cells

Cell selector

Program counter connecting addressed memory cell

Control

Program counter

Input

Output

Accumulator

OP code | Address

Instruction register

Instruction being loaded from memory to instruction register

1. Fetch instruction from memory

3. Execute instruction

Memory

Cells

Cell selector

Addressed memory cell being connected

Control

Program counter

Input

Output

Accumulator

OP Code | Address

Instruction register

Input being connected to memory

7 | PAST, PRESENT, & FUTURE OF THE COMPUTER

Where did computers come from? Probably they began with the idea of using mechanical devices to aid man in using symbols. Even to early man, fingers and toes, sticks and stones, must have seemed an inadequate way of keeping count of his animals and other belongings. "Instead of a row of stones arranged on the ground, couldn't one perhaps use beads strung on a stick?" Applying the idea, someone invented the abacus, the first of all computers. Though its origin is lost in the mists of antiquity, the abacus is known to have been in use in China some 3000 years ago.

Counting and keeping inventories were motivations for the invention of early computers. Another was the necessity for early man to plant his crops at the right time of year. This led to astronomical computers to help predict the seasons. Perhaps the most famous of these is Stonehenge, a massive grouping of great stones near Salisbury, England, which dates from about 1500 B.C. It is thought to be an astronomical observatory laid out to allow "priests" to predict the seasons, and solar and lunar eclipses as well.

Travel also was a motivation for inventing elementary computers. The Greeks, whose enterprising spirit led them to explore so many other fields, appear to have built computers to aid their sea transport. In 1900, an elaborate clockwork sort of machine was found by divers on an ancient Greek shipwreck near an Aegean island. The device dated from about 65 B.C., and it is apparently some kind of navigational computer.

However, for the ideas and devices that led directly to the modern digital computer, mankind had to wait until the beginning of the seventeenth century. About 1617, John Napier (inventor of logarithms) developed a device for multiplication. This was followed shortly afterward by the invention of the first slide rule, a device that multiplies by adding distances. In 1642, Blaise Pascal built a stylus-operated adding machine with numbered wheels geared together, somewhat as in automobile mileage indicators, so that the "carry" digit from one column could be automatically added to the next (see Fig. 10–19). Samuel Morland had a machine, based on the same principle, which could subtract as well as add. In 1694 Gottfried Wilhelm von Leibnitz introduced still another machine that could also multiply (by repeated addition), divide, and extract roots.

From an entirely different kind of endeavor came an idea that was to be critical in the development of modern computers. In the seventeenth century the Frenchman Jacquard developed a loom that could automatically weave cloth in a pattern, which, in turn, was specified by information from punched cards fed to the loom one after another. It occurred to Charles Babbage (Fig. 10–20), a mathematics professor at Cambridge University, that if

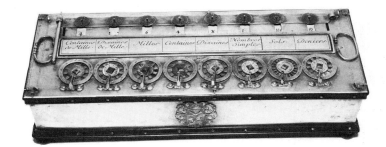

Fig. 10–19.
Pascal computer.
(Photo of original
in IBM Corporation
Antique Calculator
Collection.)

Fig. 10–20.
Portrait of
Charles Babbage.
(Photo of original
in IBM Corporation
Antique Calculator
Collection.)

the pattern of holes in punched cards could control a loom, they could also control an arithmetic calculator, causing it to carry out a planned sequence of arithmetical operations in a long computation.

In 1820 Babbage applied his theories by building a small model of a "difference engine" that evaluated polynomials (algebraic expressions consisting of two or more terms) by a method using differences between numbers. About 1833 he proposed to build an even more elaborate "analytical engine" that was enormously prophetic since it embodied all the essential features of modern electronic computers. Thus, it was to have a "store" or memory unit to hold numbers as well as the results of doing

*Fig. 10–21.
Babbage
difference machine.
(Photo of replica
in IBM Corporation
Antique Calculator
Collection.)*

arithmetic on these numbers; a "mill," or arithmetic unit, to perform the succession of steps necessary for adding, subtracting, multiplying, or dividing; and a "control" to translate and carry out instructions from the punched cards. Babbage's computer (Fig. 10–21) was not designed specifically to manipulate symbols other than numbers, but it had that basic capability.

Babbage attracted a most unlikely supporter—Lord Byron's daughter, Lady Lovelace. This titled lady, who might have been expected to lead a life of thoughtless leisure, became instead a fascinated student of mathematics. When Lady Lovelace first saw Babbage's computer, she understood what he was trying to do, and she became an enthusiastic promoter of his ambition to construct a machine that "weaves algebraic patterns just as the Jacquard loom weaves flowers and leaves." Together, she and Babbage conceived wild fund-raising schemes. They did in fact obtain some financial backing from the British government. But it was all in vain: the necessary gears and other mechanisms proved to be beyond the technology of the time. Only a very few artisans of Babbage's days could have made them accurately enough, and they appear not to have been available. Babbage died a frustrated man. For close to a century his great work would lie forgotten.

Some years later, however, machines for counting were constructed. The Hollerith tabulating machine (Fig. 10–22) was built in 1890, and was used to count census data. In fact, the Census Bureau of the United States government has provided a strong impetus for the development of computing machinery.

While we may marvel at the ingenuity of man in developing the modern digital computer, one may also wonder why it was not developed sooner. As early as 1900, telephone engineers were using the on-off properties of relay circuits in automatic dial telephony. Over the years they invented and developed circuits that could register, count, and remember, even circuits that made decisions. But nobody applied them to symbol manipulation.

Then in the thirties, possibly because of a growing need for mechanical aids of much greater capability than the desk comptometer, certain individuals began to explore the possibilities of making computers with electric circuits instead of with mechanical devices such as Babbage tried to use.

In the late 1930's, Howard Aiken of Harvard University and George Stibitz of Bell Telephone Laboratories opened the way by developing automatic calculators using telephone and other relays. The performance was impressive. During World War II, one such computer demonstrated the potential of automatic calculators by performing complex ballistic calculations. More important, these pioneer relay computers were milestones, for they were the first practical steps in the development of the modern computer.

Students of the computer at once saw that the on-off, two-valued action of telephone switches could be duplicated by electron tubes for which there existed a highly developed science and art.

Fig. 10–22.
Hollerith
tabulator,
sorter, and punch.
(*Photo of originals
in IBM Corporation
Antique Calculator
Collection.*)

At the University of Pennsylvania John Mauchly, a professor of physics, and J. Presper Eckert, Jr., a graduate student, conceived and built ENIAC (Electronic Numerical Integrator and Computer), the first automatic computer based on electronic rather than mechanical or electromechanical technology.

Completed in 1945, ENIAC could add or multiply two numbers in a fraction of a second. But it also demanded a tremendous effort to keep it working. For ENIAC's 18,000 electron tubes not only were subject to failure, but also generated an enormous amount of heat that had to be disposed of by an elaborate system of fans and air conditioning. Although ENIAC did not actually melt, it was often on the point of becoming a furnace.

Astonishing as was its performance, ENIAC had a serious limitation. For each specific problem, a plug board had to be rewired, extensively and tediously. The connections on this board were the instructions for the computation. To overcome this obstacle, the mathematician John von Neumann conceived a brilliant solution: the stored program. In 1945 and 1946 von Neumann and his colleagues proposed a computer design that could obey a program stored in its own memory and that could be changed without rewiring. The first working stored-program computers were demonstrated at about the same time in both the United States and Britain, in 1949. The first commercial electronic computer of this kind was the Eckert-Mauchly UNIVAC, put on the market in 1950.

The stored-program concept is critically important because it means that a series of instructions can be stored in the computer memory in the same way as numerical data. A program of instructions can, therefore, make modifications in another program, or even in another part of itself. A program can then be designed to modify itself automatically during the course of a computation, in a way that depends upon the intermediate results that cannot be predicted before the computer starts to work on the problem. The stored-program idea also means that the same computer can be used at various times on many different classes of problems without changing the wiring of the computer itself. Thus through the stored-program concept the computer became a truly "general-purpose" device.

At first, as a practical instrument, the digital computer continued to have several strikes against it. With its 18,000 electron tubes ENIAC weighed no less than thirty tons and demanded 15,000 square feet of floor space—a bit impractical to move or even merely to have around. Then came the transistor and the magnetic core.

Invented in 1947, the transistor (Fig. 10–23) can perform the on-off action of both relays and electron tubes. At the same time it is only a small fraction of the size of an electron tube, requires

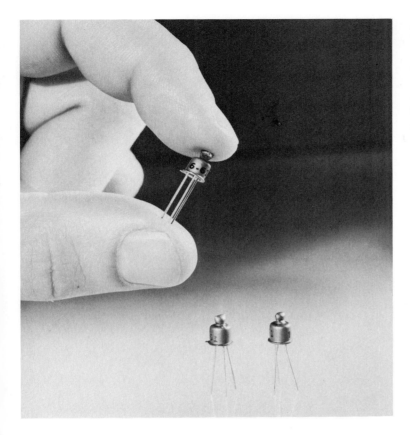

Fig. 10–23.
Transistor
in protective
casing. Transistors are
used as very fast switches
in the modern computer.
(*Courtesy of Bell*
Laboratories)

miniscule power, generates almost no heat, is more reliable and better able to withstand mechanical shock, and it can switch much faster.

At about the same time, J. W. Forrester and his colleagues at MIT developed the magnetic memory core, a minute doughnut-shaped "ferrite" ring. About the size of a period on this page, a ring can store one bit. Until this invention, computers had been supplied with memory in the form of electron tubes. Now, such tiny cores, compactly woven into a sort of cloth or mesh made of small wires, provide large computer memories.

To get an idea of the advance in computer technology brought about by the transistor and magnetic core, consider that a modern all-purpose digital computer made with relays would occupy the space of a large high school; made with electron tubes, it would occupy a gymnasium; made with transistors and magnetic memories, an all-purpose digital computer can easily be accommodated in a small single room.

More recent developments make it possible to create and connect together on a single tiny "chip" of silicon, measuring only a

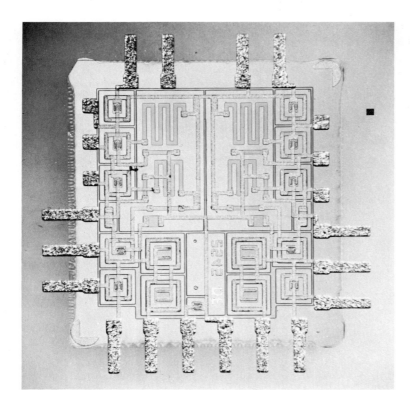

fraction of an inch across, dozens or even hundreds of components, each of which is equivalent to one transistor. These "integrated" circuits (Fig. 10–24) make it possible to build cheaply and in very little space computers of enormous logical capacity, reliability, and complexity.

The increase in computer capabilities since the 1940's and 1950's is truly remarkable. In the late 1940's, it took a typical computer between $\frac{1}{10}$ of a second and 1 second to multiply two 10-digit numbers. Today that operation can be done in less than 1 microsecond. This is a remarkable increase in speed. Note that a 600 mile-per-hour jet airplane travels about 1000 feet during the time it took the early computers to multiply two numbers. The same plane would move only about $\frac{1}{100}$ of an inch during multiplication on a modern computer. As we saw earlier, the size of memories today has grown to between a million and a billion bits, depending upon how rapidly we need to get at the stored information and how much we are willing to pay. In 1950, computer memories could hold only a few thousand bits. Reliability, too, has increased, but figures are difficult to come by and clearly the increase has not been nearly as great as for speed and memory size.

Yet, computers are reliable enough to solve large, lengthy problems. For extremely high reliability, computing facilities can be duplicated so that failure in some part is not fatal. An example is the computer complex used to control the Apollo moon-flight missions. Four large computers were checked against each other to detect incorrect results and to protect against failures.

The recent history of computers, from 1960 on, has been one of increasing penetration of all aspects of life and enterprise. Computers are used today for purposes hardly imagined a few years ago. This fact is indicated by the startling growth in the number of computers in use around the world. As Fig. 10–25 indicates, this number has grown from 250 in 1955 to 109,800 in 1970. Predictions indicate that over 272,000 will be in use by 1974. The value of these computers will grow to over $70 billion by 1974 from the $36 billion in 1970.

Any attempt to list all uses of computers would be futile. We saw earlier that computers can perform any task which can be reduced to a sequence of definite steps, that is a program. The number of such tasks is endless. Furthermore, any list of typical uses today would appear hopelessly short-sighted in only a few

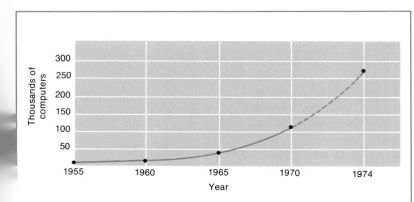

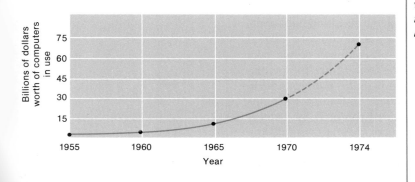

Fig. 10–25.
The number of computers in use throughout the world has doubled about every three years over the past 15. By 1974 over one quarter of a million are expected to be in operation. (These figures are for U.S. manufactured computers only.)

The worth of computers in use worldwide is expected to exceed $70 billion by 1974.

years. So, we will content ourselves with a brief review of some past and present applications which may be of importance in the future.

Computers have proved vital in space exploration. For instance, during a satellite launch the amount and direction of rocket thrust determines the orbit into which the "bird" will be projected. It is vital to know in planning such a shot what orbit will result from a particular rocket-thrust sequence. The orbit can be found from the sequence by a straight-forward calculation, but over 15 million arithmetic operations are required. A person using a desk calculator would need some 350 days to do this task if he performed one operation per second for 12 hours daily. Economically, if for no other reason, this would be unfeasible.

It might be possible to conscript some unfortunate person to undertake this job, but his chance of getting the correct answer would be almost zero. It is just beyond human capabilities to perform thousands of mechanical operations without error. On the other hand, typical computers can calculate orbits in a few minutes with almost no chance of error.

Orbits and flight paths by the hundreds were calculated before the first successful Apollo moon flight. After the launch, observations of successive spacecraft positions were fed to computers which compared the actual to the planned flight path and issued instructions for the mid-course correction to keep the ship on target. Similar calculations were vital in rescuing the Apollo 13 astronauts when a breakdown threatened to strand them in space. Computers calculated just the right maneuvers and rocket "burns" to get the astronauts back to earth. They could not have flown "by the seat of their pants" as early aircraft pilots did. The computer is literally the astronauts' compass.

A similar important computer application is in weather forecasting. Mathematical models of the atmosphere have been available since before 1900. Early in this century, scientists pointed out that such a model could be used to predict future weather by feeding in the pressure, temperature, wind velocity, and other data from many points on earth, then solving the resulting equations. The measurements to obtain the necessary data could be carried out, but solution of the equations without a computer would take so long that the forecast period would be over before the prediction was completed. Computers fast enough to produce timely predictions have been available since 1960, and since 1962 this scheme for weather prediction has been used routinely by meteorologists. About 2000 measured points are used to establish a "current" weather map, then the computer projects that map into the future (see Fig. 10–27). It takes a large computer about one hour to project one day ahead. In doing so, it solves the equations about 100,000 times. Though computer predictions are not 100%

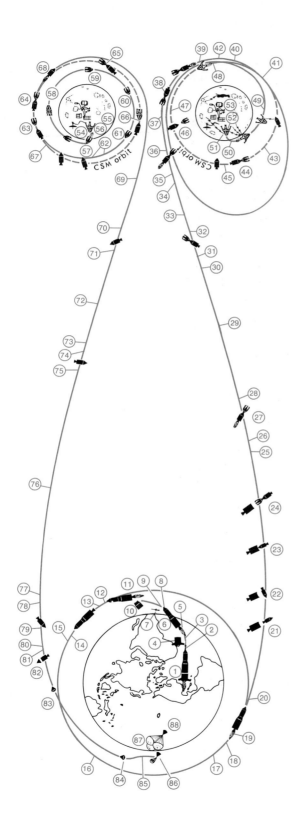

Fig. 10–26. For astronauts aboard Apollo 11, there were 88 steps to the moon and back. Computers were used in planning each one of the 88 steps in a journey to the moon. During the actual flight, data from the spacecraft were fed into the computers which compared actual to planned events and then computed corrections to keep the mission on target. (© 1970 by the New York Times Company. Redrawn by permission.)

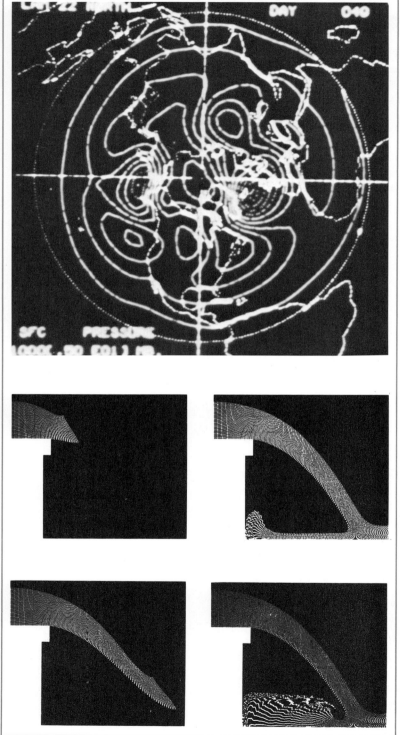

Fig. 10–27.
Weather prediction
by computer is done
by solving equations that
model the atmosphere to
project an existing weather
situation forward in time.
Typically the result is a future
weather map. The one shown at the
top tells surface air pressure. At the
bottom, a model for water flow has been
used to show water spilling over a cliff
into a pool. Notice that these pictures
are not from an artist's imagination
or impressions. They are results
computed from models of
reality in mathematical form
and displayed as pictures.
(Top photo Sol Mednick,
bottom photos University
of California, Los Alamos
Scientific Laboratory)

accurate, the information they provide has aided weathermen in increasing the accuracy of their normal forecasting methods.

These two examples illustrate tasks which are feasible only with a computer. There are other tasks which can be done either by computers or people, but computers can do some jobs more economically. Many of these are in the business and commercial world. A substantial fraction of computer usage in the United States and Europe is devoted to business applications. Examples are the preparation of monthly or weekly payrolls, keeping count of stock in warehouses, and preparation of financial balance sheets. These jobs can be done more economically by computer, since they involve repetition of the same operations time after time.

The Lockheed Missiles and Space Company has been one of the pioneers in using computers to make its operations more efficient. The Lockheed Computer Center at Sunnyvale, California, collects information from more than 200 factory stations spread over a 300-mile radius by using communication lines. The computer system records and controls the movement of more than 200,000 manufactured and stored items. Inquiry stations are also provided at key places so that information on the location of shop and purchase orders, inventory levels, and labor charges can be obtained quickly.

The whole credit card industry is based upon computer accounting techniques. In the not too distant future the use of cash will be much reduced by computer accounting methods. Money will be automatically transferred from one account to another without the passing of actual currency or even checks.

Still another class of computer applications might be called "computing for insight." Examples can be found in science and engineering. Typically computer models are used to reproduce complex natural events in laboratory situations. An interesting case is the use of computers to duplicate the speaking action of the human vocal tract. Researchers have been successful in producing intelligible speech, beginning with a printed version of what was to be said, and then using a computer to simulate the several human processes involved in turning written English into speech. It now appears feasible to build special machines to perform this task for the blind. The insight provided by the computer-based experiments was the key in this research.

Computers are also being used in manufacturing. The automobile industry in particular has used this technique for the design of new body styles. Typically a design engineer and a stylist together use a special computer console to specify the desired body shape. The computer can then issue instructions on paper or magnetic tape for machine tools to produce a full- or reduced-scale model of the new body. This automation has reduced the time

Fig. 10–28. Computers are used in manufacturing plants to assure uniform quality in the product. A paper machine at the Mead Corporation's Kingsport Division in Tennessee has a computer process-control system. The 405-foot-long machine produces rolls of paper up to 16 feet wide at the rate of 2000 feet per minute. The computer-control system monitors the process and directly controls one of the important variables, the average "basis weight" of the paper. The remote unit shown here includes a printer (right) that brings data to the operator and an input unit through which he communicates with the computer. (The Mead Corporation)

between conception of a design and the beginning of manufacture by over 50%.

Computers hold great promise for increased services to society. This possibility is of great importance because more workers in the United States today are engaged in providing services to other people than in manufacturing and agriculture. Techniques to increase the productivity of service industries are particularly important for the 1970's, therefore. For example, airline-reservation systems permit clerks to provide instantaneous confirmation of seats on future flights. This means not only efficiency in loading airline planes without overselling, but also better service for the customers.

Computers are also providing new dimensions in art and music. Artists have produced a variety of creations by programming the computer and using special output devices. Computer music, too, has been composed and played on computers. New sounds and new effects have been created.

*Fig. 10–29 a.
Artistic
drawings can
be generated by
computers using
special output printers.
This work is entitled
"Gaussian Quadratic" and
was drawn from program
instructions containing a random
element so that a number of
drawings done in succession
will each be different. Here
the random element
determines the end point
of lines along the
horizontal dimension.
(Sol Mednick)*

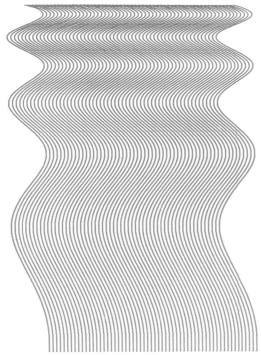

*Fig. 10–29 b.
Another artistic
drawing is "Ninety
Parallel Sinusoids with
Linearly Increasing
Period." The top curve was
specified as a mathematical
equation. This same shape
was repeated 90 times
by the computer.
(Sol Mednick)*

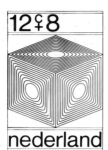
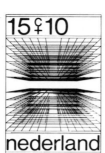

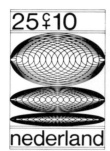
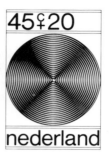

Fig. 10–30. Computer-designed stamps were produced by the Netherlands Postal Authority to aid in raising funds for Dutch charities. (Dutch Philatelic Service, designer R. D. E. Oxenaar)

There is hardly a field which computers have not touched. Certainly they are becoming the most important man-made device. In today's world, a person who does not know how to use a computer and how to control its applications can hardly consider his education complete. In the next chapter we will learn how to use computers for a number of tasks, and in the final four chapters of *The Man-Made World* we will see how computers are created.

Questions for Study and Discussion

1. Make a list of routine tasks that require the addition, subtraction, multiplication, or division of numbers, or the sorting of information, that can be more rapidly done by computers than by hand, such as sorting checks in a bank, or making out price lists from costs plus labor plus overhead, etc. Be prepared to discuss your ideas with the rest of the class.

2. How do you think the Internal Revenue System might use a large computer system? (Use a flow diagram in your answer.)

3. Count the number of times each letter of the alphabet is used on one page of this chapter. (The teacher will assign different pages to different students, so that the entire chapter will be analyzed.)

   a) Compare the frequency of the occurrence of the various letters on your page with the frequency for the entire chapter.

   b) Compare the frequency of the occurrence of the various letters on the first three pages with the frequency on the last three pages.

c) How many pages of two different books do you think would need to be counted before the frequency of occurrence of the various letters was the same for both books? Be prepared to defend your answer.

4. "Engineering is concerned with creating systems, processes, devices, and structures useful to man and his society." Would you expect engineering ideas and methods to be useful to: (a) a lawyer, (b) a teacher of biology, (c) a teacher of social science, (d) a politician?

5. Are any algebraic equations ambiguous in meaning? Defend your answer.

6. What is a major advantage of solid-state switches compared to electromagnetic relays?

7. The sentence "John made the scene" has at least three meanings. One is "John constructed the scenery for a play." State two others.

8. Discuss one instance where the computer might be used to invade personal privacy.

9. The text makes the following statement: "If a key is as long as the message and is never reused, the cipher cannot be broken without knowledge of the key itself." Defend that statement as you would to a student your age who has not read this chapter.

10. Explain the difference between the address of a memory cell and the contents of that cell. (Hint: Look at the mailbox, Fig. 10–12.) Draw a diagram of a memory showing the numbers 100 through 108 stored in memory cells whose addresses are 0 through 7.

11. Would the following program add two numbers correctly?

011
111
012
212
613
513
900

If not, why not?

Problems

1. Take a shuffled deck of playing cards. Find out how long it takes you to sort these cards into the four suits and then into the sequence of numbers from two through ace. What is the average time per sorting operation? How many errors did you have to correct during this sorting operation?

   There is a check-sorting machine in banks today that sorts checks at a rate of 1800 sorts per minute. This machine is virtually errorless. How long would it take this machine to go through the two sorting procedures that you just went through with your cards? How much faster would the machine do the routine task of sorting than you?

   If you were being paid at a rate of $2.00 per hour for such sorting work, and the sorting machine operated at a cost of $5.00 per hour of operation time, which would be the least expensive to hire? What advantages does the machine have over the human operator?

2. In modern high-speed electronic computers it takes about one-tenth of a microsecond for a switch to open or close. This switch operation can be thought to represent a "decision" on the part of the computer.

   a) How many "decisions" could such a computer make in one minute?

   b) What is the frequency of "decision making" of such a computer?

3. In encrypting a stream of binary digits (rather than letters as explained in the text), the following table can be used.

		KEY	
		0	1
INPUT (CLEAR MESSAGE)	0	0	1
	1	1	0

This is known as the Vernam Cypher.

a) If the key were a sequence of ones $(1, 1, 1, 1, \ldots)$, what effect would encryption have on a message consisting of a sequence of zeros and ones? (Take your name, encode it into ones and zeros using Table 10–3, and try this encryption scheme.)

b) Is this key effective in concealing the message? What is the relative frequency of occurrence of ones and zeros in a key stream to conceal a message? What is another necessary property of such a key stream?

4. If the loop in the flow chart for The Idiot Multiply (Fig. 10–9) were written as shown on the right, it would work most of the time.

Under what circumstances would it not work? (Hint: Start with the simplest case of multiplication first.)

5. Write a flow diagram to compute the average number of times the letter "e" appears in all the words on one page of a novel. Use only those types of operations from Fig. 10–8 which are needed for this task.

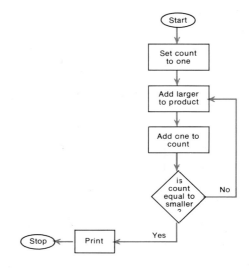

Laboratory and Projects

I | INTRODUCTION TO THE CARDIAC

CARDIAC, CARDboard Illustrative Aid to Computation, is a mechanized flow chart of the digital computer described in your textbook. In this experiment we use it for two main purposes:

a) To learn how the organization of a computer functions and produces useful results.

b) To introduce some of the basic ideas of programming.

As you work with CARDIAC following the printed instructions, it is important to notice that nowhere are you required to exercise judgment. You are only performing routine operations which can be done by a mechanism. As you experiment with the CARDIAC, you will begin to realize that, essentially, CARDIAC simulates the operation of a general purpose digital computer.

Part A

Adding Two Numbers

We begin with a simple program which reads two numbers from input cards, adds them, and punches a card giving the sum. You will follow through a series of several instruction cycles, using the following instructions:

0	Input
1	Clear and add
2	Add
5	Output
6	Store
900	Halt and reset

Using a soft pencil, copy the following program into the memory. (In the next section you will see how programs are loaded.)

Cell No.	Contents
10	017
11	018
12	117
13	218
14	619
15	519
16	900

Tuck one of the "bugs" (you have several spares) into the slot in Cell 10. This bug is the Instruction Counter; it keeps track of the instruction to be executed next.

For convenience in handling, the input and output cards are fastened together in strips. The individual cards are numbered starting at the lower end of the strip. Take one of the strips and with the special pencil write 521 on the first card and 437 on the second card. These are the two numbers to be added. Feed the strip, lower end first, into the slot above the word Input. Position it so that the first card (with 521 on it) shows through the window just under the word Input. (As CARDIAC advances through the deck of input cards the end of the strip will come out of the slit below the window.)

Take another strip and in the same manner position the blank first card in the Output window.

A few words about the accumulator are needed before you begin. The accumulator is the lowest row of four boxes. The upper two rows are only scratch pad for convenience in doing additions or subtractions. The left-most box is for the overflow digit. This box normally has a zero in it; however, if you add two numbers whose sum is greater than 999, then the overflow digit becomes a one. When the accumulator is copied into memory, only the right three boxes are copied since the memory cells can only hold a three-digit number. You will learn how to use the overflow digit in the next experiment. The sign of the accumulator is set with the slide to the left of it. It should be pointed out that you will have to perform the arithmetic operations which the program dictates. The answer is then placed in the third row of the accumulator.

Begin at the arrow labeled "Start"; that is, move the three slides so that 017 shows in the window above the box that reads "Move slides to agree with contents of the bug's cell."

Move along the colored line and follow each instruction. When you reach the "Stop," the first output card should have the number 958 on it.

Part B

Two new instructions are used in this section:

3	Test accumulator contents	(If + move the bug to next cell. If (−) jump to cell indicated by last two digits in the instructions.)
8	Jump	

The program we now use will cause the computer to generate three output cards with −9, −5, −1, and then stop. Before executing the program you must first load it into the computer. To do this you make use of another program with a short loop called a loading program which can load itself into a brand new computer, whose memory contains only one initial set of wired-in instructions. It is sometimes called a Bootstrap Routine. By proper sequence of input signals the loading program, as well as the program to be loaded and executed, is first stored in the memory section of the computer.

The Loading Program

Cell No.	Instruction	Comment
00	001	Read a card into cell 01
01	0xy	Read a card into cell xy
02	800	Jump to cell 00

If a blank card appears, the next card replaces it but the machine will halt until the "Start" button is depressed.

Here is the program which we wish to load into CARDIAC. *Look it over but do not copy it into memory as yet.*

Program to be Loaded and Executed

Cell No.	Instruction	Comment
20	126	Clear accumulator, add contents of cell 26 to accumulator
21	526	Output from cell 26
22	227	Add contents of cell 27 to accumulator
23	626	Store contents of accumulator in cell 26
24	321	Test sign of accumulator contents
25	900	Stop
26	−009	Number increased by 4 each cycle
27	004	Four

Now take one of the input strips and copy the following program into the input deck. Compare the input deck with the two programs above. Notice how the cell numbers have been used as input instructions. Place the deck in the input slot with the first card showing in the window. Put a blank strip in the output slot. The bug should be on cell 00.

Input Deck

Input Card No.	Contents	Comment
1	002 ⎫	Loading program
2	800 ⎰	
3	020	
4	126	
5	021	
6	526	
7	022	
8	227	
9	023	
10	626	Program to be loaded and
11	024	executed (with addresses)
12	321	
13	025	
14	900	
15	026	
16	−009	
17	027	
18	004	
19	(Blank)	Stop loading*
20	820	Start of execution

*CARDIAC will stop when loading is finished. Begin again at START for execution.

II | DOUBLE PRECISION ROUTINES ON A DIGITAL COMPUTER

The word "precision" is used to designate the number of digits which are computed in any numerical problem. A six-digit number is said to have twice the precision of a three-digit number. In a digital computer, the number of digits processed by the machine is limited by the original design. It is, however, possible to overcome these limitations by a process called a "double precision routine." Under normal multiplication processes, whenever we multiply two numbers of three digits each, our product will consist of a six-digit product. The "double precision routine" permits multiplication without any loss of digits in the product.

Part A

Overflow

Very often a result of an arithmetic operation produces a larger number of digits than existed in the original values. For example, the sum of 875 and 628 is 1503. Here a four-digit result appears from a set of numbers each of which is only three digits in length. The fourth digit (in the most significant position) is called the overflow digit and must not be lost. In the CARDIAC the accumulator has a position for the overflow digit but this cannot be accommodated in memory or any other register. Therefore it must be stored in two steps. The first step involves storage of the three least significant digits in a single cell. The machine is then ordered to shift the accumulator contents by three places to the right. This eliminates the three digits which have already been stored and moves the overflow into the portion of the accumulator which can be transferred into storage.

A second command to store into the next consecutive memory cell will bring this overflow into memory.

■ 1. Write a program which adds the number in location 20 to the number in location 21, puts the sum in location 22 and the overflow in location 23. Test your program on CARDIAC.

■ 2. Does it work for all possible signs of the two numbers?

■ 3. If you add two numbers having different signs, can overflow occur?

Part B

Subroutines

In writing programs, the programmer soon notices that many pieces of programs for certain basic tasks occur over and over. For example, a single program may use the sine function of different angles many times. Rather than repeat the portion of the program which calculates the sine each time it is needed, it is better to write a single program to which the main program can give the angle and get back the sine. Such a program is called a subroutine.

Whenever a program calls upon such a subroutine, a "calling sequence" is required. This calling sequence enables the machine to move from the main program to the stored subroutine and to return to the correct step in the main program when the subroutine has been completed.

Procedure

We represent a six-decimal-digit number in our CARDIAC as a combination of 2 three-digit numbers. The three most significant digits are stored in one cell and the three less significant digits are stored in the next succeeding cell of the memory. Thus the number 186324 is stored as follows:

a) In cell 95 store 186.
b) In cell 96 store 324.

We first illustrate this process with an addition subroutine:

Double Precision Subroutine

$$A \qquad + \qquad B \qquad = \text{Sum}$$

(a six-digit number)   (another six-digit number)

We store A in locations 95 and 96, and B in locations 97 and 98. We jump from the main program (call its position a, the jump step xx) to the first address of the subroutine which we have previously stored in locations 86 through 94.

When this subroutine (86-94) has been completed, we must command a return to the main program (at $xx + 1$).

Here is the subroutine program. Copy it into locations 86 through 93 of the CARDIAC Memory.

Subroutine Memory

Location	Instruction	Comment
86	199 ⎤	Prepare exit
87	694 ⎦	
88	196 ⎤	
89	298 ⎬	Add "leasts"
90	698 ⎦	
91	403 ⎤	
92	295 ⎬	Shift overflow right
93	297 ⎦	and add "mosts"
94	8($xx + 1$)	Return

1. Prepare the main program for finding the sum of $A = 186324$ and $B = 241063$. Write the machine code for the following operations placing the instructions into CARDIAC Memory beginning with location 50.

Operations	Location	Machine Code
1. Store 186 into cell 95	50	—
2. Store 324 into cell 96	—	—
3. Store 241 into cell 97	—	—
4. Store 063 into cell 98	—	—
5. Jump to address of first step in subroutine	—	—
6. Store the sum in address 59	—	—
7. Print out contents of cell 59	—	—
8. Print out contents of cell 98	—	—
9. Stop	—	—

Prepare the four-card input deck with the two numbers to be added. Place a blank strip in the output slot and start with the bug on cell 50.

2. If the two numbers used were of unlike signs, would this create a problem? Methods are available for manipulating signed numbers. Can you suggest a procedure to accomplish this?

3. If the two numbers to be added are 816,324 and 241,063, what problem would arise? Can you devise a method for such cases?

III | CIPHERS AND CODES

In Section 3, the nature and importance of ciphers and codes were discussed. You learned about two types of codes, the Polybius Code and the Vigenère Code. This activity will give you an opportunity to work with another type of code, often referred to as "Strip-Cipher."

Directions

1. The teacher will give you ten identical strips of paper containing numbers from 1–10, the entire alphabet, and assorted symbols.
2. You will also be given a card which has been ruled in such a fashion that there are ten spaces to accommodate the ten strips. This card will also have two lines running vertically down the paper. Since they have been roughly centered, we will call them "center lines."
3. Place a paper strip in each of the ten horizontal spaces of the card.
4. Now we are ready to encode and decode our first message. For example, let us see how a message, such as "ECCP is fun," can be encoded and decoded.
5. Starting at the top of the card, move the first strip so that the first letter of our message, namely "E," is between the center lines. Move the second strip until "C" is between the center lines. Continue this process until all nine letters are arranged vertically between the center lines. Since we have only nine letters in this message, the tenth strip can be removed.
6. Now we are ready to pick out the coded equivalent. You may pick out any vertical column. For example, if you had picked the third column to the left of the center lines, then the coded message would be B55MFPCRK.
7. Let us assume that you have received this message (B55MF-PCRK) and wish to decode it. Start by moving the first strip so that the first symbol "B" is located between the center lines. Do the same thing with the other eight symbols.
8. Look down the columns until you see the nine symbols which make sense. In this example, you will notice that the message "ECCP is fun" is located in the vertical column, which is three spaces to the right of the center lines.
9. You can now encode your own message, and let another student decode it while you try to decode his message. (If you have more than ten symbols in your message, encode it in groups of ten symbols, with the last group having less than ten.)

Communicating to Computers

11

I | INTRODUCTION

Without a program, a computer is a helpless collection of electronic circuitry. With a proper program, a computer can guide a satellite into orbit, operate an oil refinery, control the flow of traffic in a city, or schedule school hours to minimize classroom conflicts. In this chapter, we shall see some of the programming techniques which can turn a computer to human purposes. Becoming an expert programmer takes time and experience, and we do not expect students to achieve that in this course. However, by completing this chapter, the student will know enough of programming to use a computer for his own purposes provided he does not try to do too much too soon. Furthermore, he will be able to extend his skill on his own should he desire to do so.

We saw in the last chapter that computers execute one instruction after another blindly without leeway for error by the programmer. Thus the program must be correct in every detail. We saw a simple program for adding two numbers in the last chapter. It consisted of only seven instructions. It is often possible to write such a short program correctly the first time. Most programs are

The principal working part of genes is a complex molecule, deoxyribonucleic acid, DNA. Each DNA molecule carries a coded message which gives a part of the plans for the construction of the organism. The complete plan is carried by the chromosomes, composed of several thousand genes. This library of instructions can be likened to a program to be carried out by the growth processes. (O. L. Miller, Oak Ridge National Laboratory and Science)

longer and more complicated. Some contain thousands of instructions. It is seldom, if ever, possible to write these without error. Even relatively simple programs usually contain errors when they are first written. Thus, a vital step in preparing a program is to test it on the computer so that errors can be found and corrections made. This is a tedious process and usually requires special additional programs to help locate errors. Thus, it pays for the programmer to work carefully in the beginning to keep errors to a minimum.

Each programmer is different, and there is no universal formula or prescription to insure programming accuracy. However, there are some practices which most programmers find useful. Flow diagrams which we met earlier are one such aid. Flow diagrams show clearly the successive steps to be taken in the solution of a problem. They can be a great help in planning what a program is to do. They can help insure that a program actually solves the intended problem.

In this chapter, we use flow diagrams for outlining a program before actually writing the instructions. We find that flow diagrams can have various levels of detail. The diagram drawn in the first stages of programming a problem usually shows fewer steps than those used further along in the planning, and many fewer than the actual instructions in the final program. Two flow diagrams of differing detail and the final program for adding two numbers are shown in Fig. 11–1. Some programmers go through several stages of flow-diagram detail before actually writing instructions. Others write instructions at once. It is a matter of personal preference and experience. At each stage, however, flow diagrams must contain only operations which can be carried out by one or

Fig. 11–1.

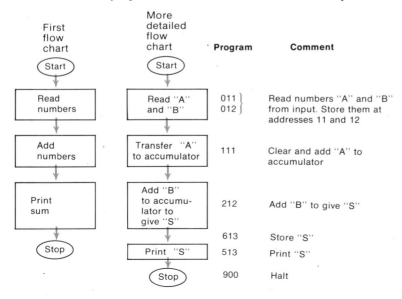

First flow chart	More detailed flow chart	Program	Comment
Start	Start		
Read numbers	Read "A" and "B"	011 012	Read numbers "A" and "B" from input. Store them at addresses 11 and 12
Add numbers	Transfer "A" to accumulator	111	Clear and add "A" to accumulator
Print sum	Add "B" to accumulator to give "S"	212	Add "B" to give "S"
		613	Store "S"
Stop	Print "S"	513	Print "S"
	Stop	900	Halt

more computer steps. In fact, accurately expanding flow-diagram operations into computer steps is crucial to making a computer do what the user wants done.

After a program has been written, the computer itself can help find simple errors and omissions in the instructions. Such simple errors are often called "bugs" to distinguish them from errors of logic in the planning of the program. To aid in debugging instructions, computers are provided with special programs which examine the instructions for inconsistencies. These "debugging packages" are usually supplied by the computer manufacturer. They look for errors and give output messages to the programmer to aid him in corrections. However, debugging packages have little to say about whether the program will perform the task intended. They only help in assuring that the instructions meet the computer's rigid requirements of form.

Though flow diagrams and debugging packages aid programmers in preparing correct sets of instructions to do the jobs intended, the major responsibility for accuracy falls on the programmer himself. It is altogether easier to debug a carefully written program than to count on trial and error to clean up a sloppy program. You will discover this for yourself later when you write your own programs.

Our purpose in this chapter is to learn the few basic programming techniques that make it feasible for people to communicate their complex desires to the computer in a form that can be acted upon. These techniques are aimed at shifting as much of the programming work as possible onto the computer, relieving the user of much tedium. In our examples, we will use the simple model computer introduced in the last chapter. There we used only six different instructions. To program realistically, four additional ones will be needed. These are introduced in the examples of this chapter. The cardboard simulator, CARDIAC, can also use these four additional instructions, so it will be of use in demonstrating their operation. Again, the programming techniques we will be discussing are all aimed at reducing the programming effort and making it easier to write correct programs.

2 | LOOPS, LOADERS, AND BOOTSTRAPS

Loops and Jumps

We saw in Figs. 10–7, 10–8, and 10–9 of the preceding chapter flow diagrams containing loops. In executing such a program, the computer runs many times through the instructions inside the loop. This feature saves the programmer from having to write out those instructions over and over again. Also, it means that the program occupies many fewer memory slots. Saving memory space is necessary for an economical and efficient computation.

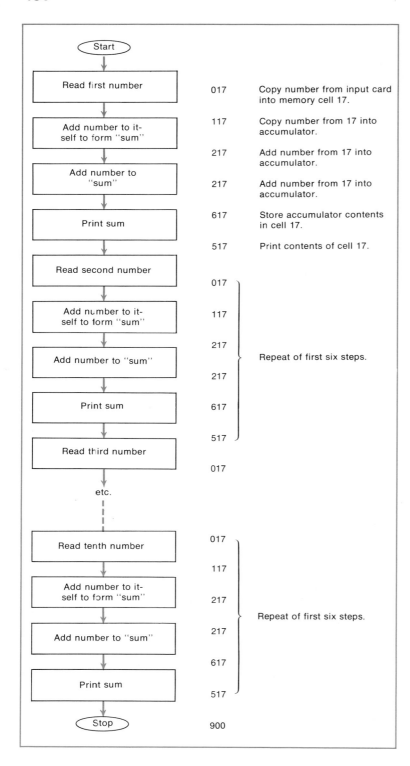

Fig. 11–2.
A program to
triple input numbers (10)
without *looping.*

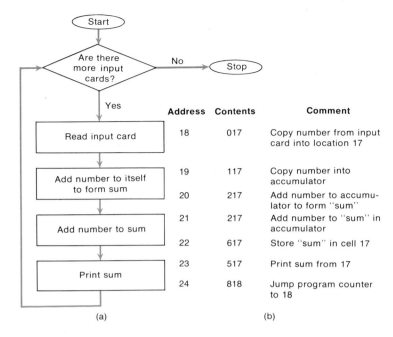

Address	Contents	Comment
18	017	Copy number from input card into location 17
19	117	Copy number into accumulator
20	217	Add number to accumulator to form "sum"
21	217	Add number to "sum" in accumulator
22	617	Store "sum" in cell 17
23	517	Print sum from 17
24	818	Jump program counter to 18

(a)   (b)

Fig. 11–3. This program will do the same task as that of Fig. 11–2, but avoids repetition of instructions by looping. In addition, this program will handle any number of input cards while the program of Fig. 11–2 works only for 10.

Let us take as an example the program to triple numbers of Fig. 11–2. It is written as though there were ten numbers punched on input cards to be read and tripled by the computer. In total, the program is 61 instructions long, and each one must be stored in the computer's memory. If there were a thousand or a hundred thousand input numbers, the program would overflow the memory of even a large computer. Notice, also, that the program consists of the same six instructions repeated over and over in the same order. This is an ideal situation for a loop to eliminate those repeated instructions.

The new flow diagram with loop is shown in Fig. 11–3a. Here a number is read from the input card stack, added to itself twice, and the sum printed out. The computer then begins the same sequence again *if* there is another card in the input stack. If not, the computer stops. Notice that this program will process *any* number of input cards *without* being informed ahead just how many there are. Thus, it does not need to be rewritten if the task is redone with a different number of input cards. This feature is another advantage of looping.

To reduce the flow diagram of Fig. 11–3a to a set of instructions, we will need an additional instruction to make the computer loop. Recall that computer control takes instructions in an order specified by the instruction counter. In a simple straight-line program such as that in Fig. 11–2, the counter is merely increased by one on each cycle. Thus, instructions are taken out of successive memory locations. To make the computer execute the same instructions many times requires some way to reset the program counter to the address of the first instruction each time the last instruction has been completed. In CARDIAC this is done by the instruction:

> 8XY* To execute this instruction, control resets the program counter to the address XY so that the next instruction executed is taken from address XY, which in the example of Fig. 11–3 is 18.

*
The letters XY as used here stand for the address numbers in the instruction. When actually used, the instruction must have the correct address substituted for XY. We will use this convention in discussing other instructions later in this chapter.

This instruction is known as the *jump* instruction since it jumps the program counter to a new number. With the jump instruction, we can write the program corresponding to Fig. 11–3a. It is shown in Fig. 11–3b. There we have assumed that the program is stored in memory cells, or locations, 18 through 24. These addresses are listed on the left and their *contents*, the instructions, on the right. In executing this program, the computer begins with its counter at 18, runs through the instructions until the counter reads 24 when it is reset to 18 by the instruction 818. Thus, the computer will loop through these seven instructions over and over.

How do we stop the computer when it is looping? The flow diagram indicates that before each pass through the instructions the card reader is examined to determine if there is an input card to be processed. If not, the computer stops. Yet, there is no instruction in our program corresponding to the decision diamond at the beginning of the flow diagram. It turns out that none is needed, for the input mechanism takes care of this action automatically. The full action of an input instruction is as follows:

> OXY Copy the word on the top card of the input-card stack into address XY in place of the previous contents of that cell, then advance (remove) the top card of the stack. *If the input card is blank or the stack is empty,* reset the program counter to 00, *set the run-stop switch to stop* and advance the input stack.

It is the italicized part of this definition that gives us the equivalent of the decision box in the flow diagram of Fig. 11–3a. This is an example of the detail that must be known about a computer's mechanism in planning its use. We will see later that all the features of OXY are important for some programs.

Not all loops can be stopped by the action of the card reader. For example, the flow diagram of the so-called Idiot Multiply (see Fig. 10–9 of Chapter 10) involves adding one number to itself repeatedly. That number is stored in a memory cell and used over and over. There is no card reading involved in that loop. To stop or break such a loop requires another special instruction. In our model computer this instruction is:

3XY If the number in the accumulator is zero or positive, take the next instruction from the next higher address in order. If the number is negative, take the next instruction from address XY by setting the program counter to XY.

This instruction is known as the *conditional jump*, because whether the program counter is jumped to a new number or not is conditional on the *sign* of the number in the accumulator.

Until now, we have said nothing about the signs of numbers in computers. However, all computers can handle both positive and negative numbers. Their signs, positive or negative, are stored usually in the first bit of each memory cell. (In our model computer the sign is stored in the first place of the memory and accumulator. A plus sign is represented by 1, a minus sign by zero. Thus an input card containing −000 is blank.) Of course, instructions have no signs even though we represent them as numbers. When stored in memory, they may have either sign, because when they are transferred to the instruction register for execution the sign is ignored. There is not even any provision for the sign bit in the instruction register.

However, when doing arithmetic, signs are important. For example, to subtract, we can add a negative number. So the accumulator has provision for keeping track of the sign. For example, if the accumulator contains the number +700 and we add −725 to it from some memory cell, the accumulator will then read −025. The conditional jump depends upon the sign of the accumulator at the time the instruction is executed.

How is the conditional jump used? As an example, let's see how to make the computer count off the last ten seconds before a rocket launch. The flow diagram is shown in Fig. 11–4a. The number −011 is put into the accumulator, then 001 is added giving −010 in the accumulator. This number is then printed out. The computer then repeats, successively printing out −009, −008, −007, and so on until the accumulator reads 000.* At that point, the computer takes the lower branch to a STOP or HALT instruction.

The program itself is shown in Fig. 11–4b. We have assumed that the numbers −011 and +001 are punched on input cards and are in the card reader with −011 first. These two numbers are read into locations 17 and 16, as indicated in Fig. 11–4b. The action then proceeds as described.

As the programs of Figs. 11–3 and 11–4 show, both the *jump* and the *conditional jump* change, or jump, the count in the program counter. The difference between the two jumps is that 8XY always changes the count while 3XY only changes the count when the accumulator sign is (−). The conditional jump is used to stop the computer from looping. It is also used to make the future course

*
Note that our computer takes the same branch when the accumulator reads zero as when its sign is (+) (see definition of 3XY instruction). Also note that in our computer the accumulator will have a sign even when its content is 000, since each bit in any register must have a value, 0 or 1, + or −; it cannot be blank. However, the signs of +000 or −000 can be ignored.

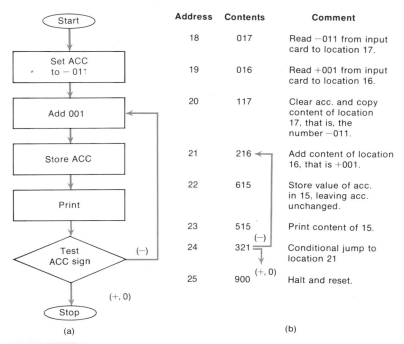

Address	Contents	Comment
18	017	Read −011 from input card to location 17.
19	016	Read +001 from input card to location 16.
20	117	Clear acc. and copy content of location 17, that is, the number −011.
21	216	Add content of location 16, that is +001.
22	615	Store value of acc. in 15, leaving acc. unchanged.
23	515	Print content of 15.
24	321	Conditional jump to location 21
25	900	Halt and reset.

Fig. 11–4. Automatic count-down for space-rocket launch.

(a) (b)

of a computation depend upon the results obtained up to that time. Thus the conditional jump is the instruction to be used at all decision points in a program. As we saw earlier, these decision points, sometimes called *branch points*, are shown in flow diagrams as diamonds to set them off from the "straight-line" instructions.

Loops are a very useful part of the programmer's bag of tricks. In fact, loops are so useful that there are few, if any, programs written without them. The reason, as we have seen, is that loops permit the same instructions to be used many times. Typically in a computation, a computer performs many more operations than the number of instructions in its memory. This relieves the programmer of having to write an instruction for each step in the computation, and it also economizes on the number of memory cells needed for storing the program. Looping is only one programming technique for making the most of a small sequence of instructions. There are others which we will see in this chapter.

An Example of Looping: Multiplication

We saw in the last chapter one way of multiplying by successive additions. We called the scheme The Idiot Multiply. Actually, when the multiplier is a single digit, this technique is as good as any other for a computer whose processor cannot multiply. To review the scheme for multiplying two numbers, the flow diagram of Fig. 10–9 in the last chapter calls for storing the larger of the two numbers in the accumulator. Then that number is added to itself *n* times where *n* is one less than the smaller number. For example, to multiply 8×20, 20 is put in the accumulator and added to itself 7 times. A somewhat different flow diagram using both instructions 3 and 8 is shown in Fig. 11–5b. The program in Fig. 11–5b assumes that the multiplication is of the form $A \times BC$ where A is the positive single digit multiplier and BC is at most a two-digit multiplicand. BC is punched on the first card and A on the fourth in the card reader. (+000 is punched on the second card to clear a memory cell to hold the product, and −001 is punched on the third card.)

This program will work for any values of A and BC. If BC were allowed to be larger, say a three-digit number, the multiplication might result in a number too large to be held in memory. For example, $5 \times 500 = 2500$. This answer would "overflow" any cell in memory since each can hold only three decimal digits plus sign. That is, the most significant digit, two, would be lost because there is no room in the memory cell to hold it.

Note that A becomes the flow diagram's *n*, and is tested in each loop to find out when it becomes negative. The loop is broken when that occurs. This method is known as looping with an index.

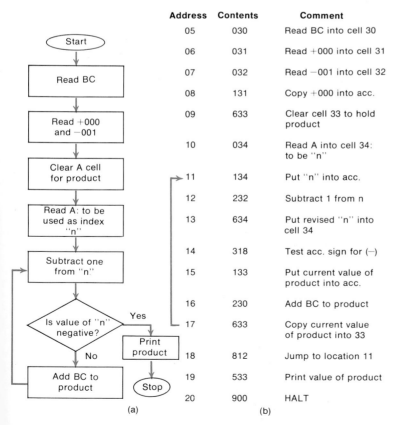

Address	Contents	Comment
05	030	Read BC into cell 30
06	031	Read +000 into cell 31
07	032	Read −001 into cell 32
08	131	Copy +000 into acc.
09	633	Clear cell 33 to hold product
10	034	Read A into cell 34: to be "n"
11	134	Put "n" into acc.
12	232	Subtract 1 from n
13	634	Put revised "n" into cell 34
14	318	Test acc. sign for (−)
15	133	Put current value of product into acc.
16	230	Add BC to product
17	633	Copy current value of product into 33
18	812	Jump to location 11
19	533	Print value of product
20	900	HALT

(a) (b)

Fig. 11–5. Program for multiplying $A \times BC$.

The index is *n*, and it merely counts the number of times the loop has been completed. The same technique can be used for any program in which a loop must be repeated *n* times. Notice that the value of *n* is supplied as data, so the program does not have to be changed if the problem is to be rerun with a different *n*.

There are many different ways to program any given problem. There is no one correct program. For example, there are many plausible ways of carrying out the $A \times BC$ multiplication correctly. Some require more steps or use more memory or both than others. Thus, a programmer may find a shorter and faster way of solving a problem by considering several alternatives. Also, the length of a program and the time it takes to run to completion depends on the instructions available on the computer at hand. For example, it is much faster to have special circuits for multiplication in the processor than to use a program. Multiplication can then be done in one step in response to one instruction. However, multiplication circuits are expensive and many small computers do not have them, as is the case with our model computer. So most computers are designed as a compromise between having large numbers of complex operations (expensive) and a few essential operations (less expensive) that can be used to program more complex ones. Our model computer has only ten instructions, probably as few as any actual computer. We have now reviewed eight of the ten.

Shifting

An example of an instruction which makes programming easier is the shift instruction. In the example of Fig. 11–5, it provides an easier way of clearing a memory cell for the multiplication product.

The shift instruction is: shift any number in the accumulator to the left X number of places, then to the right Y number of places. The values of X and Y are specified by the second and third digit of the instruction. (Note: this is the only instruction in our model computer where these digits *do not* specify a memory address.)

Digits that are shifted off either end of the accumulator are lost forever. For example, if the accumulator read 0682 (the accumulator has four decimal places plus sign) and we gave the instruction 423, (shift 2 places to left then 3 to the right) the accumulator would read 0008 after the instruction was completed. The sign of the accumulator is not changed by the shift operation. The zeros in this result may be surprising at first. They appear because there is no such thing as a blank space in the accumulator or any other computer register or memory cell. When a digit is moved out, it is immediately replaced by a zero at the opposite end of the accumulator. For instance, after the first step in the above example (the left shift by 2 places), the accumulator will read 8200.

Let us see how to use one shift instruction to replace two instructions in Fig. 11–5b. The combination of the instructions in locations 06 and 08 is to clear the accumulator so that a store instruction (location 09) can be used to clear a memory cell for the product. (The cell containing "Product" must be set to zero before the first pass through the loop.) The instructions located in 06 and 08 can be replaced by a single instruction, 404, which will clear out the accumulator. This shift instruction can be put at 08, the one at 06 replaced by the one at 05. The program is now one instruction shorter and will operate correctly as before. The first few lines of the program now read:

06	030	Read BC into cell 30.
07	032	Read −001 into cell 32.
08	404	Clear accumulator.
09	633	Clear cell 33 to hold product.
10	034	Read "A" into cell 34.

The remainder of the program reads as in Fig. 11–5b. Of course, the card with +000 on it will no longer be needed either.

We have now introduced all of the instructions for our model computer with the operations they perform, except for one. That one is a subtract instruction, 7XY. It says to subtract the word at address XY from the contents of the accumulator leaving the word in memory address XY unchanged. Except for subtracting, this instruction works exactly like the 2XY add instruction.

Bootstraps and Loaders

To this point, we have assumed that all programs are in the computer memory and are executed from there. How do they get into memory? Programs must be loaded through an input device. In our model computer, we assume this device is a card reader, as it is on many computers. Also, each instruction must be steered to its proper address. These two requirements call for a so-called "bootstrap" and a loading program. The bootstrap lets us put a program into memory *without* one being there already. The loading program steers the instructions to the proper locations.

The bootstrap is needed because when a computer is turned on, its memory and registers contain nothing but nonsense. As we said before, there can be no blanks in a computer. Each memory core, transistor bit, or other circuit must always be in one state or the other, 0 or 1. Which they contain when the machine is turned on is a chance occurrence, so the resulting words are meaningless.

Before a program can be loaded, some register or registers must be set to appropriate values. Many computers have "reset" buttons to clear the program counter and accumulator. Words can then be inserted in these by a row of input switches, and then stored in memory by other manual operations. Getting a computer started is much like handcranking an old fashioned automobile. Whatever the technique, some minimum information is inserted in the computer which then proceeds to use this to get further information from the card reader and so on. Actually, the computer appears to lift itself by its own bootstraps. So, the process of using that initial information is called *bootstrapping*.

In our model computer the special bootstrapping information is the word +001 permanently stored in cell 00. When the "run" button of CARDIAC is pushed, the program counter always reads 00, so the first instruction to be executed is taken from cell 00 and is 001. This instruction says "Read the first card and store its information in cell 01." As usual, the program counter is then increased by one to read 01. Thus the next instruction executed is the one just read off the first card. That instruction is 002. Thus the computer then reads the next card, which has 800 punched on it, into cell 02. The program counter is then advanced to 02 and the instruction 800 is executed. This merely resets the instruction counter to 00. We now have stored in the memory:

Location	Content	Comment
00	001	Wired in: read top card into 01
01	002	Read next card into 02
02	800	Jump to 00

This is a loop in which the second instruction is read from an input card. Above, we have described what happens on the first pass through the loop; namely, it has loaded itself into memory. This is the bootstrap program.

The loading program is merely a *set of cards* listing the *addresses* where the program instructions are to be stored in memory. These cards are interleaved with the cards specifying the instructions to be stored at those addresses.

To see how this works, assume that we want to load the program of Fig. 11–4b. First, punch a set of cards specifying the instructions to be loaded. These are listed in the *content* column of Fig. 11–4b. Then punch a set of cards consisting entirely of read instructions whose address part specifies the locations in memory where the instructions are to be stored. These locations are listed under the *address* column of Fig. 11–4b. For example, the first instruction reads 017 and is punched on one card; its address, 018, is punched on another card. Thus two cards are punched for each instruction in the program to be loaded. They must be arranged as shown in Fig. 11–6 in the card reader with card 1 on top.

The computer reads the first card, proceeds to set up its bootstrap, and then loads the instructions on the even-numbered cards

Card	Content of card		
1	002 ⎫	BOOTSTRAP	
2	800 ⎭		
3	018	Address ⎫	017 to be loaded into address 18
4	017	Instruction ⎭	
5	019	Address ⎫	016 to be loaded into 19
6	016	Instruction ⎭	
7	020	Address ⎫	117 to be loaded into 20
8	117	Instruction ⎭	
9	021	Address ⎫	216 into 21
10	216	Instruction ⎭	
11	022	Address ⎫	615 into 22
12	615	Instruction ⎭	
13	023	Address ⎫	515 into 23
14	515	Instruction ⎭	
15	024	Address ⎫	321 into 24
16	321	Instruction ⎭	
17	025	Address ⎫	900 into 25
18	900	Instruction ⎭	
19	Blank	Signals end of program to be loaded	
20	818	Transfer Control to program just loaded	
21	−011 ⎫	Data to be read by program after loading	
22	+001 ⎭		

Fig. 11–6.
Card stack
for loading program
of Fig. 11–4 b.

beginning with 4 and ending with 18. The addresses on the odd-numbered cards fill the 02 cell successively on each pass through the loop. A flow diagram for the bootstrap loop is shown in Fig. 11–7. The beginning of the loading sequence is indicated in Fig. 11–8. Note that the number on each input address card is used as the second instruction in the bootstrap loop. This makes it possible to steer each instruction to its correct memory location.

When the computer arrives at the blank card (Fig. 11–6, 20th card), it *stops* as specified by the flow diagram Fig. 11–7. The program counter automatically is set to 00 when the computer halts. The program of Fig. 11–4b has been loaded into memory slots 18 through 25. How then do we arrange to reset the program counter so that the computer will start executing the instructions beginning at memory location 18? Clearly, a jump instruction (818) will do this job. This instruction must be punched on an input card and placed just after the blank card of Fig. 11–6 in the card reader. Then, to run the loaded program, push the "run" button. The computer will read 818 into memory slot 01, then execute 818 on

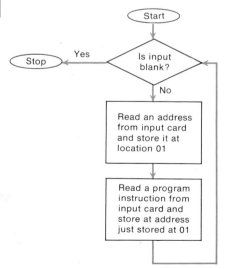

Fig. 11–7.
Bootstrap
program for
CARDIAC.

Program Step		Memory location	Instruction word	Action resulting
0	(wired into memory)	00	001	Read top card into address 01 (Top card is 002)
1	the program	01	002	Read top card into address 02 (Top card is 800)
2		02	800	Jump back to address 00
3	2nd time through the loop	00	001	Read top card into address 01 (Top card is 018)
4		01	018	Read top card into address 18 (Top card is 017)
5		02	800	Jump back to address 00
6	3rd time through the loop	00	001	Read top card into address 01 (Top card is 019)
7		01	019	Read top card into address 19 (Top card is 016)
8		02	800	Jump back to address 00
9	4th time through the loop	00	001	Read top card into address 01 (Top card is 020)
10		01	020	Read top card into address 20 (Top card is 117)
11		02	800	Jump back to address 00

and so on for 9 passes through the loop.

Fig. 11–8. Steps in loading program of Fig. 11–4b.

the next cycle. This sets the program counter to 18 and the computer will execute the program as we have described previously.

This action of the 818 jump instruction sets the computer so that its operation is controlled by the program beginning at location 18. In computer jargon, we say that "control has been passed to the missile-count-down program." This technique of passing control from one program (the bootstrap) to another (the missile-count-down) is essential to save programming effort by putting together larger programs from smaller existing subprograms. We shall see in the next section how this is done without halting the computer between subprograms.

3 | BUILDING BLOCKS FOR PROGRAMS: SUBROUTINES

Once a program has been written, it can of course be used by persons other than the original programmer. If the program performs an extremely useful task, it is likely to be used many times. The most useful programs turn out to be those which can be used as parts of larger programs. There are certain common tasks that must be done in many different programs. Some of these are programs to compute square roots, cosines, compound interest, or the roots of equations. This fact was realized in the early days of computing. As a result, programmers began to create libraries of programs which could be used as building blocks for larger programs. These building blocks are known as *subroutines*.

A subroutine is simply a program written so that it can be used as part of a larger program. In most computer centers there is a library of such subroutines. Subroutine libraries have catalogs which are updated and reissued often. These are used often by programmers since subroutines can save tremendous amounts of programming effort.

To use a subroutine, it is not necessary that its addresses be interleaved with those of the full program. Rather, the subroutine can be stored in a different part of the memory and called into action at the appropriate point in the main program. This is done by what is known as a *calling sequence* which calls the subroutine into action by means of a jump instruction. Similarly, a jump instruction is used to pass control back to the main program after the subroutine has finished its task. Thus, once the subroutine is written, it need not be changed in order to be used over and over again. A subroutine can even be used several times in the same main program.

As an example, let us consider a subroutine to add two 6-digit numbers. Recall that our model computer can store only one 3-digit number in each of its memory cells. (Its accumulator has an extra digit which will be useful to the 6-digit subroutine.) So in its direct add operation, our computer cannot handle 6-digit numbers. We say its *precision* is limited to 3-digits. We use the word *precision* because in giving a numerical value of any kind, the number of digits determines the accuracy or precision. For example, the person who says that he weighs 175.617 pounds is being twice as precise as a person saying he weighs 175 pounds. In some computer operations it is necessary to use precision greater than can be handled by the word length of the memory and processor registers. Fortunately, computers can be programmed for double, or even triple, precision operations. This can be done only at the expense of many more instruction steps than single precision operations. Increased precision means slower and more expensive performance.

Basically, the method used for handling 6-digit numbers in our computer is to store the three most significant digits in one location and the three least significant in the following location. For example, when storing 175.617, 175 would go in one cell and 617 in the following cell. When adding two 6-digit numbers, the *least* significant digits are added first, then the most significant digits *including* any carry digit from the least significant. A double-precision subroutine for adding two 6-digit numbers is indicated in Fig. 11–9. Also shown is a main program calling the subroutine and printing out its results.

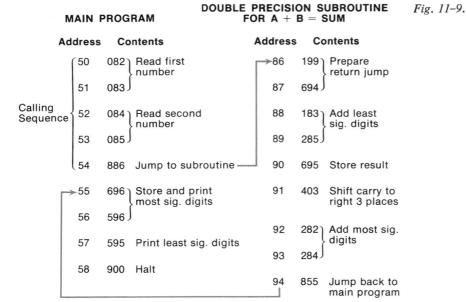

Fig. 11–9.

There are several features of Fig. 11–9 worth noting:

1. The calling sequence for the subroutine consists of five instructions. The first four of these read the numbers to be added into addresses already written into the subroutine. In this case, the addresses are 82 through 85. The subroutine will always add the numbers at those addresses (unless it is rewritten). So the main program must always load the numbers to be added into the correct locations. The final instruction in the calling sequence transfers control to the subroutine.

2. The transfer of control to the subroutine involves an action in addition to resetting the program counter. The jump instruction is arranged so that this additional action is taken automatically by the computer. The complete action of a jump instruction is:

> 8XY *Replace the second and third digits (the address component) of the word stored at address 99 by the address in the program counter.* Then reset the program counter to XY.

In our example, the italicized action means that after the transfer of control, the word at 99 reads 855 (the program counter is incremented *before* the jump is executed). The op code at 99 is permanently wired to 8. The first two instructions in the subroutine take this instruction from 99 and store it in 94 so that it can provide the return jump to the correct point in the main program.

3. After the subroutine is completed, the main program finds its results in certain specified locations. In our case, these are address 95 (least significant digits) and in the accumulator (most significant digits).

All these actions are necessary to communicate data, instructions, and results between the subroutine and main program. Programming these steps, however, is well worthwhile since they make it possible to use subroutine libraries without adapting the subroutines to each individual main program.

4 | PROGRAMS FOR WRITING PROGRAMS

As the preceding program examples show, one of the major problems in writing programs is keeping track of what information is stored where in memory. Another problem for the programmer is just remembering what the op codes mean. If computers could be arranged to take care of these details, it would be a great help. In fact, special programs have been written to permit computers to do just that. For example, instruction 11 of Fig. 11–5b can be written CLA n (clear and add n) rather than 134. The special program, called an *assembler*, can be arranged to translate such commands into numerical instructions for the computer, assigning storage locations in the process. Assemblers also attend to calling sequences and exit addresses for subroutines. Assemblers are a great aid to programming. No modern computer is without one.

To use an assembler, the "symbolic" commands, such as CLA n, are punched on cards. These cards plus the assembler program are put into the computer. The assembler processes the symbolic program and produces a conventional program which can then be used for its intended purpose. This two-step process is indicated in Fig. 11–10.

Assemblers, then, are programs for writing programs. They take a program written in assembly language and rewrite it in the more fundamental alphabet of machine language. Thus, assembly language is on a higher level than machine language in the sense that its meaning is more easily understood by humans. Happily, the process of writing a program to translate a higher into a lower level language can be carried one step further with compilers.

Compilers are used to rewrite programs of a still higher level than assemblers into assembly language. They go considerably further than assemblers towards eliminating programming drudgery. Whereas assembler programs necessitate writing a separate instruction for each machine-language instruction to be produced, compilers can translate one compiler-language instruction into

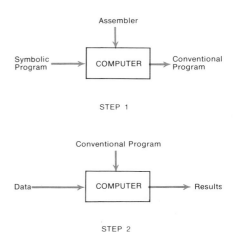

Fig. 11–10.
Using an assembler.

several assembly-language instructions. Thus, the number of statements a programmer must write is vastly reduced.

One of the best known and most widely-used compilers is FORTRAN, short for FORmula TRANslator. Its language resembles that of algebra. With FORTRAN, a programmer wishing to solve an equation such as $J = K + M - N + 2$ would write the equation exactly that way: $J = K + M - N + 2$. The FORTRAN compiler would then translate that statement into suitable assembly-language instructions. For CARDIAC, these instructions would be:

```
CLA   K
ADD   M
SUB   N
ADD   2
STO   J
```

The assembler would next translate these into machine-language instructions and assign them locations in the memory like this:

Address	Contents
20	151
21	252
22	753
23	270
24	650

At this point, the computer would cause a set of punched cards containing the complete program to be produced. If requested to do so, it would simultaneously print out the program in all three languages so the programmer could check it. The printout would also include a table of temporary storage assignments and a table of constants as follows:

Address	Contents
50	J
51	K
52	M
53	N

Table of Constants

70	002

If any routine errors were made, diagnostics might also be included in the printout. These are comments on commonplace programming errors such as the omission of brackets, or the use of undefined symbols.

Now that the programmer knows where things are and, to some extent, what (if anything) is wrong, he can go to work debugging his program. Once he locates his errors, he simply pulls the bad cards out of his deck, has new ones punched, and then slips the edited program into a card reader. He may have to repeat this process two or three times before all the bugs are gone, but, once they are, the computer will execute his program.

INSTRUCTIONS FOR MODEL COMPUTER   *Fig. 11–11.*
(CARDIAC)

Op code

Instruction		Action
0XY	*Read*	Read an input card and store its contents in memory at location XY.
1XY	*Address* *CLA*	Clear the accumulator and then add (transfer) the contents of cell XY to it.
2XY	*APD*	Add the contents of cell XY to the accumulator.
3XY	*Bran i* *XY*	Test the accumulator. On zero or positive take the next instruction from the next higher address. On negative, take the next instruction from address XY.
4XY	*X to Left* *Y to right*	Shift the accumulator contents X places to the left, then Y places to the right.
5XY	*print*	Print the contents of cell XY on an output card.
6XY	*sto*	Store the accumulator contents in cell XY.
7XY	*sub*	Subtract the contents of cell XY from the accumulator.
8XY	*Jump*	Take next instruction from address XY. *Jump*
9XY	*HLT*	Halt.

5 | BILLIARD-TABLE SIMULATION

The following illustrates how computer programs can simulate real situations. We shall see how to simulate the path of a ball on a billiard table (Fig. 11–12). The ball is projected at 45° from the lower lefthand corner of the table, which is 11 units wide and 15 units long. We assume that the ball travels back and forth across the table, always rebounding at 45° from each edge. This program determines the position of the ball on the table at the end of a specified time, assuming that the ball travels at a constant speed on the table.

With horizontal and vertical directions represented by the coordinates x and y, the position of the ball at a particular time is defined by the values of x and y. The values of x and y at the end of, say, 50 intervals of time are $x = 6$, $y = 10$.

Let us follow the path of the ball in terms of x and y, starting at 0. At the end of the first time interval $x = 1$ and $y = 1$. At the

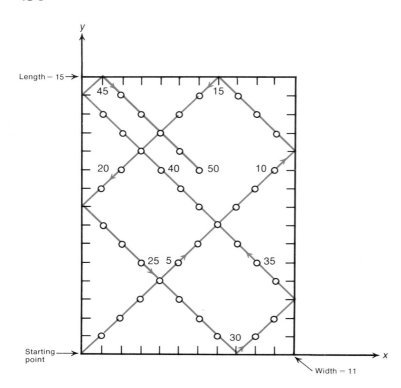

y

Length = 15 →

45   15

20   40   50   10

25 5   35

30

Starting point →

Width = 11

x

Fig. 11–12.
Idealized journey of
a ball on a billiard table.
Ball is projected at 45°
from the lower left
corner. Circles and
associated numbers
denote time
intervals.

end of the second interval $x = 2$ and $y = 2$. Thus, in each interval we add 1 to x and 1 to y.

At the end of 11 intervals $x = 11$ and $y = 11$: the ball strikes the edge of the table and rebounds. The ball now begins to travel back across the table in a direction that makes x smaller (y continues to increase as before). Later (when $y = 15$) the ball strikes and rebounds from the top of the table. Thereafter, the ball's direction is such that both x and y decrease with the passage of time.

We must program the computer to detect when x becomes equal to the width of the table and then to modify the program to subtract 1 from the value of x, rather than adding as before. When y becomes equal to the length of the table, the computer must then make the same modification for y. A flow chart for accomplishing this billiard-table calculation is shown in Fig. 11–13.

We first read in the table length (15 units), the width (11 units) and the total time (50 intervals), and we set x, y, and the interval counter to 0. We set two instructions A and B (shown toward the bottom of the flow chart) to *add* 1. These are the instructions that can be modified to *subtract* as the values of x and/or y require.

Looking at the flow chart, let's see what happens when $x = 5$ and $y = 5$. A test of x at the first test point shows that x is greater than 0. We therefore bypass the instruction modification and go

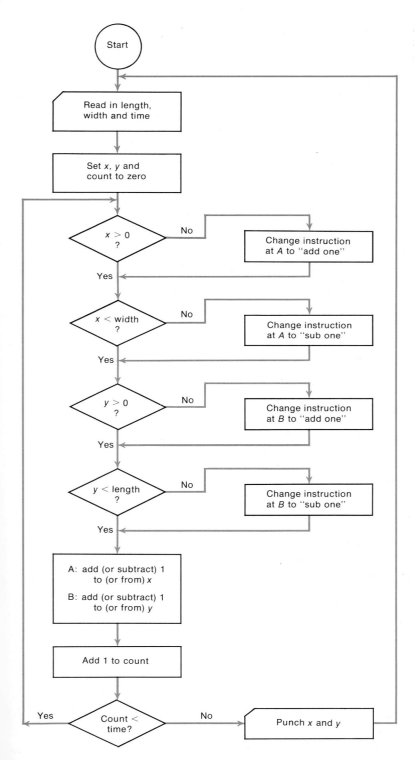

Fig. 11–13.
Flow chart for
billiard-table simulation.

to the second test point, which shows that x is less than the width. Therefore we again bypass to a similar pair of tests for y, which discloses that y is also greater than 0 and that it is less than the length. Thus, the two instructions A and B add 1 to x and y, respectively.

Next let us examine the situation where $x = 11$ and $y = 11$: the first point of rebound from a table edge. Now, the first test point shows that x is again greater than 0, so we again take the bypass to the second test point. Here, since x is now equal to the width, we do not bypass the instruction modification, and thus change the instruction at A so that it *subtracts* from x. Testing the y value, we find that no instruction modification is necessary, and that y is incremented by one.

Continuing to follow the ball's upward journey, we reach the top edge where the ball's rebound causes x and y to enter a phase in which both coordinates decrease. Thus, step by step, we make the flow chart picture the ball's journey. In Fig. 11–14 we show the problem coded as a computer program.

In this program we note that instruction 36 (A) covering x and 39 (B) covering y causes the computer to add 1 to the values of x and y in the following way. The instruction (155) at address 18, loads the word 200 (the contents of address 55) into the accumulator. Then the instruction (636) at address 19 copies 200 (the contents of the accumulator) into location 36. Similarly, 639 (the contents of address 29) stores the same information into location 39.

Now the word 200 says to add to the accumulator the contents of address 00. And in our computer the contents of address 00 is 001. Thus we add 1 to the accumulator contents, that is, to x or y.

In contrast, if the operation calls for "subtract 1," instruction 23, namely word 156, brings the word 700, which calls for "subtract 1" from the accumulator. This instruction is now copied into addresses 36 and 39 which previously called for "add 1."

The foregoing program can be varied in many ways to match different conditions. For example, it may be varied to follow a ball shot from the lower right corner, or to take into account the effect of "English" on the ball's path and recoil pattern, or the energy-sapping effects of friction. Since these physical factors are represented by numerical quantities in the memory, we can make the computer "model," or simulate, as many possible physical conditions as we please. And since a computer can execute all these instructions, computer-simulation programs can be made to represent a speeded-up advance view of a real-world situation before it happens. Such high-speed advance simulations—made possible in part by stored programs and modifiable instructions—are indispensable in space-flight experiments.

Address	Instruction	Comment
08	0 49	Read in *length*
09	0 50	Read in *width*
10	0 51	Read in *time*
11	4 40 ⎫	
12	6 52 ⎪	*Initialize x, y* and *count* to 0
13	6 53 ⎬	
14	6 54 ⎭	
15	4 40 ⎫	
16	7 52 ⎬	Create 0 − *x* and
17	3 20 ⎭	test it; if 0 < *x* go to address 20
18	1 55 ⎫	If 0 ⩾ *x*, get instruction "ADD ONE" and store it
19	6 36 ⎭	at address 36
20	1 52 ⎫	Create *x-width* and test it; if *x* < *width* go to
21	7 50 ⎭	address 25
22	3 25 ⎫	
23	1 56 ⎬	If *x* ⩾ *width*, get instruction "SUB ONE" and store
24	6 36 ⎭	it at address 36
25	4 40 ⎫	
26	7 53 ⎬	Create 0 − *y* and test it; if 0 < *y* go to address 30
27	3 30 ⎭	
28	1 55 ⎫	If 0 ⩾ *y*, get instruction "ADD ONE" and store
29	6 39 ⎭	it at address 39
30	1 53 ⎫	Create *y-length* and test it; if *y* < *length* go to
31	7 49 ⎭	address 35
32	3 35 ⎫	
33	1 56 ⎬	If *y* ⩾ *length*, get instruction "SUB ONE" and
34	6 39 ⎭	*store it at address 39*
35	1 52 ⎫	*Get x* and
(A) 36	0 00 ⎬	either add or subtract 1 from it, and
37	6 52 ⎭	store the result as the new value of *x*
38	1 53 ⎫	Get *y* and
(B) 39	0 00 ⎬	either add or subtract 1 from it, and
40	6 53 ⎭	store the result as the new value of *y*
41	1 54 ⎫	
42	2 00 ⎬	Add 1 to *count*
43	6 54 ⎭	
44	7 51 ⎫	Subtract *time* from *count* and test it; if
45	3 15 ⎭	*count* < *time* go back to address 15
46	5 52 ⎫	If *count* ⩾ *time*, print out the value of *x*
47	5 53 ⎬	and of *y*,
48	8 08 ⎭	and loop back to start
49	0 00	(*length*)
50	0 00	(*width*)
51	0 00	(*time*)
52	0 00	(*x*)
53	0 00	(*y*)
54	0 00	(*count*)
55	2 00	(instruction "ADD ONE")
56	7 00	(instruction "SUB ONE")

Fig. 11–14.
Program for
billiard-table simulation.

500

Suppose, for instance, that the space coordinates of the orbital motion of a space capsule are computed by rapidly solving equations in an appropriate computer program. If the computation proceeds quickly enough, the computer will know where the capsule will be long before it actually gets there. Computation which proceeds this rapidly is called *real-time computation*, although perhaps a better name would be "in-time" computation. Computation in real time is important in a variety of applications, such as the guidance of interplanetary probes and the control of production plants.

6 | FINAL COMMENT

We have seen in this chapter some of the techniques useful in making computers "jump through a hoop" of our own choosing. These techniques are all pointed at reducing the amount of human effort necessary to have the computer do what is wanted. Looping, subroutining, and the use of assemblers and compilers are the aids which make a computer user's life livable. There are still other aids that modern computers can and do supply, but they are all in this spirit.

Questions for Study and Discussion

1. List the ten basic operations of the computer discussed in this chapter and explain briefly how each is used.

2. Analyze the accompanying program and describe what it does.

Memory address	Word stored
22	028
23	129
24	228
25	628
26	528
27	822
28	000
29	005

3. What is a subroutine? Explain how a subroutine is used to reduce the number of stored instructions and how information and control are passed from the main program to the subroutine and back again.

4. Analyze the accompanying program and describe what it does.

Memory address	Word stored
20	029
21	030
22	130
23	729
24	631
25	531
26	130
27	629
28	821
29	000
30	000
31	000

5. The following program is one that might be used to find out whether or not a number A is larger than another number B. The top input card contains A, the second input card contains B. The answer "yes" is printed out as $+001$, the answer "no" is printed out as $+000$.

a) How many tests are required to determine if $A > B$?

b) If the question were "is $A \geq B$," how could this program be made shorter?

c) If the contents of the accumulator are positive at the time of the execution of instruction 23, what is the next instruction to be executed?

d) What does this program do if A is a negative number?

Memory address	Word stored
17	028
18	128
19	028
20	728
21	326
22	700
23	326
24	500
25	900
26	529
27	900
28	000
29	000

6. Explain how a loop conserves programming steps.

7. Explain how a bootstrap loader works. What else is necessary to be able to load a program and data into memory?

8. Explain how assemblers and compilers make programming easier.

9. If computers are as error free as suggested in these two chapters, how do we account for the numerous errors we hear about being made by department-store computers, bank statements done by computer, etc.?

Problems

1. What single machine code instruction would you write in order to have the computer do each of the following?

a) Read the top input card and put its contents into address (memory location) 34.   034

b) Add a copy of the contents in address 52 to the accumulator.   152

c) Clear the accumulator and bring a copy of the contents in address 95 into the accumulator.   195

d) Jump to the instruction stored in cell 24.   800

e) Copy the contents of the accumulator into cell 42.   642

f) Subtract a copy of the contents in cell 33 from the contents of the accumulator.   733

g) Shift the contents of the accumulator first one place to the left and then two places to the right.   412

h) Test the contents of the accumulator. If the content is negative, go to memory cell 13.   313

i) Print the contents of cell 19 onto an output card.   519

2. What is the meaning of each of the following instructions written in machine code? Write out the meaning of each in a complete English sentence.

a) 042   d) 440   g) 713   j) 516
b) 403   e) 672   h) 215   k) 900
c) 171   f) 819   i) 341   l) 309

3. If the top input card has the number 473 printed on it, and the second card has the number 052, what will each of the following programs do with these two numbers? (Assume that the top instruction is executed first.)

Memory address	Word stored
56	063
57	064
58	163
59	264
60	664
61	564
62	900
63	000

(a)

Memory address	Word stored
28	036
29	136
30	036
31	736
32	736
33	636
34	536
35	900

(b)

473
053
526

4. The contents of the accumulator are changing most of the time during any calculation. These changes in the accumulator are important. In each of the short programs below, tell what is in the accumulator after the execution of each instruction. (Assume +000 is in the accumulator before the first instruction is executed.)

Memory address	Word stored	Contents of accumulator
55	162	000
56	263	003
57	324	005
58	440	000
59	664	000
60	564	000
61	900	
62	008	
63	003	
64	000	

(a)

Memory address	Word stored	Contents of accumulator
27	134	
28	735	
29	735	
30	326	
31	636	
32	536	
33	900	
34	329	
35	127	
36	000	

(b)

5. Write as brief a program as you can (in machine code, starting at address 53) which will find and print out the value of M-N where M > N and M is positive. (Read M and N from punched cards into memory.)

6. Write a machine-code program for finding the value of (M-5N) as in Problem 5. Before writing the program, make a flow diagram for the problem.

7. Write a machine-code program that will print out any three numbers A, B, and C (on input cards) in descending order and then return for more input. Test this out on CARDIAC.

8. Draw a flow diagram and corresponding machine-code program which will examine a large set of numbers on input cards (a blank card marks the end of the set) and which will print out only those cards which have on them non-zero integers. Test this out on CARDIAC.

9. Write a flow chart and machine program which will print out only those input cards which have on them positive odd integers. (Suggestion: By a "433" instruction delete all but the right-most

digits of the numbers.) A blank card marks the end of the set of input integers.

10. Write a program which will use shifts to multiply two numbers read in from punched cards. Explain how this program is more efficient than repeated addition, considering the number of instructions executed.

11. What does the following program do?

Memory address	Word stored	
21	036	Read 36
22	136	LDA 36
23	431	× 100
24	638	STO 36
25	136	LDA 36
26	423	÷ 10
27	410	× 10
28	637	STO 37
29	136	LDA 36
30	413	÷ 100
31	237	ADD 37
32	238	ADD 38
33	636	STO 36
34	536	Writ 36
35	821	Go go 21
36	000	Read 00
37	000	Read 00
38	000	read 00

12. A survey is being made of the number of students in Johnson High School. The research committee has set up a list of the number of students in each section of each grade, and this was punched onto successive cards with −001 used to separate the grades and −002 used to indicate the end of the list. Write a subroutine which adds up the number of students within each grade and prints the total, and a mainline section which accumulates the number of students in the school grade by grade, and finally prints out the total number in Johnson High School.

Logical Thought and Logic Circuits

12

1 | INTRODUCTION

In the preceding two chapters, we have seen that a digital computer requires both input data and a program of instructions in order to produce its results. The major subunits in a computer are input, output, memory, processor, and control. All these work together automatically to help users solve problems so complicated that they could not be done in any other way. Perhaps the most startling aspect of computers' performance is that all their feats are done by simply repeating a cycle of events over and over. It is equally surprising that all computers' internal subunits, memory, processor, and control, are made from binary (two-valued) elements. In early computers, these were switches and relays. Today they are transistors, diodes, and magnetic cores. Of course, these elements, whatever they are, must be arranged, or organized, in very special ways to do the intended jobs.

First the binary elements are organized into simple electronic or electrical circuits. Several of these circuits are then combined into larger units according to certain laws. In turn, several of these are combined into still larger units until memory, processor, and control units are built. These are then organized to form the computer. This process of building in stages, understanding at each stage just how to combine simpler units into more complex ones,

*Switches
are elements
which are in one
of two states, off or on,
0 or 1, true or false, or any
other opposites. Here is a fluid
switch. Its input from the top is
switched into one of the two output
channels at the bottom. Fluid in
the horizontal arms controls
which of the two paths
the fluid in the switch
takes. (William
Vandivert)*

is a very powerful technique. It permits man to construct machines that will perform exactly as designed. Indeed, many man-made systems are constructed this way. However, this degree of understanding is rare in the world in general. Not nearly as much is understood about natural systems, such as the atomic nucleus, the human brain, or even so apparently simple a structure as molecular water.

Building in stages is illustrated by the composite picture of Fig. 12-1. The overall picture is of a gargoyle on Notre Dame Cathedral in Paris. This picture is made from a large number of small symbols which give the proper shading. The symbols, in turn, are each made from tiny letters, and finally, each letter is made of black dots. This kind of a "multilevel" structure is sometimes called *hierarchical*. In the picture of Fig. 12-1, we can see that creating each level requires that elements from the next lower level be arranged, or organized, properly. At each level, new properties emerge. This technique is basic not only to computers and other man-made systems, but also to such fields as language and chemistry. However, in each field the elements and the laws for combining them are different. Hierarchical structures are a peculiarly human creation. It is worth seeing in some detail how computers are created using this technique. This we shall do in this chapter and in the next three. In the end, we will have created a simple computer, a little slower (one million times), and with a smaller memory (one thousand times) than its multimillion-dollar electronic relatives. Yet, our computer will be complete and representative in every detail.

The first step in this process is organizing simple off-on (binary) elements into circuits according to rules of logic. The circuits are known as *logic circuits*. Logic circuits, like a simple light bulb, are always in either one of two conditions, current on or current off, 1 or 0. They behave in a simple, predictable manner. They follow rules of logic, just as a person does when he "thinks logically." (Sherlock Holmes and Perry Mason are famous practitioners of the art of logical thought: given the facts of a situation and the rules by which the facts are to be connected, they develop a logical pattern and follow it through to the inescapable logical conclusion.) Following a logical pattern is characteristic of the entire computer as well as of its individual components. We call our study of the logic circuits and their combinations "logical design."

Now we can be a little more specific about logical thought and its circuit equivalence. As an example of logical thought and its logical conclusion, we take a simple example that could occur in everyday life. Consider the statement: "I'll go to the beach if the weather is nice AND I get permission to use the family car." Here the conclusion is that the speaker will indeed go to the beach if both conditions are met: if it is true that the weather is nice

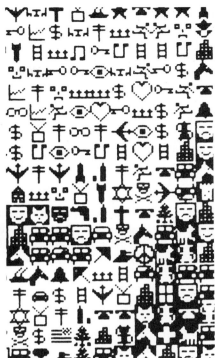

Fig. 12–1.
Gargoyle
on Notre-Dame
Cathedral in Paris.
(*Courtesy of Bell
Laboratories*)

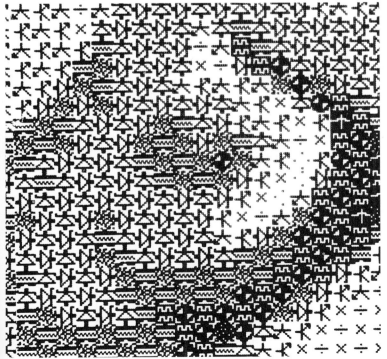

AND if it is also true that he obtains the family car. If either condition (or both) should be false, then the inescapable conclusion is that he will not go.

A less strong statement would be: "I'll go to the beach if the weather is nice OR the surf is up." Here only the first OR the second condition need be met, not both (although the combination of both might be acceptable, too). We can diagram the logical connection of logical conditions (Fig. 12–2).* Thus for an AND connective we need two True's to make the statement true, while for an OR connective we need only one (or more) True's to make the statement true. Any such logical combination is again either true or false. It can therefore be used as a logical condition within a more complex condition.

To view NOT as a logical connective may seem odd at first, but it acts much like AND and OR in combining logical conditions: "If a condition 'X' is true, 'NOT X' will be false, but if 'X' is false, 'NOT X' will be true." "X" and "NOT X" are said to be "complementary."

How, then, do we model such logical combinations with electrical circuits? Clearly, such a "logic" circuit must be able to model simple phrases which are either true or false. Consider the common light switch: it is either on or off. If we agree that a switch may represent the clause or condition "the weather is nice," then a light connected to this switch will turn on if the switch is thrown to "on," representing "true," and the light will go off if the switch is thrown to "off," representing "false" (i.e., it is not nice weather). Combinations of such simple logic circuits implementing the logical connectives AND and OR are developed in the next section.

2 | HOW TO MAKE ELECTRIC CIRCUITS SAY "AND" AND "OR"

The AND Circuit

Logic circuits—electric circuits we can switch on or off— are so much a part of our daily life that we never give them a second thought as we turn on lights, dial telephone numbers, operate automatic elevators, or make use of pedestrian-operated traffic lights.

When a person walks into a dark room, he flicks the light switch from "off" to "on." He knows that he has succeeded when the light appears. What he has done is to establish a metallic connection from a source of electricity to the light bulb. The actual source of this energy may be a generator owned by the electric company, to which the wall outlet in the room is connected, or it might be a battery. In either case, the source has two terminals which we label "+" and "−" for identification purposes, and the lamp must be connected between these (Fig. 12–3a) in order to be lighted. The connection is usually a copper wire and it conducts

*The diagrams of Fig. 12–2 may imply that the conditions must be satisfied in sequence. Actually, of course, the conditions are determined simultaneously.

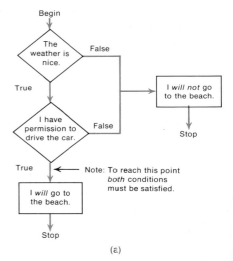

(a)

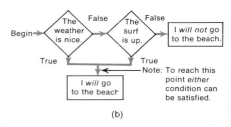

(b)

*Fig. 12–2.
Diagramming
logical conditions.*
a) *The diagram and flow chart for logical AND.*
b) *The diagram and flow chart for logical OR.*

electric current from one terminal of the battery or plug through the lamp and back to the other terminal.

For our immediate purpose, the only fact of importance is that an unbroken metallic connection from a source through a lamp and back to the source is necessary to light the lamp. Thus, to turn the light on we complete the path, and to turn it off we interrupt the path. These actions can be accomplished by inserting a switch, as shown in Fig. 12–3b. When the switch is *operated* (on), a pair of metal points (the pair of points is called a *contact*) press against each other. This action completes the path for the current. When the switch is *released* (off), the metal points no longer press against each other and the path is interrupted. In this arrangement the switch contact is said to be *in series* with the lamp.

It is possible to control a light with two switches in such a way that the light is on only when both switches are operated. The circuit in Fig. 12–4 shows how this is done using a *series* arrangement of the contacts in the two switches. This arrangement is sometimes called an AND circuit since the light is on only when switch A *and* switch B are both operated.

The operation of the AND circuit is summarized in Table 12–1. Another way of describing this circuit is to concentrate on the condition of the two contacts and the condition of the resulting path through both of them in series. If we let *0 stand for an interrupted (open) contact or path and 1 stand for a completed (closed) contact or path*, the *table of combinations* in Table 12–2 lists all possible combinations of conditions controlled by the two switches.

It is easy to imagine an important application for our AND circuit. It could be used, for example, to activate the firing circuit for a rocket when both of two operators in the blockhouse must agree that the firing should take place.

We can also use the AND circuit to represent a logical connection such as the one used in testing the "going to the beach"

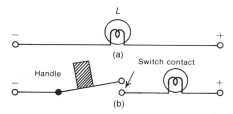

Fig. 12–3.
An electric light controlled by a switch.
a) A lamp with connecting wires.
b) A lamp in series with a switch contact.

Fig. 12–4.
The series (or AND) configuration of two contacts. A lamp controlled by two contacts.

Switch A	Switch B	Light
Released	Released	Off
Released	Operated	Off
Operated	Released	Off
Operated	Operated	On

Table 12–1.
Description of the operation of the AND circuit.

Path through contact on switch A	Path through contact on switch B	Path through the contacts in series
0	0	0
0	1	0
1	0	0
1	1	1

Table 12–2.
Combinations for the AND circuit.

condition: let switch A represent the simple condition "the weather is nice," and let B represent "having permission to use the family car." Then the "going to the beach" light will light only if *both* simple conditions are true, i.e., if both switches are on (operated). In a "truth table" (Table 12–3) for this use of the circuit, "False" entries correspond to "released," or 0, "True" to "operated," or 1. From here on the shortest table of combinations, Table 12–2, will be used to represent all three logically equivalent ways of looking at the AND circuit (the three descriptions of operation in terms of switches, in terms of combinations, and in terms of truth values). In general, the other circuits yet to be developed will be similarly represented.

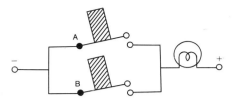

Fig. 12–5.
The parallel (or OR)
configuration of two contacts.
A lamp controlled by two
contacts in parallel.

Table 12–3.
Truth table for the
AND circuit.

C_1	C_2	C_1 AND C_2
F	F	F
F	T	F
T	F	F
T	T	T

The OR Circuit

Another basic way of connecting contacts from two switches is to put them in *parallel* with each other, so that the operation of either switch will complete a path through the circuit. In Fig. 12–5 it can be seen that there is a path for electric current if either switch A *or* switch B is operated (or if *both* are operated). This parallel connection of contacts is, therefore, commonly called an OR circuit.

As in the case of the AND circuit, the OR circuit can be summarized by a tabular description (see Table 12–4). Using 0's and 1's as before, we can write a table of combinations (Table 12–5) which shows how the condition of the parallel configuration of

Table 12–4.
Description of the operation
of the OR circuit.

Switch A	Switch B	Lamp
Released	Released	Off
Released	Operated	On
Operated	Released	On
Operated	Operated	On

Table 12–5.
Combinations for the
OR circuit.

Path through contact on switch A	Path through contact on switch B	Path through the contacts in parallel
0	0	0
0	1	1
1	0	1
1	1	1

contacts depends upon the condition of the individual contacts. The corresponding truth table is shown in Table 12–6.

C_1	C_2	C_1 OR C_2
F	F	F
F	T	T
T	F	T
T	T	T

Table 12–6.
Truth table for the
OR circuit.

As an application of the OR circuit, we can imagine that each of two astronauts lying in the capsule on the top of the rocket has a switch which he controls. The first astronaut operates his switch (A) whenever he thinks there is a fuel leak. The second astronaut, also on the lookout for fuel leaks, operates his switch (B) when he thinks he has found one. The light, controlled by the OR circuit, will go on whenever either astronaut senses trouble. (How would an astronaut feel if he knew that the warning light was controlled by an AND circuit?)

The AND and OR Circuits Combined

In a different example, we assume that the pilot of a high-performance jet airplane has just lost a large portion of a wing. He must eject, but two distinct steps are necessary. First, the cockpit canopy must be blown off; second, the seat (and he) must be ejected by another explosive charge. A straightforward method would be to fire the canopy charge with one switch and the seat ejection charge with another. But then in the confusion of the emergency, he might fire the seat charge first, and it and he would then be projected through the closed canopy—an undesirable result. How can the switches be arranged so that regardless of the order in which the switches are operated, the first will fire the canopy charge and the second will fire the seat charge? There is a way, perhaps already guessed, since it uses only a combination of the AND and OR circuits. However, a slightly different type of switch is required.

In the switches discussed so far, a single contact is controlled by a single handle or lever. To solve the pilot's problem we need a switch in which *two* contacts are controlled by a single lever. The two contacts are linked together by non-metallic material that prevents the flow of electricity from one to the other, so that they can be used in two electrically separate circuits. When the switch is operated, *both contacts close simultaneously*, and when the switch is released, *both contacts open simultaneously*.

Now let us return to the pilot's problem. We can restate it in terms of AND and OR. The canopy blowoff charge should be fired when either switch A OR switch B is operated. The seat ejection charge should be fired when switch A AND switch B are operated. Consequently, the combination of the two circuits in Fig. 12–6 does the required job. When only one switch is operated, no matter which one it is, the canopy blows off. Only when the second switch is also operated is the seat ejected. (The two contacts controlled by switch A have both been marked "a," and the two contacts controlled by switch B have both been marked "b." The reason for the distinction between a switch and its contacts will become apparent later.)

Let us close this section with a summary of the terms we have used to refer to switches. *Switches* may be in one of two *states: released* or *operated*. A *contact* controlled by a switch may also be in one of two states: *open* or *closed*. It is also true that a *path* through a set of interconnected contacts is open or closed, whether the path is through a single contact or through a complex configuration of contacts. At a given time a path is either interrupted (in the open state) or it is completed (in the closed state). As already pointed out, it is useful to use the pair of symbols 0 or 1 to stand for "open" and "closed" respectively, in speaking of the condition of a contact or a more complex path. A major source of difficulty can develop from failure to remember that 0 and 1 are here merely a pair of *arbitrary symbols* and that *they are not numerical values*. Switch contacts can be connected in series or parallel. The logical connective AND corresponds to a "series" connection, and OR corresponds to a "parallel" connection.

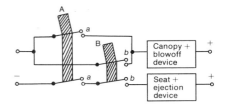

Fig. 12–6.
A circuit with contacts both in series and in parallel.

3 | THE MAJORITY-VOTE PROBLEM

Three legislators wish to vote on a number of issues and have their votes anonymous. Each one is to control a switch which is labeled "No" in the released position and "Yes" in the operated position. A lamp indicating that the majority vote is favorable is to be lighted whenever two or three of the members vote "Yes." One way of restating these requirements as a "compound" logical condition composed of simple conditions (C1, C2, and C3) and logical connectives is:

> The control circuit is to be closed if,
> (C1) switches A AND B are operated,
> OR (C2) switches A AND C are operated,
> OR (C3) switches B AND C are operated.
> (If switches A AND B AND C are all operated, then conditions 1, 2, and 3 are all satisfied at once, in which case the votes are no longer anonymous.)

To design the circuit, we remember that AND calls for a series connection and OR calls for a parallel connection. Thus, condition C1 corresponds to a series connection of contacts *a* and *b* on switches A and B. Similarly, condition C2 requires contacts *a* and

c to be placed in series, and C3 requires a series connection of contacts b and c. In addition, conditions C1, C2, and C3 are connected by OR's; thus, the above three series arrangements should themselves be connected in parallel, as shown in Fig. 12–7a. (Switch contacts have been represented by the symbol —×— in this figure. Representations of the switch handles have been omitted for the sake of clarity.)

To check that the circuit really performs as advertised, we must trace through the circuit with all possible combinations of switches open and closed to find when the overall circuit is closed and when it is open. First, we construct a table as shown in Table 12–7, listing the switches A, B, and C across the top and their various possible conditions: 1 for *Operated* and 0 for *Released*. For example, the first row, 0, 0, 0, means that all three switches are released; the second row, 0, 0, 1, means that A and B are released but C is operated. Now make an additional column corresponding to conditions C1, C2, and C3 above. We put a 1 in that column wherever the compound condition (C1, or C2, or C3) is satisfied, i.e., wherever A and B, or A and C, or B and C are 1 in the table; otherwise we put 0. Next, we construct a column corresponding to the state of the circuit just designed. We enter a 0 whenever the circuit is open and a 1 whenever it is closed. For example, with

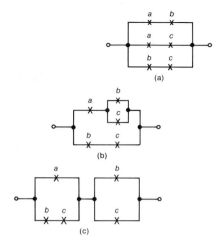

Fig. 12–7.
Three logically equivalent majority circuits.

Table 12-7.
Combinations for the majority-vote problem.

A	B	C	Logical conditions	State of network
0	0	0	0	0
0	0	1	0	0
0	1	0	0	0
0	1	1	1	1
1	0	0	0	0
1	0	1	1	1
1	1	0	1	1
1	1	1	1	1

A, B, and C all operated (1, 1, 1) the circuit is closed, 1, since all three series branches are closed. When A and B are operated and C released (1, 1, 0), the circuit is also closed, 1, since the uppermost branch is closed (the lower two branches are open). This column agrees with the previous one unless there is an error in the circuit design.

The logical statement of circuit requirements at the beginning of this section is only one of several ways to state those requirements. Another is:

The control circuit is to be closed if,
(C1) A is operated AND either B OR C is operated,
OR (C2) B AND C are operated.

An OR connective leads to two circuits in parallel, one of which (C1) is a series circuit, etc. Thus we get the circuit in Fig. 12–7b. Another way is:

The circuit is to be closed if,
(C1) A is operated, OR B AND C are operated,
AND (C2) B OR C is operated.

The resulting circuit is shown in Fig. 12–7c. There are many other ways of connecting contacts controlled by switches A, B, and C to achieve the desired result; namely, to indicate a majority vote. All, however, have the same table of combinations indicated in Table 12–7, since they must all give the same result when a particular set of switches is operated. It is always true that two or more circuits with the same table of combinations are logically equivalent. One circuit may be more "efficient" than another, however. For example, circuits b and c in Fig. 12–7 each have one fewer contact than a. In a sense we have "factored out" one contact.

It is interesting and often helpful to express these switch combinations in another way (using Boolean algebra which was introduced in Chapter 2). To do this, we express the condition AND by the symbol $\cdot$ (the sign of multiplication), and the condition OR by the symbol $+$. Then C1 of the original list leads to the representation $a \cdot b$ (or just ab); C2 gives ac, and C3 is expressed by bc. Now since these three conditions are connected by OR's, one can write, as a shorthand for the circuit, $ab + ac + bc$. The second way of stating the logical requirements (A AND either B OR C, OR B AND C) becomes $a(b + c) + bc$. This expression reduces to the previous one if the parentheses are removed as if this were an ordinary algebraic expression, but in the form shown it represents the circuit of Fig. 12–7b. This way of describing the circuit makes it clear that "factoring out" a contact has a real meaning. The circuit of Fig. 12–7c corresponds to still a different Boolean expression.

4 | HOW TO MAKE AN ELECTRIC CIRCUIT SAY "NOT"

We have seen that AND and OR can be represented by series and parallel connections of switch contacts within a circuit. There is a third basic connective of logic—the word NOT. How can we represent NOT in circuits having only switches and their contacts? The answer requires that we look at another picture of a switch to see how a handle or lever can control the opening and closing of contacts.

A single switch, A, controlling two different contacts is shown in Fig. 12–8. Whenever the upper one is closed, the lower one is open, and vice versa. Thus these contacts behave in the *comple-*

mentary way which we associated with NOT. When the upper contact is closed, the lamp L_1 is lighted and the lower contact is open so that the lamp L_2 is not lighted. On the other hand, when the upper contact is open and the lamp L_1 is not lighted, the lower contact is closed and the lamp L_2 is lighted.

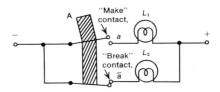

Fig. 12–8.
A single switch controlling complementary contacts.

We saw in the ejection problem that a single handle or lever could be used to control one or more contacts in an electric circuit. Now we also see that a single switch handle or lever can be made to close one group of switch contacts, while simultaneously opening the contacts of a complementary group. With this arrangement, shown in Fig. 12–8, the upper and lower contacts are always in the opposite states. Thus, if we call the upper contact *a*, the lower one will always be "not *a*."

The entire circuit of Fig. 12–8 is sometimes used in photographic darkrooms to warn people coming to the door that film is being developed inside and that the door cannot be opened without ruining the film. Needed in this situation is a red warning lamp outside the door that goes *on* when the lamp in the darkroom goes *off*. The red lamp might be L_2 and the darkroom lamp L_1.

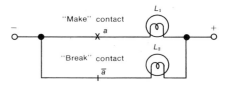

Fig. 12–9.
Two complementary types of contacts, controlled by the same switch A.

This notion of upper or lower contact is clearly too cumbersome to represent in a diagram, so we use a more natural notation. We speak of a contact as being either the "make" or the "break" type. In Fig. 12–8 the upper contact is the make contact and the lower contact is the break contact. The diagram shows the switch in the *released state:* in that state *the make contact is open and the break contact is closed.* Depressing the lever in the diagram puts the switch into its *operated state:* in that state *the make contact is closed and the break contact is open.* If the switch is labeled with a capital letter A, the *make contacts* associated with that switch are labeled with the corresponding *lower-case letter a* and the *break contacts* are labeled with that letter with a bar over it, $\bar{a}$. The *make contact* in circuit diagrams (see Fig. 12–9) is denoted by a *cross* (—×—); the *break contact* is denoted by a short perpendicular *bar* (—|—). (It helps to remember that both "bar" and "break" begin with the letter *b*.)

There is a common circuit which requires both make and break contacts. It is the circuit used to turn a light off and on by either of two switches. This is exactly the arrangement needed to control a single light from either end of a hall or set of stairs. Changing the state of one switch always changes the condition of the light, regardless of the state of the other switch. In other words, if you are at the top of the stairs in the dark, you just flip the switch and the stairs are lighted, regardless of which switch turned off the light.

The circuit given in Fig. 12–10 shows how contacts on the switches A and B can be used to control the hall light. The contact network consists of two circuits in parallel and each of these consists of two contacts in series. There will be a path through the contact network when either

> (C1) switch A is operated AND switch B is NOT operated (released),
>
> OR (C2) switch A is NOT operated (released) AND switch B is operated.

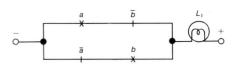

Fig. 12–10.
Hall light circuit.

Tables 12–8 and 12–9 list the properties of the circuit in terms of the operation of the switches and, equivalently, in terms of the states of the make contacts of the two switches. If a person stands at switch A (Table 12–8), he can turn the hall light off or on, whether switch B is operated or released. For instance, if he is at A, and switch A is released and the light is off (this corresponds to the first row of the table of combinations), he can operate switch

Switch A	Switch B	Path	(Light)
Released	Released	Open	(Off)
Released	Operated	Closed	(On)
Operated	Released	Closed	(On)
Operated	Operated	Open	(Off)

Table 12–8.
Description of the operation of the circuit.

A	B	Path
0	0	0
0	1	1
1	0	1
1	1	0

Table 12–9.
Combinations for the circuit.

A and the light will go on (row 3, Table 12–8). If the light had been off with switch A operated (row 4), he would have released switch A (row 2). We can verify that the same thing can be done from the opposite end of the hall by using switch B.

If we compare the table of combinations of this circuit (Table 12–9) with that of the OR circuit (Table 12–5), we note that they differ only in the last row. For OR, both switches may be operated (both conditions true) to complete the circuit (compound condition is true), but this is not allowed for the hall light circuit. If we did use an OR, then if switches A and B were both operated, the light would be on, and neither A nor B *by itself* could turn it off. The hall light circuit corresponds to an OR condition in which condition 1 and condition 2 but *not both* can be true, and we call this the *exclusive OR*. It contrasts with the normal or *inclusive OR* (either A or B or, inclusively, both):

Inclusive OR	Exclusive OR
I will go to the beach if	John will get married if
the weather is nice	he chooses Joan OR
OR the surf is up.	he chooses Linda.*

*
That is, if John lives in America, he'd better marry only one at a time.

Another way of looking at the exclusive OR circuit is that it is closed whenever just *one* of the switches is operated, and is open whenever exactly *none or two* of the switches are operated. It is possible to generalize this idea and design circuits, having three or more switches, which are *closed whenever any odd number of switches are operated* and open whenever any even number of switches are operated. A circuit of this kind is called an *odd-parity circuit*, and it will be met again in the design of a circuit that counts, for use in our computer.

5 | ANALYSIS VS. SYNTHESIS

Until now we have stated a problem and then looked at a circuit (and its table of combinations) to see that this circuit did indeed satisfy the conditions of the problem. This approach is called circuit *analysis*, but it requires that someone give us a circuit to check out. The circuit *synthesis* approach is more realistic in that we ourselves design the appropriate circuit according to the logical conditions of the problem and then check it out. As an example of this technique, we now synthesize an odd-parity circuit using three switches, A, B, and C. This circuit should be closed whenever 1 or 3 switches are closed, and open whenever 0 or 2 switches are closed. The easiest way to synthesize circuits is to rephrase the logical conditions of the problem in terms of a table of combinations, and then to translate the table into a circuit which satisfies it.

Step 1. We form all the rows of the table of combinations, starting, by convention, with 0, 0, 0, where 0 corresponds to *released* or *False*, 1 to *operated* or *True*.

Switch A	Switch B	Switch C
0	0	0
0	0	1
0	1	0
0	1	1
1	0	0
1	0	1
1	1	0
1	1	1

(There is a very regular pattern of the zeros and ones as they exhaust all possible combinations: the rightmost column alternates, the middle one goes by twos, the leftmost by fours.)

Step 2. Now we want our circuit to be completed (1) whenever one or three switches are closed (1): for those rows which have a single 1 or three 1's. We inspect each row and put down a 1 in a new column if it meets these conditions.

			3-switch odd-parity
A	B	C	
0	0	0	
0	0	1	1
0	1	0	1
0	1	1	
1	0	0	1
1	0	1	
1	1	0	
1	1	1	1

Step 3. The circuit should not be completed if there are zero or two switches closed. We consider each row and put down a 0 if it meets these conditions.

A	B	C	Circuit
0	0	0	0
0	0	1	1
0	1	0	1
0	1	1	0
1	0	0	1
1	0	1	0
1	1	0	0
1	1	1	1

We have filled all the holes in the previous stage of the table and, by so exhausting all possibilities and their corresponding results, have prescribed the actions of our desired circuit exactly. Now that the requirements have been stated in tabular form, all that remains is to interpret these requirements in circuit form.

Step 4. We translate the table of combinations by restating those rows for which the circuit is completed (rows that result in 1) in terms of simple logical conditions and their circuit equivalents. The table, in effect, tells us that a path through this particular circuit should be completed if switch A is released AND switch B is released AND switch C is operated; OR it should also be completed if A is released AND B is operated AND C is released, etc. To summarize all rows, we can say that the circuit path should be completed *if and only if*

> A released AND B released AND C operated,
> OR A released AND B operated AND C released,
> OR A operated AND B released AND C released,
> OR A operated AND B operated AND C operated.

Using parallel circuits for OR connectives, series for AND, we can draw our circuit as in Fig. 12–11. If one of these four conditions should not hold, then our circuit is not completed, a conclusion which checks with the 0 entries in the table of combinations. The same sequence of steps can be used to synthesize circuits corresponding to any logical condition.

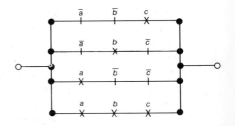

Fig. 12–11. Circuit model of 3-switch odd-parity circuit taken directly from the table of combinations.

We saw earlier that there are several circuits which satisfy a logical condition. Some of these have fewer contacts and so are simpler than others. In fact, there is a large collection of techniques for simplifying logic circuits. Simplification for relay circuits means reducing the number of contacts necessary to satisfy a table of combinations. These techniques have been accumulated through experience over the years, and we will not go into them here. However, the circuit of Fig. 12–11 can be simplified to that of Fig. 12–12, which has only 8 contacts. Yet it still satisfies the truth table for the 3-switch odd-parity circuit.

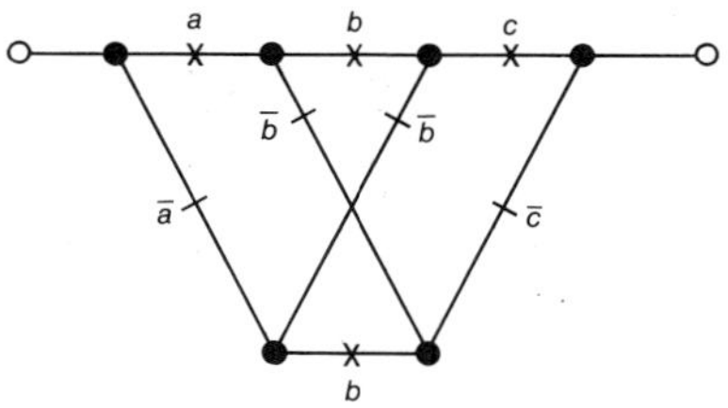

Fig. 12–12.
Final simplification of the 3-switch odd-parity circuit.

6 | ADDITIONAL TOPICS ON LOGICAL THOUGHT AND CIRCUIT MODELS

Review of Logical Thought

Centuries ago, mathematicians founded a branch of mathematics which they called *logic* (from a Greek word meaning "reasonable"), believing that all problems of the real world could be settled by applying its rules. Today we know this is an impossible feat. Nevertheless, even today, the ability to draw conclusions which are consistent with facts or theories (sometimes called *postulates*, as in geometry) is considered a mark of intelligence. Often on intelligence tests there are questions such as "Decide if the conclusion 'All zryks sneeze continuously' can be deduced from the postulates 'All three-headed zryks sneeze continuously' and 'All zryks have three heads.'" After the initial shock of such a question has worn off, we recognize that whether or not the premises are reasonable, possible, or foolish is not at issue. The question seeks to determine if the person being tested can decide what conclusions can legitimately be deduced from even the most unlikely or mysterious premises. This type of nonsense question is purposely designed so that experience with the subject of the question, or lack of it, will neither hinder nor help in deducing a legitimate conclusion. Not everyone agrees that questions of this sort really do test intelligence, but many people seem to think so. As a matter of fact, some of the conversations you hear at school may sound suspiciously like these questions.

Regardless of this issue, there is a *correct* answer to such questions: an answer which can be derived by applying a set of rules. Furthermore, these rules can be stated using the logical connectives AND, OR, and NOT which were incorporated into the logic circuits described earlier in this chapter, provided that the facts can be represented as circuit elements. This ability to deduce logical conclusions rapidly and accurately is vital in a variety of complex situations. Such situations exist in air-traffic control, in automatic observation of hospital patients, and in the operation of petroleum refineries, for example.

Basically, the rules of logic are concerned with the truth or falsity of compound statements, such as "The wind is from the northeast and it is raining." There are two component statements in this composite statement, the two being joined by the connective *and*. In logic, the truth of this statement does not depend upon how accurately we measure wind direction or whether we actually observe the rain. Logic specifies only that if both statements are true, the compound statement is logically true; otherwise the compound statement is logically false. In other words, logic deals with the truth of the *relationship* between statements and not with the subject matter of the statements themselves—just as the AND circuit deals with the state of switch contacts and their series connection, not with the name plates on the switches, the size and shape of the contacts, or for that matter any other irrelevant facts about the contacts. To emphasize this point, logicians often use nonsense statements, such as "Gorgs can play the cello" and "Zryk feathers make fine pillows."

Since the fundamental rule in logic is that the only allowable truth-values for statements are *true* or *false* (which correspond to contacts and networks of contacts having two *possible* states, *closed* or *open*), words like *maybe* and *perhaps* (useful in everyday conversation) are not useful in making the definite statements necessary for logical thought.

But a system of logic based on statements that can be only true or false is not complete without *complementary statements*. A complementary statement is one whose truth-value is opposite to that of the original statement. For example, the statement complementary to "Francis smokes cigars" is "Francis does *not* smoke cigars." The statement complementary to "Gorgs can*not* play the cello" is "Gorgs can play the cello."

It is always possible to express the complementary statement by using *not*, but often we have other English words which we interpret the same way: "The play was good," "The play was bad (not good)"; "The integer is odd," "The integer is even (not odd)"; "That man is honest," "That man is dishonest (not honest)."

A word of warning is appropriate, however. We may be used to thinking that certain pairs of words have complementary meanings when they do not. The only safe way to express a complementary idea is to use the word "not": full, not full; empty, not empty; fast, not fast; walk, not walk; and so forth.

Finally, to find out what the logical meaning of a logical connective is, it is necessary to draw the defining truth table.

A Model for a River-Crossing Puzzle

A logic circuit using contacts is ultimately nothing more than a representation of the equivalent logical word condition. One type of application in which this correspondence is quite direct is the representation of the rules for certain kinds of logical puzzles. A traditional problem is the following:

A boatman must carry a wolf, a goat, and a cabbage across a river in a boat which is so small that he can carry, at most, only one of them with him in it at a time. Moreover, whenever the wolf and goat are together, he must also be present to keep the goat from being eaten. Neither can he leave the goat alone with the cabbage. How can he carry all of them from the south bank of the river to the north bank, and not have the goat or the cabbage eaten?

A key initial step for this sort of puzzle is to find a representation for the situation at any given time. Here we let four switches, M (man), W (wolf), G (goat), and C (cabbage), represent the four main characters in the cast. When one of them is on the south shore, the associated switch will be released, and when on the north shore the switch will be operated. The puzzle then becomes one of how to start with all switches released, and then to operate all of them without violating the rules which have been set down.

The circuit designed later in this section does not itself solve the problem, but it serves as a model of the situation so that various tentative approaches to the problem can be tested without actually getting a wolf, goat, cabbage, boat, and a conveniently wide river. The circuit is used to turn on a warning light when one of the conditions of the problem has been violated.

One condition is that on either river bank the wolf should never be with the goat without the man being present. Thus, when M is operated (man is on the north shore) and W and G are both released (wolf and goat are on the south shore), the warning light should be on. It should also be on when M is released, and W and G both operated. The other condition is that the goat and the cabbage should never be together without the man being present. Thus, when M is operated and C and G are both released, or when M is released and C and G are both operated, the warning light should be on.

The two preceding pairs of conditions (four in all) lead to a contact network with four corresponding sets of contacts in series, all of these to be placed in parallel with each other. These are:

$$\begin{aligned} \text{C1} \quad & \bar{w}, \bar{g}, m \\ & w, g, \bar{m} \\ \text{C2} \quad & \bar{c}, \bar{g}, m \\ & c, g, \bar{m} \end{aligned}$$

The corresponding circuit is given in Fig. 12–13a. A simplified circuit which is equivalent to that one is given in Fig. 12–13b. The circuit is readily derived by using Boolean algebra. The stated conditions lead to the expression:

$$\bar{w}\,\bar{g}\,m + w\,g\,\bar{m} + \bar{c}\,\bar{g}\,m + c\,g\,\bar{m}$$

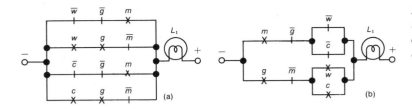

Fig. 12–13. Circuit models for the "river-crossing" problem. a) First form of the circuit. b) Alternate form of the circuit.

We can combine the first and third terms to yield $\bar{g}\, m\, (\bar{w} + \bar{c})$ when factored; in the same way the second and fourth give us $g\, \bar{m}\, (w + c)$. Since these are still in parallel the final expression is $\bar{g}\, m\, (\bar{w} + \bar{c}) + g\, \bar{m}\, (w + c)$, which is the shorthand for the network of Fig. 12–13b. The reader should verify that the two circuits are equivalent by examining each of the sixteen possible combinations in which the four switches can be operated or released.

The rules of the puzzle tell us that the boatman can travel alone between the banks of the river or can transport, at most, one of the others with him. If he goes alone from the north shore to the south shore, this is represented on our model by moving the switch M from the operated to the released position. If he takes the wolf to the north shore from the south shore, this is represented by operating both of the switches M and W.

The puzzle now becomes one of starting with all switches released and of operating or releasing the switch M (and at most one of the other switches) so that eventually all of the switches are operated, and so that in the process the warning light does not go on. Furthermore, to go from one situation or *state* (i.e., a given row in the table of combinations describing where all four participants are) to another, the player may change the M switch (0 to 1 or 1 to 0), meaning the man has crossed the river alone, or two switches (M rowing the boat, and one other) for the crossing of two participants, but never more than two switches. One possible sequence of switch states which solves the problem is shown in Table 12–10. That solution is:

1. The man crosses to the north shore with the goat.
2. He goes back to the south shore alone.
3. The man crosses to the north shore with the cabbage.
4. He goes back to the south shore with the goat.
5. The man crosses to the north shore with the wolf.
6. He goes back to the south shore alone.
7. He crosses to the north shore with the goat.

There is another slightly different solution. Can you discover the other solution? How many steps are necessary to develop the alternate solution?

W	C	G	M	Light	Solution
0	0	0	0	0	
0	0	0	1	1	
0	0	1	0	0	
0	0	1	1	0	
0	1	0	0	0	
0	1	0	1	1	
0	1	1	0	1	
0	1	1	1	0	
1	0	0	0	0	
1	0	0	1	1	
1	0	1	0	1	
1	0	1	1	0	
1	1	0	0	0	
1	1	0	1	0	
1	1	1	0	1	
1	1	1	1	0	

Table 12–10. Combinations and a solution of the "river-crossing" problem.

7 | FINAL COMMENT

This chapter shows how electric circuits made up of on-off elements (contacts controlled by switches) are put together to represent the basic connectives of logic: AND, OR, and NOT. Furthermore, these basic contact circuits can be combined to represent, or model, logical situations such as the river-crossing puzzle, or to perform logical operations such as in the majority circuit. Logic circuits are at the root of automatic aids to human thought and action. Thus, any problem or operational requirement (as in the hall-light problem) which can be stated in logical form can be met by a logic circuit.

Logic circuits are not the only aids to thought. There are many others, but none relieves us of the obligation to think. For example, reducing a real problem to logical terms may be a mental challenge as great as or greater than the logical manipulations themselves. Furthermore, a mistake in this reduction can easily result in an inappropriate answer. Nevertheless, aids to thought put us a step ahead.

Also, we have progressed a significant step toward understanding digital computers. We have seen that switch-controlled contacts properly organized into circuits can perform useful tasks. We will see in the next chapter how these circuits can in turn be assembled into larger units to form the basic subunits of a computer.

Questions for Study and Discussion

1. How will the firing circuit for a rocket be controlled when three, or more, operators in the blockhouse must all concur that it should be fired?

2. How would you draw a circuit to model the second "going-to-the-beach" problem in Section 1?

3. Describe one or more situations which might require the operation of three contacts in series.

4. Describe one or more situations which might require the operation of three contacts in parallel.

5. Define *odd-parity* and *even-parity* as used in contact network analysis.

6. Describe one or more situations where a majority circuit would be appropriate.

7. Sam Smith, an explorer, discovers two strange tribes in the jungle. One of these tribes consists of people who never tell the truth, but the other tribe consists of people who never lie.

On one of his journeys Sam encounters a tall and a short tribesman walking together. He recognizes that each is a member of a different tribe but he cannot tell which of these people tells the truth and which one of them lies, so he asks the tall tribesman "Do you tell the truth?"

The tall tribesman answers "Groom."

The short tribesman who speaks English says "He says yes, but he is a liar."

Which tribesman is the truthful one? Justify your answer.

Problems

1. Draw a diagram of a circuit for the following:

   a) $a \cdot b + c$

   b) $a \cdot (b + c)$

   c) $a \cdot (\bar{a} + b)$

   d) $b \cdot [c + (\bar{a} \cdot c)]$

   e) $(a \cdot b \cdot c) + (a \cdot \bar{b}) + (\bar{a} \cdot \bar{b} \cdot c)$

   f) $(\bar{a} \cdot c) + (\bar{a} \cdot \bar{c}) + (\bar{b} \cdot \bar{c}) + (a \cdot b \cdot \bar{c}) + (\bar{a} \cdot b \cdot \bar{c})$

2. Design two more equivalent three-switch majority-vote circuits, using the Boolean algebra method.

3. Design a four-switch even-parity circuit.

4. A doorlock is to be operable only when time switch T and manual switch M are both activated. Draw the circuit from the components shown below.

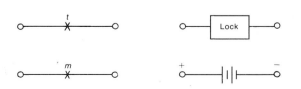

5. A single house electric bell is to be operated when either the front- or rear-door push buttons are operated. Draw the wiring diagram.

6. Review the seat-ejection problem in Section 2. Complete the table of combinations for the circuit in Fig. 12–6.

A	B	Canopy blowoff charge	Seat ejection
0	0		
0	1		
1	0		
1	1		

7. The figure below shows a circuit to be analyzed.

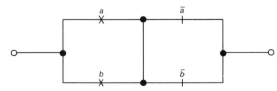

a) Construct and complete a table of combinations for the network.

b) Compare the table with that of Problem 6.

c) Which of the following is a correct description of the circuit: AND, OR, odd-parity, even-parity?

8. Construct and complete a table of combinations for the network shown below.

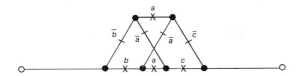

9. A board of trustees for the Last National Bank consists of four voting members. Any loan must be approved by at least three of the board members before it is accepted. The members wish to vote in secret but wish to know if any three or more members voted yes. Below you will find a contact network for this "at least three out of four" vote problem.

a) What three contacts should be placed in the bottom branch of this network so that the network is completely specified?

b) Are any other branches in parallel necessary? If so, why?

c) Draw two other networks that have fewer contacts and which will do the same job.

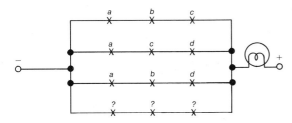

10. Complete the table of combinations for the contact network shown below.

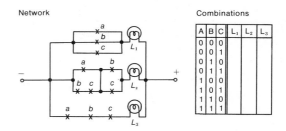

a) How many switches must be operated in order to light only L_1?

b) How many switches must be operated in order to light only L_1 and L_2?

c) How many switches must be operated in order to light all three lamps?

d) Does the order in which the switches are operated determine which lamps will be lighted?

e) Can you think of a real-life situation in which this circuit might be used?

11. Districts I, II, and III combined to form a regional school with each district having two members on the Board of Education. Action by the board requires a majority vote by district. A negative vote by one representative of a district acts as a veto on a positive vote by the other representative. Design a circuit which will permit secret voting by individuals. Use a lighted lamp to indicate affirmative action by the board.

Laboratory and Projects

I | INTRODUCTION TO THE LOGIC CIRCUIT BOARD

The Logic Circuit Board has all of its components except the power supply on the top of the board. The top panel displays four lamps, four slide switches, and four relays. Identify each of these units. Compare the arrangement of these elements on the panel with Fig. 4.

Notice that four sets of double eyelets marked with a minus sign are available at the center of the panel. Electrical energy required to operate any unit on the board can be obtained by inserting the tapered pin at the end of the connecting wire into any of these eyelets, and then inserting the free end of the wire into the eyelet connected to the element which is to be operated. To prevent breakage always grasp the tapered pin by the plastic grip and insert or remove the wire with a twisting motion. Otherwise, the wire will be broken away from the pin.

Each terminal is connected by two eyelets. This is sufficient to connect any desired circuit since an arbitrary number of terminals can be connected together by a chain of jumpers. (As a convenience, extra eyelets are supplied for the negative power supply terminal since it is used so much.)

To aid you in wiring circuits, the eyelet wiring field (Fig. 4) is labeled with the same symbols used in the circuit diagrams. Although the notation used may seem strange at first, you will quickly find as you become familiar with it that it has many advantages over the more pictorial notation sometimes used.

Before long you will be able to analyze a complex problem and represent it electrically on the Logic Circuit Board.

Slide Switches

Figure 1 shows a slide switch cut away so that you may see how it works. In Fig. 1 the switch is in the "0" position. The metal bridge carried by the insulating handle is connecting the terminal *b* to the terminal *c*. Since the lamp is wired to the battery through terminals *m* and *c*, there is a gap in the metallic path between *m* and *c* and the lamp is not lit.

In Fig. 2 the switch has been operated; the slider is in the "1" position. Now the metal bridge connects terminals *m* and *c*; since there is a complete metallic path from the battery through the lamp, it is lit.

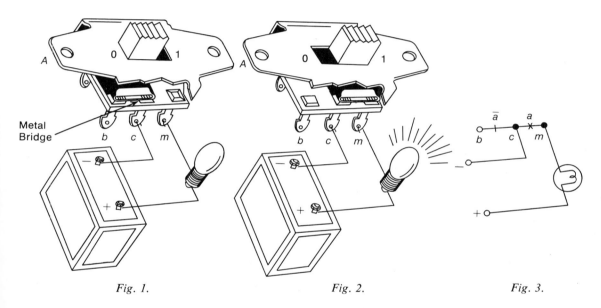

Metal Bridge

Fig. 1.   *Fig. 2.*   *Fig. 3.*

Figure 3 is a circuit diagram arranged to show how the circuit symbols correspond to the switch, lamp, battery, and wires in the pictures. Between terminals *m* and *c* we have a make contact on switch *A*. This is shown in the diagram by the symbol $\times^{a}$. Remember that a make contact is closed only when the switch is operated ($A = 1$). The break contact is drawn with the symbol $+^{\bar{a}}$. This of course is closed only when the switch is not operated ($A = 0$).

Notice further that the circuit diagram (Fig. 3) does not indicate whether the switch *A* is actually operated. It can be in either state. A circuit diagram shows only the logic of the connections, *not the state of the switches.*

For clarity the details of a slide switch are drawn for only one make-break contact on the switch. These switches each have four sets of such contacts (often called a four-pole double throw, or 4-PDT switch.) The four sets are insulated from each other but are logically connected; whenever the slider is operated, all four make contacts are closed, and whenever the switch is returned to a non-operated position, all four of the break contacts are closed.

Figure 5 shows Fig. 3 redrawn. The break contacts are omitted since they are not used. This circuit is similar to the light circuits in your house which are controlled by a single switch.

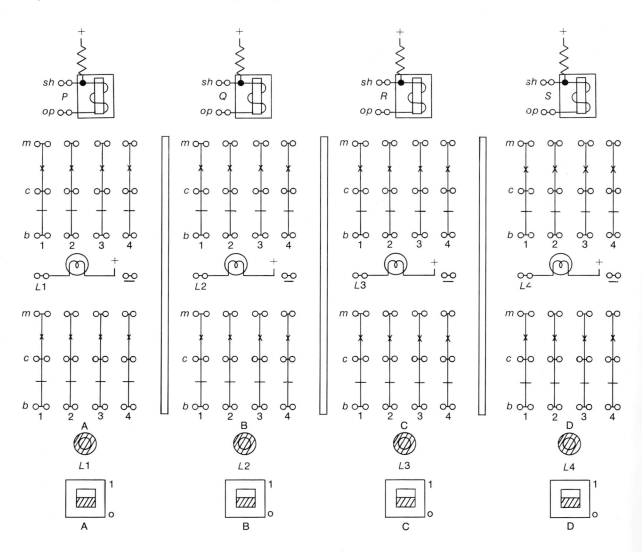

Fig. 4.
Logic Circuit Board.

In your first learning experience with the LCB, wire the circuit shown in Fig. 5. We wish to control lamp L1 with make contacts on switch A. Use the section 1 contacts controlled by switch A. To do this connect a wire from lamp L1 to the m (make) eyelet of section 1 of A. The diagram (Fig. 5) indicates that the other

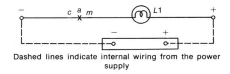

Dashed lines indicate internal wiring from the power supply

Fig. 5.

A	L1
0	0
1	1

Fig. 6.

contact of section $A1$ must connect to the minus terminal for electrical energy. The c (center) eyelet represents the other contact of the switch. This eyelet, therefore, should be connected to any of the minus ($-$) eyelets on the panel. Check your circuit by operating switch A. Now shift the tapered pin from the minus terminal on the panel to any other negative terminal and observe the action of the switch.

The operation of this circuit is described by the Table of Combinations (Fig. 6). Column A represents the state of switch A and Column $L1$ represents the state of lamp $L1$. Thus a zero under A designates "Switch A is not operated"; while a one in Column A designates "Switch A is operated." Correspondingly, a zero under $L1$ represents "Lamp $L1$ is not lighted" and a one in Column $L1$ states "Lamp $L1$ is lighted." Check the Table in Figure 6.

Switch to Open One Lamp Circuit While Closing Another

Connect lamps $L1$ and $L2$ and two contacts of switch A as shown in Fig. 7. What happens as switch A is operated? Explain the result.

■ 1. Copy the Table, Fig. 8, into your notebook in the way shown in Fig. 6.

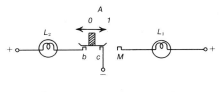

Fig. 7.

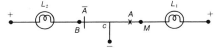

A	L1	L2
0		
1		

Fig. 8.

2. A street intersection has 8 lamps (Fig. 9) RN (Red North), RE, RS, RW, and GN (Green North), GE, GS, GW. These lamps are all controlled by contacts on a single switch A. If $A = 1$ means East-West traffic can flow, show the complete circuit diagram.

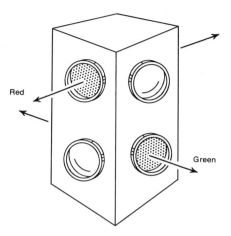

Fig. 9.
Traffic-control signals.

Red

Green

AND Circuits

Connect $L1$ and one contact from each of switches A and B as shown in Fig. 10.

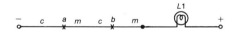

Fig. 10.

A	B	$L1$
0	0	
0	1	
1	0	
1	1	

Fig. 11.

3. Operate switches A and B, observe lamp $L1$, and complete a table like the Table of Fig. 11. Why is the circuit called an AND circuit? How many ones are in the output column of the Table?

4. How many ones would be found in the output column of a Table for A AND B AND C?

■ 5. In order to initiate a call, the receiver in a telephone booth must be lifted and a coin deposited to obtain a dial tone. Each of these acts operates a switch. Prepare a diagram of a circuit that meets these requirements.

OR Circuit

Connect $L1$ and contacts of A and B as shown in Fig. 12. What is the effect on $L1$ of operating (1) A only, (2) B only, (3) A and B?

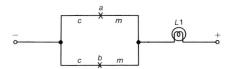

Fig. 12.

■ 6. Complete the Table of Fig. 13. Why is this circuit called an OR circuit? How many ones in the output column of the Table?

Fig. 13.

A	B	L1
0	0	
0	1	
1	0	
1	1	

■ 7. How many ones has the output column of a Table of Combinations for A OR B OR C?

■ 8. The fire-alarm system of a school has a number of switches located throughout the building. Prepare a diagram showing how any one of four switches may be used to sound the alarm.

A Problem

Here is a circuit problem for you to solve. You are given switches A and B and lamps $L1$ and $L2$. Review the problem of the jet pilot from the text and devise a circuit so that lamp $L1$ represents the canopy and lamp $L2$ represents the seat ejection— then using switches A and B and the two circuits for AND and OR from the first part of this experiment complete the problem. One additional constraint to remember is that the two switches may not be thrown simultaneously.

■ 1. Which circuit controls the canopy ($L1$); which circuit controls the seat ($L2$)?

II | THE ODD-PARITY AND MAJORITY CIRCUITS

In our last experiment we examined an OR circuit which said in effect "Either *A* or *B* or both." Some of you may remember that this has a special name in mathematics, it is called the "Inclusive Or." Let us now examine a circuit which will say "Either *A* or *B*, but *not both*." This too has a special name, the "Exclusive Or."

As circuits become more complicated, the wiring of a circuit directly from its diagram becomes difficult. An intermediate step which helps is a list of the terminals that are to be connected. Such a list is called a "Running List." To make the list we need a notation for the terminals to be connected. First we number the various contacts on a switch or relay. We label a contact terminal in three parts:

> First a capital letter for which switch or relay;
> Next a number for which contact;
> Finally *m*, *c*, or *b* for which terminal.

Relay coil terminals are labeled with a capital letter followed by *op* or *sh*. The negative power terminal is *Neg*.
Examples. The terminals of make contact *a2* (on switch *A*) are *A2c* and *A2m*. The terminals of break contact *ā2* are *A2c* and *A2b*. The terminals of break contact *b̄1* (on switch *B*) are *B1c* and *B1b*. From now on we will give contact numbers in the diagrams. You will find them an aid in finding wiring errors.

A sample running list for the odd-parity circuit is:

> *A1m, B1b; A1c, Neg;*
> *A1b, B1m; L1, B1c.*

The first line of this running list is read:

> Connect terminal *1m* of switch *A* to terminal *1b* of switch *B*.
> Connect terminal *1c* of switch *A* to the negative power supply terminal.

■ 1. Write two more sentences for the second line.

Another name for the "Exclusive Or" circuit is the odd-parity circuit. Why would the term odd parity be appropriate for this circuit? After you have wired the circuit found in Fig. 14, complete a table like Fig. 15.

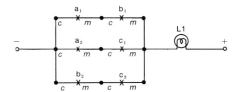

Fig. 14.
An odd-parity circuit.

A	B	a	$\bar{a}$	b	$\bar{b}$	$a \cdot \bar{b}$	$\bar{a} \cdot b$	L1
0	0							
0	1							
1	0							
1	1							

Fig. 15.

The following circuit, Fig. 16, is an odd-parity circuit with three variables.

■ 2. Before wiring it prepare a running list. Check that it agrees with Fig. 16.

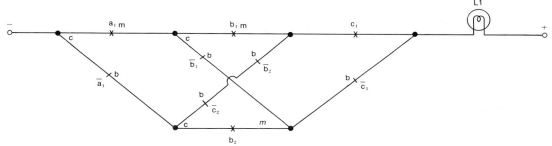

After you have wired the circuit, operate the switches and summarize your results in the Table of Combinations of Figure 17.

Fig. 16.
Three-variable
odd-parity circuit.
(*Lower case script represents make, break, and center.*)

A	B	C	L1
0	0	0	
0	0	1	
0	1	0	
0	1	1	
1	0	0	
1	0	1	
1	1	0	
1	1	1	

Fig. 17.
Table of Combinations for three-variable odd-parity circuit.

3. The following two majority circuits of Fig. 18 are equivalent. Test this by making two tables of combinations like Fig. 19.

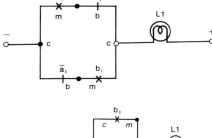

Fig. 18.
Three-variable majority circuits.
(*Lower case script represents make, center, and break.*)

A	B	C	L1
0	0	0	
0	0	1	
0	1	0	
0	1	1	
1	0	0	
1	0	1	
1	1	0	
1	1	1	

Fig. 19.
Table of Combinations for the majority circuit.

III | CIRCUITS WHICH MODEL LOGIC PROBLEMS

In this experiment we see how a circuit can help solve a problem by serving as a model of a situation. Using the model, various approaches to the problem can be tried without setting up the actual situation. The river-crossing problem is an example of this method of problem solving.

Part A

The River-Crossing Problem

A boatman must carry a wolf, a goat, and a cabbage across a river in a boat which is so small that he can carry, at most, one of them with him in it at a time. Moreover, whenever the wolf and goat are together, he must also be present to keep the goat from being eaten. Neither can he leave the goat with the cabbage. How can he carry all of them from the south bank of the river to the north bank?

Use a table of combinations to list all situations that might occur. The warning light must go on when a dangerous situation appears.

	A Boatman	B Cabbage	C Wolf	D Goat	Warning light
a.	0	0	0	0	0
b.	0	0	0	1	0
c.	0	0	1	0	0
d.	0	0	1	1	1 (light on)
e.	0	1	0	0	0
f.	0	1	0	1	1
g.	0	1	1	0	0
h.	0	1	1	1	1
i.	1	0	0	0	
j.	1	0	0	1	
k.	1	0	1	0	
l.	1	0	1	1	
m.	1	1	0	0	
n.	1	1	0	1	
o.	1	1	1	0	
p.	1	1	1	1	

Note: A zero indicates items that are on one side of the river; a one represents those items that are on the other side.

1. On a separate sheet of paper complete the table for the entries *i* through *p*. The total number of conditions which will produce a warning light is six. Does your table reveal them all?

Represent the boatman by switch A, the cabbage by switch B, the wolf by switch C, and the goat by switch D. Remember that a zero indicates a break set of contacts and a one indicates a make set of contacts. Assume that the zero position of the switches represents the south bank of the stream and the one position the north bank.

Under these conditions the warning light which should go on for line *d* uses a path given by $\bar{a}\,\bar{b}\,c\,d = 1$. Figure 20 shows this for lines *d*, *f*, and *h*.

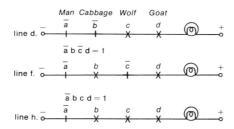

Fig. 20.

■ 2. Draw the similar paths for the warning conditions on the lower half of the table.

Note that if the wolf and goat have been transferred to the north side of the river while the man and the cabbage are on the south side, the lamp will light. Examine all of the six conditions which are dangerous and make sure that the circuits indicate a lighted lamp when the goat and the wolf or the goat and the cabbage are together on either side without the presence of the man.

The network for the three conditions listed in Fig. 20 can now be constructed:

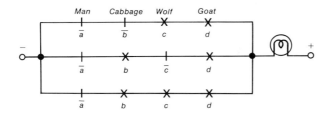

Fig. 21.
Combined paths from Fig. 20.

These three branches will not give warning for all possibilities because there are three additional conditions that must be satisfied.

■ 3. Draw the circuit for the remaining conditions.

Part B

Simplification of the Logic Network

This network uses more contacts than necessary. We can use Boolean algebra to improve it. Remember that we represent series contacts with products and parallel contacts with sums. The circuit of Fig. 2 can be written as:

$$\bar{a}\,\bar{b}\,c\,d + \bar{a}\,b\,\bar{c}\,d + \bar{a}\,b\,c\,d = 1$$

Factoring $\bar{a}\,d$ from each term gives:

$$\bar{a}\,d\,[\bar{b}\,c + b\,\bar{c} + b\,c] = 1$$

Further factoring b from the last two terms leaves:

$$\bar{a}\,d\,[\bar{b}\,c + b(\bar{c} + c)] = 1$$

The term $(\bar{c} + c)$ may be dropped since it always equals 1, that is, there is a conducting path regardless of the state of switch C. The equation then becomes:

$$\bar{a}\,d\,[\bar{b}\,c + b] = 1$$

From our use of a sum to represent a parallel arrangement and a product to represent a series arrangement, the combined circuit is represented as:

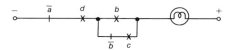

Fig. 22.
Simplified paths from Fig. 21.

■ 1. Combine and simplify the circuit you designed for the remainder of the Table of Combinations.

■ 2. Prepare a running list for the entire circuit.

Check this running list against the combined and simplified circuits.

Now wire your LCB and simulate the problem by moving the slide switches. Remember the zero (0) position of the switches represents the south bank of the stream and the one (1) position represents the north bank, and that you can move the A switch (boatman) from one side to the other either alone or with only one other switch.

■ 3. Starting with all zeros, prepare a sequence of safe moves which leads to all switches in the one position.

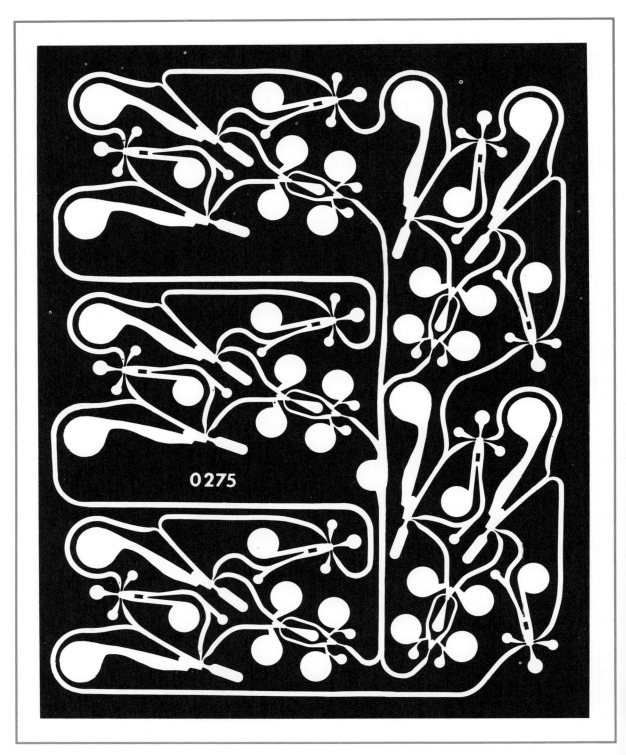

Logic Circuits as Building Blocks

13

I | INTRODUCTION

Among other uses, the symbols "1" and "0" are convenient as shorthand to show that a logical statement is true or false as we saw in Chapter 12. Similarly, they can indicate that contacts, or a network of contacts, are closed or open. The fact that 1 and 0 can be used in both these ways makes it possible to use contacts, alone or in networks, to represent statements in logic. Two such networks, developed in Chapter 12, were the majority circuit and the odd-parity circuit. In this chapter we will see how these can be used to construct a larger circuit which adds numbers. Our "adder," however, will work for numbers only if they are expressed as a sequence of 0's and 1's, for only in this form can there be a direct relation between the individual digits of the numbers and the switch contacts. Thus, even though decimal numbers may be more familiar, binary numbers are more convenient for logic circuits and for computers.

This fluid logic circuit performs the operation of dividing by 10 in an all-fluid digital computer: for every 10 input pulses the circuit delivers one outout pulse. Input pulses enter from above the plane of the circuit through the circular, bumplike hole attached to the straight channel running from top to bottom just to the right of center. The 10 identical logic elements, or modules, are arranged in a series of five pairs, three at left and two at right.
(Willam Vandivert)

2 | THE DECIMAL AND THE BINARY NUMBER SYSTEMS

In the decimal system, a number is expressed as an *ordered* sequence of digits, such as 85263. Each digit can have one of ten values, 0 through 9. The *position* of each digit in the sequence determines its value in units, tens, hundreds, etc. The rightmost digit gives the units and is called the least significant digit, while the leftmost digit is the most significant and in our example represents the number of ten thousands.* The values (we might call them *weights*) in a decimal number are all powers of 10. For instance, the value of the 8 in the number 85263 is $10^4 = 10 \cdot 10 \cdot 10 \cdot 10 = 10,000$. The 8 is "weighted" by 10,000; it really stands for 80,000. Remembering that $10^3 = 1000$, $10^2 = 100$, $10^1 = 10$, and $10^0 = 1$, we know that the number 85263 represents $8 \cdot 10^4 + 5 \cdot 10^3 + 2 \cdot 10^2 + 6 \cdot 10^1 + 3 \cdot 10^0 = 8 \cdot 10000 + 5 \cdot 1000 + 2 \cdot 100 + 6 \cdot 10 + 3 \cdot 1$. A decimal fraction is so called because its digits are ordered in just the same way, but each one is multiplied by 10 raised to a negative power. The decimal point is used merely as an indicator or a marker to show where the negative exponents begin.

A similar system is used for binary numbers.† A binary number is expressed as an ordered sequence of binary (two-valued) digits, such as 10111. The value of each of these digits is a power of two, and the digits are arranged so that the least significant digit is at the right and the most significant at the left, as before. Therefore, the binary number 10111, when expressed as a decimal number, is $1 \cdot 2^4 + 0 \cdot 2^3 + 1 \cdot 2^2 + 1 \cdot 2^1 + 1 \cdot 2^0 = 1 \cdot 16 + 0 \cdot 8 + 1 \cdot 4 + 1 \cdot 2 + 1 \cdot 1 = 23$. Any number in the binary system can be converted to a decimal number by this technique of expanding in powers of 2. The list of numbers below, written in both the binary and decimal systems are examples for practice. Each of the binary numbers can be evaluated in the way that the first one is, and thus be shown to be equivalent to the corresponding decimal number.

*This positional notation, which we now take for granted, did not always exist. In some ancient systems, such as the Roman in which the number 1984 would be written MCMLXXXIV, computation with numbers was so difficult that only a learned scholar could handle even the simplest problems in arithmetic. We use the Hindu system, which was developed from the earliest positional notation we know of — that of the Babylonians and Sumerians.

A significant advantage of a positional notation is that after learning only a few rules which apply to the numbers represented by single digits, one can use these same rules to add, subtract, multiply, and do other computations on numbers of any size whatsoever.

†Any integer *could* be used as a number base. From time to time advocates of a base-12, or "duodecimal," system appear. They claim that this base would have many advantages over the decimal system because 12 is exactly divisible by four smaller integers — 2, 3, 4 and 6 — rather than only by the two — 2 and 5 — which are possible in the decimal system. The duodecimal system would require twelve different symbols, for example, 0, 1, 2, 3, 4, 5, 6, 7, 8, 9, α, and β.

Binary representation	Decimal representation
$1 \cdot 2^4 + 1 \cdot 2^2 + 1 \cdot 2^0 = 10101$	$2 \cdot 10^1 + 1 \cdot 10^0 = 21$
111001	57
1001101	77
1110111	119
1111111	127
11111000000	1984
1010	10
1100100	100
1111101000	1000
11110100001001000000	1000000

All modern computers deal with binary rather than decimal numbers in their internal operations. However, they usually convert binary to decimal before printing out results, since decimal numbers are more familiar and easier to read.

One of the most useful of all internal units in any computer is an adder for binary numbers. Adding binary numbers is very similar to adding decimal numbers. To add in either system we start with the least significant digits of the numbers being added. We determine their sum, write it in the least significant place, then add the next most significant digits including any carry-over from the least significant digits, and so on. For example, adding two decimal numbers,

```
CARRY DIGITS → 1 1 1 1 0
             5 6 7 5 0 ← FIRST NUMBER
             6 7 3 5 9 ← SECOND NUMBER
           1 2 4 1 0 9 ← SUM
```

Similarly, for two binary numbers (remembering that in binary $1 + 1 = 10$; in decimal this corresponds to $1 + 1 = 2$, two in binary is 10):

```
CARRY DIGITS → 1 1 1 1 0 0 0 1
             1 1 0 1 1 0 0 1 ← FIRST NUMBER
             1 1 1 1 0 1 0 1 ← SECOND NUMBER
           1 1 1 0 0 1 1 1 0 ← SUM
```

In the rightmost column, $1 + 1 = 0$ with a 1 carry to the next column to the left. There, $0 + 0 + 1 = 1$, 0 carry; then $0 + 0 + 1 = 1$ with 0 carry; then $0 + 1 + 0 = 1$ with 0 carry, and so on. Thus, the rules for binary addition are as follows:

1. In each column there are three numbers to be added (after the first column), one from each number and one carry from the position to the right. The sum digit for each column is:
   - 0 if there are no or two 1's in the column.
   - 1 if there are one or three 1's in the column.
2. The carry digit to the next column is:
   - 0 if there is no or one 1 in the column.
   - 1 if there are two or three 1's in the column.

Note that *sum* digit is the *odd parity* of the 1's in a column. That is, if the inputs to an odd-parity circuit are the three digits in a column to be added, the output will be the sum digit for that column. The *carry* digit is the *majority* of the digits in a column. That is, if the inputs to a majority circuit are the three digits to be added, the output will be the carry digit for the next column.

So, we can now draw a symbolic diagram of an adder based on logic circuits studied in the preceding chapter. The diagram is

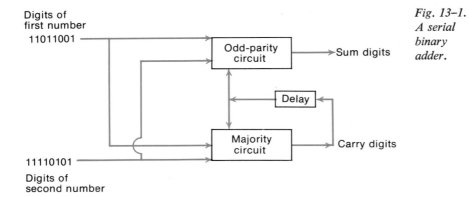

Digits of
first number
11011001

Odd-parity circuit → Sum digits

Delay

Majority circuit → Carry digits

11110101

Digits of
second number

Fig. 13–1.
*A serial
binary
adder.*

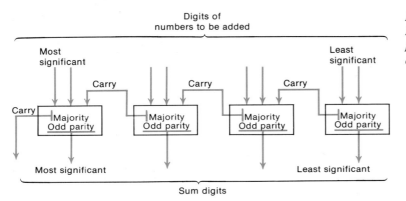

Digits of
numbers to be added

Most significant

Least significant

Carry   Carry   Carry

Carry

Majority
Odd parity   Majority
Odd parity   Majority
Odd parity   Majority
Odd parity

Most significant   Least significant

Sum digits

Fig. 13–2.
*Four-digit
parallel
adder.*

shown in Fig. 13–1. At the left, digits of each number are fed to the adder one at a time. The least significant digits of each number are put in first. As successive pairs of digits are put in, the carry digit from the preceding pair is supplied also to the two circuits. The sum digits appear at the right, the least significant first. An adder of this kind is known as a *serial* adder since it adds one column after another in series.

Another kind of adder operates on all columns simultaneously. Such adders are known as *parallel* since they work on all columns in parallel. Parallel adders can be faster than serial adders, but require more equipment, as is indicated in Fig. 13–2. There, the digits of the numbers to be added are all supplied at the same time from the top to individual adding units, each composed of a majority and an odd-parity circuit. The carry digits are generated almost instantaneously, so the correct sum appears at the bottom after a short delay.

Recall that the inputs to logic circuits in the last chapter were manual switches which could be either operated or released. The states of these switches must correspond to the digits of the numbers to be added in both Figs. 13–1 and 13–2. A different situation exists for the carry digits, however. In both adders, the output of the majority circuit drives an *input* of the majority and odd-parity circuits for the next column. In other words, the output of one logic circuit must be able to operate or release switch contacts in another logic circuit. This can be done by a device known as a *relay*. Relays are a key to building larger computing elements from smaller ones in a hierarchical fashion. They also make possible the automatic operation of an organized collection of logic circuits.

3 | AN AUTOMATIC PARALLEL BINARY ADDER

Parallel adders can be created by organizing majority and odd-parity circuits appropriately. The organization needed is shown in Fig. 13–2. The automatic carry operation indicated there, however, requires that the switching action in one circuit control switch contacts in an entirely independent circuit. This can be done by a relay, diagrammed in Fig. 13–3.

In this diagram, we note that switch A, at left, has no electrical connection with contacts r and $\bar{r}$, at right. So contact a may be in one logic circuit while r and $\bar{r}$ may be in another. By operating switch A we can operate contacts r and $\bar{r}$. The secret lies in the electromagnet shown at center.

An electromagnet consists of a rod of iron, or other magnetic material, around which an insulated wire is wrapped. When electric current is sent through the wire, the rod becomes strongly magnetic, so that a nearby piece of iron is strongly attracted. In the diagram we show such a piece of iron labeled "armature." It is in the form of a lever that is free to swing on a pivot but is pulled away from the electromagnet by the "spring." In the condition shown in Fig. 13–3a, switch A is released. No current flows; the electromagnet is not energized; contact r is open and contact $\bar{r}$ is closed.

If now we operate switch A so as to close contact a, current energizes the electromagnet. In turn, the electromagnet strongly attracts the armature which then swings on its pivot toward the electromagnet, Fig. 13–3b. This motion of the armature causes contact r to close and contact $\bar{r}$ to open. Thus, by operating contact a we have reversed the states of contacts r and $\bar{r}$ even though there is no electrical connection between them and contact a. Thus, the state of switch A in one logic circuit can control the state of contacts in another.

Since an entire contact network, such as a majority circuit, can be either open or closed, it too can control contacts through a relay. This slight extension of relay control is indicated in Fig. 13–4. In fact, this is the way we use relays for carry digits in the

Fig. 13–3.
Diagram of a relay.

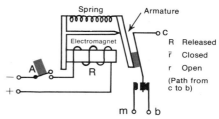

a) *Released.*

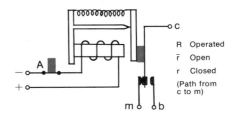

b) *Operated.*

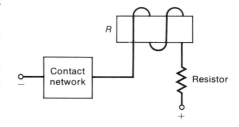

Fig. 13–4.
Contact network controlling operation of relay R. (*Make and break contacts of relay* R *not shown.*)

X₁	Y₁	Sum	Carry
0	0	0	0
1	0	1	0
0	1	1	0
1	1	0	1

(a)

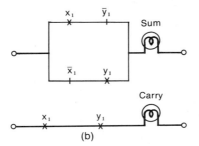

(b)

Fig. 13–5. Circuits for adding two binary digits, X and Y. The upper circuit is an exclusive OR, which is the same as an odd-parity circuit for two inputs. Similarly the lower circuit is an AND, which is the same as a majority circuit for two inputs. The sum appears at the right through indicator lamps which are "on" for 1 and "off" for 0.

automatic parallel adder. The armature of a relay can operate several contacts at once rather than only two as indicated in Fig. 13–3.

Majority circuits, odd-parity circuits, and relays are all we need for a binary adder. To see how they work together, let us first review what we do when we add in binary. Consider first the task of adding two digits X_1 and Y_1. In Fig. 13–5a we show the combinations involved in adding two digits X_1 and Y_1. In Fig. 13–5b we show two circuits that perform these combinations when operated by two switches X and Y, each of which controls two contacts. With reference to the top circuit, when we add 1 to 0 or 0 to 1, it is required that the sum be 1. Correspondingly, the "sum" lamp in the circuit lights. When we add 1 to 1 the sum must be 0 and we must carry 1. Correspondingly, the sum lamp does not light. Instead the "carry" lamp in the lower circuit lights.

To add binary numbers of more than one digit requires that we use a relay to transfer the carry digits. As an illustration, let us consider a collection of circuits for adding 11 to 11 to yield the sum 110. Actually these circuits are able to add any two binary numbers of two-digit length. The $11 + 11$ example is convenient for illustration because it exercises the adder fully. This slightly more complex task of adding 11 to 11 in order to yield the sum 110 requires that we use a relay for transfer of the carry digits.

In adding 11 to 11 we must do the following things:

1. We add 1 to 1. This means we put down 0 and carry 1 to the next column.
2. In the next column we add 1 to 1 to the carried digit 1. This gives us three 1's. We put down 1 and carry 1 to the next column.
3. Since there are no digits in the third column, we merely put down the carried digit 1.

To perform these operations, we need five circuits as follows:

1. A circuit to combine 1 and 1 to register 0.
2. A circuit to carry 1 from the first to the second column.
3. A circuit to combine 1 and 1 and 1 and then register 1.
4. A circuit to carry 1 to the third column.
5. A circuit to register the carried 1.

In Fig. 13–6 we show the required five circuits. We note that the circuits in stage 1 are identical to those in Fig. 13–5 except the "carry" lamp is replaced by a relay. Each of the circuits on the right handles the digits in a particular (binary weight) column. Each feeds into a lamp that registers the sum for the column by lighting for 1 or remaining off for 0. The relay circuits on the left do the job of carrying a digit as required to the next column. Let us now follow the switching action of the circuits in Fig. 13–6 as they perform the operation of adding 11 to 11 (a problem chosen to emphasize the carrying function).

We operate the switches X_1 and Y_1 to represent the two 1's to be added in the first column and the switches X_2 and Y_2 to represent the two 1's in the second column. As a result, no less than 21 contacts are called into play.

For the first column, as a result of operating switches X_1 and Y_1, we affect the contacts x_1, $\bar{x}_1$, y_1, and $\bar{y}_1$ as follows: In the circuit connected to the lamp L_1, x_1 closes but $\bar{y}_1$ remains open; also, although y_1 closes, $\bar{x}_1$ remains open. Consequently, no current flows and L_1 does not light and thus registers the required 0.

Meanwhile in the circuit connected to the relay C_2 the closing of contacts x_1 and y_1 energizes the relay. As a result, relay C_2 carries the action to the circuits for the second column, closing all contacts labeled c_2 and opening C_2.

Examining the circuit connected to L_2, we see that since contacts x_2, y_2, and c_2 are all closed (X_2, Y_2, and relay C_2 are operated),

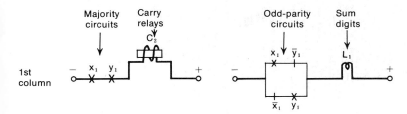

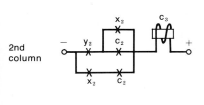

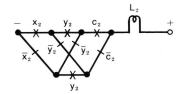

Fig. 13–6. A two-digit binary adder.

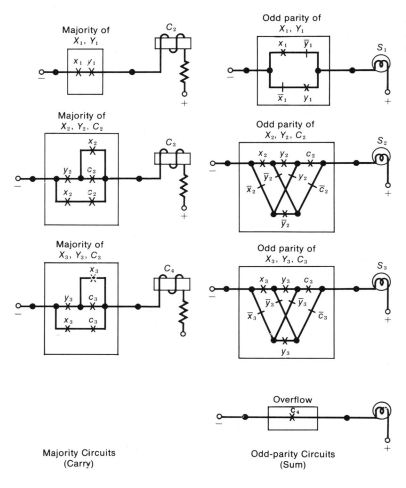

Majority of X_1, Y_1

C_2

$x_1 \ y_1$

Odd parity of X_1, Y_1

$x_1 \quad \bar{y}_1$

$\bar{x}_1 \quad y_1$

S_1

Majority of X_2, Y_2, C_2

x_2

$y_2 \quad c_2$

$x_2 \quad c_2$

C_3

Odd parity of X_2, Y_2, C_2

$x_2 \quad y_2 \quad c_2$

$\bar{y}_2 \quad y_2$

$\bar{x}_2 \quad \bar{c}_2$

$\bar{y}_2$

S_2

Majority of X_3, Y_3, C_3

x_3

$y_3 \quad c_3$

$x_3 \quad c_3$

C_4

Odd parity of X_3, Y_3, C_3

$x_3 \quad y_3 \quad c_3$

$\bar{y}_3 \quad \bar{y}_3$

$\bar{x}_3 \quad \bar{c}_3$

y_3

S_3

Overflow

C_4

Majority Circuits
(Carry)

Odd-parity Circuits
(Sum)

*Fig. 13–7.
Binary adder
circuits.*

current flows in L_2 which lights, registering 1. Meanwhile, in the circuit connected to relay C_3, since the contacts y_2 and c_2 are closed, the relay C_3 operates, closing contact c_3 which in turn causes L_3 to light up, registering 1 and thus completing the third column of the addition.

This particular example of addition illustrates the basic functions of an adder even though we limited it to two digits and explained it for two specific binary numbers, 11 and 11. Of course, any number of digits can be handled by adding additional circuits. A three-digit is shown in Fig. 13–7. This figure is drawn to emphasize the role of the odd-parity and majority circuits previously covered. Also, these adders carry out all the operations necessary for summing any two binary numbers with the suitable number of digits. That is, our adders perform the general logical rules for addition as presented in Section 2.

4 | NUMBERS WITH A SIGN

Until now we have avoided the problems of handling numbers with a sign ($+101$, -1101, etc.). How do we represent these in a computer? A common method is to give the sign by the leftmost digit of the binary number. This digit is called the *sign-bit*. *If the sign-bit is "1" the number is positive*, and *if the sign-bit is "0" the number is negative.* For example:

1000000000101 represents $+5$, and
0000000000101 represents -5.

In this example there are 13 bits in each number. Since the sign occupies the first bit, there are only twelve bits available to represent the magnitude. Thus, numbers as large as $2^{12} - 1 = 4095$ can be represented. The representation for $+4095$ is 1111111111111, and that for -4095 is 0111111111111.

If we wanted to add two numbers with signs, it would be simple if we had to add only positive numbers, because the adder of the previous section would do. But sometimes we must subtract a larger number from a smaller, and in any case, the situation is complicated by the fact that either sign or both may be negative.

To compute the sum of numbers with signs that can be either positive or negative, we must treat the signs and the magnitudes separately. From the signs we determine the proper arithmetic operation, which we then perform on the magnitudes. We determine the sign of the result by comparing the magnitudes.* The rules which must be used can be described by the following table.

*
The magnitude of a number, X or Y for example, is denoted as $|X|$ or $|Y|$.

Addition:
Compare signs
   if same   : add $|X| + |Y|$, use sign of X
   if different: compare magnitudes
       if equal: SUM = 0
  if $|Y| < |X|$: subtract $|X| - |Y|$, use sign of X
  if $|X| < |Y|$: subtract $|Y| - |X|$, use sign of Y†

†
Subtraction circuits operate in much the same way as adders do.

The following numerical examples, numbered to match the operations listed above, are set down to show how the rules work.

1. $X = 5$, $Y = 5$: $5 + 5 = 10$.
   $X = -5$, $Y = -5$: $5 + 5 = 10$, X negative, Sum $= -10$.
2. $X = 5$, $Y = -5$: $5 + (-5) = 0$.
3. $X = -5$, $Y = 4$: $5 - 4 = 1$; X negative: $S = -1$.
4. $X = 4$, $Y = -5$: $5 - 4 = 1$; Y negative: $S = -1$.

Clearly addition requires comparisons of the signs and of the magnitudes of the numbers X and Y in order to decide which of several things are to be done next. The two-variable odd-parity circuit may be used to compare the sign-bits of two numbers. If both bits are 1 or both are 0, then the resulting signal is "zero"; if one is 1 and the other is 0, then the result is "one." A procedure and a circuit for comparing the magnitudes of two numbers is presented in the next section.

5 | A CIRCUIT WHICH COMPARES THE MAGNITUDES OF TWO NUMBERS

The magnitudes of two binary numbers (that is, binary numbers without a sign) can be compared in the following way to determine which is larger. Starting with the leftmost bit position (after the sign-bit), we compare the corresponding bits of the two numbers.* If these digits have the same value, we compare the next pair of digits to the right. This process is continued until an unequal pair is found or until there are no more digits. If the value of the digits is the same in every position, the numbers are equal; if an unequal pair is found, the number which contains a "1" at this position is larger.

The procedure is illustrated in Table 13–1, and Fig. 13–8 shows a circuit that compares one pair of corresponding bits (a and b) in two binary numbers (A and B). The completed path through the circuit is different for three conditions: $A = B$; $A = 1$, $B = 0$; and $A = 0$, $B = 1$. Four of these basic comparison circuits can be combined as in Fig. 13–9 to compare four-bit numbers. The comparison is done in steps as indicated in Table 13–1, where A_4, A_3, A_2, A_1 are switches corresponding to the digits in the number A, and B_4, B_3, B_2, and B_1 correspond in the same way to digits in the number B.

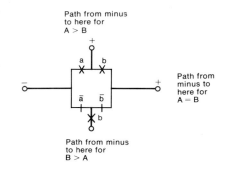

Path from minus to here for $A > B$

Path from minus to here for $A = B$

Path from minus to here for $B > A$

Fig. 13–8. Circuit to compare single-bit binary numbers A and B.

$$
\begin{array}{rcccc}
A = & 1 & 0 & 0 & 1 \\
B = & 1 & 0 & 1 & 1
\end{array}
$$

Step 1. Compare ——
Result: Equal
Step 2. Compare ——
Result: Equal
Step 3. Compare ——
Result: Digit of $B >$ digit of A
Conclusion: The magnitude of B is greater than the magnitude of A.

Table 13–1. Comparison of magnitude of two binary numbers, 1001 and 1011.

$A = 1001$ (9 in decimal) is represented by $A_4 = 1$, $A_3 = 0$, $A_2 = 0$, $A_1 = 1$.

$B = 1011$ (11 in decimal) is represented by $B_4 = 1$, $B_3 = 0$, $B_2 = 1$, and $B_1 = 1$.

The operated switches, then, are A_4, A_1, B_4, B_2, and B_1. The others are released. Under these conditions the heavily lined path is completed to lamp L_3, indicating that $A < B$. A path would be

*For this scheme to produce correct results, each number should contain the same number of bits; any difference in bit length can be eliminated by placing a sufficient number of 0's to the left of the shorter number, as is indeed done in a computer.

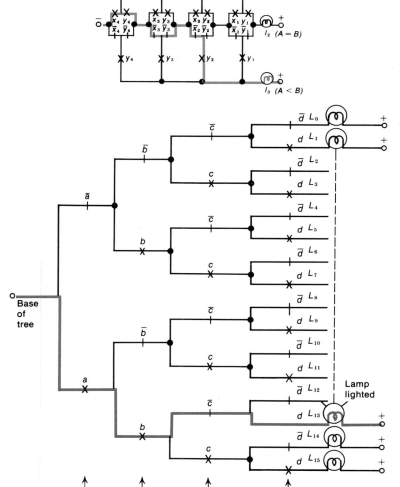

Fig. 13–9.
Circuit to compare
four-digit binary numbers.
(Note that the subcircuit to
compare each bit pair of
numbers is the same
as the circuit in
Fig. 13–8.)

Fig. 13–10.
A selection
"tree" circuit
set to 1101. (The
heavy line indicates
the selected
path.)

completed to L_1 if A were greater than B, and to L_2 if A were equal to B.

We see that the subcircuit to compare each pair of bits of the two numbers is the same as the circuit in Fig. 13–8. Hence, numbers of any bit length can be compared simply by adding more circuits, or stages, of exactly the same type.

6 | A "TREE" CIRCUIT

A circuit that converts binary to decimal numbers is shown in Fig. 13–10. Known as a selection-tree circuit, it plays an important role in the computer for another purpose as well. One of sixteen different lights (numbered L_0 through L_{15}) is lit by operating four switches (A, B, C, and D). The state of the switches represents the binary number, and the lighted lamp the corresponding decimal number. For instance, if switches A, B, C, and D are set to 1, 1, 0, and 1, respectively, lamp #13 is turned on since the contacts a, b, $\bar{c}$, and d are closed. When these contacts are closed, the *only* path completed from the base of the tree to a lamp is the one which goes to lamp #13. Thus, the binary number is converted to decimal.

Tree circuits are commonly used in computers to select a designated element (a particular lamp in Fig. 13–10, for example) so that it can be connected to some other part of the computer. For example, this tree circuit can be used to select a memory cell containing information that is to be transferred to another part of the computer, or a memory cell in which we wish to store information brought from somewhere else.

7 | FINAL COMMENT

In this chapter we have studied circuits vital to the operation of any digital computer. The selection-tree circuit permits any one of many elements to be singled out and connected to some other selected element. The adder circuit permits binary numbers to be combined according to the rules of addition. The comparison circuit permits the signs of numbers to be taken into account. These circuits were constructed according to the logic developed in Chapter 12, using switches and contacts. In order to automate the operations, one additional element was required, namely, the relay. Relays transfer the state of a contact or a contact network automatically to a different network. For example, in the adder relays were used to transfer the carry digit from one digit network to the next. These circuits and elements take us a large step toward understanding computers. They are "literal-minded" and do not temper their logic with common sense, but they are tireless in their operation, and they rarely make an error within their logical framework.

Logic circuits, as we have discussed them, respond to the settings of switches and relays and reach a static condition. In a computer, however, one event is followed by another in sequence. Furthermore, the nature of each event is modified, or even determined entirely, by the outcome of previous events, implying that some element or elements record or remember previous outcomes. Thus, time and memory, in addition to logic, are vital elements in computers. In the next chapter we will explore these matters.

Questions for Study and Discussion

1. In the game "Twenty Questions," to determine the age of a person who is at least 0 but less than 64, what is the largest number of questions one would ever have to ask (using the best system) to get the correct answer? (Hint: Study Fig. 13–10.)

2. Explain the function of

   a) the first lefthand section of the binary adder in Fig. 13–7.

   b) the first righthand section of Fig. 13–7.

3. Name and discuss one application of a tree circuit.

4. Study Fig. 13–10. Can you find a closed path between lamps 6 and 8 for any state of the switches? Between any other pair of lamps?

5. Why is a relay needed for constructing a binary adder?

6. How can the circuit of Fig. 13–8 be used to compare the signs of two numbers?

7. In Chapter 10, Table 10–3 shows a six-bit alphanumeric code. How could a tree be used to translate from the code to the alphanumeric characters?

Problems

1. Convert the following binary numbers to decimal form.

11	10000
101	110010
110	11010
1011	1100100
1010	1111101000

2. Convert the following decimal numbers to binary form.

1	15
8	16
4	31
2	32
9	27

3. Perform the following binary addition. Check your work by converting the numbers to decimal form.

$$1010 + 110010 =$$
$$1011 + 11010 \ =$$

4. Perform the following binary subtractions on unsigned (positive) numbers. Check your work as in Problem 3 above.

$$100 - \ \ \ 11 =$$
$$1010 - \ \ \ 101 =$$
$$110011 - 11010 =$$

5. Perform the following subtractions on unsigned (positive) numbers, and check your work.

$$1100100 - 11010 \ =$$
$$11010 - 110100 =$$

6. Refer to Section 5 and Fig. 13–9. For the following pairs of numbers, which lamp will light? Why?

   1. $A = 1010$ $B = 111$
   2. $A = 1010$ $B = 1100$
   3. $A = 1010$ $B = 1010$

7. a) What is the greatest number of questions necessary to guess someone's age known to be between 0 and 63? (See Question 1.)

   b) Assume the age is known to be between 0 and 31 and repeat part (a).

   c) What is the relation between the number of guesses and the range in ages?

8. a) What is the largest number representable by four lamps of the LCB?

   b) How many bulbs would be necessary to give the binary equivalent of 31?

   c) What is the decimal equivalent of each of the four lamps when they are used to display binary numbers?

9. A tennis elimination tournament among eight players proceeds as shown below. Explain how this process may be considered an application of a selection tree.

10. How many binary digits are necessary to identify uniquely (to "address") each of approximately 65,000 books in a library?

11. Figure 13–9 indicates a method to compare two n-digit binary numbers. Design a schematic diagram of a circuit to compare three n-digit binary numbers, using L1, L2, and L3 to indicate which is the largest number. (In case of a tie you may light all the bulbs of highest value.)

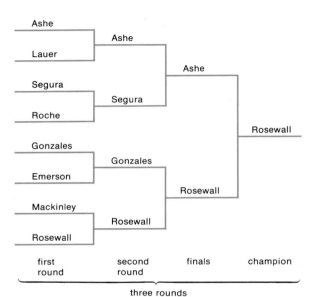

Ashe			
	Ashe		
Lauer			
		Ashe	
Segura			
	Segura		
Roche			
			Rosewall
Gonzales			
	Gonzales		
Emerson			
		Rosewall	
Mackinley			
	Rosewall		
Rosewall			
first round	second round	finals	champion

three rounds

Laboratory and Projects

EXTENDING OUR SENSES USING
THE LOGIC CIRCUIT BOARD

The Relay

As more complicated circuits are developed we need a method of controlling one set of contacts from a network of other contacts without using a person to operate a switch. We use the relay such as shown in Fig. 1 to do this. Fig. 1a shows the relay P in the released state. Since the wire between the negative power supply terminal and the op terminal is open, there is no current through the relay coil. The spring pulls on the iron armature which is hinged above the coil as shown. Mounted by an insulating block to the armature is a spring strip connected to the terminal c. The armature forces this strip against the terminal b. Thus there is a metallic path between the terminals c and b.

In Fig. 1b the wire has been connected from the negative power supply to the relay op terminal. Current now flows through the relay coil. This causes a magnetic field which attracts the iron armature to the coil, stretching the spring. When the armature moves, it breaks the connection between terminal b (break) and the terminal c (center) and makes a new connection between terminal m (make) and terminal c.

A circuit diagram (Fig. 1c) shows the correspondence between the symbols and the picture. An advantage of the symbols is that we are not required to draw the contacts p or $\bar{p}$ adjacent to the coil P. Thus, a circuit can show the logic more clearly. We can always recognize which contacts and coils are associated by the lettering.

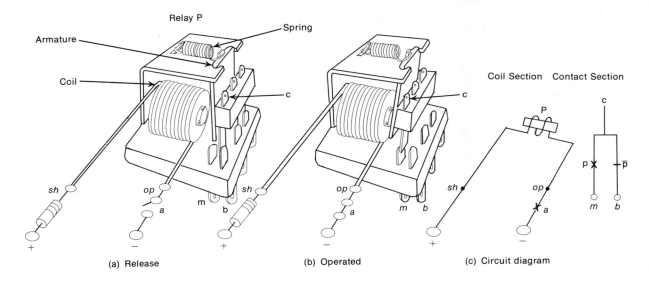

(a) Release (b) Operated (c) Circuit diagram

Each relay can have several sets of contacts electrically insulated from each other, but logically in identical states.

The connection from the terminal *sh* through the resistor to the positive side of the power supply is internally made in the logic circuit board for each of the relays.

To become familiar with the operation of the relay, connect terminal *op* of relay *P* through a make contact of switch *A* to $(-)$ as shown in Fig. 2. Operate and release switch *A* several times and note the movement of the armature of relay *P*.

Fig. 1.
Relay.
a) Release.
b) Operated.
c) Circuit diagram.

Fig. 2.

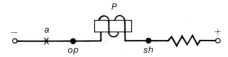

Connect lamp *L*1 through a make contact of relay *P* to $(-)$, and lamp *L*2 through a break contact of *P* to $(-)$ (Fig. 3). Operate *A* and observe the results.

Fig. 3.

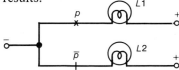

◼ 1. Summarize your results in a table like Fig. 4.

Fig. 4.

A	*L*1	*L*2
0		
1		

Another method of controlling a relay is shown in Fig. 5. The relay is released when the contact b is closed and is operated when contact b is open. This is the opposite of the circuit of Fig. 2.

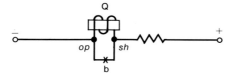

Fig. 5.

2. Wire the circuit shown in Figs. 5 and 6 and compare it with the circuit of Figs. 2 and 3.

Summarize your results in a table like Fig. 7.

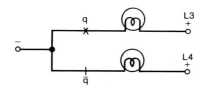

Fig. 6.

Fig. 7.

B	L3	L4
0		
1		

3. A relay coil is part of a burglar alarm operating on **dry** cells; which coil control method would be preferable? Suggest other situations where these considerations would be important.

II | BINARY NUMBERS AND THE BINARY ADDER

Part A

For the first part of this experiment combine some of the logic circuits that you have studied in Chapter 12. Make a circuit that can take as its input two binary numbers of two digits each. Switches A and B are one input and switches C and D are the other. The binary output (represented by the lights on the logic board) is the sum of the input numbers. In the second part of the experiment extend this type of circuitry to add two binary numbers of four digits each. Refer to the odd-parity circuits and the majority circuits from Chapter 12 and review your text on binary numbers.

Detailed Example of Binary Addition

Consider the following two binary expressions:

$$2^4 2^3 2^2 2^1 2^0$$

	Carry	1 1 1 0
Expression A 1 0 1 1		1 0 1 1
Expression B 1 1 1 0	Addition	1 1 1 0
	Sum	1 1 0 0 1

1. Is there a majority of 1's in the first column from the Right? No, and so no carry is involved.
2. Is there a majority of 1's in the second column from the right? Yes, and so a 1 is carried to the third column.
3. Is there a majority of 1's in the third column? (Remember that this includes any 1's you may have carried.) Yes, and so a 1 is carried to the fourth column.
4. Is there a majority of 1's in the fourth column? Yes, and so a one is carried to a fifth column that is the 2^4 digit.
5. Is there a majority of 1's in the fifth column? No, and so no carry is involved.
6. Go back to the first column from the right, the 2^0 column. Is the number of 1's odd or even? Odd, write a 1 in the sum row in the 2^0 column.
7. Is the number of 1's odd or even in the 2^1 column? Even, write a 0.
8. Is the number of 1's odd or even in the 2^2 column? Even, write a 0.
9. Is the number of 1's odd or even in the 2^3 column? Odd, write a 1.
10. Is the number of 1's odd or even in the 2^4 column? Odd, write a 1 in the 2^4 column. The problem is completed.

1. Prepare a running list and then wire the binary adder circuit shown in **Fig. 8.**

2. Test your circuit by adding the following pairs of numbers.

$$\begin{array}{ccc} 10 & 01 & 11 \\ +10 & +11 & +10 \\ \hline \end{array}$$

How many different pairs of numbers can this circuit add?

Part B

The circuit for the binary adder using two circuit boards is but an extension of the two-digit binary adder—here two logic circuit boards are used—the four switches on one board make up one input while the four switches on the other make up the other input. Place the boards side by side. Call the left one X and the

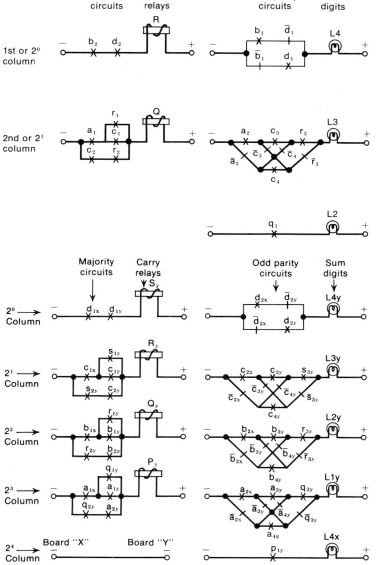

Majority circuits   Carry relays   Odd parity circuits   Sum digits

Fig. 8.
Two-digit binary adder.
The inputs are A, B, and C, D . . . B
and D are the 2⁰ inputs while A and B
are the 2¹ column inputs.

Fig. 9.
Binary adder for
two binary numbers of four digits each;
two L. C. B.'s required.

right one *Y*. Be sure to connect the *neg.'s* of the two boards. The teacher's manual has a running list for this circuit.

III | A COMPARATOR CIRCUIT FOR TWO 4-DIGIT BINARY NUMBERS

The following circuit compares two 4-digit binary numbers. Review Section 5 of Chapter 13 to see how this circuit is used in a computer. As in the four-digit binary adder, two LCB's must be used, board X and board Y. Remember that there must be a wire from a (−) of board X to a (−) of board Y. Note that $L1$ means $X > Y$, $L2$ means $X = Y$, and $L3$ means $X < Y$.

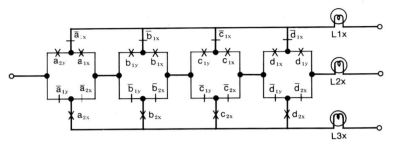

Fig. 10.
Circuit to compare two
4-digit binary numbers.

Wire the circuit of Fig. 10. Switches A, B, C, D on board X represent one number, and switches A, B, C, D on board Y represent the other number.

1. Test your circuit by comparing the following pairs of numbers: 1101, 1100; 1101, 1101; 1101, 1110; 0000, 0000; 1111, 1111; 1010, 0101; 0001, 0000.

2. How many different pairs of numbers can this circuit compare?

IV | TREE CIRCUITS

A circuit that by various settings of switches connects a base point to any one of a group of other points is known as a tree circuit. The setting of switches determines which point is connected to the base. N switches permit the selection of one of the 2^N terminals. For example, three switches can result in a tree with nine branches.

1. Prepare a running list for the circuit of Fig. 11.

2. Wire the circuit of Fig. 11 and complete a table of combinations like Figure 12.

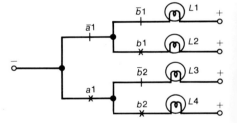

Fig. 11.
Complete two-stage tree.

A	B	$L1$	$L2$	$L3$	$L4$
0	0				
0	1				
1	0				
1	1				

Fig. 12.
Table of combinations for
two-stage tree.

3. Is there at any time a continuous path from one lamp to any other lamp of the tree circuit?

Figure 13 shows a partial four-stage tree. We have changed the order of the contacts in some branches so that it will fit on the switches. This tree can be used to convert binary numbers to decimals. If we represent a binary number on switches A, B, C, and D in the usual order (A has weight 8, B weight 4, etc.), then the lamps can be labeled 0 through 9, so that the lamps give the decimal equivalent of the binary number on the switches.

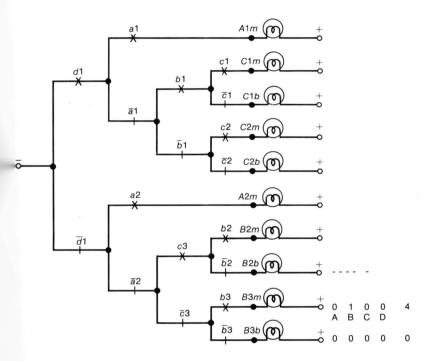

Fig. 13.
A partial tree for
binary to decimal translation.

Study Fig. 13, then label the readout lamps with their appropriate binary numbers* in the column adjacent to the lamps. Complete the second column with the equivalent decimal number.

Since the LCB has only four lamps, we must test this circuit in steps.

*Some of the lamps will light for 4 binary numbers; in every case only one of the numbers is less than 9.

4. Prepare a running list for decimal numbers zero through three for the circuit of Fig. 13. To assist you in preparing this list the output terminals just ahead of the lamps have been marked with a heavy dot and labeled as *Alm*, *Clm*, etc.

Obtain ten small squares of masking tape. Label them 0, 1, 2, 3, etc. Place the squares marked zero through three adjacent to the corresponding readout lamps of the LCB.

5. Wire and test your circuit, then complete the table of Fig. 14 for this portion of the tree circuit.

N_{10}	A	B	C	D	L0	L1	L2	L3	L4	L5	L6	L7	L8	L9
0	0	0	0	0										
1	0	0	0	1										
2	0	0	1	0										
3	0	0	1	1										
4	0	1	0	0										
5	0	1	0	1										
6	0	1	1	0										
7	0	1	1	1										
8	1	0	0	0										
9	1	0	0	1										
10	1	0	1	0										
11	1	0	1	1										
12	1	1	0	0										
13	1	1	0	1										
14	1	1	1	0										
15	1	1	1	1										

6. Prepare a running list for that portion of the tree circuit necessary for reading out 4, 5, 6, 7. Relabel the readout lamps with these digits. Wire and test this section, then complete the appropriate section of the table.

Fig. 14.
Table of combinations for circuit of Fig. 13.

7. Prepare the running list to give the decimal numbers 8 and 9. Relabel the readout lamps with these digits, wire and test this section, then complete the appropriate section of the table.

8. The wiring diagram of Fig. 13 represents a folded tree, for switches B and C appear on both the third and fourth vertical sections of the diagrams. Why is this folding necessary? ˛

9. With four switches we should be able to obtain 2^4 or 16 combinations representing the numbers 0 through 15.

a) Can you give reasons for restricting this tree circuit to ten readout devices?

b) Can you modify the tree circuit so that no lamps will light when the switches are set to represent 1011 through 1111? (Completion of the lower section of the Table of Combinations in Fig. 14 should provide you with a clue.)

V | SIMPLE CARD READER FOR THE LOGIC CIRCUIT BOARD

From the examples given in Section 6, Chapter 10, we become familiar with how punched-card reading equipment is used to transfer symbolic information from cards to a computer. We show this transfer of information, from punched cards to lighted lamps, using two logic circuit boards and the card reader as follows:

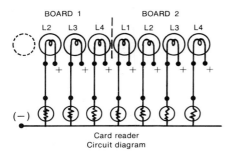

Fig. 15.

(Helpful hint! Be sure to connect one negative from one board to a negative on the other.)

We may use the first three positions on our reader to denote an address, while the last four could contain information. We may also punch out cards to represent any of the characters from the six-bit alphanumeric code found in Table 3, Chapter 10.

Machine Memory

14

I | INTRODUCTION

In the previous two chapters, we have seen how to combine on-off switches into circuits that represent logical statements. With the aid of relays, these can be coupled to form some of the basic subunits of computers, for example, the parallel adder. To complete the inventory of subunits, an additional basic on-off element is needed, namely, a *memory* element. A memory element is useful not only in the memory bank of computers but also in the instruction register, the program counter, and in the processor (the accumulator is a memory register).

Memory circuits are used widely not only in computers but also in everyday life. There are circuits at the telephone exchange which remember the individual digits of a telephone number one by one as they are dialed, until the dialing procedure is completed. When you push the "Up" or "Down" button to call an automatic elevator to your floor, there is no need to keep on pressing it because there are circuits which remember that the button was pushed. Once inside, you push another button for the floor you want. When that floor is reached, the elevator remembers to stop and to open the door, even if it had to make several other stops first. At many pedestrian crossings there are traffic lights which can be changed by pushing a button. The circuits for these are often designed so that a pedestrian can push the button and cause the light to change shortly to "Red" for the automobile traffic. If another pedestrian pushes the same button a fraction of a minute after the light has gone "Green" again, the circuit has been

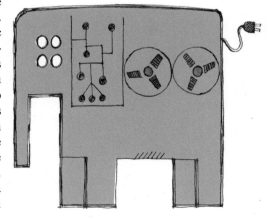

Computer operator controlling the computer. Note in the diagram on the face of the computer that the memory is a vital part of the system. While computer memory is often more readily accessible than human memory, man's ingenuity and creativity are vital in controlling the machine.

designed to remember that the pedestrian button was pushed less than, say, two minutes ago. When the interval is over, however, the circuit remembers that the button has been pushed a second time, and it again turns the traffic light to "Red."

None of the logic circuits discussed so far has a memory of this kind. The lamp controlled by the majority-vote circuit lights or does not light depending on the present positions of its switches and is not affected by what settings these switches might have had in the past. The hall-light odd-parity circuit is affected only by the present status of its switch positions and not by what happened to these switches previously. The adder circuit delivers its answer independent of the past history of its use. Memory, as we are using the word, means storing the evidence of past events.

Human memory is complicated and subtle. The state of the brain seems to depend upon the chemical and electrical states of its billions of neurons. But no one has been able to identify any specific area of the brain as the repository for memories of past events. It has not been demonstrated that any specific memory trace has been destroyed by cutting into the brain.

Computer memory is very much easier to understand. Each memory cell stores one bit concerning one past event. But since it is we who have defined how the computer elements are interconnected, we know exactly how the states of the computer elements are determined and how they can be changed. Furthermore, these states are binary in nature: switches and relays are operated or released, contacts are closed or open, and even entire networks are open or closed. Therefore, as we have seen, a table of combinations gives complete information about the state of networks, based solely on the binary states of their component elements.

There are many different memory devices that could be used as our basic memory element. In Chapter 10, we studied briefly magnetic cores as elements in an addressed computer memory. Transistor memory is being used more and more in computers and computer-related equipment. Here, however, we will use relays as our memory elements because they are convenient and it is easy to see how they work as memories. In the last chapter, we used relays in one logic circuit to control switch contacts in another. However, no memory was involved. To make a memory, a relay must control a switch contact *in its own circuit*. We shall see such an arrangement and how it results in memory in Section 2.

2 | THE BASIC RELAY MEMORY ELEMENT

The operation of the basic relay memory circuit is shown in Fig. 14–1. In Fig. 14–1a, relay P is initially released, as are also the two input switches, A and B. The a and b make contacts are open. Note that the p contact is controlled by the relay P, and the

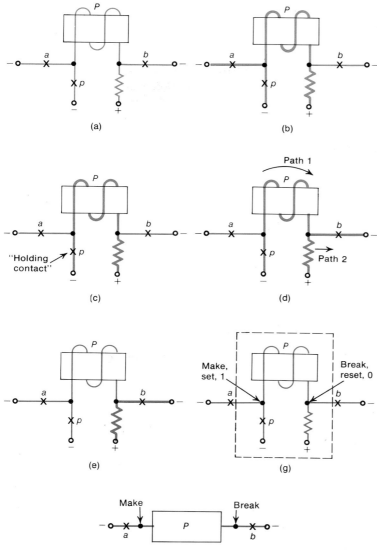

*Fig. 14–1.
Operation
of basic
memory circuit.
(a)* P, A, B *released.
(b)* A *operated,* P
*operates, ''p'' closes.
(c)* A *released,* P *remains
operated. (d)* B *operated,* P
about to be shunted. (e) P
*shunted through ''b''
make. (f) Memory
labeled with functions.
(g) Symbolization of
one-bit memory
circuit.*

two are in the same electrical network. Still in Fig. 14–1a, P is released, the p make contact is open. Since a and p are open, the
relay remains released. But if switch A is now operated, contact a
closes and P's coil is connected to minus. Current starts, the armature is attracted, and within a fraction of a second, the p make
contact closes (Fig. 14–1b).* Therefore P is now also connected to
minus through its own make contact which can supply P with
current independent of the operation of A (Fig. 14–1c). Thus A
need only be operated momentarily and then released to cause P
to become (and to stay) operated. Hence the circuit can remember

*
Closed paths and operated relays are shown
by heavy shading.

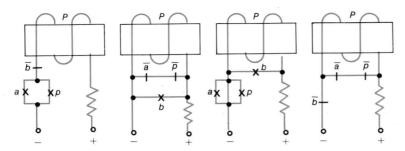

Fig. 14–2.
Equivalent memory
circuits.

indefinitely that switch A was operated. It does so through its own contact p, which is called a *holding* contact. This is in contrast to the simple relay circuit of Fig. 13–3, Chapter 13, which stays operated only as long as input switch A is operated.

To clear the memory relay, we need to be able to reset it. In the case of the elevator, for instance, this happens after the correct floor has been reached. To clear the memory, we operate switch B (Fig. 14–1d) which immediately causes the current to bypass the coil (Fig. 14–1e).* (Since path 2 offers no "resistance" to the current, current through path 1 almost ceases because the coil in this path has more resistance.) As the current in the coil diminishes, the magnet grows weaker; soon it is weak enough that P releases, the armature swings back, and p opens. B may now be released and P will remain released (Fig. 14–1a). Since A causes the relay to operate and B causes it to be released, we refer to the corresponding sides of this relay circuit as the make and break sides, respectively (also set and reset).

Since we symbolize the operation of a relay or switch by "1" and its release by "0," the make side may be viewed as the "record-a-1-in-memory" side, and the break side as the "record-a-0-in-memory" side. These conventions are symbolized in Fig. 14–1f. Figure 14–1g is a black box representation of the one-bit memory circuit.

Our memory circuit requires one holding contact and two input contacts. These can be organized quite flexibly, however, as the four equivalent memory circuits shown in Fig. 14–2 illustrate. Each circuit can be analyzed in terms of the cycle studied with our standard circuit: (a) A, B, and P released; (b) A operated; (c) A released; (d) B operated; (e) B released.

What would happen if A and B were operated simultaneously? The electrical properties of the memory circuit are such that a resetting or break action (switch B) overrides a setting or make action (switch A). In other words, resetting the relay causes it to be released regardless of its previous state (operated or released), and regardless of the make side input.

This relay memory circuit can be a basic building block for large units. A cluster of a dozen such circuits can be arranged to

*This is the point at which the need for the resistor in the relay circuit becomes clear. Without this resistor contact b shortcircuits the power supply.

remember a whole 12-bit number. Then we can imagine hundreds, even thousands of copies of the cluster—but the cost of the building in which to house all those relays, and the cost of the power to operate them, becomes too high. This, of course, is one reason why large memories are not built with relays, but with magnetic cores or transistors according to the same principles.

The same very simple memory element, when provided with a minor modification of the input wiring, enables us to make a shift register, crucial in multiplication. And an apparently trivial change in the shift register gives us, in turn, a counter which is essential to the control function of the computer. The memory circuit turns out to be very powerful, very adaptable.

3 | AN ADDRESSABLE MEMORY

The simple circuit of Fig. 14–1 can store a "0" or a "1." When switch A is operated and released, a "1" is stored. When switch B is operated and released, a "0" is stored. In Fig. 14–3 a light bulb has been added in parallel with the relay winding so that we can read the state of the relay by seeing if the bulb is lighted or not. If the relay is operated (and A and B released), the holding contact p is closed, $\bar{b}$ is closed, and the lamp is connected directly to minus. It will light, indicating that a "1" is stored. If the relay is released (A and B released), the a and p make contacts are open. The lamp circuit cannot be completed and the lamp will not light.

To make a computer memory, we now need as many copies of the dashed box circuit of Fig. 14–3a as we have binary digits (bits), and a means for selecting each of these memory circuits for storing and reading. We use the tree circuit discussed in Section 6, Chapter 13, to do the selecting, but just one tree is inadequate. We need to select both the make and the break sides of a given circuit, so that we can record either "1" or "0."

Thus, for four-memory circuits (or cells) we have the circuit of Fig. 14–4. The make and break trees are identical and are set simultaneously by the C and D switches from whose contacts they

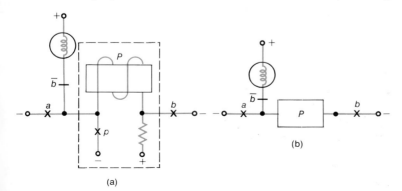

(a)

(b)

Fig. 14–3.
One-bit
memory with
readout.
a) Detailed circuit.
b) Symbolic
diagram.

568

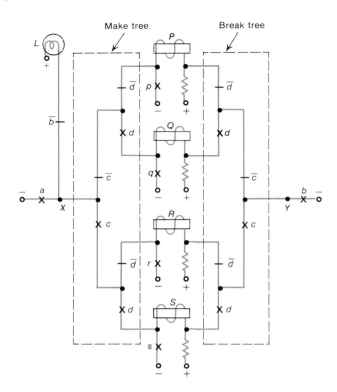

Fig. 14–4.
Four-cell
memory with make
and break
trees.

are formed. To record in a given memory circuit, a two-step process is followed:

1. Select the desired circuit. Switches C and D are appropriately set. The make tree selects (is connected to) the make side of the specified relay, while the break tree selects the break side of the specified relay.

2. Set the selected relay to 1 (or 0) by connecting the root of the make or the break tree (marked X or Y) to minus.

EXAMPLE: Recording and reading cell Q

Step I. Select Q by releasing switch C and operating switch D. Two paths are now set up, one in each tree, from root X to one terminal of Q, from root Y to the other terminal, Fig. 14–5a. No connection to minus has yet been made, so that the relay has not yet been affected.

Step II. Set Q by connecting the root of the make tree to minus. This is done by closing make switch A. Thus, current flows through Q's coil to operate Q. Q then remains operated. Figure 14–5b shows the completed path accentuated and circled. Once Q operates, the make tree is no longer needed in the relay circuit, since the Q make side is connected through q to minus; thus A may be released.

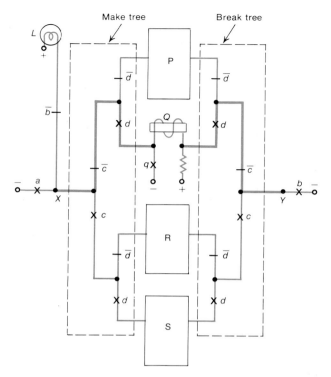

Make tree   Break tree

*Fig. 14–5a.
Operation for
four-cell memory.
(a) Select Q by
operating D and
releasing C.*

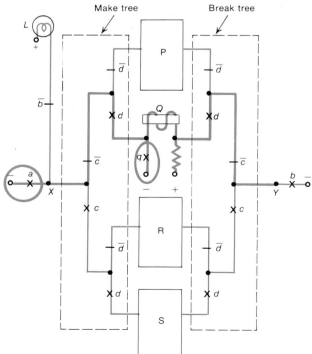

*Fig. 14–5b.
Operation of
four-cell memory.
(b) Set Q to 1 by
operating A.*

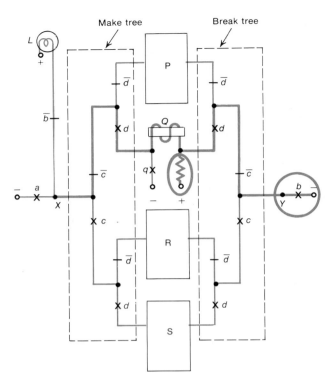

Fig. 14–5c.
Operation of
four-cell memory.
(c) Set Q to 0 by
operating B.

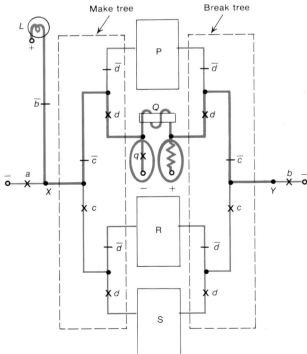

Fig. 14–5d.
Operation of
four-cell memory.
(d) Reading out
of Q = 1.

If we had wanted to record a 0 in Q instead of a 1, Step I would have been the same; only Step II would be different. To record a 0, we connect the break tree root Y to minus by closing switch B. The break side of Q (and no other break side) is now connected to minus, the relay Q releases, and remains released. Figure 14–5c shows the completed path accentuated and circled. As before, once Q releases, the break tree is no longer needed and B may be released.

To read the contents of any memory circuit, we simply select it by setting switches C and D. The lamp L, connected to the root X of the make tree, "reads out" the contents of the selected memory circuit. Thus to read the Q memory circuit, it is selected as in Fig. 14–5a. If Q is operated, X connects through the make tree to minus and the lamp lights (Fig. 14–5d). If Q is released, X is "at plus" (i.e., it is connected to plus through the lamp, but it has no minus connection); therefore the lamp does not light.

Through proper settings of switches C and D, any one of the four memory circuits can be selected. Hence the lamp can read the contents of any cell. Each memory cell is said to have an "address" that corresponds to the setting of the C and D switches to select it. Memory cell P has the address 0, (00); Q has the address 1, (01); R has 2, (10); and S has 3, (11). "Addressing" a cell is the act of setting C and D to the proper values to select that cell.

4 | SHIFTING AND SHIFT REGISTERS

An important operation in any computer is shifting bits in a register to the right or left. In our model computer we found it useful to shift numbers held in the accumulator register. There was a shift instruction that caused the accumulator contents to be shifted a specified number of digits to the right or left.

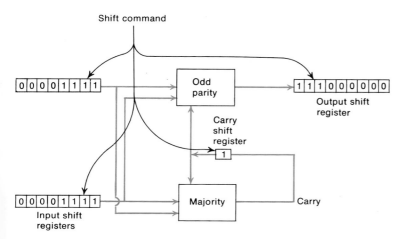

Fig. 14–6. Serial adder with input, output, and carry shift registers summing $11111111 + 11111111$ as pictured. The four least significant digits have been added. The four least significant digits of the sum appear in the first four places of the output shift register.

Shifting in a computer is done by a circuit with memory called a *shift register*. Shift registers have many uses in a computer, other than in the accumulator. For example, they are needed for multiplication by adding partial products which are shifted just as in ordinary manual multiplication. Also, shift registers are needed to complete the serial adder, as indicated in Fig. 14–6. They are used to store and shift input numbers on command to the adder, a digit at a time. An output shift register stores the sum digits as they are produced, and a single stage shift register holds one carry digit. On each shift command, a digit is shifted from each register on the left into the odd-parity and majority circuits. At the same time, the carry digit from the preceding addition is also shifted into the circuits while the resulting sum digit is shifted into the first place of the output shift register.

As pictured in Fig. 14–6, the two 8-digit numbers, 11111111 and 11111111, have been shifted four times resulting in 00001111 in each input register and 111000000 in the output sum register. (Note that it has one more place than the input registers to hold the last carry.)

Shift registers are built from a number of nearly identical component circuits. The basic circuit for one component is shown in Fig. 14–7a. Notice that it is composed of two parts, a memory circuit and a recording circuit. (See inset 14–7a.) The contact p is the holding contact to give the relay memory. The contact a determines whether a 1 or a 0 is stored, and operation of S causes the shift to take place.

As in the one-bit memory circuit, we record a 1 or 0 depending on whether we connect the make or the break side to minus. The complementary a and $\bar{a}$ in the recording network ensure that only one side is so connected at any given time. When switch S is operated, the negative terminal of the electrical source is connected to the make terminal of the relay if $a = 1$, or to the break terminal of the relay if $a = 0$. Therefore, when S is operated, the memory circuit records the state (operated or released) of the switch A.

When the switch S is released the contacts a and $\bar{a}$ are in series. Regardless of the state of switch A, therefore, no path can be completed through both of these contacts; when one is closed the other is open. Thus, when switch S is released, the relay is controlled by its own holding contact p. After switch S is released, switch A can no longer affect the state of P. The value of the bit present in the relay depends, therefore, upon the previous state of switch A.

One complete stage of a shift register requires two circuits of this kind (see Fig. 14–7b). One is exactly like Fig. 14–7a. The second has contacts p and $\bar{p}$ controlled by relay P. Let us suppose that S is released. The relay P is operated or released depending upon the value of the bit stored there. The state of Q is the same

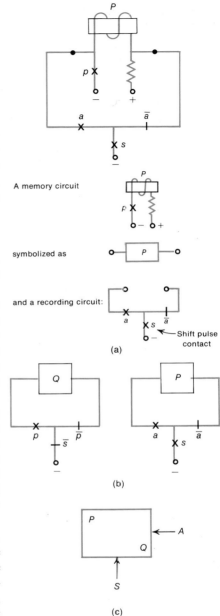

A memory circuit

symbolized as

and a recording circuit:

Shift pulse contact

(a)

(b)

(c)

Fig. 14–7.
The shift register.
a) Basic shift-register circuit. b) Complete shift-register stage. c) Complete stage symbolized.

as the state of P, because with the contact $\bar{s}$ closed, the state of the p and $\bar{p}$ contacts determines the state of Q. So long as S remains released, we may set the switch A to either state without affecting the states of either P or Q.

We now set A to some new value, say 1, which is to be stored on P and then shifted to Q. With A in its "new" state, we operate S. This action causes P to take the same state as A (Fig. 14–8a). With S operated, however, the contact $\bar{s}$ in the Q half of the circuit is open and the relay Q retains its former state, although its recording circuit takes on the same state as P. Now we release S once again. The value stored in the relay P is passed on to the relay Q (Fig. 14–8b). After switch S is released, the state of A can be changed once more without influencing the state of P (Fig. 14–8c). The original state of switch A has been shifted through the P relay to the Q relay.

A complete shift register can be made by cascading several copies of the circuit shown in Fig. 14–7c. The circuits are connected as suggested by the diagram in Fig. 14–9. The Q relay of each stage controls the recording contacts of the next P relay. The shifting signal S actuates all of the stages. Each time it changes from 0 to 1 and then back to 0, the bit value stored in each stage is shifted one stage to the left.

In a four-stage shift register containing 0110, the detailed operation during one shift with input A (0 or 1) is as follows:

	Q_4	P_4	Q_3	P_3	Q_2	P_2	Q_1	P_1	
Original:	0	0	1	1	1	1	0	0	
Operate S:	0	0	1	1	1	1	0	0	A
becomes	0	1	1	1	1	0	0	A	
Release S:	0	1	1	1	1	0	0	A	
becomes	1	1	1	1	0	0	A	$A = 110A$.	

The more detailed circuit diagram of Fig. 14–10 shows four stages of the shift register. There it can be seen that the q contacts of one stage control the next stage in the same way that the a contacts are used in the first stage.

Registers to shift numbers to the right instead of to the left are sometimes needed. It is easy to imagine how this could be done. If the circuit of Fig. 14–9 is built backward (Stage 4 becoming Stage 1, etc.), its shifting action is reversed in exactly the way

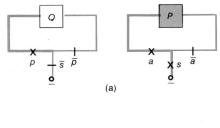
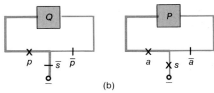
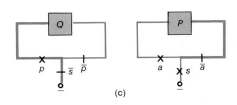

Fig. 14–8.
Shift-register stage in use.
a) A = 1 transferred to P half of stage (S operated).
b) A = 1 transferred to Q half of stage (S released).
c) A reset, holding contact retains P unchanged

Fig. 14–9. Description of a shift register—a cascade connection of four shift-register stages.

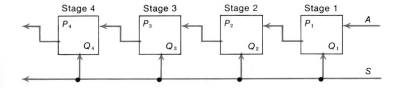

desired. In practice, there is no need to construct a whole new circuit; it is better to have a second set of recording contacts which can be switched in, and the usual set switched out, by another relay.

5 | CIRCUITS THAT COUNT

In carrying out a program a digital computer behaves like a man performing a series of errands in planned order. As he performs each errand, the man counts it off in his mind or checks it off on a list—one, two, three, and so on. In a computer, the program counter keeps count of the list of instructions so that each is performed at the proper point in the sequence. The counter contains the address of instructions, counting through them in order as explained in Chapter 10. Here we study how circuits with memory are made to count. Later we will learn how to construct such circuits.

In counting from 0 to 7 in binary, for example, a counter must have three stages corresponding to the three binary weights 2^0, 2^1, and 2^2. On the count of one, the first stage, 2^0, must register 1 while the others remain 0. On the count of two, the second stage, 2^1, must register 1 while the others are 0. This counting process must continue as indicated in Fig. 14–11. Note that the 2^0 stage must alternate between 0 and 1 on each count, while the 2^1 stage must alternate only every second count, and so on. There is a circuit with memory that can be the basic unit of a counter. Several of them connected together have the alternating property needed for a counter.

The basic counter stage is shown in Fig. 14–12a. The P and Q memory circuits have each other's contacts in their recording networks. The resulting action gives the counter stage a behavior quite different from the shift register stage which looks similar at first glance.

The contacts a and $\bar{a}$ are the counter input. That is, the count is increased by one when switch A is operated and then released. The state of relay Q corresponds to the 2^0 digit of the counter.

To see how this comes about, look at the right of Fig. 14–12a. Relay P is operated if its holding contact p is closed, or if A is operated and Q released. In either case, P remains operated as long as Q is released. A similar statement can be made about Q. Let us see what happens in the stage as we turn A on and off. Assume that, at the start, relays P and Q and switch A are all released. Figure 14–13a shows the paths that are completed. Next we operate A to count 1. Current appears in P which operates (Fig. 14–13b). Next we release A. This time current appears in Q which operates (Fig. 14–13c). A complete operate-release cycle of A has thus caused both P and Q to store 1's.

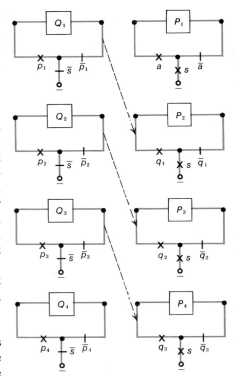

Fig. 14–10 Completed four-stage shift register.

		Binary Count	
	2^0	2^1	2^0
0	0	0	0
1	0	0	1
2	0	1	0
3	0	1	1
4	1	0	0
5	1	0	1
6	1	1	0
7	1	1	1

Decimal Count

Fig. 14–11.

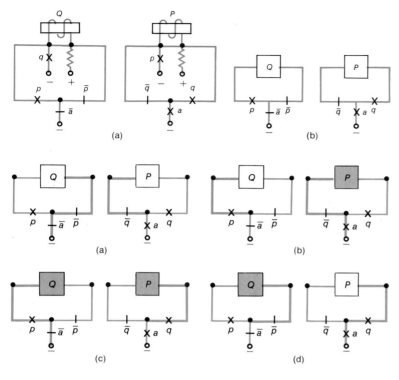

Fig. 14–12.
The counter.
a) Counter
stage circuit.
b) Counter stage
symbolized
(memory plus
recording
circuit).

Fig. 14–13.
Two A pulses operate
the counter stage.
a) A = 0, P = 0, Q = 0.
b) A = 1, P = 1, Q = 0.
c) A = 0, P = 1, Q = 1.
d) A = 1, P = 0, Q = 1.

Now we run through another cycle of A. First we operate A.
P releases (Fig. 14–13d). Next we release A. This releases Q (Fig.
14–13e). We are now back to the same state as in Fig. 14–13a.
We can summarize the actions of the counter circuit as A is operated and released by listing the successive states in a table (Table
14–1). Notice that it requires two cycles of A (0-1-0-1) to get
one cycle of Q (0-0-1-1). This is exactly the relation between successive digits of a counter as indicated in Fig. 14–11.

Step:	(0)	1	2	3	4	5	6	7	8	9	10	11	· · ·
A:	0	1	0	1	0	1	0	1	0	1	0	1	
P:	0	1	1	0	0	1	1	0	0	1	1	0	
Q:	0	0	1	1	0	0	1	1	0	0	1	1	

Table 14–1.
Summary of
the action of one
stage of the
counter.

A multistage counter can be formed by cascading stages identical to the one discussed above. The arrangement is shown in
Fig. 14–14. Note that each Q relay controls contacts in the next
stage's P and Q circuits. The actual connections are shown in Fig.
14–15 for a three-stage counter. (The reader should ignore the
dashed part of the circuit for now.)

The opening and closing of the q_1 contacts in the second stage
influence that stage exactly as the opening and closing of the a

Fig. 14–14.
A three-stage
counter illustrating the
influence of one stage of the
counter upon the next.

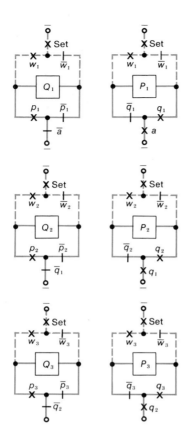

Fig. 14–15.
*Complete
three-stage
counter.*

contacts influence the first stage. In the third stage the q_2 contacts serve the same purpose.

In a multistage counter, two openings and closings of the q contacts of one stage cause the q contacts in the next stage to open and close just once. Thus, in Fig. 14–15 the values of q_3, and q_2, and q_1 are a sequence of three-digit binary numbers (reading from bottom upward) which starts at 000 and progresses (each time A is released) through 001, 010, 011, 100, 101, 110, and 111. In effect, the circuit "counts" how many times the A switch has been pulsed (on-off) up to 7. The Q relays can provide the counter output. A counter with M stages can count from 0 through $2^M - 1$. For example, a ten-stage counter can count to $2^{10} - 1 = 1023$.

In many applications, counters must be set to some starting number other than zero, or the number must be jumped to a lower or higher value. Both these operations must be done with the program counter in a computer. The dashed circuits in Fig. 14–15 can set the counter to any desired value, from which it can proceed by counting in the normal fashion. For instance, suppose we want to set the counter to 5, but we previously had counted to 7: q_3, q_2, and q_1 are all 1. The relay states and recording paths are shown

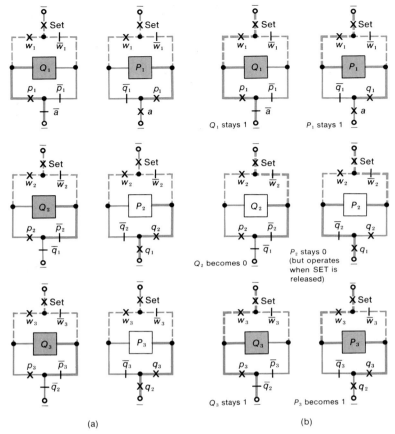

(a)

(b)

Fig. 14–16.
Setting
a counter.
a) Counter set
to 7 (i.e., 111).
b) Counter set
to 5 (i.e., 101).

in Fig. 14–16a. To set the counter to 5, we first set $W_3 = 1$, $W_2 = 0$, $W_1 = 1$ and then pulse the SET switch. The resulting relay states are indicated in Fig. 14–16b.

In the automatic operation of a computer, as explained in Chapter 10, the address in the program counter *selects* that memory cell so that the instruction there can be brought into the instruction register. This action is indicated in Fig. 10–18, Step 1, of Chapter 10. This operation is carried out by using the program counter to control a "tree" circuit. The diagram in Fig. 14–17 demonstrates how the q contacts from three stages of a counter can control a tree. When the switch A is operated and released, the count is increased by one and the completed path in the tree is connected from the "root" of the tree to the terminal path on the right, which has a number one greater than the previous count. To jump to any particular memory location, set the counter to the appropriate number. This is one way that memories of computers are addressed. Another use of counter-controlled trees is examined in the next section.

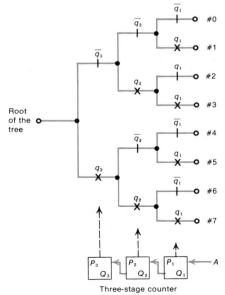

Fig. 14–17.
A counter-controlled
tree.

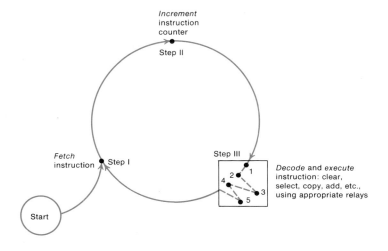

Increment
instruction
counter

Step II

Fetch
instruction · Step I

Step III

Decode and execute
instruction: clear,
select, copy, add, etc.,
using appropriate relays

Start

Fig. 14–18.
Instruction-cycle
sequence showing
when the instruction is
actually executed.

A	●	—			
B	—	●		●	●
C	—	●		—	●
D	—	●		●	
E	●				
F	●	●		—	●
G	—	—		●	
H	●	●		●	●

Fig. 14–19.
Morse code
representation of
A *through* H.

6 | TURNING A NUMBER INTO ACTION

In the preceding sections we have seen how various elements of a computer can be made from switches and relays. Adders and shift registers are commonly used in computer processors. Memories are of course used in constructing addressable storage. We have even seen how one element, the counter, can drive another, the addressable memory, so that information can be stored or read out from each cell independently. This is exactly how the program counter in a computer sets the memory so that the appropriate instruction can be brought into the instruction register. Recall that in the computer's automatic cycle this was Step 1, "fetch an instruction." A simplified reminder of the events in the cycle is shown in Fig. 14–18.

Looking further at this cycle, note that during "execution," a number of steps must occur one after another automatically. For example, in executing an "add" instruction, the following steps must take place: (1) the memory cell containing the number to be added must be connected to the adder in the processor; (2) the number in the accumulator must be transferred to a register that is connected to the adder input; (3) the output of the adder must be connected to the accumulator, so that the sum will appear there; (4) the adder must be actuated to produce the sum; and (5) all these connections must then be broken. Remember, too, that all these successive steps are done in response to an instruction which appears merely as a number (2XY, that is, op code plus address) in the instruction register. So the control mechanism must contain the necessary logic circuitry to turn that op code plus address into the correct sequence of steps *automatically*. This requires a combination of a counter, a selector tree, and several logic networks. In this section we will not see all details of this

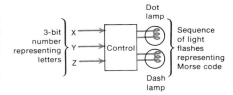

3-bit
number
representing
letters

X

Y

Z

Control

Dot
lamp

Dash
lamp

Sequence
of light
flashes
representing
Morse code

Fig. 14–20.

Letter	3-bit code X Y Z			Morse code			
A	0	0	0	●	—		
B	0	0	1	—	●	●	●
C	0	1	0	—	●	—	●
D	0	1	1	—	●	●	
E	1	0	0	●			
F	1	0	1	●	●	—	●
G	1	1	0	—	—	●	
H	1	1	1	●	●	●	●

Fig. 14–21.
Three-bit code
for the first eight
Morse code
letters.

Letter	Binary code			Morse code output							
	X	Y	Z	Time interval							
				0	1	2	3	4	5	6	7
A	0	0	0				●		—		
B	0	0	1		—		●		●		●
C	0	1	0		—		●		—		●
D	0	1	1		—		●		●		
E	1	0	0				●				
F	1	0	1	●			●		—		●
G	1	1	0		—		—		●		
H	1	1	1	●			●		●		●

Fig. 14–22. Timing the Morse code output.

control mechanism. Rather we will consider how a number can be turned into a simpler sequence of actions, namely the sending of a Morse code signal. This example illustrates the principle of computer control.

In Morse code, letters are each represented by a different sequence of dots and dashes. The code for the first eight letters of the alphabet is shown in Fig. 14–19. When a Morse code operator sends a message, he keys out dots and dashes one after another to spell the message. Here we want to use switches and relays to generate the correct dots and dashes when a letter to be sent is chosen. This process is indicated in Fig. 14–20, where a three-digit binary number, representing the letter, is the input and the correct sequence of dots and dashes is the output. In terms of computer control, the binary numbers are the op codes and the Morse code signals are the execution steps. The corresponding numbers and codes are listed in Fig. 14–21.

Let us assume that each Morse signal is to be sent during an 8-second interval. The 8-second time interval is shown in Fig. 14–22 cut into 8 one-second intervals. In each interval one of three things is to happen: a dot is sent (the dot lamp is lighted), a dash is sent (the dash lamp is lighted), or nothing happens. (Spaces are necessary to keep consecutive dots and dashes from being confused.) During time intervals 0, 2, 4, or 6, nothing need be done. On the other hand, during interval 3, for instance, the dot lamp should be lighted for letters A, B, C, D, E, F, and H, or the dash lamp should be lighted for letter G. This relation can be represented by a logic circuit as shown in Fig. 14–23a. Notice that the contacts in these networks are open or closed depending on whether switches X, Y, and Z are operated or released as listed in Fig. 14–22. Similar networks can be arranged for the other seven time intervals of Fig. 14–22. These are shown in Fig. 14–24. (Remember for intervals 0, 2, 4, 6 neither light need be lit; no network is necessary.) The final problem then is to connect each dot-dash network pair to minus at the appropriate interval in the cycle. We can use a counter circuit to count from 0 through 7 (and then back to 0). It

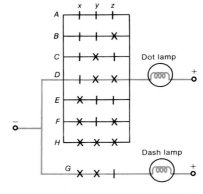

Fig. 14–23. Circuit diagram for the Morse coder at time intervals.

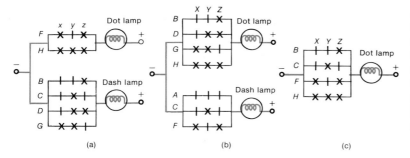

Fig. 14–24.
Circuit
diagrams for
the Morse coder
at other active time
intervals.
a) Interval 1.
b) Interval 5.
c) Interval 7.

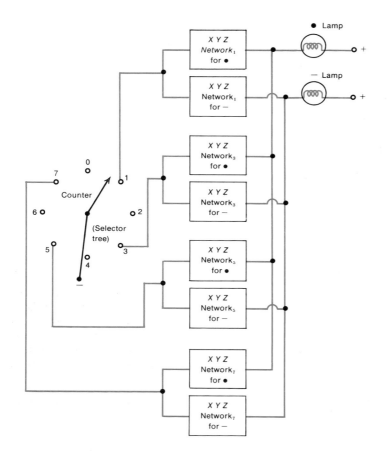

Fig. 14–25.
Schematic
diagram of
Morse
coder.

can drive a selection tree to select networks for intervals 1, 3, 5, and 7 in a timed sequence. This action is illustrated in **Fig. 14–25.** The selection of a · or a – in any time interval is shown as a two-step selection process: first the counter must be at the proper time interval (1, 3, 5, or 7); second, although a path is made from minus through the counter tree to a network, at the same time the network must itself be completed in order to light the lamp. In other words, the first half of a completed path is determined by the time interval, the second by the state of switches X, Y, and Z.

It is exactly in this way that computer control is arranged so that op codes can be translated into appropriate series of actions to carry them out.

7 | FINAL COMMENT

Earlier we saw how relays could be used to couple logic circuits by having a relay in one circuit control contacts in another. In this chapter, we have seen that by allowing relays to control contacts in their own circuit, a whole new family of computer building blocks results. These have memory and are a vital part of any computer that stores and processes information. The actual circuits presented are practically all that is needed to create an elementary, but representative, computer. In addition, we have now studied all the principles necessary in understanding digital computers and what they can and cannot do.

Questions for Study and Discussion

1. In this chapter, relays were discussed as a memory device. Name some other devices which could be used to remember an event.

2. Discuss the meaning of the term *address*. How many addresses can be accessed using a three-variable binary code; a three-variable decimal code?

3. The telephone company uses a ten-variable decimal code for your "address." What is the maximum number of telephone numbers that can be assigned if there are no other restrictions on the assignment?

4. Discuss the function of the shift register in the arithmetic unit of a digital computer.

5. Explain the operation of the two circuits below.

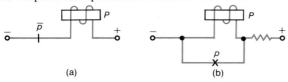

(a)   (b)

Do these circuits have any practical applications?

6. Refer to Fig. 14–4 of the text. Recall that 0's and 1's can be stored by operating and releasing switches A and B, and that the states of operation of C and D determine the address at which data are stored or sensed. For all parts of this problem assume that all switches and relays are initially unoperated.

   a) What and where is information stored when the switches D and A are operated (in that order) and switch A is then released?

   b) What and where is information stored when C is operated, A is operated and released, and B is operated and released (in that order)?

   c) What sequence of operations is necessary to store a 1 in the relay S?

   d) What sequence of operations is necessary to store a 0 in the relay R?

7. What is one advantage that machine memory has over human memory?

8. What is one advantage that human memory has over machine memory?

9. Describe the sequence of operations that are necessary to turn on the light in the following circuit. Give the reason for each operation.

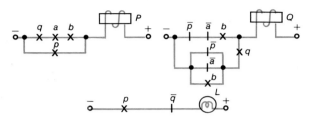

10. Discuss what conditions are necessary to turn the lamp L on and off in the following two circuits.

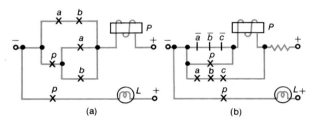

(a) (b)

11. Discuss how to operate the switches A and B to turn the lamp L on and off in the next circuit. If the relay and switches are initially unoperated what is the shortest sequence of operations which will turn the lamp on?

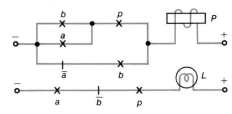

12. The switches in the following circuit are operated in the following sequence:

	1	2	3	4	5	6	7	....	13	14
a:	0	0	1	1	0	0	1	....	0	0
b:	0	1	1	0	0	1	1	....	0	1

Describe how the relays P and Q operate and release, and how the number of times they operate relates to the number of operations of switches A and B.

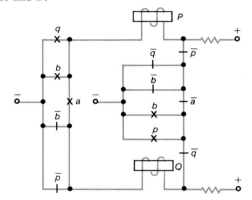

Problems

1. a) In the two circuits below, discuss what happens when, starting with the condition that relay P is released, the switch A is operated.

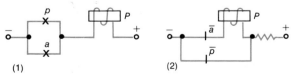

(1) (2)

b) What happens when, starting with condition that P is operated, A is released?

2. Analyze the operation of the following two circuits, that is, describe step-by-step, starting with all switches and the relay unoperated, what happens when:

a) switch A is operated and then switch B is operated;

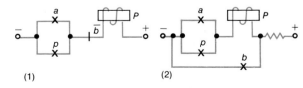

(1) (2)

b) switch B is operated and then switch A is operated.

3. In each of the three circuits below analyze what happens, starting with both switches and the relay unoperated, when

   a) switch A is operated and then released, then switch B is operated and then released;

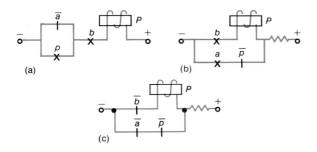

(a)   (b)

(c)

   b) switch B is operated and then released, then switch A is operated and then released.

4. In the circuit below, switch A has been released for a long time, and then it is operated. What are the possible resulting states of operation of the relay P? Explain.

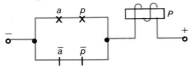

5. The following circuit is used to determine which of two contestants in a TV quiz show operates his switch first. Discuss the operation of this circuit assuming that switch C is released. What is the probable purpose of switch C?

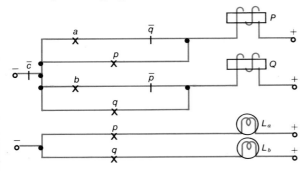

6. a) Describe the shortest sequence of operations of switch A which will cause the lamp L to be turned on. Assume that all switches and relays are initially unoperated.

   b) How can the lamp be turned off again?

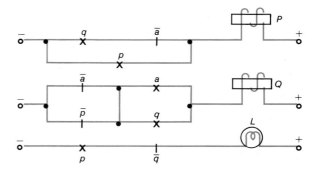

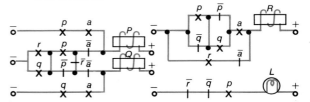

7. Describe step by step what happens in the following circuit when switch A is alternately operated and released.

Laboratory and Projects

I | CIRCUITS WITH MEMORY

If the present state of a circuit depends on what its inputs were at some past time, we call it a *sequential circuit*. Such a circuit needs memory. How can we make a memory circuit with a relay?

Figure 1 illustrates two methods of controlling the current in the relay coil. Fig. 1a shows a make contact in series with coil P; in Fig. 1b a make contact is in parallel (or shunt) with coil Q.

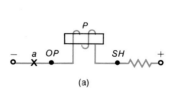
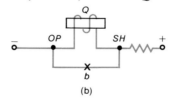

(a)   (b)

Fig. 1.

Memory

Wire the circuits shown in Fig. 1a and 1b. Note the movement of the metal armature of relay P as switch A is operated; of relay Q as B is operated. To determine the state of the relays more readily add the circuits shown in Fig. 2a and 2b to your present hookup.

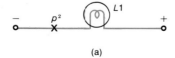

(a)   (b)

Fig. 2.

1. When switch A is operated, what is the state of $L1$? Explain.

2. When switch B is operated, what is the state of $L2$? Explain.

■ 3. Summarize your results by preparing and completing a table similar to Fig. 3a and 3b.

Fig. 3.

A	P	L1
0		
1		

(a)

B	Q	L1
0		
1		

(b)

Now modify the circuit of Fig. 1a so that the operations of the relay will influence the state of the relay; that is, the relay's own contacts will be in the control path. This process is known as feedback—some of the output signal is used as input to the circuit. Wire the circuit of relay *P* as shown in Fig. 4. Operate, then release switch *A*.

Fig. 4.

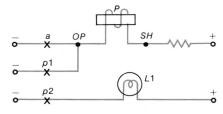

■ 4. Explain your results.

A circuit which remains in a given state unless influenced by external signals is known as a *stable circuit*. The circuit shown in Fig. 4 is stable.

An unstable circuit continually changes from one state to another. Two such circuits are shown in Figs. 5a and 5b. Note that Fig. 5a shows the control switch in series with the coil while Fig. 5b illustrates control of a relay by a switch in parallel with the coil. Wire the circuits of Figs. 5a and 5b. These circuits are known as buzzers.

Fig. 5.

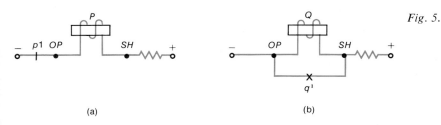

(a)

(b)

■ 5. Which circuit buzzes faster, 5a or 5b?

■ 6. Why should there be a difference in the frequency at which the circuits buzz?

Usually buzzing is an indication that a circuit has been incorrectly wired or that a component is defective. There are, however, situations where periodic pulses are desired, as in clock or timing circuits.

It is also possible for the contacts of one relay to control a second relay. Thus two or more relays can control each other and give either stable or unstable behavior.

Examine Fig. 6. Analyze the circuit shown.

■ 7. Does Fig. 6 represent a stable or unstable arrangement? Explain. Wire the circuit and verify your prediction.

Change the circuit of Fig. 6 by replacing the $\bar{s}$ contact controlling relay P to an s make contact.

■ 8. Is this new arrangement a stable or an unstable circuit? Explain.

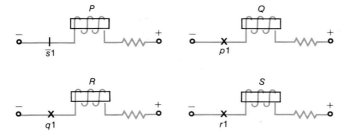

Fig. 6.
A circuit to analyze and test for stability.

Memory with Control

By rearranging the circuits of Fig. 4 and using a make contact $b1$ in parallel with the coil, we obtain the circuit shown in Fig. 7.

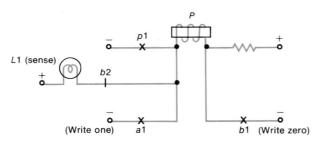

Fig. 7.
One-bit memory with control contacts and sensing lamp.

A break contact on switch B ($\bar{b}_2$) is essential to prevent current from following a *sneak path* between (+) and (−) via the relay coil and the lamp when we "write" a zero. Lamp $L1$ is the readout lamp indicating the state of relay P, while $p1$ is the holding contact for relay P.

Since the relay P has two stable states, we call this circuit a one-bit memory. We say that when P is operated (and holding itself in), the circuit is storing a "one." When P is released, it is storing a "zero." It is then natural to call switch A the "write one" switch since momentarily operating A will set P to "one." In the same way switch B is the "write zero" switch. We can "read" the memory circuit by looking at lamp $L1$; if the lamp is lighted, the circuit is storing "one," otherwise, "zero." Later when we talk of storing information on cards, a relay punch will replace the lamp.

◼ 9. Prepare a running list for the circuit of Fig. 7.

◼10. Wire the circuit. Write a "one." Does $L1$ remain lighted when switch A is released?

◼11. Write a "zero." Can you write a "one" while switch B is in the operated condition?

◼12. Remove the $\bar{b}_2$ contact from the circuit, then operate switch B. Explain your result.

Memory with Four Cells Each Storing One Bit

The central memory for even a small computer contains many thousand bits. In order to use such a memory we must have a way of selecting the relatively few bits desired at any instant. This is called the *access circuit* of the memory. The control signal to the access circuit is a binary word which is called the Address of the selected part (or *cell*) of the memory. We use two copies of a tree circuit for our access circuit. Figure 8 shows the circuit diagram of the 4 × 1 (four cells, each one bit) memory which you are using in this part. The copy of the tree on the left side of the diagram steers both the "write one" signal and the sense lamp to the selected one of the four memory relays P, Q, R, and S. The copy of the tree on the right side steers the "write zero" signal to the selected memory relay.

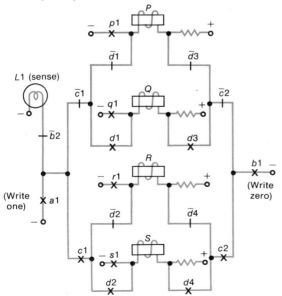

Fig. 8.

Wire the circuit of Fig. 8, set switch $C = 0$, switch $D = 0$; this is the address of the cell using relay P. Operate switches A and B momentarily several times to successively write "ones" or "zeros" in cell 00. Leaving cell 00 with a "one," go to cell 01 ($C = 0$, $D = 1$: relay Q). Successively write several "ones" and "zeros" into this cell. Leaving cell 01 with a stored "zero," go back to cell 00 and check that it has not been disturbed by writing into cell 01; that is cell 00 should still contain a "one." Test all the cells in a similar fashion until you feel your understanding of this circuit is complete.

■13. What is the address of the cell using relay R? Relay S?

II | THE SHIFT REGISTER

Introduction

Another of the important sequential circuits which involves memory or storage elements is the *shift register*. This circuit has many applications and is of particular importance in the process of multiplication in the computer. It can also be used to clear the contents of the accumulator in the computer.

Multiplication in the binary system is particularly simple since only shifting and summing or adding are required. Thus to multiply the number 1010 by 101 we merely rewrite and shift the multiplicand by one place to the left for each 1 in the multiplier and shift without a rewrite of the multiplicand for each 0 in the multiplier. We then sum the columns as in decimal multiplication, but we add in accordance with the arithmetic of the binary system. The above problem would then be solved as follows:

```
      1010
       101
      1010 ← rewrite multiplicand multiplied by 1
     0000  ← shift multiplicand and multiply by zero (0)
    1010   ← shift multiplicand and multiply by 1
    1      ← carry digit
  110010 ← product
```

Our shift register will consist of relays connected in a manner capable of shifting the state of any relay to its immediate neighbor.

In Fig. 9 a relay P is arranged to store either a 1 or a 0 by manipulating switch A. If a 1 is to be stored, switch A is first

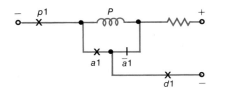
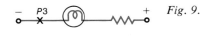

Fig. 9.

operated. The shift switch D is now operated to energize the coil P through contacts d_1. The coil will "lock" into an operated condition through contacts p_1.

The storage of information should be considered as represented by the state of relay p. When it is operated, a 1 has been stored and when it is released, it represents a 0. Our operation of switch A followed by the operation of switch D can be considered to have accomplished the storage of a 1 in two steps. In the first step the movement of switch A may be considered as the "read-in" of the digit into the particular part of the circuit, and in the second step the movement of switch D can be considered as the "shift" of the digit into storage at relay p. This method of examination of the circuit will make the explanation of the shifting operation easier to understand. We simply keep in mind the fact that the contacts in the position occupied by $a1$ and $\bar{a}1$ produce a read-in of the digit, while the operation of the contacts in the position occupied by $d1$ will produce the actual storage of the digit.

In Fig. 9, therefore, we:

1. Operate switch A to read in a digit 1. (If we do not operate switch A we have the equivalent of a read-in of a 0.)
2. Operate switch D to store the digit in relay P.

In Fig. 10 we have a full stage of a shift register. It should be noted that each half is designed in the same fashion: it contains a set of read-in contacts across the coil of the relay, a set of storage contacts in series with the coil, and a set of shift contacts which shift the information from the read-in to the storage contacts.

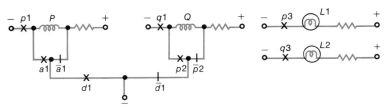

Fig. 10.

In Fig. 10 the read-in contacts for relay Q ($p2$ and $\bar{p}2$) are set for a 1 whenever relay P stores a digit 1. They do not read-in a digit 1 when a 0 state is represented at relay P.

The shifting operation from relay P to relay Q can now be traced as follows:

1. Read-in a digit 1 into stage P by operating switch A.

2. Store this in P by operating switch D.

3. If a zero is now to be read into this stage, the switch A must be returned to its unoperated state. As long as it remains in its operated state, each movement of the switch D stores a 1 into relay P.

4. Storage of digit 1 at relay *P* automatically reads-in the information into stage *Q* through contacts *p2* and *p̄2*. Notice that although the 1 is *read* into stage *Q*, it cannot be *stored* at *q1* because contacts *d̄1* are open when contacts *d1* are closed.

5. Shift switch *D* is now returned to its unoperated position. This permits current to energize coil *Q* and storage takes place. We have thus shifted the digit 1 from *P* to *Q*.

We can extend the shift register to as many stages as we wish. Figure 11 represents a two-stage (four relay) shift register.

■ 6. Prepare a running list for this circuit and wire this on your LCB.

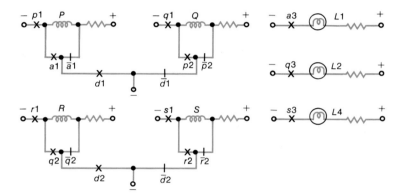

Fig. 11.
Two-stage shift register.

■ 7. If time permits wire a shift register of four stages with two LCB's.

Remember to connect a negative (−) terminal on one board to a negative (−) terminal on the other to serve as a return path for the signal.

III | THE COUNTER CIRCUIT

In this experiment, the *counter*, a frequently used type of sequential circuit, is studied. The name of the circuit (counter) describes one of the major applications of the arrangement. The circuit counts the number of pulses that are applied to its input. But the counting process is actually a *scaling down* process. After a fixed number of pulses have been applied to the input, the circuit produces a readout which may be a glowing lamp, or an electrical pulse which can actuate a second stage of the counter.

This experiment involves the wiring of a *binary counter*; an arrangement in which at each stage one count or pulse is produced at the output for two input pulses. The input is thus scaled down by the factor of two for each stage. With two stages the count is scaled down by four; for three stages by eight, etc.

The Single-Stage Counter

A single-stage counter is illustrated schematically in Figure 14a. The pulses required to operate this counter stage are produced by manipulating the manual switch A. When switch A is operated, the electrical current flows through relay coil P. When a is brought back to the 0 position, the current through the a make contacts is zero. Operating A once more causes current to flow again. Thus this back and forth motion of switch A produces a series of pulses of current much as illustrated in Figure 12.

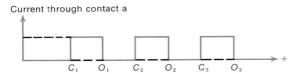

Current through contact a

C_1 O_1 C_2 O_2 C_3 O_3

Fig. 12.
Note: C_1, C_2, C_3, etc., represent contact closure.
O_1, O_2, O_3, etc., represent contact opening.

By studying Fig. 14a, you can verify that: (1) operating switch A causes relay P to operate and hold, (2) releasing switch A then causes relay Q to operate and hold (this operates lamp 4), (3) once again operating switch A causes relay P to be released, and (4) when switch A is released, relay Q is also released, causing lamp 4 to go to the unoperated state. Figure 13 illustrates the relationship between the pulsing of switch A and the operation of lamp 4.

Fig. 13.

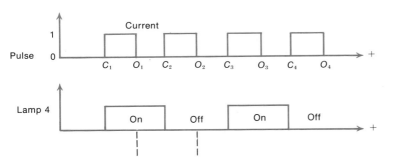

Current

1

Pulse 0

C_1 O_1 C_2 O_2 C_3 O_3 C_4 O_4

Lamp 4

On   Off   On   Off

Wire the single-stage binary counter (Fig. 14a) and check to see if it agrees with the analysis given above and with the information on Fig. 13.

Do not remove any of the wires from this single-stage counter, but use relay contacts q_3 as a new pulser and complete Fig. 14b. This two-stage counter is a combination of two single-stage counters; the pulses produced by the first stage for operating lamp 4 may now be used to produce input pulses for the second stage.

1. Operate the binary counter and complete a copy of the following chart.

Switch A	0	1	0	1	0	1	0	1	0	1	0	1	0
Relay P													
Relay Q													
Relay R													
Relay S													
Lamp 4													
Lamp 3													

We see that counter circuits are actuated by pulses. Our counter, because of the physical limitations of the LCB, was restricted to counting from binary 0 0 to 1 1. The circuit may be extended, using additional LCB's to give greater range. If time is available, try to extend the range using more than one LCB.

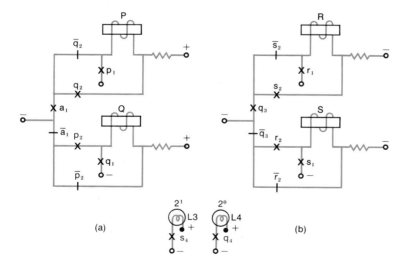

Fig. 14.
Two stages of a binary counter.

IV | AN AUTOMATIC MORSE CODE TRANSMITTER

In this experiment we examine a switching circuit in which a series of operations is performed in a sequential manner. The device itself, a Morse code transmitter, is described because it displays many of the elements we have already discussed and in addition gives some insight into the logical arrangement of these elements to produce a desirable sequence of operations. Computers and other devices are required to produce a specified series of operations at definite time intervals and make use of similar but considerably more complex switching arrangements.

We consider our device as a typewriter, which produces a series of lamp flashes when any key is depressed. Three lamps are used. The first flashes on as an indicator that the machine is ready to interpret into Morse code any letter key which is depressed. A second lamp will flash to indicate a dot and a third lamp will flash to indicate a dash. In actual practice, therefore, if lamp L_1 is glowing to indicate that the machine is ready and we depress typewriter key A, lamp L_2 will flash on to indicate a dot; it will then be extinguished and lamp L_3 will flash on to indicate a dash and then it too will be extinguished as lamp L_1 comes on again to indicate that the machine is ready for the translation of the next letter into Morse code.

Our LCB will permit only the first eight letters of the alphabet to be set by the switches. The principles we use can be extended with a greater number of switches. The Morse code for the first eight letters of the alphabet is as follows:

$$A = \bullet \,—\qquad\qquad E = \bullet$$
$$B = —\,\bullet\,\bullet\,\bullet\qquad F = \bullet\,\bullet\,—\,\bullet$$
$$C = —\,\bullet\,—\,\bullet\qquad G = —\,—\,\bullet$$
$$D = —\,\bullet\,\bullet\qquad\quad H = \bullet\,\bullet\,\bullet\,\bullet$$

With three switches we can secure eight combinations so that each combination can be used to represent a single letter of our group. We shall represent each letter by a switch arrangement as shown in Fig. 15.

Fig. 15.

Letter	Switch Operation		
	B	*C*	*D*
A	0	0	0
B	0	0	1
C	0	1	0
D	0	1	1
E	1	0	0
F	1	0	1
G	1	1	0
H	1	1	1

Thus when switch B is not operated, but switches C and D are operated, the letter D is to be given by the machine as a flash of lamp L_3 followed by two flashes of lamp L_2 to indicate a dash followed by two dots.

On our LCB we set these switches manually, but in the type-writer arrangement of a single depression of the typewriter key marked "*D*" would be arranged to set the equivalent of the LCB switches inside of the machine. We must also remember that the lamps do not flash on and off at the same time, so that some form of timing mechanism or clock must be present to produce timing pulses which turn on the dot and dash lamps at the proper instant of time (Fig. 16). On our LCB the timing pulses will be produced manually.

The Morse code transmitter which we will build thus involves a series of two major steps for its operation. The first step requires that we set switches *B*, *C*, and *D* for the letter we require, an action that would normally be obtained by striking the typewriter key; and the second step requires that we produce a series of timed pulses so that the dots and the dashes for the required letter are produced in the required order which would normally be developed in the typewriter by some type of timing mechanism built into it.

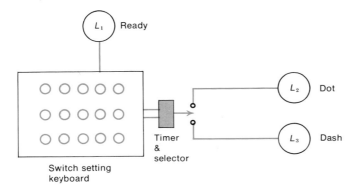

Fig. 16.

In Fig. 17 the lamp operating sequence is shown for each of the letters to produce the appropriate Morse code output. The chart is divided into 8 instants of time and for each instant of time the operated lamp is shown as a dot or a dash. Intervals 0, 2, 4, and 6 are instants when no lamp is flashed and these indicate a space between two consecutive Morse Code symbols.

Figure 20a, b, and c is the complete circuit for the transmitter. In Fig. 20c the switching arrangement for operating the dot and dash lamps is drawn schematically to the right of the dotted line. The circuit may appear to be quite complex but is essentially a tree circuit for selection of the appropriate lamp at the appropriate time once the switches *B*, *C*, and *D* are set. We will examine one method by which such selection circuits can be devised later. But for the moment we should observe that the switches to the left of the dotted line in Fig. 20c are operated by the clock or timing mechanism. If, for example, a dash is required at time interval 1, the clock must operate contacts a_3 and leave unoperated contacts

	0	1	2	3	4	5	6	7
A				•		—		
B		—		•		•		•
C		—		•		—		•
D		—		•		•		
E				•				
F		•		•		—		•
G		—		—		•		
H		•		•		•		•

Fig. 17.

$\bar{q}_4$ and $\bar{s}_2$. If the switches in the selection circuit *to the right of the dotted line* have been properly set, lamp L_3 (dash) will flash at this instant. We examine the operation of the complete Morse code transmitter in a series of subdivisions.

Dot-Dash Lamp Selector

The design of the lamp selector circuit can be illustrated by examining the design of the dash lamp switches at "time 1." Figure 17 indicates that the dash lamp should be "on" for the letters *B, C, D,* and *G* and "off" for all the remaining letters in our group of eight. This necessitates that the switches to produce these letters must first be set manually and then the pulses produced by switch *A* will flash the dash lamp at time "interval 1."

Figure 17 then informs us that the following Boolean expression must be true for this condition:

$$\overline{B}\overline{C}D + \overline{B}C\overline{D} + \overline{B}CD + BC\overline{D} = 1$$

Factoring the first and third term as well as the second and fourth term gives:

$$\overline{B}D\,(\overline{C} + C) + C\overline{D}\,(\overline{B} + B) = 1$$

By our postulate that any variable plus its negation must equal 1, the above expression reduces to:

$$\overline{B}D + C\overline{D} = 1$$

The circuit for this expression is then:

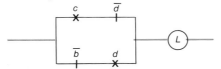

Fig. 18.

As can be seen in Fig. 20c, this is the circuit which will be energized to produce a dash when the switches A, Q, and S are properly adjusted to select the path which is marked "time 1." DOT and DASH circuits for the other letters can be designed.

Clock or Counter Circuit

Recall that the action of the counter unit (Fig. 20a) depended on the sequential energizing of two relays. The sequence of operations can be indicated with a diagram as shown in Fig. 19.

If the second stage (R and S) is operated from the pulses at relay Q, Fig. 19 will display the operation of the relays. It should be noted that the relative time sequences are similar to Fig. 19,

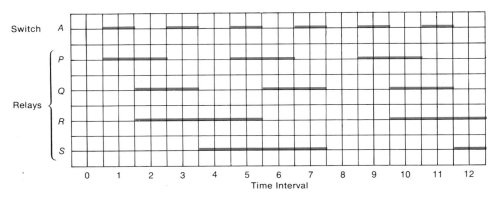

Fig. 19.
Switch relay timing
sequence.

but that time lengths of operation are twice as long as in the first stage. During "time 1" switch A is shown in an operated state while relays Q and S are unoperated. Reference to Fig. 20c indicates that under these conditions a path for current to the dash lamp circuit does exist from the negative terminal.

At "time 2" the unoperated state of switch A opens the path to the lamp circuit and neither lamp flashes. The student should trace the circuit connections for "time 3," etc.

Start Lamp

The lamp L_1 will flash when Q, S, and A are all released. If any letter is set-up for interpretation in Morse code, the absence of a glow in L_1 will indicate that the machine is not ready for pulsing to produce the required Morse code output. Pulsing of the machine should begin only when the lamp L_1 is bright.

We have thus discussed the three major units required to produce a device in which a series of desired outputs in a sequential order must be produced. A running list for this device is given below. Wire the LCB, set the switches, and pulse the counter to check the Morse code output for each letter.

Design a transmitter to produce a Morse code output for letters I and J. The Morse code for I is • • and for J is • — — —.

Running List for Automatic Morse Code Transmitter

a) Counter: *Neg, A3c, A2c, A1c, P1c, Q2c, Q3c, R1c; A1m, P1m, Pop, Qop; A2m, Q1c; Q1m, Psh; Q1b, Qsh; Q3m, S1c; Q2m, R1m, Rop, Sop; S1m, Rsh; S1b, Ssh; Q2b, S4c; S4b, A4c; A4b, L1.*

b) Tree: *A3m, Q4c; Q4b, S2c; Q4m, S3c.*

c) Networks: *S2b, C1c, D1c; C1m, D3b; D1m, B1c; S3b, C2c; C2m, B2c; B2m, B3b, D2c; S2m, B3c; B3m, C3c; C3b, D3m; S3m, D4c; D4b, C4m; D4m, C4b, B4m; C4c, B4b, L4, D3c, D2b, B1b; L3, D2m, C3m, C2b, B4c, B2b, B1m.*

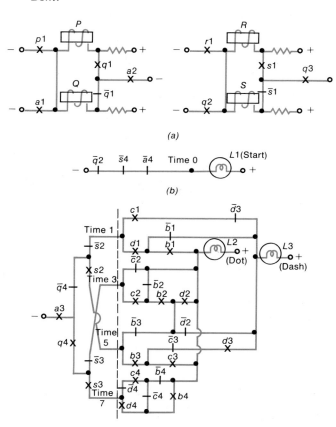

Fig. 20.
a) Counter circuit.

b) Starting indicator.

c) Automatic Morse code transmitter.

If you were to design a similar automatic Morse code transmitter for all 26 letters:

▮ 1. How many input code switches would you need?

▮ 2. How many stages would your counter require?

▮ 3. How many lamps would you need?

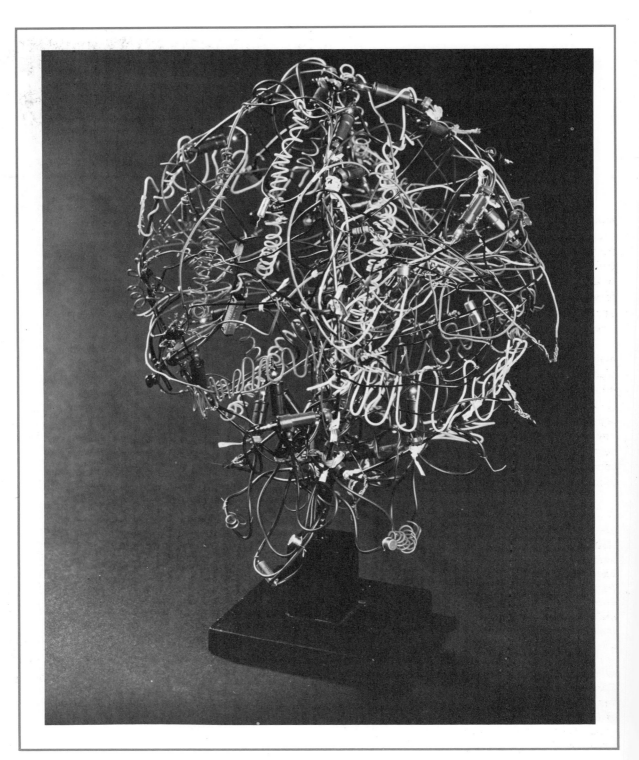

A Minimicro Computer 15

I | INTRODUCTION

There is a world of difference between looking at a computer from the outside and looking at it from the inside. When we view a computer from the outside, we consider the different attitudes people have toward it, the different kinds of information it can handle, and the varieties of uses to which it can be put. We may be dazzled by its capabilities. In contrast, the inside view is straightforward and prosaic. Internally, a computer does one and only one thing, it works with on-off devices and signals in response to a program which must be correct in every detail.

In Chapters 10 and 11, we looked at the computer as a symbol manipulator and studied how to program it to do our bidding. Next we looked inside and explored the building blocks for computers, examining how logic circuits using AND, OR, and NOT actually manipulate information, representing, storing, adding, counting, and shifting. Then, we saw how a control mechanism might execute computer instructions by connecting appropriate components inside the computer at the proper time. The Morse-code transmitter, as a primitive model for control, showed how numbers representing instructions might be converted to a sequence

Here is Thomas Adamatz's (11 years old) work of art representing his idea of a computer. It is made from Christmas-tree lights and discarded electronic components, and was awarded a prize in Data General Corporation's computer art contest. Actually computers are not nearly as chaotic as they appear to the uninitiated. They are highly organized and constructed by assembling successive levels of building blocks. (Computer Decisions, *June 1970*)

of small actions to perform the desired operations. In this chapter we will see how the computer as a whole is organized, and "what makes it tick."

In the process, we will actually "construct" a small, but "general-purpose," computer, using *only* components we've already discussed in previous chapters. This computer is called "general purpose" because in principle it can carry out any task which can be programmed, any task which its million-dollar relatives could carry out. Our minimicro relay computer typifies the same fundamental principles as do transistorized ones.

The fascinating fact is that any computer is somehow much "greater" than the sum of its parts. Each of its components carries out only one logical function, whereas the computer as a whole can carry out any task specified logically and in detail. This over-all capability results from the particular organization of components which was pioneered by Babbage and von Neuman.

The block diagram showing the organization of our minimicro computer is very similar to that we studied in Chapter 10, except that we use many of the components we have studied in Chapters 12, 13, and 14. To be specific, our minimicro computer deals with information encoded as 13-bit binary numbers (12 bits-plus-sign) stored and processed by binary selection trees, memory cells, counters, adder, instruction decoder, and so on.

2 | TRANSFERRING BINARY INFORMATION BETWEEN COMPONENTS (COPYING)

Basic to the operation of any computer is the transfer of binary information from one logic circuit to another. The computer must copy information into its processor, or memory, or instruction register without erasing it from its original locations—we want to preserve the information for further use.

The copying mechanism must make the "states" of the receiving relays the same (regardless of prior setting), relay for relay, as the sending relays (Fig. 15–1). We used copying without mentioning it in the shift register of Chapter 14 to copy the states of P relays onto Q relays within a stage, and to copy the states of Q in one stage to P in the next. Figure 15–2 shows another example of copying a single bit from relay P to relay Q. Note that we can impose a new state on a memory relay by choosing to connect either its make or break side to minus (Fig. 15–2a), or we can release the relay first, and then connect its make side only through a make contact to minus (Fig. 15–2b). The advantage of the second method is that it requires that the P relay control only one contact in the Q circuit. On the other hand, it takes two separate operations to record: clear, and then copy. We make use of both methods in our computer.

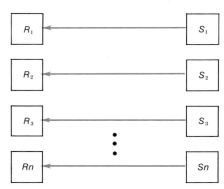

Fig. 15–1. Transferring a binary number from sending relays (or switches) $S_1, S_2 \ldots S_n$ to receiving relays $R_1, R_2 \ldots R_n$.

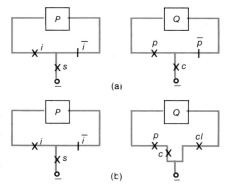

Fig. 15–2. Setting P and then copying P onto Q.

a) Set input switch I, operate and release switch S; then operate and release copy switch C.

b) Same as (a), except Q is cleared with switch CL before C is used.

To summarize, though we may speak loosely about transferring or moving information from one circuit to another, bits don't physically move—they are copied by making states of receiving relays identical to states of corresponding sending relays.

3 | INPUT, OUTPUT, AND MEMORY

Input

As explained in Chapter 10, information can be put into computers by typewriters, light pens, and many other devices. The most common, however, is a punched-card reader which senses holes in the cards and transfers this information to memory inside the computer. Here we see how this process works.

A punched card is effectively divided into areas, each of which can either have a hole in it or not, punched or unpunched. Our computer's simple card is divided into thirteen columns to represent a 13-bit binary number. A punched column represents a 1; an unpunched column represents a 0, as shown in Fig. 15–3a.

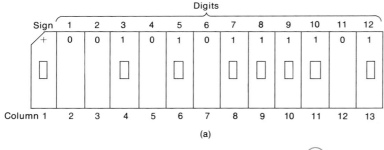

(a)

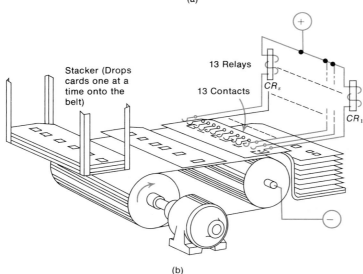

(b)

Fig. 15–3.

a) *Input devices. A 13-bit punched card for our machine (represents the number +0010101111101).*

b) *Punched card reader.*

Furthermore, the leftmost column represents the sign: unpunched for $-$, punched for $+$.*

Holes in cards are difficult for people to read. For that reason, most punched cards have the corresponding digits printed at the top of the card so that they are easily "man-readable." The computer reads the card by sensing the holes (Fig. 15–3b). A deck of cards is put into a stacker, then the motor-driven belt moves them one at a time under a set of thirteen contacts. At the instant that the contacts are lined up over the row containing the punched holes, a pulse of current is sent through the electrical circuits which run between the $+$ and $-$ terminals. Each of the 13 circuits will be complete only if the brush makes contact with the drum through a punched area on the card. Thus only those relays (CR_s and CR_1 through CR_{12}) which correspond to columns containing punches will be operated. The states of the relays now match the states of the columns, and we can say that the information has been transferred, copied, or read into the relays.

Output

In a similar way, information can be transferred from relay contacts to punched cards by means of the card punch shown in Fig. 15–4.

The card punch has 13 rectangular hole punches, each driven by an electromagnet similar to that in a relay, though stronger. During operation, a motor moves a blank card from the supply stack to the punches. When the punches are lined up with the digit positions on the card, the 13 electromagnets are operated according to the binary information stored on other switches or

*Thus a blank card denotes "$-000 \ldots 0$," an unacceptable number, whereas "$+000 \ldots 0$," i.e., plain zero, is represented by a single punch in column 1. In this way zero and a blank card can be distinguished.

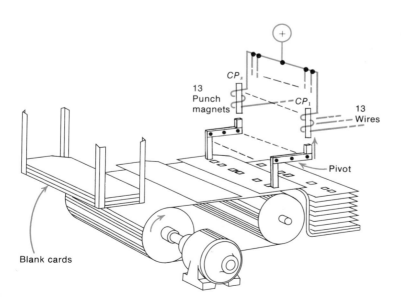

Fig. 15–4.
Card punch.

relays. This action causes the desired binary number to be punched on the card. Then the motor moves the card to the output stacker.

Information can also be stored in "machine-readable" form on punched paper tape or magnetic recording tape, or any one of many other media. All operate in essentially the same fashion as the punched card except for details of the recording mechanisms involved.

Memory

The unit of Fig. 14–4, Chapter 14, is an addressable four-cell memory. This unit is reproduced in simplified form in Fig. 15–5. Each cell can store one bit in its relay circuit. Tree contacts on relays C and D can select one of the four cells so that 1 or 0 can be written into that cell. Each cell has an address. Relay P is selected when $C = 0$, $D = 0$. Its address is then 00. The address of Q is 01. R's address is 10, and S's is 11. To store not a single bit in each cell, but a full 13-bit number, we must have 13 "copies" of this basic four-cell one-bit memory. The selection trees in each copy are identical and are controlled by the same relays, C and D.

A setting of the C and D relays selects the relay in the same position in each of the 13 circuits simultaneously. For instance, if we release C and operate D, we select address 01, the second relay from the top in each circuit. So 13 relays are selected. The 13 relays selected by a single address are designated by a letter with subscripts (P_s, P_{12}, P_{11}, . . . P_1). Each group of 13 relays stores a single 13-bit number with its sign. This memory then can store four 13-bit words.

To store the 13-bit number $+000000000101$ into cell Q (Q_s, Q_{12}, Q_{11}, . . . Q_1), we set 01 in the selection-tree relays. Then we connect X_s to minus, Y_2 through Y_9 to minus, X_{10}, Y_{11}, and X_{12} to minus respectively, thereby setting all Q relays appropriately.

Now, if we wished more than four words (i.e., more storage positions), we should expand the trees to include contacts on three relays, C, D, and E, for instance. Three relays provide 2^3 possible paths, so that each tree would then select from eight memory cells. Nine tree relays and 13 circuits provide 512 13-bit cells. ($2^9 = 512$)

Input and Output Connected to Memory

For purposes of illustration, let us see how we might connect our card reader, card punch, and a 4-cell 13-bit memory. Figure 15–6 shows a block diagram of the circuit. We have labeled the boxes with general names but have indicated in parentheses the actual hardware used in this example. The solid lines with arrows running between boxes show the "direction" in which binary signals are transmitted from box to box. For instance, a number can be transferred from the card reader to a memory cell, but not from the memory cell to the card reader. Relays I (Input) and O

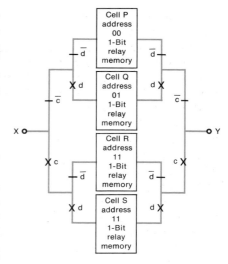

Fig. 15–5.
A four-cell
addressable memory;
each cell stores
one bit.

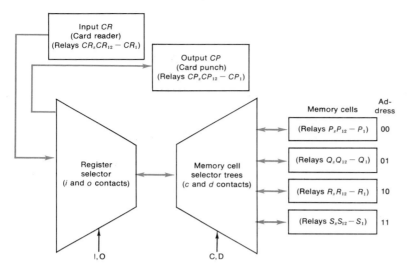

Fig. 15–6.
Block diagram of
a four-cell memory
connected to a card
reader and a card
punch. (The hardware
associated with each
block is shown in
parentheses.)

(Output) are used to connect card reader or card punch at the appropriate time to the selected cell in memory.

The actual circuit for one digit of Fig. 15–6 is shown in Fig. 15–7. It consists of one 4-cell memory connected to two relay contacts on the card reader and to one card-punch electromagnet. The memory inputs, X and Y, are connected to the card reader contacts which are labeled cr. The punch electromagnet is CP. In addition, the selection contacts on relays I and O are shown. The operation of the memory part of the circuit shown enclosed in dotted lines is identical to that of Fig. 15–5.

The objective of reading a number into memory is to transfer the states of the card to the stages of the selected memory relays. Thus, this card read into location 01

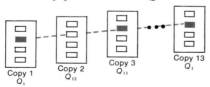

would produce this relay pattern in cell Q

Note that the first memory-circuit copy thus represents the sign, Q_s, not the most significant digit, Q_{12}.

To read this single 13-bit word from the card reader into the specific memory cell, each bit must be handled separately by a circuit like the one of Fig. 15–7. Each one uses a two-step select-copy process. First, we must select the address of the cell by selecting a path in the trees, and then read the first bit, 1 or 0,

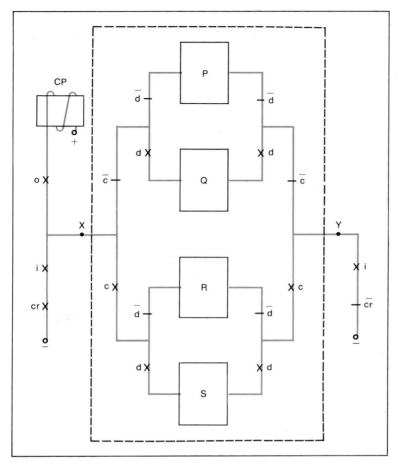

Fig. 15–7.
Card reader
and card punch
connected to
memory.

into the selected relay on the first circuit, at the same time the second bit into the corresponding relay on the second circuit, etc. We read the 13-bit word by simultaneously copying the states of 13 card-reader relays onto 13 cell relays. Thus, the action is:

1. Clear: Clear the CR relays and run the motor on the card reader to move a new card into the reading position.
2. Select: Choose a single memory cell by setting relays C and D to that cell's address.
3. Copy: Operate relay I to close all i contacts.

In order to read out from memory, we connect our memory to the card-punch electromagnet. The connection is made through the contacts of relay O. Then, to punch a card:

1. Clear: Clear the CP relays and run the motor to bring a new card under the punches.
2. Select: Set an address by relays C and D.
3. Copy: Momentarily operate relay O.

4 | ARITHMETIC UNIT

A major part of most computer processors is an arithmetic unit. It often is designed to add, subtract, multiply, and divide. Here for illustration we will concern ourselves only with addition. Our arithmetic unit, then, is simply a binary adder with 13-bits.

Figure 15–8a shows a signless 12-bit adder. The most significant bit is in place 12, the least significant is in place 1.

The A and B inputs come from relay registers which contain the numbers to be added. The sum appears in another register labeled S. This relationship is indicated in Fig. 15–8b. Note that the thirteenth overflow bit is stored in the S register, but cannot be copied into memory from S. The sign bits are handled by additional logic circuits which are independent of the adding circuits themselves.

Let us see how we use the arithmetic unit to add two numbers. On first thought one would suggest to "fetch" the first number from memory and copy it into the A register, then fetch the second one and put it into B, after which they could be added in the conventional way. Though this would work well for two numbers, how would one add five numbers? The actual mode for computer adders is to preserve the partial results of previous calculations in the S register, and then to add a new number to these "accumulated" results. For this reason, we often refer to the entire adder as the "accumulator." The accumulator is connected through two selector trees to memory so that it can have access to the cell containing the number to be added. This arrangement is shown in Fig. 15–9, along with the other units discussed so far. The

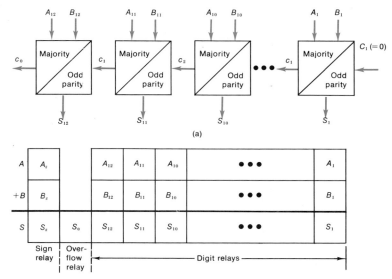

(a)

(b)

Fig. 15–8. Symbolic representations of the binary adder.

a) A 12-bit signless binary adder (a circuit for determining resultant sign S_s based on A_s and B_s is not shown.)

b) Signed binary adder symbolized as S register = A register + B register.

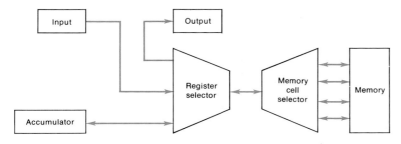

Fig. 15–9.
Block diagram of
connected input, output,
accumulator, and
memory.

"register selector" is a tree that permits either input, output, or accumulator to be connected to memory.

5 | INSTRUCTION CYCLE AND CONTROL UNIT

Let us now illustrate the flow of information and control in our minimicro computer by showing in detail how it executes one instruction. In our discussion, we will use decimal numbers rather than binary as we did with our model computer and its simulator, CARDIAC, in Chapters 10 and 11. This difference does not change the basic operation of the computer, but makes it easier to understand. The instruction to be executed is:

Stored in location	Instruction	Meaning
04	201	"Add contents of cell 01 to the accumulator."

To execute this instruction, we need the control mechanisms discussed earlier.

Figure 15–10 shows the block diagram of Fig. 15–9 with the necessary additions: a control box, an Instruction Register, and an Instruction Counter. The instruction register, a register of standard memory relays, always stores the instruction which is currently being executed, while the instruction counter, a resettable counter of the type discussed in Chapter 14, keeps track of which instruction should be executed next. Note that both the instruction register and the instruction counter can be selected by the register selector tree when it is properly set up. This is done by control.

To execute the instruction 201, we must fetch it from memory location 04 when it is stored, and bring it to the instruction register. Then we must increment the instruction counter to 05, the address of the next instruction, and finally we must properly interpret "201" as "add from cell 01." These are the steps in the automatic-instruction cycle discussed earlier and illustrated in Fig. 14–18, Chapter 14.

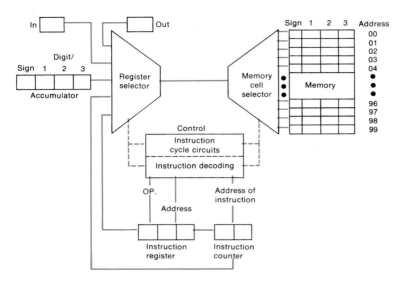

Fig. 15–10.
Block
diagram of a
three-decimal-digit
computer, with a
one-digit operation code
(op), two-digit address, 100
three-digit-plus-sign memory
cells (addresses 00 through
99). (Broken lines mean
control of information
flow; solid lines
mean flow of
information.)

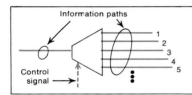

This component functions like a *switchboard*: the control signal determines which of the multiple lines is to be connected to the single line to form a completed path through the selector. Thus the selectors are used by Control to "direct traffic" in the computer.

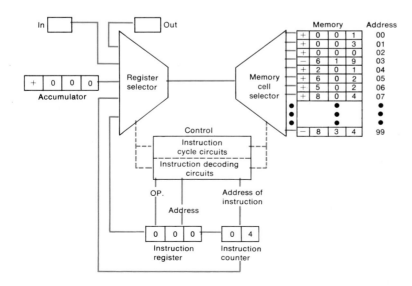

Fig. 15–11.
Memory loaded,
instruction counter
has been set to
address of first
instruction.

Now let us look at this entire process in somewhat greater detail. Figure 15–11 shows our block diagram, with a program loaded in locations 04 through 07. The data word +003 is stored in location 01. Note that the accumulator, instruction register, and instruction counter all have initial values; in particular, the instruction counter has been set to the address (04) of the first instruction (201).

When we now throw a "run-stop" switch to the run position, the computer is ready to start its first instruction cycle. We present a diagram of the outcome of each of the three steps of this cycle to give a clear picture of the action. In Fig. 15–11 we see that the computer is ready to do Step I. Control must cause the instruction in address 04 to be fetched and copied in the instruction register (where it will be decoded during Step III).

Step I. FETCH (Fig. 15–12):

In order to fetch and copy the instruction, Control must perform the standard select-copy process:

Select: Control's instruction circuits operate selection-tree relays so that address 04 is set up in the memory-cell selector, and the "address" of the instruction register (*IR*) is set up in the register selector (note the dotted control lines).

Copy: Now that a completed path has been established, the number 201 is copied into the instruction register. (Remember, this action erases any previous information in the instruction register, and that this is being done as instruction # 4. Before the information in the instruction register can be used, the instruction cycle must be incremented.)

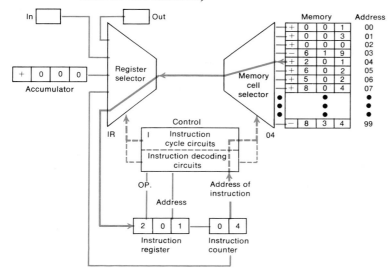

Fig. 15–12. Step I of instruction cycle "fetch instruction from memory."

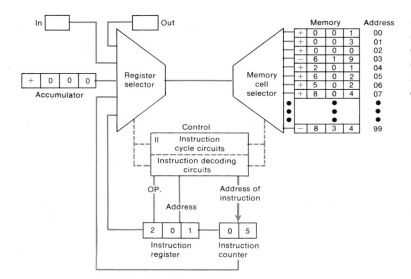

Fig. 15–13.
Step II
of instruction
cycle "increment
instruction
counter."

Step II. INCREMENT (Fig. 15–13):

In Step II the instruction counter must be increased by one to the "next" instruction's address. This is done automatically by a pulse from control. Figure 15–13 shows that the contents of the instruction counter have been increased from 04 to 05.

Step III. EXECUTE (Fig. 15–14):

We have now arrived at the crucial step of the instruction cycle, the execution of the current instruction contained in the instruction register. Control's instruction-decoding circuit now takes over to execute its timed sequence of actions. Just as the Morse coder's action sequence depended upon the setting of *XYZ* input switches,

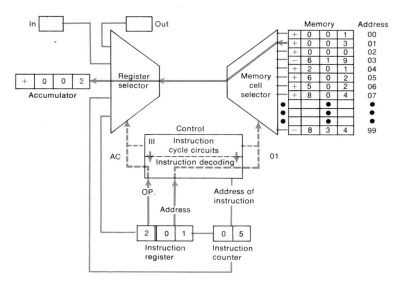

Fig. 15–14.
Step III of
instruction cycle
"decode and execute
instruction."

the action sequence for instruction-decoding and execution depends on the setting of the instruction-register relays.

In particular, the first digit "2" is interpreted as the operation code (op code) for "Add to the accumulator the contents of the cell specified in the two-digit address portion of the instruction." The setting of "2" on the op-code relays causes the register selector to be set to accumulator (AC), while the setting of "01" on the address portion of the instruction-register relays causes the memory-cell selector to be set to 01. Thus a path is established from memory cell 01 to the accumulator, with the result that $+000$ (old AC contents) and $+003$ (01's contents) are added to accumulate the sum $+003$ in the AC. Since numbers are only copied, it is clear that 01's contents are not changed by this operation.

Having successfully caused the execution of Step III, Control starts another instruction cycle by fetching the instruction in 05. The computer continues to run through its cycle until it comes to a halt instruction.

6 | FINAL COMMENT

We have completed our study of digital computers. Looked at from the outside, computers appear to be immensely complicated and able to perform astounding feats. Yet, as we have seen, they are made of simple binary elements combined into units that are in turn combined into larger units, and so on until a complete computer emerges. At each stage, laws and rules are used to combine units into larger units. By building this way, engineers can create computers containing millions of transistors and other components. This approach to the creation of complex machines is one of the unique features of the man-made world. In this creation, man does not copy nature. He creates and builds by using detailed, but simple, techniques. There was once a story going around among students that the man who invented the desk-top adding machine went insane after the conception. The story implies that the concept of a complex system comes like a revelation and exhausts a person's creative ability forever afterward. This view is romantic, but untrue. The concept of the computer came from a long series of inventions, insights, and experiments. Many people built upon the contributions of others, and finally the modern ideas came into being.

This process of creation means that any man-made systems or devices are landmarks for further progress based upon the technique and knowledge on which they rest. More exciting and significant, creation added to creation opens an endless frontier of accomplishment for man's benefit. In Columbus' time, sailing vessels made the discovery of the "new world" possible. Who knows what new world will be discovered as the result of modern technology. Of course, technology can be misused to man's detriment, and the computer as one of man's most impressive creations is a good candidate. Some people fear computers. Fear of *people* misusing computers is more realistic. This can be prevented only by people who are informed about both computers and humanity. It is such people who can prevent computers from robbing men of their privacy, freedom, and dignity and see to it that they make man's life easier, safer, and more humane.

Questions for Study and Discussion

1. Explain the contents of the instruction counter. Explain the difference between an address in the program counter and the address *part* of the relevant instruction.

2. How does the instruction register affect the instruction cycle? Does it control any of the actions of the other components of the computer?

3. List the three steps of the instruction cycle and explain what happens in each. Relate each step to the operations of CARDIAC.

4. Explain in your own words how the control circuit and the clock pulses serve to make the computer execute its instructions in proper order.

5. Is a shift circuit used in the accumulator when it is performing an addition? Explain the method used to add in the accumulator, including any circuits involved.

6. Is there any inherent reason why the op-code part of the instruction precedes the address part?

7. Point out any differences in the card reader and the card punch (Figs. 15–3b and 15–4). Could these be combined into one machine?

8. Examine the computer block diagram in Fig. 15–10. Are there any parts which can be eliminated? Can you think of any necessary function not performed by any of the components?

Problems

1. During each instruction cycle of a computer:

   a) How many times is the program counter incremented?

   b) How many times is the instruction register changed?

2. Elaborate Fig. 15–1 into a detailed diagram of how "sending" relays may be copied into "receiving" relays.

3. Construct (draw) a logic circuit which, when properly connected to a card reader of the kind shown in Fig. 15–3b, will turn on a light when a blank card is being read.

4. Would −003 in address 01 of Fig. 15–13 change the program or the output? Would −602 in address 05 change anything? Could you have predicted the effects without knowing in what program these words were to be used?

5. Suppose the instruction in cell 04 (Fig. 15–13) were 701: "Subtract the word at address 01 from the contents of the accumulator." How would this change the sequence of operations explained in Section 5? Explain in detail how 701 would be executed during the three steps of the appropriate instruction cycle.

Laboratory and Projects

I | A MINIMICRO COMPUTER

Although the minimicro computer studied in this chapter was "general purpose" in that it could carry out any task which could be specified by a program of instructions that the computer could execute, the construction of such a computer might involve more laboratory equipment and material than we would have at our disposal. Our minimicro computer will involve only the logic circuit boards that you already have, and will require only those circuits that you have previously wired and examined, such as the binary counter, binary adder, and memory circuits. When your computer is finished, you will be able to punch out the necessary instructions and information on your data cards, feed this information through your card reader and associated circuits into your memory section, instruct your computer to operate, read the output on the lamps of the adder (accumulator), and by punched-card instruction cause the computer to halt and reset to zero.

The circuits involved make this a "special purpose" computer in that it limits our program to one that will add two binary expressions with decimal values of up to fifteen each, however, an excess availability of address and instruction circuits as well as five unused control switching circuits limit its possible functions only to your ingenuity. The proposed computer involves the use of ten logic circuit boards—of these, three (boards X, Y, and Z) will be used as a binary adder; another three (boards U, V, and W) will function as a sequential control unit. Two boards, M and N, will be used as four-bit memory cells (incidentally the number of memory cells may be increased by making only a slight circuit change). One logic circuit board, board I, will be used as an information input board from the card reader. Board A, receiving information also from the card reader, will function as an address and control board for your computer.

To minimize the number and the length of connecting wires the following basic layout was one of the most successful.

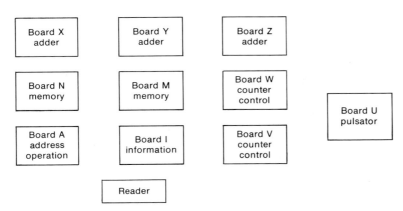

Fig. 1.
A minimicro computer board configuration.

The use of this number of logic circuit boards demands some adaptations generally not required with the less complex circuits we have previously completed. One requirement is the need to release all of the relays, even though holding contacts are engaged and relay shunts are inaccessible, without turning off all of the 110v A.C. line switches. To enable us to do this, we employ the following technique. First, *none* of the hand-operated switches (A, B, C, or D) are used on *any* board. Make certain that all of the switches are in the "Off" or "Unoperated" position. We will then use the B switch section of boards X, Y, Z, M, N, V, and W to provide us with a common negative, called K, that we can control. Wired as shown in each of the parts of this exercise we will have nine (9) common negatives available, all unique from the board's own negative or minus (−) terminal.

Examination of the suggested configuration for boards reveals the fact that existing leads, in light of your previous experience, will not be long enough. This is where the other unused hand-operated switch sections A, C, and D become very valuable. One lead to an eyelet on common allows the other common eyelet, and both of the break eyelets on that section to be at the same electrical potential, this is your junction or wire splicer. Using this technique, you have available over 120 unique junction points and the length of the existing leads is satisfactory. An actual count of a rather minimized circuit layout for the computer revealed the use of fifty-five (55) long green leads, forty-five (45) medium length red leads, and approximately one hundred (100) short black leads required to complete the circuit.

Although the "Minimicro Computer" circuit may most certainly be wired in one sitting—the unit has been divided into four parts—each relatively independent of the other. Each of the first three parts will take the best part of a class period to complete and test. The *TEST* is vital because any component failure would cause the final circuit to fail. Part D will allow us to bring them all together and start operation—from that point on your imagination will control what direction your computer may take.

Part A

Part A is the computer "Control Unit" and is composed of a form of runway circuit to provide pulses to operate a three-stage binary counter. The binary counter output is used as an addressing unit to look at first one memory cell and then another memory cell, and route the output of these memory cells to the proper registers of the binary adder (our accumulator). Examination of the three-stage binary counter relay contacts q_w, s_w, and s_v reveal the following sequence:

q_w	s_w	s_v	
0	0	0	
0	0	1	- - - - - - - - $\ast$ $\ast$
0	1	0	
0	1	1	
1	0	0	
1	0	1	
1	1	0	- - - - - - - $\ast$ $\ast$
1	1	1	- - - - - - - $\ast$ $\ast$

The first two values marked with the asterisk, $\overline{q_w}$, $\overline{s_w}$, s_v and q_w, s_w, $\overline{s_v}$ will be used in a later part to provide the controlled negative value K to the relay contacts of boards M and N and hence enable the transfer of the values of the relays (operated or released) of boards M and N to the binary adder. It might be mentioned at this time that although our unit is very crude the transfer time is approximately 1.5 second. To stop the counter and pulsator after the counter has reached the value of decimal seven, or binary 1 1 1, we use switch contacts q_{3w}, s_{4w}, and s_{4v} to provide a negative K to the shunt of relay Q_v which effectively stops the pulsator circuit.

When wiring the three boards U, V, and W for Part A, omit the use of the three relay contacts from board A, these are shown connected from a K to $\overline{q_{3a}}$ to r_{3a} to s_{3a} to relay Q_v operate. This part will be completed in Part D of this exercise.

To test Part A after completion use temporary leads which will be removed after the test, and connect a negative (−) to one of the K terminals. Using another single lead, go from K on board V to the operate terminal of relay Q_v. The pulsator circuit should

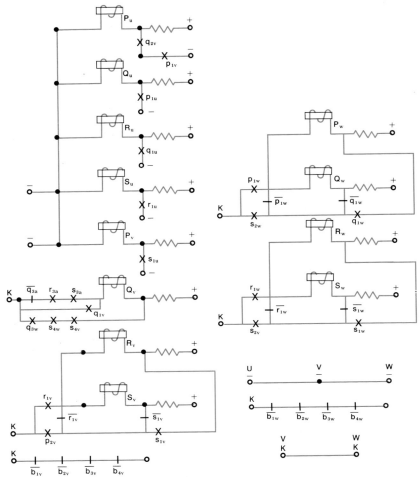

Fig. 2.
A minimicro computer,
Part A.

start and the counter unit should sequence through a complete cycle and halt. Try this two or three times to assure yourself that the circuit is correct and then remove the temporary leads and start on Part B.

Part B

Part B will be the computer input, control instruction, memory, and addressing unit. It is also that section that provides the controlled negative value K. Access to the two memory boards M and N is gained through the relay contacts of relays Q_a, R_a, and S_a. Since only one address may be used on the punched card at one time, the value of the relays P_i, Q_i, R_i, and S_i is copied into the correct memory board and stored there using holding contacts

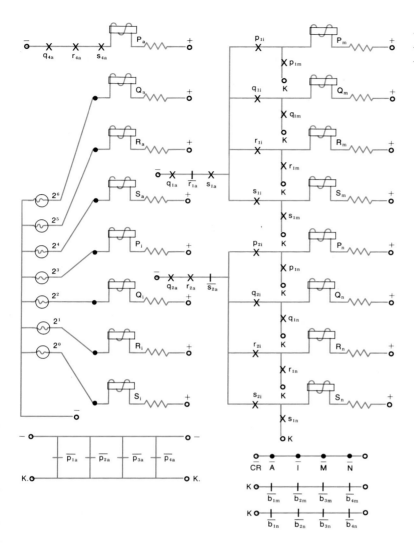

Fig. 3.
A minimicro computer,
Part B.

such as p_{1m} or q_{1n}. In testing, this memory may be removed by using the binary address 1 1 1 on the punched card as this energizes P_a and removes the negative value K. Examination of the small circuit in the lower left-hand corner of the schematic for Part B reveals that all of the four break contacts are wired together. This is to provide enough current carrying capacity for relay P_a. You may wish to include at this time "readout" lamps on the memory boards as a check on your card-reader input. The two circuits on the right-hand side of the schematic on Part D allow you to do this, and in no way interfere with either this or later sections. No temporary wiring is needed to test this section as it stands on its own. When you are satisfied with its action, start on Part C, the binary adder circuit.

618

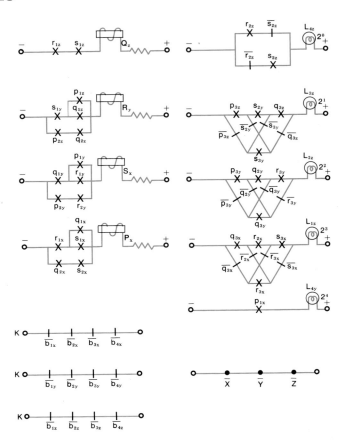

Fig. 4.
A minimicro computer,
Part C.

Part C

Part C is the familiar binary adder circuit—the one possible change is the use of relays and their contacts as the input values to the registers rather than the hand-operated switches. First make certain that a negative lead from each board is connected to the others, then do the same for the K negative value, even though it is not used on this part of the circuit itself. Proceed to connect relay Q_z and Lamp 4_z in the schematic and immediately test to make certain it is correct. The test may use once again the temporary wiring of leads to operate relays R_z and S_z. If this much of the circuit is correct, remove the temporary leads and proceed to the next section of the circuit, connect relay R_y and Lamp 3_z in their respective circuits, test as before using temporary leads and operating relays S_y, P_z, or Q_z. This procedure is repeated for all five sections of the binary adder. When you are convinced that this portion of the circuit is in working order, move on to Part D which will complete your minimicro computer.

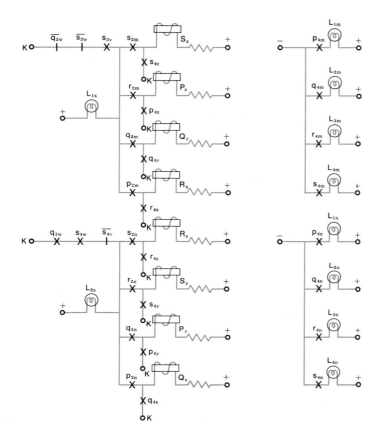

Fig. 5.
A minimicro computer.
Part D.

Part D

Completion of the schematic of Part D, and interconnecting all of the boards previously wired will complete your Minimicro Computer. One order of wiring this final section that has been successful is: Make certain that all boards are common as far as the negative or minus $(-)$ value is concerned, this will involve *one* wire from a negative on one board to the negative on the next. The next order of operation is to make certain that the switched value of the negative K found on board A is connected to each board where it is required. Once again one wire running from one board to the next is sufficient. The lamps L_{1x} and L_{2x} enable you to determine whether the control unit from Part A is providing the necessary negative value K to boards M and N, so that it can be fed to the binary adder registers. Your completion of this circuit completes your computer. Try a few punched inputs for addition and see whether decimal eleven plus decimal value fifteen equals decimal value twenty-six.

EPILOGUE

ARE ALL THINGS POSSIBLE?

Engineering, science, and technology have brought forth many marvels. People of an earlier age could scarcely have believed the modern world. To the Elizabethan or even Victorian man, a conversation between a person in London and one in Tokyo, breakfast in Rome and dinner in Los Angeles on the same day, a man walking on the moon, hearts transplanted from one person to another, energy obtained from the nucleus of an atom, or eating fish in Paris that had been delivered from South Africa would have seemed sheer fantasy. Yet today we take such things for granted.

In contrast to an often stated view, modern man does appreciate how fantastic the realities of today's world are. Because of them, he has almost unlimited expectations. Can we rocket to Alpha Centauri, can we stop the aging processes in man so that life will become eternal, can we replace defective eyes and ears, can we control heredity, can we transport materials by electric currents, can we accurately predict the future? The answer often heard is: if not today, then certainly in five years, or 10, or 20, or 50 years!! Regardless of the truth of this statement, modern man is scarcely surprised at any new feat of engineering, science, or technology.

Yet there are impossibilities. These arise from many causes, some technical, some political, some economic, and some arise from human limitations. The answer to the title of this section, then, is "all things are *not* possible." Nature provides certain limits on what can be done, and these limits cannot be exceeded. Within these limits, engineering and science are continually opening new possibilities, new options for people and societies. In fact, technology opens many more possibilities for man than he can or should pursue. The various options must be held up against the yardstick of human values before any one is adopted as a goal to be achieved. Many things are possible technically that are not possible humanely. Making life better and easier for people and nations lies in dodging the impossible; it is the "art of the possible" that counts. This art requires knowledge and technique, plus human values and feelings.

But let us return for a moment to the impossible. Are there really impossibilities? Yes, there are laws of nature which cannot

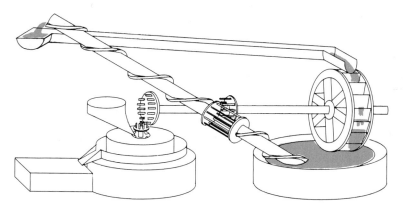

Fig. 1.
This perpetual motion
mill was proposed by Robert
Fludd in 1618. The waterwheel
is propelled by a waterfall. Both a
pump to keep the water flowing and the
millstone are powered by the wheel.
No outside source of energy
is required. This mill
will not work for it
violates the Law of
Conservation
of Energy.

be violated. For example, a "closed-cycle mill" as proposed by Robert Fludd in 1618 is shown in Fig. 1. It was said to be closed-cycle because it required no outside source of power. It was to be used in areas that lacked flowing streams to power mills and factories. The falling water turns a waterwheel which drives a pump to recirculate the water and thereby keep the water falling. The wheel also drives a mill to do some useful work, such as grinding wheat to produce flour. This is one example of a perpetual motion machine. Many different versions of such machines have been proposed even in recent years. Some are extremely complex and ingenious. However, we do not need to take such proposals seriously for they violate a fundamental law of nature, namely, The Conservation of Energy.

There are several ways of stating this law. One version says that energy can neither be created nor destroyed. Another says that the total energy in the universe is constant.* In either case, the waterfall and mill of Fig. 1 requires a source of energy if it is to do any useful work. Because there is no source of energy to drive the system, it will not operate as illustrated. Useful work aside, no mill can pump all its water supply back to the uphill starting point because some energy is lost as heat due to friction. Even if we started the mill operating as shown, all the available work energy would soon be converted to heat and the mill would stop.

How do we know that the energy is *always* conserved? How do we know that the "law" cannot be violated? One answer is that in repeated observations and measurements over 150 years, no one has found a confirmed violation. More importantly, if we *assume* that the law *is valid*, we can *prove* logically many well-known properties of nature. If the law did not hold, the world would be different from the one we live in. Thus, we rely on both observation (experiment) and inference when we assert that conservation of energy is a valid law of nature.*

---

*
A full discussion of energy conservation can be found in modern physics textbooks. Here we use it as an illustration without discussing it in detail.

*
Experiment and inference can occasionally be deceiving. In the early 1960's, Lee and Yang, two physicists, found a violation of the "conservation of parity" law which had been thought to hold in nuclear interactions. Their finding overturned the supposed law, and they won the Nobel Prize for their discovery. However, long established laws which have been confirmed myriads of times are not likely to be overturned.

There are many such "laws" and each underlies some observable part of the man-made or natural world. Newton's Law of Gravity is one. Another is a law, first stated by C. E. Shannon, which sets the limits on the maximum amount of information which can be sent over a communication line in any interval of time. We know from Shannon's work that it is useless to try to send more information than his limit. Another example of lawful limitations is the speed of light. It is useless to try to transmit energy through space at a speed faster than that of light. In modern computers, this limit has become an issue, for it may well take longer to transmit a binary pulse from memory to processor than it does to close or open switches in the processor. Yet there is no way to speed up the transmission. The minimum time is set by the speed of light and the length of wire in the memory-processor path.

Some limits are much more widely restrictive than others. The speed of light, the law of gravity, and the conservation of energy have very broad implications. Shannon's communication limit is based upon a particular model of communication channels. So Shannon's limit applies only to those situations which fit his model. It turns out that many channels do fit reasonably well, and the model can be modified to fit many others. (Similar limits result.) However, laws and limits must be examined for their relevance before they are taken literally to apply to a situation. Serious errors have been made by scientists and engineers not following this caution. For example, around 1930 a mathematical model was used to "prove" that frequency modulation broadcasting was impossible. The mathematics was correct, but the model did not describe the situation adequately, and today FM broadcasting is widespread. (FM is used also in many communication systems other than broadcasting.)

So, though many laws of nature are very well established indeed, we must keep our minds open to change. In most cases, it is not too hard to determine when some fundamental principle is being responsibly challenged, but there are questionable cases. Two very prominent ones concern the existence of extrasensory perception and flying saucers. Many responsible scientists have examined both, but the public in general remains unconvinced that true answers have been found. Have the scientists been too skeptical or have the lay advocates been too uncritical of the evidence for these phenomena? In the end, people as individuals must review the evidence, and perhaps make observations and experiments of their own before they can have a responsible opinion about such matters. However, should any theory or "happening" violate a well-established law of nature, it makes that event or theory much less believable. ESP and flying saucers do seem to violate such laws and that is enough to make many people reject them as either false or illusory. Perhaps this is what Einstein had in mind when he reportedly remarked about flying saucer sightings, "These people have certainly seen something, but I don't care what it was."

In any case, these laws of nature are by far the most robust knowledge in man's storehouse, and they are much prized. Yet many other principles of a less absolute sort are useful and vital, and give great insight. For example, we can count the number of times each word appears in a text and make a list of the words and the number of times each appears. We can then rank order the words putting the most frequent first and ending with the least frequent. If we now plot the number of times each appears against the rank order, as shown in Fig. 2, we find a straight-line relation. The slope of the line is such that, for example, the tenth word in order is used only one-tenth as many times as the most frequent word. It turns out that this relation holds for a surprisingly large range of documents, magazines, and papers. As indicated in Fig. 2, both James Joyce's *Ulysses* and newspapers conform. This relation is commonly associated with the name of G. K. Zipf, and is known as *Zipf's Law*. A similar law, known as *Lotka's Law of Productivity*, holds for scientific authors. The number of authors publishing ten papers, during their careers, in scientific journals is $1/10^2$ (one one-hundredth) the number publishing only one paper.

Laws of this kind clearly *can* be violated, particularly if a person sets out to do so. They do not necessarily hold in all cases, but they do define what we might call "normal" behavior. If a book or a group of authors violates one of these laws, we might be interested in why, but we wouldn't be as shocked as if we had found a violation of energy conservation. So, despite the fact that laws of this kind do not hold rigidly, they do give insights. However, they are in no way comparable to the laws of nature in ruling certain feats as impossible.

So much for the *impossible*. More important is the *possible*. What is possible? Just because some goal is not impossible — does not violate any law of nature — does not mean that it is possible in any human sense. Prevention of cancer in human beings may not violate any law of nature, but we don't know how to do that today. We say that such possibilities are "beyond the state-of-the-art." Advancing the state-of-the-art so that new things become possible in a real sense is what the country's technical enterprise is all about. Research, development, testing, and experimentation are all devoted ultimately to improving man's lot by advancing the state-of-the-art. This quest has proved altogether more productive and rewarding for mankind than efforts to find loopholes in the laws of nature. So despite the public's fascination with reports of unexplained happenings, such as unidentified flying objects, the welfare of the nation and mankind hinges on new achievements by people working within the state-of-the-art and extending it. "Mi-

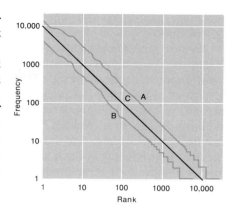

Fig. 2.
Zipf's law says that in "normal" text, the tenth most common word (rank) occurs one-tenth as often as the most common word. Shown here are word counts (frequency) plotted against rank order for James Joyce's Ulysses (A), a newspaper article (B), and an ideal curve (C). In the passage from the newspaper, for example, the most frequent word was used about 7,000 times, the tenth most frequent word about 700 times, and the thousandth most frequent word about 7 times.

raculous" happenings are sensational but not very significant in this picture.

Even advances in the state-of-the-art require devoted cultivation if they are going to work to man's benefit. Developing a new treatment, product, or service using state-of-the-art technology requires a substantial investment of money. Putting these in forms that are desired by the people and organizations that will use them is a problem of marketing. Marketable items and services must be distributed and delivered, and service must be available to correct deficiences and failures, which are inevitable. Finally, performance must be studied so that the product or service can be improved and adapted to changing conditions and demands. Failure on any of these fronts can change a potential benefit to a menace.

Even given all of these resources, new approaches may be politically unacceptable. For example, a computer data bank containing basic personal, business, and professional information about United States residents is certainly possible. It might even be quite valuable for planners and sociologists. However, there has been strong political and public opposition to such a data bank because of the fear that the information it contains might be misused. People could be subjected to intimidation and might lose their privacy. Such issues must be faced before the newly possible is brought into being and has its impact on people.

There are indeed many hurdles in bringing new products and services to public use. New products and services must be based on sound technology which works reliably. They must be economically feasible. That means people must be willing to pay for them. They must be backed up by efficient service and marketing organizations. They, and their side effects, must be politically and humanly acceptable. All of these things and more are needed to turn new possibilities into realities. Yet it is through this complex process that change will come. The world will be different tomorrow.

Furthermore, the direction of change will not be determined by the laws of nature, though of course we cannot violate them. Rather change will be shaped by advances in state-of-the-art technology steered by adventuresome people. If the world is to be a better place, more and more people must take part. We cannot leave our destinies unquestioningly in the hands of "experts." Yet relatively few can, or ought to be engineers, scientists, or entrepreneurs. So society needs "fans," observers who can cheer or boo as the players perform. To be effective, fans must know the rules of the game and be able to pick out the stars and the goats as new realities are being brought into being. Fans must be technologically literate. That is what *The Man-Made World* is all about. Whether you will be a player or a fan, *The Man-Made World* should help you follow the action and have a voice in shaping the world of the future.

INDEX